High-Resolution Electron Microscopy

High-Resolution Electron Microscopy

Fourth Edition

John C. H. Spence
Department of Physics and Astronomy
Arizona State University/LBNL California

OXFORD
UNIVERSITY PRESS

Great Clarendon Street, Oxford, OX2 6DP,
United Kingdom

Oxford University Press is a department of the University of Oxford.
It furthers the University's objective of excellence in research, scholarship,
and education by publishing worldwide. Oxford is a registered trade mark of
Oxford University Press in the UK and in certain other countries

© John C. H. Spence 2013

The moral rights of the author have been asserted

First Edition published in 1980
Second Edition published in 1988
Third Edition published in 2003
Fourth Edition published in 2013
Previously editions were entitled Experimental high-resolution electron microscopy

Impression: 1

All rights reserved. No part of this publication may be reproduced, stored in
a retrieval system, or transmitted, in any form or by any means, without the
prior permission in writing of Oxford University Press, or as expressly permitted
by law, by licence or under terms agreed with the appropriate reprographics
rights organization. Enquiries concerning reproduction outside the scope of the
above should be sent to the Rights Department, Oxford University Press, at the
address above

You must not circulate this work in any other form
and you must impose this same condition on any acquirer

Published in the United States of America by Oxford University Press
198 Madison Avenue, New York, NY 10016, United States of America

British Library Cataloguing in Publication Data

Data available

Library of Congress Control Number: 2013938574

ISBN 978–0–19–966863–2

Printed and bound by
CPI Group (UK) Ltd, Croydon, CR0 4YY

Links to third party websites are provided by Oxford in good faith and
for information only. Oxford disclaims any responsibility for the materials
contained in any third party website referenced in this work.

To
Vernon, Penny and Andrew,
and in memory of John Wheatley

Preface to the Fourth Edition

The history of science can be viewed as progress in our efforts to measure things—as Kelvin commented: 'Unless you can measure it, you cannot improve it'. And so it has been in the recent history of high-resolution electron microscopy, whether in structural biology, in the analytical and imaging modes of the scanning transmission (STEM), or the full-field transmission electron microscope (TEM). The decade since the last edition of this book has been characterized more than in any other way by advances in the accurate quantification of the various signals from these instruments, advances in the power and speed of computer algorithms for simulating them, and the drive toward three-dimensional imaging at atomic resolution. This drive to quantify imaging and spectroscopic signals, partly due to improved computers and detectors, has finally moved us from the realm of the physicists to the level of specialists in materials science and structural biology. At the same time, the entire agenda of modern science which is relevant to microscopy has shifted since the first edition of this book over 30 years ago, for example toward nanoscience, rather than the study of defects in bulk materials, toward a more quantitative biophysics, and toward protein dynamics, in addition to structure. A wide range of competing new microscopies has also developed in this period, as reviewed in the two volumes of *Science of Microscopy* edited by P. Hawkes and J. C. H. Spence (Springer, 2006).

Since the basic theoretical diffraction physics on which electron microscopy is based had been fairly well worked out by the late 1960s, apart from some aspects of the theory of plasmon excitation from small particles in energy-loss spectroscopy, and linear dichroism for electron beams, subsequent dramatic progress has resulted mainly from developments in instrumentation. So the last decade has been the decade that has seen the widespread adoption of aberration-corrected machines, the development of entirely new direct-injection detectors, major advances in three-dimensional imaging at near-atomic resolution, and the first commercially available electron monochromators for energy-loss spectroscopy in STEM. I have added coverage of the relevant electron-optical theory for aberration correction, and also the fascinating analysis of electron Ronchigrams now used for automated aberration correction (which has its origins in work at ASU in the late 1970s). The aberration correctors have in turn provided greater space for incorporation of new types of energy-dispersive X-ray detectors with collection solid angles approaching 1 sr, so that, in combination with the now widely accepted field-emission electron sources, the detection of inner-shell X-ray fluorescence from a single atom is possible, with a similar achievement also reported using energy-loss spectroscopy. Atomic resolution has been achieved with the SEM. The discovery of graphene has similarly been a heaven-sent gift to TEM and image quantification, since, at a stroke, it has solved the long-standing problems of thickness measurement, multiple-scattering perturbations, and sample charging, while providing an ideal conductive substrate, or even an envelope for a liquid cell. Most importantly, images of graphene agree closely with simulations, perhaps finally settling the issue of the 'Stobbs factor', and ushering in an era of truly quantitative atomic-resolution electron microscope imaging. Yet, with the more widespread use of STEM and energy-filtered imaging, radiation damage remains a more serious problem than ever, and we are now seeing a new generation of machines built to operate below the knock-on threshold for materials like graphene. (We eagerly anticipate learning whether one solution to the problem of damage—the use of short pulses as developed with X-ray lasers—can be applied to electron microscopy.) Terminology and techniques have changed greatly since the first edition of this book—the universally used expression 'lattice image' seems quaint nowadays (a lattice is a mathematical abstraction) and has been avoided (mostly) in this edition, while the declining use of film has taken longer than materials scientists might have anticipated, since biologists do not need a large dynamic range to record their faint cryo-EM images, and prize the large numbers of pixels offered by film above all else. (Since the excellent image plates require a film transport mechanism, I have not deleted all references to this type of image recording.) But the new direct-injection detectors with their fast recording and noiseless readout will surely be a major advance for all fields in the battle against radiation damage and sample movement. With the development of transistors based on carbon nanotubes, the discovery of the nanotube by TEM gains in importance with time, as does work on graphene, plasmonics, and catalysis using nanoparticles, among many other areas. In the Plates I have tried to collect some of high-resolution images which have had the most impact on their fields. Because of the power of web-based search engines, I have deleted the section on applications of high-resolution electron microscopy, relying instead on the web, and the examples used to illustrate points of technique throughout the book. The final acceptance, after more than three decades, of the STEM mode by the materials community reminds us that energy-loss spectroscopy in STEM, combined with its atomic-resolution imaging capability, remains an unrivalled spectroscopy with the highest spatial resolution in all of modern science.

'Rosedale', Orinda, CA and Tempe, AZ   J.C.H.S.
April 2013
(ASU and LBNL)

Preface to the Third Edition

The appearance of this third edition, fifteen years after the second, comes at a time of great excitement in high-resolution electron microscopy (HREM). The structure of the ribosome (the large biomolecule which makes proteins according to DNA instructions—'life itself') has recently been determined for *E. coli* by single-particle tomographic cryo-electron microscopy. The structure of tubulin, also, has now been determined by the electron crystallographic methods pioneered earlier for purple membrane and other proteins which are not easily crystallized. And tubulin has since been fitted to the tubule structure (which carries proteins about the cell) by the same single-particle cryomicroscopy methods, and the location and function of the anticancer drug Taxol identified. In materials science, Iijima's discovery of the nanotube in 1991 has done much to establish the new field of nanoscience, described by Raul Hoffman as 'the next industrial revolution', following Iijima's earlier first observation of the buckyball in 1980 (see the Plates). The atomic-scale study of interfaces by HREM has finally succeeded in supplying three-dimensional interface structures as the basis for electronic structure calculations, which, together with energy-loss spectra now fulfil a thirty-year old idea that HREM may ultimately contribute to a complete description of the ground state electronic structure of the defects in materials which control their properties, and their excitations. More recently we have seen the first convincing images of individual substitutional dopant atoms in crystals (by ADF-STEM), and the first useful energy-loss spectra obtained from individual columns of atoms. These last results were obtained by STEM, a sure sign that STEM, the method first used to see atoms by electron microscopy in 1970, has emerged from being a physicists' plaything to a useful tool for materials characterization at the atomic scale. The possibilities of direct-write inorganic lithography by STEM continue to be explored, as does the 'SCALPEL' lithography program for certain applications. The purchase, around the turn of the century, of 60 field-emission STEM instruments by the semiconductor industry recognizes that, with 1.2 nm width gate oxides, you cannot now see a transistor without one.

These recent spectacular achievements—the development of cryomicroscopy, the buckytubes and STEM—should not mask the enormous amount of new science that has come from straightforward HREM in the last decade. From understanding the role of interfaces in limiting current flow in superconductors, the magnetoresistance manganates and the strength of composite materials, to measurements of the quantized resistance of nanowires formed *in situ*, to identification of the shape changes in the proteins which pump protons through the cell membrane in photosynthesis, the range and depth of this work has been remarkable. Ultra-high vacuum TEM has continued to develop, and of the many recent papers entitled 'Gabor's dream realized', one did in fact achieve exactly that—improvement of resolution in electron microscopy by holography. *In situ* and environmental HREM have produced remarkable atomic-scale images of catalytic reactions at pressures up to 5 Torr, the MOCVD process, and even the liquid–solid interface. Inspired by the example of the nanotube, this may initiate an era of materials discovery by atomic-scale search procedures combined with *in situ* combinatorial materials synthesis. By comparison with other particles used as probes of matter, the electron, for which we have the brightest sources in all of physics, the strongest interactions, and highly efficient parallel detectors, together now with corrected lenses and monochromators, has continued to demonstrate its worth as an imaging, spectroscopy and diffraction tool for science. And electron scattering, in the CBED geometry, where experimental parameters can be accurately quantified and extinction errors are absent, has proven to be more accurate (at the 0.025% level) for measurement of X-ray structure factors than X-ray crystallography itself in many cases, allowing chemical bonds to be seen clearly in oxides for the first time.

Over the past fifteen years at least two high voltage HREM machines have been built with a resolution close to one angstrom, while image processing techniques and aberration correctors in both the TEM and STEM mode have brought the same resolution (or better) to medium voltage machines. The point has been made that aberration correction became possible when computer speeds made automated alignment possible in a time shorter than the stability time of the microscope, and this auto-tuning capability will eventually lead to much more user-friendly machines with remote-user capabilities. It seems unlikely that we will see much further improvements in resolution, or that they will be needed. The important challenges in the field now lie with atomic resolution tomography (recently achieved in both STEM and HREM), *in situ* studies, holography of fields at higher resolution, and further development of ELS in STEM, which can give us the equivalent of soft X-ray absorption spectra from regions of atomic dimensions.

Advances in instrumentation have generated as much excitement as I can remember since the beginning of the field. Field emission sources, imaging energy-analysers, monochromators and aberration correctors, CCD cameras and Image Plates, together with far more powerful computers have transformed the field in the last decade. All this has confirmed the crucial importance of the microscope environment, since aberration correction to the angstrom level is worthless without comparable stabilities. The most successful atomic-resolution instruments of the future will certainly be those in the quietest locations, the most favourable environments.

This new edition reflects the above developments. Three new chapters (on STEM, on super-resolution methods and image processing, and on HREM in biology) have been added, and entirely new sections added or replaced on field-emission guns, detectors, aberration correction, three-fold astigmatism, radiation damage, organic crystals, holography, autoalignment, environmental factors, energy-loss spectroscopy, and environmental cells. A thorough revision of the entire text has resulted in many other changes throughout. The previous edition included a chapter reviewing applications of HREM in materials science—the literature on this is now so vast, and web-search engines so efficient, that no attempt is made to systematically review applications in this edition. The new section on imaging of defects in crystals does, however, provide recent case studies in crucial areas, and references are provided to several excellent recent reviews of HREM.

Electron microscopy remains important because it is the only technique which can provide real-space images at atomic resolution of the defects within materials that largely control their properties. In addition, the STEM, with its fine probe and multitude of detectors for spatially-resolved spectroscopy, is the ideal instrument for the study of inorganic nanostructures. In the past it has been limited by the projection approximation, and the poor agreement of ELNES with theory. With the recent appearance of tomographic images of crystals and defects at atomic resolution, the achievement of accurate quantification of diffraction data in the CBED mode, and the excellent agreement now possible in ELNES calculations for known structures in many cases, the situation has drastically changed. The integration of these methods is now proceeding rapidly. All this, together with the need to understand the host of useful new nanostructures being synthesized and the further development of cryomicroscopy, guarantee an exciting future for the field.

'Rosedale', Orinda, CA
March 2002
(ASU and LBL)

J.C.H.S.

Acknowledgements

The bulk of the first edition of this book was written during the year (1975) following my term as a post-doctoral fellow at the Department of Metallurgy and Materials Science at the University of Oxford. I therefore owe a great deal to all members of that department for their support; in particular Professor P. B. Hirsch, Dr M. J. Whelan, and Dr C. J. Humphreys for their encouragement and optimistic conviction that the methods of high-resolution electron microscopy can be useful in materials science and elsewhere. I am also grateful to Dr J. Hutchinson for his advice and encouragement.

I also wish to thank members of the Melbourne University Physics Department (Drs A. E. Spargo and L. Bursill in particular) for sharing their experience of practical and theoretical high-resolution work with me during 1976. My debt to the Australian groups generally will be clear from the text.

The quality of technical support is particularly important for high-resolution work. I have been fortunate both at Oxford (with Graham Dixon-Brown) and at Arizona State University (with John Wheatley) to work with skilled and enthusiastic laboratory managers who have each taught me a great deal.

I have also learnt a great deal since coming to Arizona State University from Dr Sumio Iijima, to whom I express my thanks, and from John Barnard on visits to Cambridge.

I owe a particular debt of gratitude to Dr David Cockayne for his interest in the book and painstaking review of some of its chapters.

My debt to Professor John Cowley will be obvious to all those familiar with his work. His major contributions to high-resolution imaging have laid the important theoretical foundations in this subject and I have used many of his results. Drs Peter Buseck and Leroy Eyring have freely discussed their experiences with high-resolution work. Harry Kolar and Dr Michael O'Keefe have each also been kind enough to assist in the book's production. I am grateful to Drs A. Glauert and D. Misell for their detailed comments on an early draft of the first edition manuscript.

In preparing this fourth edition in 2012/13 I have been particularly grateful to Prof Ernst Bauer at ASU, and to John Barnard in Cambridge for his suggestions and corrections to the text, especially on experimental technique. The affectionate support of my wife Margaret, allowing me to find time for this large project, continues to be the greatest blessing of my life.

Contents

Symbols and abbreviations	xvii
1 Preliminaries	1
1.1 Elementary principles of phase-contrast TEM imaging	2
1.2 Instrumental requirements for high resolution	8
1.3 First experiments	10
References	11
2 Electron optics	13
2.1 The electron wavelength and relativity	13
2.2 Simple lens properties	16
2.3 The paraxial ray equation	22
2.4 The constant-field approximation	24
2.5 Projector lenses	25
2.6 The objective lens	28
2.7 Practical lens design	29
2.8 Aberrations	31
2.9 The pre-field	37
2.10 Aberration correction	38
References	43
Bibliography	45
3 Wave optics	46
3.1 Propagation and Fresnel diffraction	47
3.2 Lens action and the diffraction limit	50
3.3 Wave and ray aberrations (to fifth order)	55
3.4 Strong-phase and weak-phase objects	61
3.5 Diffractograms for aberration analysis	63
References	65
Bibliography	66
4 Coherence and Fourier optics	67
4.1 Independent electrons and computed images	69
4.2 Coherent and incoherent images and the damping envelopes	70
4.3 The characterization of coherence	76
4.4 Spatial coherence using hollow-cone illumination	79
4.5 The effect of source size on coherence	81
4.6 Coherence requirements in practice	83
References	86
Bibliography	87

5 TEM imaging of thin crystals and their defects — 88
- 5.1 The effect of lens aberrations on simple lattice fringes — 89
- 5.2 The effect of beam divergence on depth of field — 93
- 5.3 Approximations for the diffracted amplitudes — 96
- 5.4 Images of crystals with variable spacing—spinodal decomposition and modulated structures — 102
- 5.5 Are the atom images black or white? A simple symmetry argument — 104
- 5.6 The multislice method and the polynomial solution — 106
- 5.7 Bloch wave methods, bound states, and 'symmetry reduction' of the dispersion matrix — 107
- 5.8 Partial coherence effects in dynamical computations—beyond the product representation. Fourier images — 113
- 5.9 Absorption effects — 115
- 5.10 Dynamical forbidden reflections — 117
- 5.11 Relationship between algorithms. Supercells, patching — 122
- 5.12 Sign conventions — 125
- 5.13 Image simulation, quantification, and the Stobbs factor — 126
- 5.14 Image interpretation in germanium—a case study — 129
- 5.15 Images of defects and nanostructures — 134
- 5.16 Tomography at atomic resolution—imaging in three dimensions — 143
- 5.17 Imaging bonds between atoms — 145
- References — 146

6 Imaging molecules: radiation damage — 154
- 6.1 Phase and amplitude contrast — 154
- 6.2 Single atoms in bright field — 157
- 6.3 The use of a higher accelerating voltage — 165
- 6.4 Contrast and atomic number — 169
- 6.5 Dark-field methods — 171
- 6.6 Inelastic scattering — 174
- 6.7 Noise, information, and the Rose equation — 177
- 6.8 Single-particle cryo-electron microscopy: tomography — 180
- 6.9 Electron crystallography of two-dimensional crystals — 188
- 6.10 Organic crystals — 190
- 6.11 Radiation damage: organics and low-voltage EM — 192
- 6.12 Radiation damage: inorganics — 195
- References — 197

7 Image processing, super-resolution, and diffractive imaging — 204
- 7.1 Through-focus series, coherent detection, optimization, and error metrics — 204
- 7.2 Tilt series, aperture synthesis — 210
- 7.3 Off-axis electron holography — 211
- 7.4 Imaging with aberration correction: STEM and TEM — 212
- 7.5 Combining diffraction and image data for crystals — 215
- 7.6 Ptychography, Ronchigrams, shadow images, in-line holography, and diffractive imaging — 219
- 7.7 Direct inversion from dynamical diffraction patterns — 226
- References — 226

8 Scanning transmission electron microscopy and Z-contrast — 233
- 8.1 Imaging modes, reciprocity, and Bragg scattering — 233
- 8.2 Coherence functions in STEM — 240
- 8.3 Dark-field STEM: incoherent imaging, and resolution limits — 243
- 8.4 Multiple elastic scattering in STEM: channelling — 249
- 8.5 Z-contrast in STEM: thermal diffuse scattering — 251
- 8.6 Three-dimensional STEM tomography — 257
- References — 260

9 Electron sources and detectors — 264
- 9.1 The illumination system — 265
- 9.2 Brightness measurement — 268
- 9.3 Biasing and high-voltage stability for thermal sources — 270
- 9.4 Hair-pin tungsten filaments — 274
- 9.5 Lanthanum hexaboride sources — 274
- 9.6 Field-emission sources — 275
- 9.7 The charged-coupled device detector — 276
- 9.8 Image plates — 281
- 9.9 Film — 282
- 9.10 Direct detection cameras — 283
- References — 286

10 Measurement of electron-optical parameters — 289
- 10.1 Objective-lens focus increments — 289
- 10.2 Spherical aberration constant — 291
- 10.3 Magnification calibration — 293
- 10.4 Chromatic aberration constant — 295
- 10.5 Astigmatic difference: three-fold astigmatism — 295
- 10.6 Diffractogram measurements — 296
- 10.7 Lateral coherence width — 299
- 10.8 Electron wavelength and camera length — 302
- 10.9 Resolution — 303
- 10.10 Ronchigram analysis for aberration correction — 306
- References — 312

11 Instabilities and the microscope environment — 315
- 11.1 Magnetic fields — 315
- 11.2 High-voltage instability — 318
- 11.3 Vibration — 319
- 11.4 Specimen movement — 319
- 11.5 Contamination and the vacuum system — 321
- 11.6 Pressure, temperature, and draughts — 323
- References — 323

12 Experimental methods — 324
- 12.1 Astigmatism correction — 325
- 12.2 Taking the picture — 326
- 12.3 Recording atomic-resolution images—an example — 328

xvi *Contents*

12.4	Adjusting the crystal orientation using non-eucentric specimen holders	335
12.5	Focusing techniques and auto-tuning	337
12.6	Substrates, sample supports, and graphene	340
12.7	Film analysis and handling for cryo-EM	343
12.8	Ancillary instrumentation for HREM	344
12.9	A checklist for high-resolution work	345
	References	346
13	**Associated techniques**	**348**
13.1	X-ray microanalysis and ALCHEMI	348
13.2	Electron energy loss spectroscopy in STEM	357
13.3	Microdiffraction, CBED, and precession methods	363
13.4	Cathodoluminescence in STEM	372
13.5	Environmental HREM, imaging surfaces, holography of fields, and magnetic imaging with twisty beams	376
	References	380
Appendices		388
Index		403

Symbols and abbreviations

Symbols

Δ	Chromatic damping parameter (see eqn (4.9))
Δf	Defocus. Negative for weakened lens. Special cases—Scherzer focus (eqns (6.16) and (4.12)), stationary phase focus (eqn (5.76)), Gaussian focus
θ_c	Beam divergence
C_s	Spherical aberration constant. Positive
C_e	Chromatic aberration constant. Positive
$\chi(u)$	Phase shift due to lens aberrations
Φ	Complex wave amplitude
ϕ	Magnitude of Φ
v_g	Fourier coefficients of crystal potential (in volts)
U_g	Similar (in Å^{-2}). $U_g = \sigma v_g/\pi\lambda = 1/\lambda\xi_g = 2m|e|v_g/h^2$
S_g	Excitation error. $S_g = -\lambda g^2/2$ in axial orientation
ξ_g	Extinction distance
F_g	Structure factor
$f(\theta)$	Atomic scattering factor
θ	Scattering angle. $\theta = 2\theta_B$
θ_B	Bragg angle
λ	Relativistic electron wavelength (see eqn (2.5))
m	Relativistically corrected electron mass
m_0	Rest mass of electron
V_0	Accelerating voltage of electron microscope
$\phi(\mathbf{r})$	Crystal potential (in volts)
W	Total relativistic energy of electron ($W = mc^2 = \gamma m_0 c^2$)
V_r	Relativistically corrected accelerating voltage
γ	Relativistic factor; $\gamma = m/m_0$
$\gamma^{(j)}$	Eigenvalue of dispersion matrix. May be complex
$\mathbf{k}_i, \mathbf{k}_e$	Three-dimensional vacuum electron wavevectors
$\mathbf{k}_0$	Three-dimensional electron wavevector inside crystal corrected for mean potential
$\mathbf{k}^{(j)}$	Labelling electron wavevector for Bloch wave j inside crystal
$\mathbf{K}$	Two-dimensional scattering vector $\hat{\mathbf{K}} = u\hat{\mathbf{i}} + v\hat{\mathbf{j}}$, with $(u^2 + v^2)^{1/2} = |K| = \theta/\lambda$
$\mu^{(j)}$	Absorption coefficient for Bloch wave j.

The 'crystallographic' sign convention, which takes a plane wave to have the form $\exp(-i\mathbf{k}\cdot\mathbf{r})$ is used throughout this book, except in Section 5.7. This convention, and its relationship to the 'quantum-mechanical' convention, is discussed in Section 5.12.

Abbreviations and acronyms

ADF, annular dark-field
ADT, automated diffraction tomography
ALCHEMI, atom location by enhanced channelling microanalysis
AR, antireflection
ART, algebraic reconstruction
BS, beam stop
CBED, convergent-beam electron diffraction
CCDs, charge-coupled devices
CL, cathodoluminescence
CMOS, complementary metal-oxide semiconductor
CMR, colossal magnetoresistance
cryo-EM, cryo-electron microscopy
CTF, contrast transfer function
DQE, detective quantum efficiency
DADF, displaced-aperture dark-field
DFT, discrete Fourier transform
DM, discrete tomography
EDX, energy-dispersive X-ray (spectroscopy)
EELS, electron energy-loss spectroscopy
ELNES, electron energy-loss near-edge spectroscopy
EMCD, electron energy-loss magnetic circular dichroism
EXAFS, extended X-ray absorption fine structure
EXELFS, extended energy-loss fine structure
FCC, face-centred cubic
FEG, field emission gun
FOM, figure-of-merit
FWHM, full width at half maximum
HAADF, high-angle annular dark-field
HiO, hybrid input–output
HREM, high-resolution electron microscopy
HRTEM, high-resolution TEM
HVEM, high-voltage electron microscopy
HOLZ, higher-order Laue zone
ILST, iterative least-squares method
KCBED, kinematic CBED
MAL, maximum-likelihood [algorithm]
MAPS, monolithic active pixel device
MOFS, metal-oxide framework structures
MTF, modulation transfer function
OTF, optical function
PAM, parabola method
PCD, projected charge density
PIE, ptychographical iterative engine
PL, photoluminescence

PSF, point spread function
RMS, root mean square
SALVE, sub-angstrom low-voltage electron microscope [project]
SCEM, scanning confocal (electron) microscopy
SIRT, simultaneous iterative reconstruction
STEM, scanning transmission electron microscopy
STM, scanning tunnelling microscope
TDS, thermal diffuse scattering
TEM, transmission electron microscope/microscopy
TIDF, tilted-illumination dark-field
UHV, ultrahigh vacuum
UPS, ultraviolet photoelectron spectroscopy
XANES, X-ray absorption near-edge structure
XPS, X-ray photoelectron spectroscopy
YAG, yttrium aluminium garnet
ZOLZ, zero-order Laue zone

1
Preliminaries

This chapter is intended to re-orient an electron microscopist familiar with conventional transmission electron microscope (TEM) imaging at relatively low resolutions to the specialized requirements of atomic-resolution electron microscopy. Scanning transmission electron microscopy (STEM) is considered in Chapter 8, but much of the following also applies to STEM.

The conventional TEM bears a close resemblance to an optical microscope in which image contrast (intensity variation) is produced by the variation in optical absorption from point to point on the specimen. Most electron microscopists interpret their images in a similar way, taking 'absorption' to mean the scattering of electrons outside the objective aperture. Unlike X-rays or light, electrons are never actually 'absorbed' by a specimen; they can only be lost from the image either by large-angle scattering outside the objective aperture or, as a result of energy loss and wavelength change in the specimen, being brought to a focus on a plane far distant from the viewing screen showing the elastic or zero-loss image. This out-of-focus 'inelastic' or energy-loss image then contributes only a uniform background to the in-focus elastic image. Thus image contrast is popularly understood to arise from the creation of a local intensity deficit in regions of large scattering or 'mass thickness' where these large-angle scattered rays are intercepted by the objective aperture. The theory which describes this process is the theory of incoherent imaging (see, e.g., Cosslett (1958)).

By comparison, the high-resolution TEM (HRTEM) is a close analogue of the optical phase-contrast microscope (Mertz 2009). These optical instruments were developed to meet the needs of biological microscopists who encountered difficulty in obtaining sufficient contrast from their thinnest specimens. Nineteenth-century microscopists were dismayed to find the contrast of their biological specimens falling with decreasing specimen thickness. More than a century later, improvements in specimen preparation techniques have allowed us to see exactly the same behaviour in electron microscopy when observing specimens just a few atomic layers thick. Since the transmission image is necessarily a projection of the specimen structure in the beam direction (the depth of focus being large—see Section 2.2) these low-contrast ultra-thin specimens must be used if one wishes to resolve detail at the atomic level.

Broadly then, while low-resolution biological microscopy is mainly concerned with the electron microscope used in a manner analogous to that for a light microscope, this book is mainly devoted to the use of the TEM as a phase-contrast instrument for high-resolution studies. For the STEM, we will see that an incoherent mode of atomic-resolution imaging is also possible.

2 Preliminaries
1.1. Elementary principles of phase-contrast TEM imaging

Fresnel edge fringes, images showing the structure of a thin crystal, and images of small molecules are all examples of phase contrast. None of these contrast effects can be explained using the 'mass thickness' model for imaging with which most microscopists are familiar. All are interference effects. The distinction between interference or phase contrast and conventional low-resolution contrast is discussed more fully in Section 6.1. In this section, however, a simple optical bench experiment is described which will give the microscopist some feel for phase contrast and anticipate the results of many electron-imaging experiments. An expensive optical bench is not needed to repeat this experiment—good results will be obtained with a large-diameter lens (focal length, f_0, about 14 cm and diameter, d, about 5 cm) and an inexpensive 1.5 mW He–Ne laser. The object-to-lens and lens-to-image distances U and V could then be about 17 and 80 cm, respectively, as indicated in Fig. 1.1.

The very thin specimens used in high-resolution electron microscopy can be likened to a piece of glass under an optical microscope. An amorphous carbon film behaves rather like a ground-glass screen for optical wavelengths. In a typical high-resolution electron microscope experiment this 'screen' is used to support a sample of which one wishes to form an image. Figure 1.2 shows the image of a glass microscope slide formed using a single lens and a laser. The laser beam has been expanded (and collimated) using a ×40 optical microscope objective lens. The presence of a small molecule or atom is difficult to simulate exactly. However, the glass slide shown contains a small indentation produced by etching the glass with a small drop of hydrofluoric acid. This indentation will produce a similar contrast effect to that of an atom imaged by an electron microscope. Figure 1.2 shows a 'through-focus series' obtained by moving the lens slightly between exposures. These through-focus series are commonly taken by high-resolution microscopists to allow the best image to be selected (see Chapter 5). In particular, we notice the lack of contrast in Fig. 1.2(b), the image recorded at exact Gaussian focus when the object and the image fall on conjugate planes. *Exactly similar effects are seen on the through-focus series of electron micrographs*

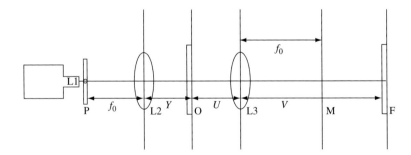

Figure 1.1 The optical bench arrangement used to record the images shown in Figs. 1.2–1.4. Here L1 is a ×40 microscope objective lens at the focus of which is placed a pin-hole aperture P. Lenses L2 and L3 have a focal length of $f_0 = 14$ cm. The object is shown at O and the film plane at F; distances are $Y = 30$ cm, $U = 17$ cm, and $V = 80$ cm. The pin-hole aperture is used as a spatial filter to provide more uniform illumination. Back-focal plane masks may be inserted at M.

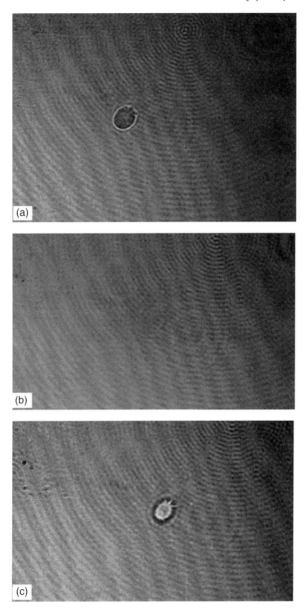

Figure 1.2 Optical through-focus series showing the effect of focus changes on the image of a small indentation in a glass plate (phase object). The image at (a) was recorded under-focus, that is, with the object too close to the lens L3. It shows a bright fringe surrounding the indentation similar to that seen on electron micrographs of small holes; image (b) is recorded at exact focus and shows only very faint contrast; image (c) is recorded at an over-focus setting (object too far from L3) and so shows a dark Fresnel fringe outlining the indentation. The background fringes arise in the illuminating system. Compare these with the electron micrographs in Fig. 6.7.

4 *Preliminaries*

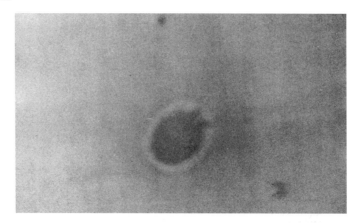

Figure 1.3 An image recorded under identical conditions to that shown in Fig. 1.2(a), with the laser source replaced with a tungsten lamp focused onto the object (critical illumination). The faint contrast seen is due to the preservation of some coherence in the illumination introduced by limiting the size of lens L2. This contrast disappears completely if a large lens is used. Variations in the size of this lens (or an aperture near it) are analogous to changes in the size of the second condenser aperture in an electron microscope.

of single atoms shown in Fig. 6.7. Figure 1.3 shows the same object imaged using a conventional tungsten lamp-bulb as the source of illumination to provide 'incoherent' illumination conditions. Despite wide changes in focus, little contrast appears.

If this analogy between the indented glass slide and a high-resolution specimen in electron microscopy holds, this simple experiment suggests that our high-resolution specimens will be imaged with strong contrast only if a coherent source of illumination is used and if images are recorded slightly out of focus. These conclusions are borne out in practice, and the importance of high-intensity coherent electron sources for high-resolution microscopy is emphasized throughout this book. Two chapters have been devoted to the topic—Chapter 12 deals with experimental aspects and Chapter 4 describes the elementary coherence theory.

To return to our optical analogue. Given coherent illumination the question arises as to whether methods other than the introduction of a focusing error can be used to produce image contrast. During the last century many empirical methods were indeed developed to enhance the contrast of images such as that in Fig. 1.2(b). These included: (1) reducing the size of the objective aperture as in Fig. 1.4(a); (2) introducing a focusing error as in Fig. 1.2(a); (3) simple interventions in the lens back-focal plane as in Fig. 1.4(b), where Schlieren contrast is shown (the back-focal plane is approximately the plane of the objective aperture for an electron microscope); and (4) the use of back-focal plane phase plates, similar to the Zernike phase plate used in optical microscopy.

All of these methods have their parallel in electron microscopy. The first is not useful, since image resolution is necessarily limited. The second is the usual method of high-resolution electron microscopy and is discussed in more detail in Chapters 3 and 12. Methods based on interventions in the lens back-focal plane have appeared from time to time (see,

Elementary principles of TEM imaging 5

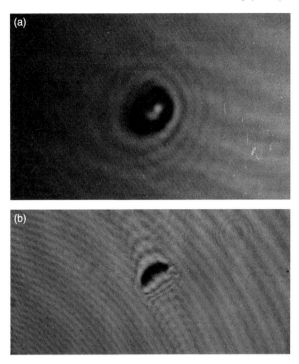

Figure 1.4 Using a laser source, both these images were recorded at exact focus (exactly as in Fig. 1.2(b)). Yet in both, the outline of the indentation can be seen. In (a), a small aperture has been placed on the axis at M, severely limiting the image resolution. On removing this aperture the image contrast disappears. The use of a small aperture at M (the back-focal plane) is analogous to the normal low-resolution method of obtaining contrast in biological electron microscopy. In (b) a razor blade has been placed across the beam at M, thus preventing exactly half the diffraction pattern from contributing to the image. The resulting image is approximately proportional to the derivative of the phase shift introduced by the object taken in a direction normal to the edge of the razor blade. Notice the fine fringes inside the edge of the indentation arising from multiple reflection within the glass slide.

e.g., Spence (1974)), while there have been steady improvements in the construction of Zernike phase-plates over the years for electron microscopes (corresponding to the use of back-focal plane phase plates), as reviewed by Hall et al. (2011). Carbon films or applied electric fields across the back-focal plane have been used—in one scheme, for example, an infrared laser beam, focused onto the central spot of the diffraction pattern, will cause the required 90° phase shift in the central undiffracted electron beam (Muller et al. 2010).

In both electron and optical microscopy the reasons for the lack of contrast at exact focus are the same—these thin specimens ('phase objects') affect only the phase of the wave transmitted by the specimen and not its amplitude. That is, they behave like a medium of variable refractive index (see Section 2.1). It is this variation in refractive index from point to point across the specimen (proportional to the specimen's atomic potential in volts for electron microscopy) which must be converted into intensity variations in the image if we are to 'see', for example, atoms in the electron microscope. Through the reciprocity theorem, discussed in Section 8.1, all these considerations apply to the imaging of single atoms in the STEM in bright-field mode.

We can sharpen these ideas by comparing the simplest mathematical expressions for imaging under coherent and incoherent illumination. For the piece of glass shown in Fig. 1.2, the phase of the wave transmitted through the glass differs from that of an unobstructed reference wave by $2\pi(n-1)/\lambda$ times the thickness of the glass, where n is the refractive index of the glass. If $t(x)$ is the thickness of the glass at the object point x, the amplitude of the optical wave leaving the glass is given from elementary optics (see, e.g., Lipson and Lipson (1969)) by

$$f(x) = \exp(-2\pi i n t(x)/\lambda) \qquad (1.1)$$

The corresponding expression for electrons is derived in Section 2.1. Optical imaging theory (Goodman 2004) gives the image intensity, as

$$I(x)_c = |f(x) * S(x)|^2 \qquad (1.2)$$

for coherent illumination and

$$I(x)_i = |f(x)|^2 * |S(x)|^2 \qquad (1.3)$$

for incoherent illumination. In these equations the function $S(x)$ is known as the instrumental impulse response, and gives the image amplitude which would be observed if one were to form an image of a point object much smaller than the resolution limit of the microscope. Such an object does not exist for electron microscopy; however, the expected form of $S(x)$ for one modern electron microscope can be seen in Fig. 3.8. This function specifies all the instrumental imperfections and parameters such as objective aperture size (which determines the diffraction limit), the lens aberrations, and the magnitude of any focusing error. The asterisk in eqns (1.2) and (1.3) indicates the mathematical operation of convolution (see Section 3.1), which results in a smearing or broadening of the object function $f(x)$. It is not possible to resolve detail finer than the width of the function $S(x)$. Using eqn (1.1) in eqns (1.2) and (1.3) we see that under incoherent illumination the image intensity from such a phase object does not vary with position in the object, since

$$|\exp\{-2\pi i t(x) n/\lambda\}|^2 = 1 \qquad (1.4)$$

Only by using coherent illumination and an 'imperfect' microscope can we hope to obtain contrast variations in the image of a specimen showing only variations in refractive index. *In high-resolution electron microscopy of thin specimens the accurate control of illumination coherence and defect of focus are crucial for success.* The amount of fine detail in a high-resolution TEM micrograph increases dramatically with improved coherence of illumination, while completely misleading detail may be observed in images recorded at the wrong focus setting.

Elementary principles of TEM imaging 7

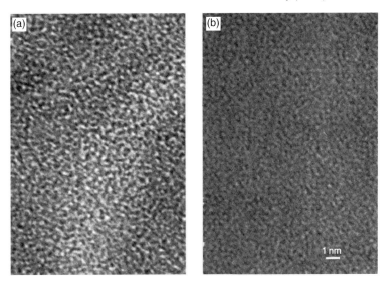

Figure 1.5 Two electron microscope images of amorphous carbon films recorded at the same focus setting but using different condenser apertures. In (a) a small second condenser aperture has been used, resulting in an image showing high contrast and fine detail. This contrast is lost in (b), where a large aperture has been used. This effect is seen even more clearly in optical diffractograms (see Chapter 4).

To demonstrate the effect of coherence on TEM image quality at high resolution, Fig. 1.5 shows images of an amorphous carbon film using widely different illumination coherence conditions but otherwise identical conditions. The loss of fine detail in the 'incoherent' image is clear. In practice, for a microscope fitted with a conventional hair-pin filament, the illumination coherence is determined by the size of the second condenser lens aperture, a small aperture producing high coherence. With a sufficiently small electron source (such as a field-emission source) the illumination may be almost perfectly coherent and so independent of the choice of second condenser (illuminating) aperture.

In most cases of practical interest the imaging is partially coherent. By this we loosely mean that for object detail below a certain size X_c we can use the model of coherent phase contrast imaging (see Fig. 1.2) while for detail much larger than X_c the imaging is incoherent. The distance X_c is given approximately by the electron wavelength divided by the semi-angle subtended by the second condenser aperture at the specimen, when using a hair-pin filament. In Section 10.7 a fuller discussion of methods for measuring X_c is given which includes the effect of the objective lens pre-field. Since the width of Fresnel edge fringes commonly seen on micrographs is approximately equal to X_c, we can adopt the rough rule of thumb that the more complicated through-focus effects of phase contrast described in Section 6.2 will become important whenever one is interested in image detail much finer than the width of these fringes.

To summarize, the main changes in emphasis for microscopists accustomed to conventional TEM imaging and considering a high-resolution project are as follows: in conventional imaging we deal with 'thick', possibly stained, specimens using a small objective aperture and a large condenser aperture. Focusing presents no special problems. In high-resolution phase-contrast imaging we deal with the thinnest possible specimens (unstained), using a larger objective aperture of optimum dimensions and a small illuminating aperture. The choice of focus becomes crucial and must be guided either by experience with computed images or by the presence of a small region of specimen of known structure in the neighbourhood of the wanted structure. This image simulation experience indicates that for the vast majority of large-unit-cell crystals imaged at a resolution no higher than 0.20 nm the choice of focus which gives a readily interpretable image containing no false detail is between 20 and 60 nm under-focus (weakened lens) for microscopes with a spherical aberration constant between 1 and 2 mm. A calibrated focus control is therefore needed to set this focus condition. Methods for measuring the focus increments are given in Section 10.1. For aberration-corrected instruments, the choice of aberration-free conditions produces an in-focus image with zero contrast, as we see from eqn (1.4). Section 7.4 describes the optimum choice of defocus and aberration constants for best resolution and contrast for these instruments. The limits on specimen thickness are discussed in Chapter 5.

1.2. Instrumental requirements for high resolution

The major manufacturers of electron microscopes offer instruments that are closely matched in performance. All are capable of giving point resolution better than 0.2 nm for bright-field images. The use of side-entry eucentric goniometer stages, of the kind used by biologists for many years, for high-resolution work greatly facilitates the procedure of orienting a small crystal, since the lateral movement accompanying tilting of the specimen is minimized by these stages. However, the drift rate due to thermal effects using these stages has been considerably higher than that for top-entry stages.

A laboratory which has recently purchased a TEM and wishes to use it for high-resolution studies should consider the following points. This list is intended as a rough checklist to refer the reader to further discussion of these topics in other chapters.

1. The microscope site must be acceptable. Mechanical vibration, stray magnetic fields, and room temperature must all be within acceptable limits. These factors are discussed in Chapter 11.

2. A reliable supply of clean cooling water at constant temperature and pressure must be assured (see Chapter 11). In installations where internal recirculating water is not available and the external supply is impure it is common to find both the specimen drift rate and illumination stability deteriorating over a period of weeks as the water filter clogs up, causing the pressure of the cooling water to fall and leading to fluctuations in temperature of the specimen stage and lens. A separate closed-circuit water supply and external heat exchanger is the best solution for high-resolution work, with the water supplied only to the microscope. Vibration from the heat exchanger pumps must be considered; these should be in a separate room from the microscope. Despite the use of distilled water, algae are bound to form in the water supply. Some manufacturers warn against the use of algae inhibitors as these may corrode the lens-cooling jacket, repairs to which are very expensive. About

all that can be done is to replace the water periodically. Thermal stage drift is one of the most serious problems for high-resolution imaging, particularly for long-exposure dark-field images. If external hard water is used *it is essential that the condition of the water filter is checked regularly and replaced if dirty*. If the microscope's inbuilt thermoregulator is used it is equally important to check every few days that it is operating within the temperature limits for which it is designed.

3. In order to record an image at a specified focus defect it will be necessary to measure the change in focus between focus control steps ('clicks') using the methods of Section 10.1. In addition, the spherical aberration constant C_s must be known for the optimum objective lens excitation. This should be less than 2 mm at 100 kV if high-resolution results are expected. Methods for measuring C_s are given in Section 10.2. Microscope calibration—the measurement of defocus increments, spherical aberration constant, and objective lens current—is therefore an important first step. Once the optimum specimen position in the lens gap has been found, the objective lens current needed to focus the specimen at this position should be noted, and the highest-quality images recorded near this lens current. This is not always possible when using a tilting specimen holder if a non-axial object point is required, since the specimen height (and hence the objective lens current needed for focus) varies across the inclined specimen.

4. The resolution obtained in a transmission image depends, amongst other things, on the specimen position in the objective lens. The optimum 'specimen height' must be found by trial and error. In reducing the specimen height and so increasing the objective lens current needed for focus, the lens focal length and spherical aberration constant are reduced, leading to improved resolution. Figure 2.12(a) shows the dependence of spherical aberration on lens excitation. A second effect, however, results from the depth of the objective lens pre-field, which increases as the specimen is immersed more deeply into the lens gap. This increases the overall demagnification of the illumination system and increases its angular magnification (see Section 2.2). The result is an increase in the size of the diffraction spots of a crystalline specimen which, for a fixed illumination aperture, allows the unwanted effect of spherical aberration to act over a larger range of angles (see Section 3.3). Finally, as shown in Fig. 2.12(a), for many lens designs the chromatic aberration coefficient C_c passes through a minimum as a function of lens excitation, and this parameter affects both the contrast and resolution of fine image detail. The complicated interaction between all these factors which depend on specimen position can best be understood using the 'damping envelope' concept described in Section 4.2 and Appendix 3. This 'damping envelope' controls the information resolution limit (loosely referred to by manufacturers as the 'line' or 'lattice' resolution) of the instrument and depends chiefly on the size of the illumination aperture and C_c. The point resolution, however, is determined by spherical aberration. A method has been described which would allow both these important resolution limits to be measured as a function of specimen position in the lens bore through an analysis of optical diffractogram pairs (see Section 10.6). A systematic analysis of this kind, however, represents a sizeable research project, involving many practical difficulties. For example, in such an analysis, diffractograms are required at similar focus defects for a range of specimen positions, yet the 'reference focus' used to establish a known focus defect (see Section 12.5) itself depends on C_s, which varies with specimen position. In addition, both the focal increment corresponding to a single step on the focus control and the chromatic

aberration constant (which affects the 'size' of diffractogram ring patterns) depend on the specimen position. For preliminary work, then, a simpler procedure is to record images of a crystal with a large unit cell at the Scherzer focus for a range of specimen heights and to select the highest-quality image, judging this by eye. Experience in comparing computed and experimental images of crystals with large unit cells shows that a useful judgement of image quality (a combination of contrast, point, and line resolution) can be made by eye. Once this near-optimum specimen position has been found, further refinements can be made using the computer image-matching technique together with diffractogram analysis. To obtain the required range of specimen heights on top-entry machines it is necessary to fit small brass washers above or below the specimen. A shallow threaded cap to fit over the specimen may also be needed to bring the tilted specimen closer to the lower cold-finger. The measured objective lens current needed to bring the image into focus is then used as a measure of specimen height, and must be recorded for each image. The specimen height and corresponding lens current needed to give images of the highest quality can then be determined and permanently recorded for the particular electron microscope. Crystals with large unit cells have the useful property that the highest-contrast images are usually produced near the Scherzer focus, so that this focus setting can be found routinely by experienced microscopists working with these materials.

5. A vacuum of 0.5×10^{-7} Torr or better is needed (measured in the rear pumping line). The simplest way to trace vacuum leaks is to use a partial pressure gauge (see Section 11.5). A microscope which is to be used for high-resolution imaging should be fitted permanently with such a gauge, while plasma-cleaning apparatus for the sample holder is also crucial. Inexpensive gauges of the radiofrequency quadrupole type are useful. In a clean, well-outgassed microscope with no serious leaks and a vacuum of 0.5×10^{-7} Torr the major source of contamination is frequently the specimen itself. Always compare the contamination rates for several different specimens before concluding that the microscope itself is causing a contamination problem.

6. The high-voltage supply of the microscope must be sufficiently stable to allow high-resolution images to be obtained. Given a sufficiently stable high-voltage supply (see Sections 11.2 and 2.8.2) this generally requires a scrupulously clean gun chamber and Wehnelt cylinder if a field-emission gun is not used. Methods for observing high-voltage instability are discussed in Section 12.2.

7. The room containing the microscope must be easily darkened completely, and a room-light dimmer control needs to be fitted within arm's reach of the operator's chair. Specimen changes can then be made in dim lighting so that the user does not lose the slow chemical dark adaptation of the eyes. The newest generation of atomic-resolution machines are controlled digitally from a neighboring room.

1.3. First experiments

This section outlines a procedure which will enable a microscopist new to high-resolution work to become familiar with the techniques of high-resolution TEM and to practise the required skills of focusing, astigmatism correction, and image examination. The test specimen used consists of small clusters of heavy atoms supported on a thin carbon film. The microscope is assumed to satisfy the conditions of Section 1.2 and the focal increments and spherical aberration constant are presumed known.

1. Specimens, consisting of small clusters of heavy atoms lying on a thin carbon substrate film, can be purchased from suppliers.

2. Once a suitable specimen area has been found, bright-field TEM images can be recorded at high magnification (400 000 or more) using a charge-coupled device (CCD) camera. Chapter 12 gives details of all necessary precautions. A microscope which has been switched on in the morning, with all cold-traps kept filled with liquid nitrogen, should be in thermal equilibrium by early afternoon when serious work can commence. The morning can be profitably spent examining possible specimens. The cold-traps must not be allowed to boil dry during a lunch break. *The thermal stability and cleanliness of the three components in the objective lens pole-piece gap (objective aperture, cold-finger, and specimen) are of the utmost importance in high-resolution electron microscopy.* The best arrangement is to leave the microscope running through the week, fitted with an automatic liquid-nitrogen pump for the cold-traps. Record images without an objective aperture inserted (so that resolution is limited by incoherent instabilities—see Section 4.2) and also using an aperture whose size is given by eqn (6.16b). The semi-angle θ_{ap} subtended by the objective aperture can be measured by taking a double exposure of the diffraction pattern of some continuous gold film with, and without, the aperture in place (see Section 10.2). A check of the following must be made before taking pictures: (1) condenser astigmatism (for maximum image intensity); (2) gun tilt; (3) current or voltage centre (see Section 12.2); (4) high-voltage stability (see Section 11.2); (5) absence of thermal contact between specimen and cold-finger (see Section 12.2); (6) cleanliness of objective aperture; (7) specimen drift (see Section 11.4); and (8) contamination (see Section 11.5).

Use the minimum-contrast condition (see Section 12.5) to correct astigmatism and take several bright-field images in the neighbourhood of the focus value given by eqn (6.16a). A through-focus series about the minimum contrast focus in steps of, say, 20 nm will show the characteristic change from a dark over-focus Fresnel fringe around an atom cluster to a bright fringe in the under-focus images.

3. Digital diffractograms of the images should be obtained as described in Section 10.6. The measured diameter of the rings seen in these can be used with the simple computer program given in Appendix 1 to find the focus setting for each micrograph and the microscope's spherical aberration constant. Alternatively, the images may be analysed using commercial software. These diffractograms reveal at a glance the presence of astigmatism or specimen movement during the exposure (drift—see Section 10.6). This immediate 'feedback' is essential for a microscopist learning the skills of astigmatism correction and focusing. With practice the microscopist will become adept at finding the minimum-contrast condition, correcting astigmatism, and resetting the focus control a fixed number of 'clicks' toward the under-focus side to obtain images of the highest contrast and resolution using these test samples.

References

Cosslett, V. E. (1958). Quantitative aspects of electron staining. *J. Roy. Microsc. Soc.* **78**, 18.
Goodman, J. W. (2004). *Introduction to Fourier optics*. Third Edition. McGraw-Hill, New York.
Hall, R. J., Nogales, E., and Glaeser, R. M. (2011). Accurate modeling of single-particle cryo-EM images quantifies the benefits expected from using Zernike phase contrast. *J. Struct. Biol.* **174**, 468.

Lipson, S. G. and Lipson, H. (1969). *Optical physics*. Cambridge University Press, London.

Mertz, J. (2009). *Introduction to optical microscopy*. Roberts and Co., Greenwood Village, CO.

Muller, H., Danev, J., Spence, J., Padmore, H., and Glaeser, R. M. (2010). Design of an electron microscope phase plate using a focussed continuous-wave laser. *New J. Phys.* **12**, 073011.

Spence, J. C. H. (1974). Complex image determination in the electron microscope. *Opt. Acta* **21**, 835.

2
Electron optics

An elementary knowledge of electron optics is important for the intelligent use of an electron microscope, particularly at high resolution. This chapter describes the important physical properties of magnetic electron lenses, such as image rotation, aberrations, focal length, and the distinction between projector and objective modes. In addition, we review the history of the successful effort toward aberration correction, define the higher-order aberration coefficients, and outline the principles of correction for later elaboration. Calculations of lens characteristics are given, showing the way in which lens aberrations depend on lens excitation and geometry. The elegant contribution of early workers such as Lenz, Glaser, Grivet, and Ramberg using simple closed-form approximations for the lens field can be traced through the electron optics texts in the references. Useful introductory accounts of electron optics can be found in the books by Hall (1966), Hawkes (1972), and especially Grivet (1965), while a comprehensive modern review and extensive bibliography is given in Hawkes and Kasper (1996). Modern lens designs are based on computed solutions of the Laplace equation (see Septier 1967 for a review) and subsequent numerical solution of the ray equation (Orloff 2007, Carey 2006, Mulvey and Wallington 1972). The accurate measurement of electron-optical parameters and aberration coefficients, both individually and as part of an automated feedback loop for aberration correction, is discussed in Chapter 10. For superconducting electron optics see Dietrich (1977).

2.1. The electron wavelength and relativity

Rather than solve the Schrödinger equation for the electron microscope as a whole, it is simpler to separate the four problems of beam–specimen interactions, magnetic lens action, detection, and fast electron sources. The first problem is a many-body problem solved by reducing it to the interaction between one beam electron and an optical potential, while the second has traditionally been treated classically. A wavelength is assigned to the fast electron as follows.

The principle of conservation of energy applied to an electron of charge $-e$ traversing a region in which the potential varies from 0 to V_0 gives

$$eV_0 = p^2/2m = h^2/2m\lambda^2 \tag{2.1}$$

where p is the electron momentum and h is Planck's constant. Thus,

$$\lambda = \frac{h}{\sqrt{2meV_0}} \tag{2.2}$$

where the de Broglie relation $p = mv = h/\lambda$ has been used. An electron leaves the filament with high potential energy and thermal kinetic energy, and arrives at the anode with no potential energy and high kinetic energy. The zero of potential energy is taken at ground potential. If λ is in nanometres and V_0 in volts, then

$$\lambda = 1.22639/\sqrt{V_0} \qquad (2.3)$$

At higher energies the relativistic variation of electron mass must be considered. Neglect of this leads to a 5% error in λ at 100 kV. The relativistically corrected mass is

$$m = m_0/(1 - v^2/c^2)^{1/2}$$

and the relativistic equation corresponding to eqn (2.1) is

$$eV_0 = (m - m_0)c^2$$

with m_0 the electron rest mass and c the velocity of light. These equations may be combined to give an expression for the electron momentum mv. Used in the de Broglie relation, this gives the relativistically corrected electron wavelength as

$$\lambda = h/(2m_0 eV_r)^{1/2} \qquad (2.4)$$

where

$$V_r = V_0 + \left(\frac{e}{2m_0 c^2}\right) V_0^2$$

is the 'relativistic accelerating voltage', introduced as a convenience. For computer calculations the value of λ may be taken as

$$\lambda = 1.22639/\left(V_0 + 0.97845 \times 10^{-6} V_0^2\right)^{1/2} \qquad (2.5)$$

with V_0 the microscope accelerating voltage in volts and λ in nanometres.

The relativistic correction is important for high-voltage electron microscopy (HVEM). If V_0 is expressed in MeV, a good approximation is $V_r = V_0 + V_0^2$, so that $V_r = 6$ MeV for a 2 MeV microscope. Instruments have been built operating at energies up to 3 MeV, and there is some renewed interest in these for fast electron microscopy, in view of the reduction in Coulomb interactions between electrons at high energy. The formal justification for these definitions of a relativistically corrected electron mass and wavelength must be based on the Dirac equation, as discussed by K. Fujiwara and A. Howie (see Section 5.7).

A method for measuring the relativistically corrected electron wavelength directly from a diffraction pattern is discussed in Section 10.8. This method requires only a knowledge of a crystal structure and does not require the microscope accelerating voltage or camera length to be known.

The positive electrostatic potential $\phi(r)$ (in volts) inside the specimen further accelerates the incident fast electron, resulting in a small reduction in wavelength inside the specimen

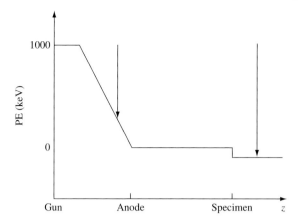

Figure 2.1 Simplified potential energy (PE) diagram for an electron microscope. The length of the vertical arrow is proportional to the kinetic energy of the fast electron and inversely proportional to the square of its wavelength. The sum of the electron's potential energy (represented by the height of the graph) and its kinetic energy is constant. Electrons leave the filament with low kinetic energy and high potential energy (supplied by the high-voltage set) and exchange this for kinetic energy on their way to the anode, which is at ground potential. As with a ball rolling down a hill, they are further accelerated as they 'fall in' to the specimen. Approximate distance down the microscope column is represented on the abscissa and the potential step at the specimen has been exaggerated.

(Fig. 2.1). Ignoring the periodic variation of potential, which gives rise to diffraction and the dispersion surface construction, the mean value of this inner potential is given by $v_0 = \phi_0$, the zero-order Fourier coefficient of potential (see Section 5.3.2). A typical value for v_0 is 10 V. The refractive index n of a material for electrons is then given by the ratio of wavelength λ in a vacuum to that inside the specimen λ', so that

$$n = \frac{\lambda}{\lambda'} = \left(\frac{1.23}{\sqrt{V_0}}\right)\left(\frac{\sqrt{V_0 + \phi_0}}{1.23}\right) \approx 1 + \frac{\phi_0}{2V_0}$$

The phase shift of a fast electron passing through a specimen of thickness t with respect to that of the vacuum is then

$$\theta = 2\pi(n-1)t/\lambda = \pi\phi_0 t/\lambda V_0 = \sigma\phi_0 t$$

as suggested in Fig. 2.2. Here $\sigma = \pi/\lambda V_0 = 2\pi m e\lambda/h$. If the approximation is then made that the exit-face wavefunction can be found by computing its phase along a single optical path such as AB in Fig. 2.2, the product $\phi_0 t$ can be replaced by the specimen potential function projected in the direction of the incident beam (see Section 3.4). The neglected contributions from paths such as CA can be included using the Feynman path-integral method as discussed in Jap and Glaeser (1978).

16   Electron optics

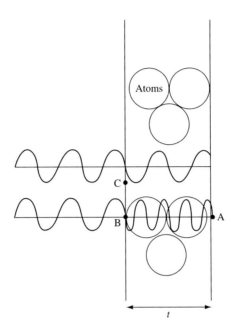

Figure 2.2 The electron wave illustrated in two cases passing through a specimen. The wave passing through the centre of an atom (where the potential is high) has its wavelength reduced and so suffers a phase advance relative to the wave passing between the atoms, which experiences little change in its wavelength. The assumption of this simplified model, used in the phase grating approximation, is that the amplitude at A can be calculated along the optical path AB with no contribution at A from a point such as C. For thick specimens this approximation is unsatisfactory.

2.2. Simple lens properties

Modern electron microscopes have many imaging lenses, of variable focal length, beyond the specimen, with the position of the object and final viewing screen fixed for the purposes of focusing. At the high magnifications usually used for high-resolution microscopy, the lens currents (which determine the focal lengths) of lenses L2, L3, and L4, for example, might be used to control the magnification, as shown in Fig. 2.3 and Table 2.1. For a fixed magnification setting, focusing is achieved by adjusting the strength of the objective lens L1 until the fixed plane P1 is conjugate to the exit face of the specimen. The properties of these lenses can be understood from the equations for ideal lens behaviour given in this section, which also provide results for later chapters.

The study of electron optics seeks to determine the conditions under which the electron wavefield passing through an electron lens satisfies the requirements for perfect image formation. For comparison purposes, it is convenient to set up a model of the ideal lens. The ideal lens is a mathematical abstraction which provides perfect imaging given by a projective transformation between the object and image space. The constants appearing in this transformation specify the positions of the cardinal planes of the lens. The six important cardinal planes are the two focus planes, the two principal planes, and the two nodal

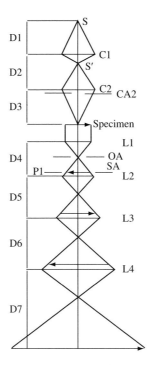

Figure 2.3 Ray diagram for an electron microscope with two condenser lenses, C1 and C2 and four imaging lenses, L1, L2, L3, and L4, operating at high magnification. A typical set of dimensions for $D1$ to $D7$ is given in Table 2.1, together with the possible range of focal lengths. These values may be used for examples throughout this book. Here OA is the objective aperture, P1 is a fixed plane, and SA is the selected area aperture.

Table 2.1 Electron-optical data for a typical electron microscope (see Fig. 2.3).

Distances between lens centres (approximate) (mm)	Focal length range (mm)
$D1 = 143.6$	$1.65 < f(C1) < 19$
$D2 = 94.3$	$30 < f(C2) < 1060$
$D3 = 251.4$	$15.4 < f((L2) < 281$
$D4 = 215.5$	$3.1 < f((L3) < 99.5$
$D5 = 44.9$	$2.06 < f(L4) < 16.4$
$D6 = 73.6$	
$D7 = 345.6$	

For magnifications greater than 100 000 the magnification is controlled by adjusting the focal length of L3 with $f(L2) = 15.4$ mm fixed and $f(L4) = 2.1$ mm fixed. The focal length of L3 is set as follows: $f(L3) = 9.8$, 7.0, 5.0, 3.1 mm for $M = 150, 200, 400, 750$ K, respectively.

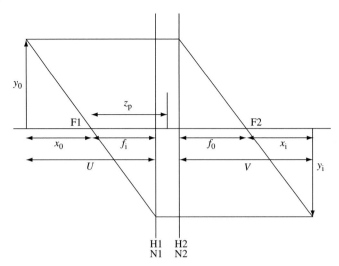

Figure 2.4 The thick lens. The nodal planes (N1, N2), principal planes (H1, H2), and focal planes (F1, F2) are shown together with the lens focal lengths (f_i, f_o) and the object focal distance z_p. For a magnetic electron lens the principal planes are crossed.

planes as shown in Fig. 2.4. For magnetic lenses the nodal planes coincide with the principal planes. The points where the axis crosses the nodal planes are called nodal points, N1 and N2. Principal planes are planes of unit lateral magnification, while nodal planes are planes of unit angular magnification. For an axially symmetric lens, the projective transformation for perfect imaging simplifies to

$$\frac{y_i}{y_o} = \frac{f_i}{x_o} = \frac{x_i}{f_o} \qquad (2.6)$$

where the symbols are as defined in Fig. 2.4. Equation 2.6 is Newton's lens equation. Figure 2.4 provides a convenient graphical construction for eqn (2.6). Here F1 and F2 are known as the object and image focus respectively with H1 and H2 the object and image principal planes. The determination of the positions of these planes is the key problem of electron optics—once they are known, the rules for graphical construction of figures satisfying eqn (2.6) can be used to find the image of an arbitrary object. The rule for a construction which gives the conjugate image point P' of a known object point P is:

1. Draw a ray through P and F1, intersecting H1 at Q. Through Q draw a ray YQ parallel to the axis extending into both object and image spaces.
2. Draw a ray parallel to the axis through P to intersect H2. From this intersection draw a ray through F2 to intersect the ray YQ at P. P' is the image of P.

As an example of this construction, Fig. 2.5 shows these rules applied to the objective lens of a modern electron microscope operating at moderate magnification (about 40 000). Note that the image formed by the objective lens is virtual, and that the principal planes

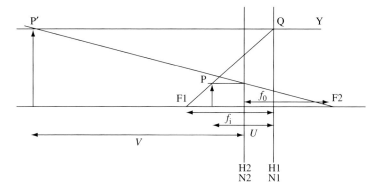

Figure 2.5 Ray diagram for the objective lens of a microscope operating at moderate magnification. The image is virtual and the principal planes are crossed. Object and image focal lengths are equal for magnetic lenses. A typical value for f_2 is 2 mm, and the magnification $M = V/U$ may be about 20.

are crossed, as they are for all magnetic electron lenses. The use of this mode in a four-lens instrument has advantages for biological specimens where radiation damage must be minimized. At this moderate magnification lens L3 is switched off. At high magnification all lenses are used. Modern lens designers use the methods of matrix optics; an elementary introduction to these techniques can be found in Nussbaum (1968).

The simple thin-lens formula can still be used if the object and image distances U and V are measured from the lens principal planes H1 and H2. Equation (2.6) becomes

$$\frac{f_i}{U} + \frac{f_o}{V} = 1$$

If the refractive indices in the object and image space are equal, as they are for magnetic electron lenses, then

$$f_i = f_o = f \tag{2.7}$$

and so

$$\frac{1}{U} + \frac{1}{V} = \frac{1}{f} \tag{2.8}$$

A construction can also be given to enable the continuation of a ray segment in the image space to be found if it is known in the object space. Note that a ray from P arriving at N1 at an arbitrary angle leaves N2 at the same angle. For a thin lens the principal planes coincide, so this is the ray through the lens origin. Equation (2.8) is quite general if the following sign convention is obeyed: U is positive (negative) when the object is to the left (right) of H1, V is positive (negative) when the image is to the right (left) of H2. Both focal lengths are positive for a convergent lens, and all magnetic lenses are convergent for electrons and positrons. From eqn (2.8) and the definition of magnification (eqn 2.9) three cases emerge:

20 Electron optics

1. $U < f$: image is virtual, erect, and magnified.
2. $f < U < 2f$: image is real, inverted, and magnified.
3. $U < 2f$: image is real, inverted, and reduced.

Some additional terms, commonly used in electron optics, are now defined.

1. *The lateral magnification M is given by*

$$M = \frac{y_i}{y_o} = -\frac{V}{U} \tag{2.9}$$

Using eqn (2.8) we have

$$M - 1 = -\frac{V}{f} \tag{2.10}$$

so that if V is fixed and the magnification is large, as it is for the objective lens of an electron microscope, the magnification is inversely proportional to the focal length of the objective lens. For high magnification, U must be slightly greater than f—both are about a millimetre for a high-resolution objective lens.

2. *The angular magnification m is, for small angles,*

$$m = \frac{\tan \theta_i}{\tan \theta_o} \approx \frac{\theta_i}{\theta_o} = \left| \frac{1}{M} \right| \tag{2.11a}$$

as shown in Fig. 2.6.

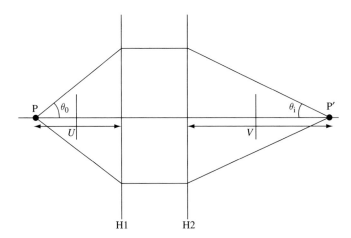

Figure 2.6 Angular magnification. The image P' of a point P is shown together with the angles which a ray makes with these points.

3. *The entrance and exit pupils* of an optical system are important in limiting its resolution and light-gathering power. These concepts also simplify the analysis of complex optical systems. The entrance pupil is defined as the image of that aperture, formed by the optical system which precedes it, which subtends the smallest angle at the object. The image of the entrance pupil formed by the whole system is known as the exit pupil. The 'aperture stop' is the physical aperture whose image forms the entrance pupil, as shown in Fig. 2.7.

4. *The Gaussian reference sphere* for an image point P is defined as the sphere, centred on P, which passes through the intersection of the optic axis with the exit pupil (see Fig. 2.7). For an unaberrated optical system, the surface of constant phase for a Huygens spherical wavelet converging toward P coincides with this reference sphere. The deviation of the wavefront from the Gaussian reference sphere specifies the aberrations of the system (see Section 3.3), while the diffraction limit is imposed by the finite size of the exit pupil, or, equivalently, the entrance pupil. This is discussed further in Chapter 3.

5. *The longitudinal magnification*, M_z, can be used to relate depth of field to depth of focus (see below). Differentiation of eqn (2.8) gives

$$\frac{\Delta V}{\Delta U} = -M^2 = M_z \qquad (2.11b)$$

Thus an object displacement ΔU causes a displacement ΔV of conjugate image planes given by this equation. For example the image planes conjugate to the upper and lower surfaces of an atom 0.3 nm 'thick' are separated by 3 m if the lateral magnification M is 100 000.

6. *Incoherent imaging theory* gives the depth of field or range of focus values (referred to the object plane) over which an object point can be considered 'in focus' as

$$Z_D = 2d/\theta = 2\lambda/\theta^2 \qquad (2.11c)$$

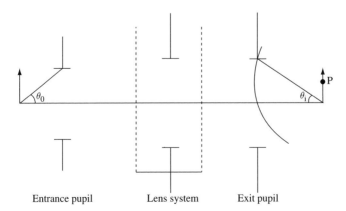

Figure 2.7 The entrance and exit pupils of an optic system. A complicated optical system consisting of many lenses can be treated as a 'black box' and specified by its entrance and exit pupils and a complex transfer function. A Huygens spherical wavefront is shown converging to an image point P.

22 Electron optics

where θ is the objective aperture semi-angle and d is the microscope resolution. However, this result cannot be accurately applied to the coherent high-resolution imaging of phase objects (see Sections 3.4 and 5.2).

The methods used by lens designers to determine the position of the cardinal planes of electron lenses are discussed in the next sections. From Fig. 2.4 it is seen that the axis crossing of a ray entering (leaving) parallel to the axis defines the image (object) focus. In electron optics the trajectory of an electron entering the lens field parallel to the axis can similarly be used to find the lens focus once the equation of motion for the electron can be solved for a particular magnetic field distribution. Real electron trajectories follow smooth curves within the lens magnetic field. In order to use the ideal lens model it may be necessary to use the virtual extensions of a ray from a point well outside the influence of the field to define the lens focus.

2.3. The paraxial ray equation

The focusing action of an axially symmetric magnetic field can be understood as follows. Figure 2.8 shows a simplified diagram of an electron lens, including a typical line of magnetic flux. The actual arrangement used for one instrument in the important region of the lens pole-pieces is shown in Fig. 2.9. The dimensions of the pole-piece are S, R_1, and R_2 as shown in Fig. 2.8. The magnetic field B is confined to the pole-piece gap, where an electron of charge $-e$ and velocity v experience a force

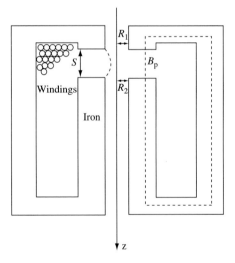

Figure 2.8 Simplified diagram of a magnetic electron lens. The dimensions of the pole-piece are indicated and the path taken for Ampère's law is shown as a broken line. A line of magnetic flux is also shown as broken, giving a qualitative indication of the focusing action of the lens (see text). The field strength in the gap far from the optic axis is B_p. In practice the windings are water cooled and the pole-piece is removable.

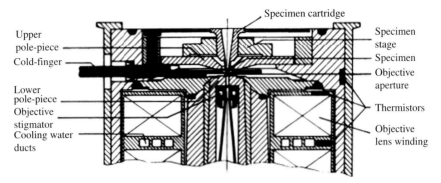

Figure 2.9 Detail of an actual high-resolution pole-piece. The upper pole-piece bore diameter $2R_1$ is 9 mm, the lower bore diameter $2R_2$ is 3 mm, and the gap S is 5 mm in this commercial design. (See Fig. 2.8 for the definitions of R_1, R_2, and S.)

$$F = -ev \times B = m\frac{d^2r}{dt^2} \tag{2.12}$$

The direction of F is given by the left-hand rule (current flow opposite to electron flow), which from Fig. 2.8 is seen to be into the page as the electron enters the field on the left (assuming the upper pole-piece is a north pole). An electron entering on the right side experiences a force out of the page. These forces result in a helical rotation of the electron trajectory. The rotational velocity component r_θ imparted interacts with the z component of the field $B_z(r)$ to produce a force towards the axis, again given by the left-hand rule. This force is responsible for the focusing action of the lens.

Using certain approximations described in most optics texts, eqn (2.12) can be simplified for meridional rays in a cylindrical coordinate system. These are rays in a plane containing the axis before reaching the lens. The simplifications include the neglect of terms which would lead to imperfect imaging (lens aberrations). The paraxial ray equation which results contains only the z component of the magnetic field evaluated on the optic axis $B_z(z)$. This is

$$\frac{d^2r}{dz^2} + \frac{e}{8mV_r}B_z^2(z)r = 0 \tag{2.13}$$

where r is the radial distance of an electron from the optic axis and V_r is the relativistically corrected accelerating voltage. Once the field $B_z(z)$ on the axis is known, computed solutions of eqn (2.13) can be used to trace the trajectory of an electron entering the field. For a symmetrical lens, a ray entering parallel to the axis will define all the electron lens parameters discussed in later sections. Note that the approximation has been made that the z component of the field does not depend on r.

Equation (2.13) is a linear differential equation of second order with two linearly independent solutions. It can be shown that the solutions describe rays which satisfy the conditions for perfect lens action. Using a computer, one can also solve eqn (2.12) to find the electron trajectories for a particular magnetic field. By comparing these true electron

24 Electron optics

trajectories with the idealized trajectories satisfying eqn (2.13) (and producing perfect imaging) it is possible to determine the aberrations of a magnetic lens. Alternatively, it is possible to use simple expressions for the various aberration coefficients which are given as functions of the solution to eqn (2.13) (see Section 2.8.1).

Similarly, the helical rotation of meridional rays can be found. The rotation of this plane is given by

$$\theta_0 = \left(\frac{e}{8mV_r}\right)^{1/2} \int_{z_1}^{z_2} B_z(z) dz \qquad (2.14)$$

Rays entering the lens in a given meridional plane remain in that plane which rotates through θ_0 as the rays traverse the lens field. Note that the total image rotation is $(180 + \theta_0)$ degrees on account of the image inversion (M is negative).

It can be seen that the lens shown in Fig. 2.8 is very inefficient, since most of the power dissipated supports a field in the z direction which produces no force on the electron entering parallel to the axis. Aberration correctors use the far more efficient hexapoles and quadrupoles.

2.4. The constant-field approximation

A review of computing methods used in the solution of eqn (2.13) is given in Mulvey and Wallington (1972) and Orloff (2007). However, eqn (2.13) is easily solved if the z dependence of the field can be neglected. It then resembles the differential equation for harmonic motion. The accuracy of this 'constant field' approximation has been investigated by Dugas et al. (1961), who found it to give good agreement with experiment for the focal lengths of projector lenses at moderate and weak excitation. It is included here for the physical insight it allows into the action of magnetic lenses and to clarify the definition of the lens focal length. Since the expression for C_s (eqn 2.32) involves derivatives of the field, we cannot expect to understand the influence of aberrations using such a crude model.

If the origin of coordinates is taken on a plane midway between the pole-piece gap (length S), then a field constant in the z direction is given by

$$\begin{aligned} B_z(z) &= B_p & \text{for } -S/2 \leq z \leq S/2 \\ &= 0 & \text{elsewhere} \end{aligned} \qquad (2.15)$$

Equation (2.13) becomes

$$\frac{d^2 r}{dz^2} + k^2 r = 0 \qquad (2.16)$$

with

$$k^2 = \left(\frac{e}{8m_0 V_r}\right) B_p^2 \qquad (2.17)$$

that is,

$$k = 1.4827 \times 10^5 B_p / V_r^{1/2}$$

with B_p in tesla and V_r in volts. The solution to eqn (2.16) is

$$r = A\cos kz + B\sin kz \tag{2.18}$$

where A and B are constants to be determined from the boundary conditions. Matching the slope and ordinate r_0 at $z = S/2$ for a ray which leaves the lens parallel to the axis gives

$$r = r_0 \cos k(z - S/2) \tag{2.19}$$

The 'constant field' strength B_p can be related to the number of turns N and the lens current I using Ampère's circuital law. If the bore of the lens D is sufficiently small that it does not disturb the magnetic circuit and the reluctance of the iron is considered negligible compared with that of the gap S, then we have, for the circuit indicated in Fig. 2.8,

$$B_p = \frac{\mu_0 NI}{S} = 4\pi \times 10^{-7} \left(\frac{NI}{S}\right) \tag{2.20}$$

with I in amperes and B_p in tesla. Note that B_p gives the flux density in any lens gap if measured sufficiently far from the optic axis.

2.5. Projector lenses

Lenses which use the image formed by a preceding lens as object, such as lenses L2, L3, and L4 of Fig. 2.3, are known as projector lenses. 'Intermediate' lenses fall in this category, to distinguish them from lenses which use a physical specimen as object.

It may happen that the image formed by, say, L2 falls within the lens field of L3. The image formed by L3 can nevertheless be found by the constructions of Section 2.2 if a virtual object is used for L3. This virtual object is the image formed by L2 with L3 removed. A similar procedure applies if the image formed by L2 falls beyond the centre of L3.

The behaviour of real electron trajectories within the lens is given for a simple model by eqn (2.19). This equation is now used to give the focal length and principal plane position of an equivalent ideal lens. The image formed by a system of lenses can then be found by successive applications of the ideal lens construction.

The projector object focus f_p may fall inside or outside the lens field. The two cases are indicated in Figs 2.10(a) and (b). Notice that the extension of the asymptotic ray direction has been used for $z \to -\infty$ to define the 'virtual' or asymptotic projector focal length f_p. The distance between the principal plane H1 and the object-focus in Fig. 2.10(a) is

$$f_p = r_0/\tan\theta = r_0 \bigg/ \left(\frac{dr}{dz}\right)_{-\infty} = r_0 \bigg/ \left(\frac{dr}{dz}\right)_{-S/2}$$

Using eqn (2.19) gives

$$f_p/S = [Sk\sin(Sk)]^{-1} \tag{2.21}$$

26 Electron optics

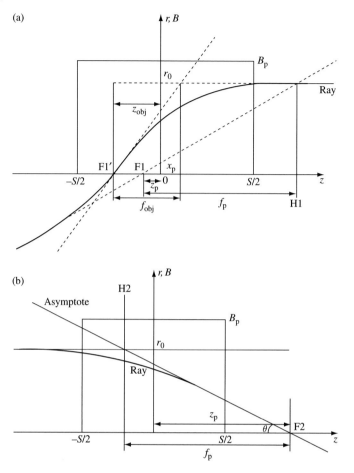

Figure 2.10 (a) Definition of the objective and projector focal lengths f_{obj} and f_p. The objective and projector focal distances z_{obj} and z_p are also shown. The specimen for an objective lens is placed near F1'. This diagram shows the lens used both as an objective (with focus at F1') and as a projector (with focus F1) if the focus lies inside the lens field. (b) Ray diagram for a projector lens if the focus lies outside the lens field, which extends from $-S/2$ to $S/2$. The ordinate represents both the field strength and the distance of a ray from the optic axis. The image focus occurs at F2.

This function is plotted in Fig. 2.11, and shows the minimum focal length characteristic of projector lenses. Lenses are generally operated in the region $0 < Sk < 2$, though the properties of 'second zone' lenses with $Sk > 2$ have been investigated for high resolution (von Ardenne 1941). Equation (2.21) gives a minimum for $f_p = 0.56S$.

Using eqn (2.20) the abscissa for Fig. 2.11 is, under the constant-field approximation,

$$Sk = 0.1862 \left(\frac{NI}{\sqrt{V_r}} \right) \qquad (2.22)$$

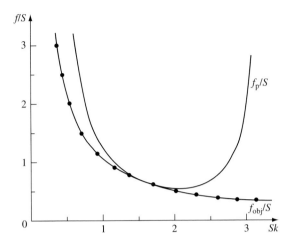

Figure 2.11 Relative focal lengths of objective and projector lenses (f_{obj} and f_{p}) as a function of lens excitation (proportional to the lens current) for the constant-field model. The abscissa is defined in the text. The projector minimum focal length occurs at $f_{\text{p}} = 0.56S$, where S is the size of the pole-piece gap.

The object focal distance z_{p} is defined as the distance between the centre of the lens and the object focus. From Fig. 2.10(a) it is seen that, together with f_{p}, this specifies the position of the principal plane. From the figure,

$$z_{\text{p}} = S/2 - r(-S/2) \bigg/ \left(\frac{\mathrm{d}r}{\mathrm{d}z}\right)_{-S/2}$$

Using eqn (2.19) gives

$$\frac{Z_{\text{p}}}{S} = \left(\frac{1}{2} - \frac{1}{kS \tan kS}\right) \tag{2.23}$$

for a projector lens. The sign is reversed for Fig. 2.10(b).

If the focal length is much larger than the lens gap ($f_{\text{p}} \gg S$) the lens is a weak lens such as C2 of Fig. 2.3. With $kS < 2$, eqn (2.21) gives

$$S/f_{\text{p}} = (Sk)^2 \tag{2.24}$$

which is the formula for a 'thin' magnetic lens when the distance between principal planes is neglected. A slightly more general form of this result was obtained by Busch (1926) in one of the first papers on electron optics. His result is obtained by integration of eqn (2.13) assuming r constant within the lens field. Then

$$\frac{1}{f_{\text{p}}} = \frac{e}{8mV_{\text{r}}} \int_{-\infty}^{\infty} H_z^2(z) \mathrm{d}z \tag{2.25}$$

which agrees with eqn (2.24) for a 'top-hat' field.

In an attempt to express the focal properties of all magnetic lenses on a single universal curve, Liebmann (1955a) found that lenses of finite bore D could be described by a modified form of eqn (2.25). His result is

$$\frac{1}{f} = \frac{A_0(NI)^2}{V_r(S+D)} \tag{2.26}$$

where A_0 is approximately constant. This is a good approximation for weakly excited lenses and can be extended to include asymmetrical lenses by taking

$$D = R_1 + R_2$$

Unfortunately, the properties of strong lenses become increasingly sensitive to the details of pole-piece geometry, making eqn (2.26) unreliable. In particular, since the focal length of a modern objective lens is not large compared with the pole-piece dimensions, these lenses cannot be treated as 'thin' lenses.

2.6. The objective lens

The objective lens, which immediately follows the specimen, is the heart of the microscope. Since angular magnification is inversely proportional to lateral magnification (eqn 2.11a), the magnification provided by this lens ensures that rays travel at very small angles to the optic axis in all subsequent lenses. We shall see that lens aberrations increase sharply with angle, so that it is the objective lens, in which rays make the largest angle with the optic axis, which determines the final quality of the image. The most important lens aberrations are specified by aberration constants C_c and C_s, which are very approximately equal to the focal length of the lens (Liebmann 1955a). For high-resolution microscopy, where the reduction of aberrations is important, the focal length and hence lens aberrations can be reduced by introducing the specimen into the field of the objective lens. This is known as an immersion lens. The field maintained on the illuminating side of the object is known as the pre-field (see Section 2.9) and plays no part in image formation. The extent of the remaining field available for image formation depends on the position of the object. Thus the position of the cardinal planes of the lens and the aberration coefficients depend on the object position z_{obj}, which becomes an important electron-optical parameter for these lenses. It is measured from the centre of the lens. By convention, z_{obj} and the objective lens focal length f_{obj} are specified for a lens producing an image at infinity (infinite magnification) so that the specimen is at the exact object focus. Thus z_{obj} is equal to the object focal distance.

For the simple lens described by eqn (2.17) it is easy to show from symmetry that the image and object focal distances are equal. These are defined by rays entering and leaving the lens parallel to the axis, respectively. The object focal distance can consequently be found from Fig. 2.10(a) as indicated on the figure. From the figure and eqn (2.19), the ordinate is zero when the cosine argument is $-\pi/2$. That is

$$\frac{z_{obj}}{S} = -\left(\frac{\pi}{2kS} - \frac{1}{2}\right) \tag{2.27}$$

The 'real' focal length f_{obj} is given by

$$f_{obj} = r_0 \bigg/ \left(\frac{dr}{dz}\right)_{z_{obj}} \tag{2.28}$$

that is

$$\frac{f_{obj}}{S} = (Sk)^{-1} \tag{2.29}$$

as shown in Fig. 2.11.

A comparison with eqn (2.21) shows that the focal length of a lens used as an objective may be shorter than its focal length when used as a projector lens. In practice a lower limit to the focal length is set by saturation of the pole-pieces of the lens.

Data for the objective lens, such as those given in Section 2.7, include a plot of lens excitation against z_{obj}, both in suitable units. This is interpreted as follows. For a specified excitation, the ordinate gives the object position required to produce an image at infinity. This will be the approximate specimen position for a lens operating at high magnification. The object is then separated from the object principal plane by the focal length, and is usually situated on the illuminating system side of the lens.

For an asymmetrical objective ($R_1 \neq R_2$) operated at high excitation, the focal length depends on which of R_1 or R_2 is in the image space. For weaker lenses this is not so. Highly asymmetrical objectives are popular in modern microscopes with a top-entry stage in which the specimen is introduced from above, since the specimen-change mechanism must be accommodated. A typical value of $2R_1$ is about 20 mm for a tilt stage, while $2R_2$ may be as small as 2 mm. The effect of pole-piece asymmetry is to increase the maximum value of $B_z(z)$ (R_1 and S held constant) and to shift the position of the maximum toward the smaller pole-piece (R_2). For $R_2 \to 0$ this maximum coincides with the vanishing pole-piece face. For real lenses $B_z(z)$ is a smooth peaked function and the important refractive effect of the lens occurs near the maximum of the field. Liebmann (1955b) was able to show from analogue computations that this maximum value for an asymmetrical lens agrees within 1% of the maximum for an equivalent symmetrical lens whose pole-pieces have diameter $D = R_1 + R_2$. The focal properties of a highly asymmetrical lens can be obtained by tracing two rays through the lens, one entering parallel to the axis and one leaving parallel to the axis. In practice a single ray traced through an equivalent symmetrical lens gives a good approximation to the focal length of the lens.

2.7. Practical lens design

Figure 2.12(a) shows the variation of measured aberration constants C_s and C_c for the JEOL UHP pole-piece ($C_s \approx 0.7$ mm) as a function of specimen position. It can be seen that there is an optimum specimen position to minimize C_c, but that C_s steadily decreases

30 *Electron optics*

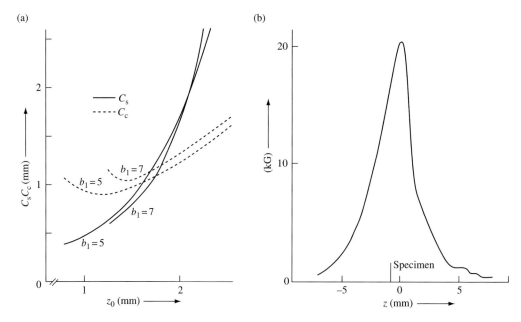

Figure 2.12 (a) The variation of C_c and C_s with specimen position for the JEOL UHP pole-piece ($C_s = 0.7$ mm). Here z_0, the specimen position, is measured from the top of the lower pole-piece. The values of C_c and C_s are experimentally measured values, courtesy of Drs Shirota, Yonezawa, and Yanaka. (b) The measured lens field distribution, obtained by the image-rotation method. The position of the specimen, where most of the refractive effect of the lens occurs, is shown. The ordinate indicates the z component of the axial field strength in kilogauss.

with z_0, the specimen height. Here b_1 is the upper pole-piece diameter (the lower diameter is 2 mm). Figure 2.12(b) shows the lens-field distribution, measured by observing the shadow-image rotation as an edge is moved down the optic axis. An upper limit to the field is set by saturation of the pole-piece iron—a typical saturation flux density is 2.4 T (24 000 G) for Permendur, a commonly used type of iron. From these and similar curves we find that the spherical and chromatic aberration constants are of the same order as the focal length of the lens, and vary with lens excitation in a similar way. The best specimen position for minimum chromatic aberration is found to be well to the illumination side of the centre of the lens, while spherical aberration is minimized by placing the specimen close to the centre of the lens. From eqns (2.29), (2.20), and (2.26) we see that a reduction in the lens gap increases the flux in the gap and reduces the focal length of the lens, unless it is otherwise limited. The overall instrumental magnification of a microscope can be increased in this way by reducing the pole-piece gap of the projector lens or by increasing the current in the lens if the additional heat dissipated can be absorbed by the lens-cooling system, thus making the instrument more suitable for high-resolution work. The very large depth of focus at high magnification means that the overall magnification can be increased by reducing the focal length of the projector without altering the strength of other lenses.

2.8. Aberrations

By about 1960, and prior to the current era of aberration-corrected electron microscopy, electronic instabilities had been reduced to the extent that the dominant aberrations affecting TEMs were third-order spherical aberration C_s, chromatic aberration C_c, and astigmatism. (The correction of these and higher-order aberrations is discussed in Sections 2.10 and 3.3.) The consequences of including higher-order terms in the derivation of the ray equation are discussed in many papers (see, e.g., Zworykin et al. (1945) for a textbook treatment). The departure from perfect imaging which results from the inclusion of these terms has components which are classified by analogy with the corresponding aberrations in light optics. Retention of the term in r^3 is analogous to the inclusion of the term in θ^3 in the expansion of $\sin\theta$ used in Snell's law applied to a glass lens. The result is that the third-order aberrations are generated, one of which, spherical aberration, is most important at high resolution.

Most high-resolution images are recorded at high magnification ($M > 80\,000$). The magnification dependence of the aberrations of a system of lenses is not discussed here (see Hawkes 1970) and we restrict ourselves to the case in which the object is near the object focus of the objective lens. At high magnification, then, the quality of the final image is determined by the aberrations of the first lens (objective lens), since here the scattering angle (θ_0 of Fig. 2.4) is largest and the spherical aberration (ray aberration) depends on the cube of this angle. Since the images formed by subsequent lenses are larger, their angular spectra (diffraction patterns) are smaller.

Another image defect important for high-resolution imaging arises from both electronic instabilities and beam–specimen interactions, and is known as chromatic aberration. This refers to the effect of a spread in energies in the beam. The distinction between ray and wave aberrations is discussed in Chapter 3; here we deal only with the geometrical-optics treatment of aberrations. The power of θ_0 is increased by one for an aberration expressed as a wave aberration or phase shift.

An elegant treatment of aberrations makes use of Fermat's principle applied to the optical path $W(P_1, P_2)$ between two points P_1 and P_2. This is, with n the refractive index,

$$W(P_1, P_2) = \int_{P_1}^{P_2} n\,ds$$

which, according to Fermat's principle, is a minimum or takes a stationary value. The function $W(P_1, P_2)$ is also known as Hamilton's point characteristic function. The expansion of this function as a power series allows terms to be identified which describe perfect imaging. Deviations from perfect imaging are expressed by terms of higher order. This theory goes far beyond the interests of high-resolution imaging; however, the following sections summarize the main results of aberration theory. Early textbook treatments can be found in Zworykin et al. (1945), Cosslett (1946), Grivet (1965), Septier (1967), and Reimer (1984), which summarizes the aberration-correction methods we discuss in Sections 2.10, 3.3, 7.4, and 10.10.

2.8.1 Spherical aberration

The effect of spherical aberration is indicated in Fig. 2.13. Rays leaving an axial object point at a large angle θ_0 are refracted too strongly by the outer zones of the lens and brought to a focus (the marginal focus for a ray which just strikes the aperture) before the

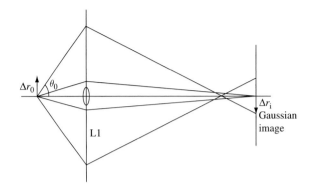

Figure 2.13 The effect of spherical aberration. Rays passing through the outer zones of the lens (far from the axis) are refracted more strongly than those paraxial rays passing close to the axis. These outer rays meet the axis before the Gaussian image plane, and meet that plane a distance Δr_i from the axis. Spherical aberration (or aperture defect) is the most important defect affecting the quality of high-resolution images.

Gaussian image plane. This is the plane satisfying eqn (2.8), where paraxial rays, obtained as solutions of the paraxial ray equation (2.13), are focused. The trajectories of aberrated rays can be obtained by the computed solution of the full-ray equation. If all rays from an axial object point are considered, the image disc of smallest diameter is known as the circle of least confusion. The distance Δr_i, in the image plane at high magnification is found to be proportional to θ_0^3, the constant of proportionality being the third-order spherical aberration constant C_s if Δr_i is referred to object space. That is

$$\Delta r_0 = C_s \theta_0^3 \quad \text{and} \quad \Delta r_i = M C_s \theta_0^3 = M^4 \theta_i^3 C_s \tag{2.30}$$

The radius of the circle of least confusion is $\frac{1}{4} C_s \theta_0^3$. The value of C_s depends on lens excitation and object position, but is usually quoted for an object close to the object focal plane, corresponding to the case of high magnification. On modern uncorrected instruments, C_s ranges between about 0.5 and 2.5 mm and sets the limit to 'interpretable' resolving power. A value for C_s can be obtained either from computed solutions of the full (non-paraxial) ray equation or through the use of an expression given by Glaser (1956). Simple recurrence relations have been given by Liebmann (1949) for the first method, expressing the deviation of the true trajectory from the paraxial trajectory at increments along the ray path. If the deviation at the point where a paraxial ray, which left the object parallel to the axis, crosses the axis is Δ, then the spherical aberration is given by

$$C_s = \Delta/\beta^3 \tag{2.31}$$

where β is the slope of the paraxial ray at the image focus.

Glaser (1956) found a useful expression for C_s as

$$C_s = \frac{e}{16m_0 V_r} \int_{z_1}^{z_2} \left[\left\{ \frac{\partial B_z(z)}{\partial z} \right\}^2 + \frac{3e}{8mV_r} B_z^4(z) - B_z^2(z) \left\{ \frac{h'(z)}{h(z)} \right\}^2 \right] h^4(z) dz \qquad (2.32)$$

Here $h(z)$ is the *paraxial* trajectory of a ray which leaves an axial object point with unit slope. The occurrence of derivatives in eqn (2.32) makes the evaluation of C_s particularly sensitive to the shape of the field $B_z(z)$. For high resolution, where resolution is limited by C_s (see Chapter 3), eqn (2.32) shows that the detailed form of $B_z(z)$, and hence the shape of the pole-piece, sensitively determines the instrumental resolution. It is for this reason that the utmost care must be taken when handling the pole-piece. An expression similar to eqn (2.32) has been given by Scherzer (1936). The fifth-order spherical aberration term C_5 may, however, also be important at very high resolution. For example, it has been found that at 100 kV, with $C_s = 0.5$ mm, the effects of C_5 cannot be neglected for spacings smaller than 0.1 nm (Uchida and Fujimoto 1986).

Simple techniques for *in situ* experimental measurement of C_s for the objective lens of an electron microscope are given in Chapter 10. Spherical aberration is discussed from the point of view of wave optics in Chapter 3, and the correction of spherical aberration is discussed in Sections 2.10, 3.3, 7.4, and 10.10.

2.8.2 Chromatic aberration

Chromatic aberration arises from the dependence of the focal length of the lens on the wavelength of the radiation used, and hence on the electron energy. With polychromatic illumination, in-focus images are formed on a set of planes, one for each wavelength present in the illuminating radiation. There are three important sources of wavelength fluctuation in modern instruments. These are, in decreasing order of importance, for the thin specimens ($t < 10$ nm) used at high resolution:

1. The energy spread of electrons leaving the filament (see Chapter 9): at a high gun-bias setting, $\Delta E / E_0 = 10^{-5}$.
2. High-voltage instabilities: a typical fluctuation specification is $\Delta V_0 / V_0 = 2 \times 10^{-6}$/min with V_0 the microscope high voltage.
3. Energy losses in the specimen.

It is convenient to include fluctuations both in the high-voltage supply and in the objective lens current in the definition of the chromatic aberration constant C_c. Differentiation of eqn (2.26) gives

$$\frac{\Delta f}{f} = \frac{\Delta V_0}{V_0} - \frac{2\Delta B}{B} = \frac{\Delta V_0}{V_0} - \frac{2\Delta I}{I} \qquad (2.33)$$

To allow for deviations from the thin-lens law (eqn 2.8), a constant of proportionality K is introduced in 2.33. The chromatic aberration constant C_c is defined as $C_c = Kf$, so that

$$\Delta f = C_c \left(\frac{\Delta V_0}{V_0} - \frac{2\Delta B}{B} \right) = C_c \left(\frac{\Delta V_0}{V_0} - \frac{2\Delta I}{I} \right) \qquad (2.34)$$

34 Electron optics

In practice the fluctuations in lens current and high voltage are unlikely to be correlated, so that the random fluctuations in focal length should be obtained from the rules given for manipulation of variance in statistical theory. Adding the fluctuations in quadrature gives the variance

$$\frac{\sigma^2(f)}{f^2} = \frac{\sigma^2(V_0)}{V_0^2} + \frac{4\sigma^2(I)}{I^2} \tag{2.35}$$

The fluctuation in focal length, expressed by the standard deviation $\sigma(f)$, is given by

$$\frac{\sigma(f)}{(f)} = \left[\frac{\sigma^2(V_0)}{V_0^2} + \frac{4\sigma^2(I)}{I^2}\right]^{1/2} = \Delta/C_c \tag{2.36}$$

The variances in I and V_0 must be obtained by measurement, or their values may be deduced either by treating Δ as a fitting parameter for the image matching experiments described in Section 5.13, or from diffractogram analysis (see Section 10.6).

The effect on the image of a small change in focal length is now determined using the methods of geometric optics as a first approximation. As shown in Fig. 2.14, a ray at slightly higher energy, for which the lens has a longer focal length, is brought to a focus distance ΔV from the Gaussian image plane where

$$\Delta r_i = \Delta V \theta_i$$

Differentiating the lens equation gives

$$\frac{\mathrm{d}V}{\mathrm{d}f} = \left(\frac{V}{f}\right)^2 \approx M^2$$

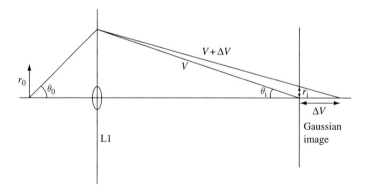

Figure 2.14 The effect of chromatic aberration. The faster electrons which have been accelerated through a potential $V + \Delta V$ are less strongly refracted than lower-energy electrons accelerated through a potential V. These higher-energy electrons are thus brought to a focus beyond the Gaussian image plane, which they pass at a distance r_i from the axis.

so that
$$\Delta V = M^2 \Delta f \tag{2.37}$$

thus
$$\Delta r_i = M^2 \Delta f \theta_i$$

Referred to the object space, the extended disc image of an axial object point has radius
$$\Delta r_0 = M \Delta f \theta_i$$

Using eqn (2.11) this is
$$\Delta r_0 = \theta_0 \Delta f$$

From eqn (2.34) we have
$$\Delta r_0 = \theta_0 C_c \left(\frac{\Delta V_0}{V_0} - \frac{2 \Delta I}{I} \right) \tag{2.38}$$

As an example, consider the image of a specimen which introduces a discrete energy loss of 25 eV. Many biological specimens show a rather broad peak at this energy. The loss image will be out of focus by an amount ΔU referred to the object plane. At high magnification, the lens law gives
$$\Delta U \approx \Delta f \tag{2.39}$$

so that the focus defect for the loss image can be obtained from eqn (2.34), neglecting the fluctuation in lens current. For $C_c = 1.6$ mm and $\Delta E = 25$ eV, this gives $\Delta U = 400$ nm at 100 keV. Since the lens has a shorter focal length for the loss electrons, with V fixed an objective lens focused on the elastic image will have to be weakened to bring the inelastic image into focus. Thus inelastic loss images appear on the under-focused side of an image where the first Fresnel fringe (see Chapter 3) appears bright and the lens current is too weak. Unfortunately the optimum focus condition for high-resolution phase contrast is also on the under-focus side (about 90 nm at 100 kV) so that any inelastic contribution appears as an out-of-focus background blur in the elastic image. For many specimens the total inelastic image contribution may exceed the elastic contribution (see Section 6.6).

In practice C_c can be evaluated once the field distribution is known (Glaser 1952) using
$$C_c = \frac{e}{8mV_r} \int_{z_1}^{z_2} B_z^2(z) h^2(z) \mathrm{d}z \tag{2.40}$$

with $h(z)$ the function in eqn (2.32). Chromatic aberration can be eliminated in principle by combining a magnetic lens with an electrostatic mirror, whose chromatic aberration is of the opposite sign.

36 Electron optics

The constant C_c can be measured experimentally from measurements of the change in focal length with high voltage or lens current. A method is described in Chapter 10. Chromatic aberration is absent from STEM instruments if no lenses are used after the specimen. The representation of chromatic aberration in terms of wave optics is discussed in Chapter 3, and it should be stressed that eqn (2.38) is a geometrical approximation, accurate for the spatially incoherent imaging of point objects at moderate defocus, that is, for ΔV not too small. For small ΔV or coherent illumination (or both) the wave-optical methods of Chapter 3 must be used. The effect of variations in C_c or Δ on high-resolution image detail is shown in Fig. 4.3.

2.8.3 Astigmatism

Severe astigmatism is caused by an asymmetric magnetic field. A departure from perfect symmetry in the lens field can be represented by superimposing a weak cylindrical lens on a perfectly symmetrical lens. Figure 2.15 shows the image of a point formed by an astigmatic lens as two line foci at different points on the optic axis. Astigmatism can be thought of as an azimuthally dependent focal length, which produces the phase shift given by eqn (3.22). *In situ* methods for measuring the astigmatism constant z_a are discussed in Section 10.5. An experienced operator using a modern instrument fitted with a stigmator for correcting astigmatism can reduce this aberration so that it does not limit resolution; however, this requires considerable practice (see Chapter 12). A stigmator is a weak quadrupole lens whose excitation is controlled electronically to allow correction of field asymmetry in any direction by introducing a compensating weak cylindrical lens of the correct strength and orientation.

A machining tolerance of a few microns in the ovality of the objective lens pole-piece is required to bring astigmatism within easily correctable limits. The homogeneity of the iron used is very important.

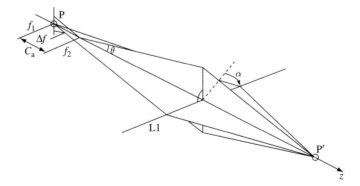

Figure 2.15 Astigmatism. The focal length of the lens depends on the azimuthal angle α of a ray leaving the object. These rays are in a plane containing the optic axis. Planes at right angles for the maximum and minimum focal length are shown, with a mean focus f. The difference between the maximum and minimum focus is the astigmatism constant C_a.

Recent improvements in resolution to the sub-angstrom level have exposed three-fold astigmatism. An example of its measurement and removal can be found in Overwijk *et al.* (1997). Two- and three-fold astigmatism are discussed from the point of view of wave optics in Chapter 3.

2.9. The pre-field

For an objective lens of minimum spherical aberration, the specimen will be placed well within the lens field. The field on the illuminating side of the specimen is known as the pre-field and has the effect of a weak lens placed before the specimen. Figure 2.16 shows a ray entering a 'constant-field' objective parallel to the axis with the lens excitation to the right of the specimen arranged to produce an image at infinity. These rays define both the pre-field focal length and the objective focal length. Expressions for the pre-field focal length, focal distance, and the demagnification of the incident beam by the pre-field have been given by Mulvey and Wallington (1972) using an analysis similar to that of Section 2.6. The significance of this for high resolution is that the divergence of the incident beam, and hence the coherence length, may be affected by the strength of the pre-field lens. In practice, pre-field effects are included in a measurement of transverse coherence width if this is obtained from measurements of diffraction spot sizes as discussed in Section 10.7. It follows that for an immersion lens the illumination conditions are affected by specimen position and objective lens focusing. The modern trend has been toward the use of increasingly symmetrical 'condenser-objective' lenses. These allow convenient switching from the high-resolution electron microscopy (HREM) mode to the probe-forming mode for convergent-beam electron diffraction (CBED) and STEM on the same specimen region (see Section 13.3).

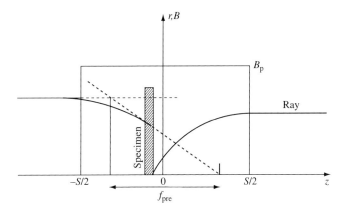

Figure 2.16 The effect of the objective lens pre-field. The specimen is immersed in the lens field and a ray entering the field parallel to the axis is bent toward the axis, providing partially converged illumination. Diffraction spots appear larger than they would in the absence of the pre-field to the left of the specimen. The focal length of the pre-field f_{pre} is shown, which depends on the specimen position. A large pre-field is commonly used for high-resolution work since the shortest possible focal length is required to reduce aberrations.

2.10. Aberration correction

Scherzer (1936) showed that for rotationally symmetric static lenses (not mirrors) forming a real image of a real object without space-charge, conductors on the axis or discontinuities in the electrostatic potential, C_s, is positive-definite, suggesting that under these conditions it could not be eliminated. Equation (2.32) and similar forms were used in the unsuccessful search for round lens fields with $C_s = 0$. The relaxation of Scherzer's conditions, in the search for reduced spherical aberration, were analysed by him in his famous 1947 paper (Scherzer 1947). This paper inspired subsequent work on aberration correction for electron microscopy, leading ultimately to success (by relaxing the rotational symmetry condition) half a century later.

Comprehensive historical reviews of aberration correction can be found in Krivanek *et al.* (2007), Rose (2008), and Hawkes (2009), and of electron optics generally in Hawkes and Kasper (1996). Early work, which could not take advantage of real-time image analysis, aberration measurement, and feedback, and which resulted in resolution improvements inferior to the best uncorrected instruments of the time, was undertaken mainly in Cambridge, Chicago, and Darmstadt, where Seeliger (1953) and later Mollendstedt (1956) implemented an early proposal of Scherzer. At the AEI company in the UK, Archard proposed quadrupole–octupole correctors (Archard 1958). In Cosslett's group in Cambridge a series of correctors were built, including the quadrupole–octupole probe-forming corrector of Deltrap (1964), the electrostatic/electromagnetic corrector of Hardy (1967), and the quadrupole–octupole corrector of Thomson (1968). The aberrations of non-round lenses, including sextupoles, and their application to aberration correction were analysed fully by Hawkes (1965). In Chicago, a quadrupole–octupole corrector was built by Beck and Crewe in 1976, followed by an important new sextupole design (Beck 1979), which culminated in a sextupole-round-lens design to correct all aberrations up to six-fold astigmatism (Shao *et al.* 1988). In retrospect the experimental failure of these early approaches can be seen to result from the large number of alignment parameters and current settings which required adjustment. As a result, it came to be believed around 1990 that the resulting 'multiparameter optimization problem' was insoluble. The situation changed with a couple of innovations. (a) Powerful on-line methods of characterizing the total aberration function of STEM and TEM instruments were developed [the Ronchigram (Section 7.6, Fig. 7.5) and the Zemlin tableaux (Section 12.5, Fig. 12.8)]. This on-line analysis was made possible by the development of the CCD camera for electronic image recording (Spence and Zuo 1988), which obviated the need for chemical processing of film, prior to aberration measurement. (b) The performance of modern computers improved sufficiently during the 1990s to allow analysis of these patterns, with feedback to the microscope, in a time less than the stability time of the microscope's electrical and mechanical systems. 'Auto-tuning' of HREM instruments by this method began around 1984 when it first became competitive with human alignment, focus setting, and stigmation (see Smith (2008) for a review), and later for STEM. The modern approach involves measurement of the total aberration coefficients using diffractograms or Ronchigrams, in order to set up a matrix relating the excitation of optical elements to the aberrations. If there are more excitation signals than aberration coefficients, this matrix can be inverted to allow a particular set of aberration coefficients (e.g. all as small as possible) to be established.

Since about 1990 progress has been rapid. It became clear that the off-axis aberrations of the octupole–quadrupole arrangement made it more suitable for scanning probe instruments (where electrons are confined to a small region around the optic axis), while the sextupole was more suited to the off-axis TEM mode with its isoplanatic requirement. Success in this field might be defined as the achievement of better resolution than that obtainable by uncorrected machines, such as the 1 Å resolution reported uncorrected by Ichinose et al. (1999). (Aberration correction has advantages other than simple improvement in resolution—for example the larger space made available in corrected lens pole-pieces has made possible the use of much larger X-ray detectors, subtending solid angles approaching a steradian, and resulting in single-atom detection.) In Germany, the first practical operating TEM corrector, based on sextupoles and a round lens, was constructed by Haider et al. (1995) (see also Rose (1990)), and this was used to demonstrate atomic resolution imaging in 1998 (Haider et al. 1998; see Fig. 7.2). This design has since become the most widely accepted design for TEM instruments. The first corrector for both C_s and C_c (Rose 1971) had been demonstrated previously for a SEM (Zach and Haider 1995). For STEM instruments, work at Cambridge University's Cavendish Laboratory in 1997 resulted in a successful first-generation C_s-corrected instrument (Krivanek et al. 1997). When combined with a cold field-emission tip this allowed far higher probe currents to be used for microanalysis, producing a modern STEM instrument using four quadrupoles and three octupoles, now capable of sub-angstrom resolution (Nellist et al. 2004) and single-atom species identification by energy-loss spectroscopy (Krivanek et al. 2010).

In general, in the presence of higher-order aberrations, the two-dimensional ray aberration Δr_0 in eqn (2.30) (and corresponding phase aberration) can be expanded in a power series, each complex term $\mathbf{C}_{n,m}$ of which depends on a different power n of the scattering angle (the order of the aberration) and on its symmetry m. Rotation by $2\pi/m$ about the optic axis brings the aberration function into coincidence with itself. A second azimuthal angle will be needed for anisotropic aberrations such as the example of 'astigmatism' shown in Fig. 2.15 (more correctly 'two-fold astigmatism of the first order' $C_{1,2}$) for which $m = 2$, since the aberration function (shown later in Fig. 3.6) has the shape of a saddle. In terms of this general notation, the third-order spherical aberration of Section 2.8.1 is $C_s = C_{3,0}$ (or C_3 in the notation of Haider and Rose). Erni (2010) provides a tabulated comparison of notations. We use here the notation of Krivanek et al. (2007). Other important aberrations include axial coma $C_{2,1}$, three-fold astigmatism $C_{2,3}$, and four-fold astigmatism $C_{3,4}$. Current aberration correctors have corrected aberrations up to fifth order. The degree of an aberration refers to its power dependence on energy-spread in the beam, and the rank is the sum of the order and the degree. Geometrical aberrations have degree zero. Aberrations which depend only on scattering angle, not object coordinate (as in the isoplanatic approximation), are known as aperture or axial aberrations, otherwise as off-axis aberrations. In the mid-nineteenth century Ludwig Seidel analysed the symmetry-allowed geometric aberrations of a round lens up to third order—these are known as the Seidel aberrations (spherical aberration, coma, field astigmatism, field curvature, and image distortion). We will see in Section 3.3 that the phase shift associated with an aberration depends on scattering angle to a power one greater than the order of the ray aberration Δr_0.

Just as in conventional HREM it is common to choose a focus setting which compensates to the greatest extent possible for the effects of spherical aberration in the round objective

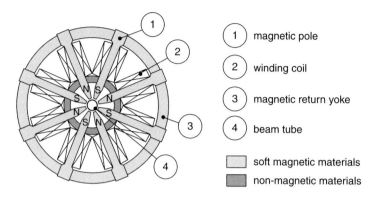

Figure 2.17 Cross-section of magnetic octupole lens. (From Krivanek et al. (2007).)

lens ($m = 0$), the essential idea of aberration correction is to add additional electron-optical elements (which may not be round), whose aberrations cancel those of the objective lens. Figure 2.17 shows an example of a magnetic octupole. For such a multipole, the symmetry m is half the number of poles ($m = 1$ for a dipole). Because all of the field in a multipole is used to refract the electrons, the power consumption of multipoles is far less than that of round lenses, with a consequent reduction in cooling needs. As a simple example, we consider qualitatively the operation of the quadrupole–octupole corrector. This is one of the two main successful corrector designs, the other being the sextupole corrector (Crewe 1984). A multipole with symmetry m introduces an aberration of type $C_{m-1,m}$ or four-fold astigmatism $C_{3,4}$ for the octupole shown in Fig. 2.17. Unlike a round lens, for which a single current fixes both the aberrations and the Gaussian focal length, the aberration coefficients $C_{n,m}$ of a multipole are proportional to the current through it, and may have opposite sign to that of a round lens.

The simplest aberration correctors therefore are those which correct astigmatism (the 'stigmator' on a microscope). To correct for two-fold ($C_{1,2}$) or three-fold ($C_{2,3}$) astigmatism we need only add either a quadrupole or sextupole.

To correct spherical aberration $C_s = C_{3,0}$, a quadrupole–octupole corrector may be used, and it should be noted that a design similar to this is also capable of correcting for chromatic aberration (Haider et al. 2008). For one form of this type of corrector for use in STEM the elements are arranged as shown in Fig. 2.18 (Krivanek et al. 2007, Delby et al. 2011). Quadrupole 1 (Q1) converts the round beam into an elliptical beam in octupole 1 (O1), Q2 changes it back into a round beam in O2, Q3 changes it into an elliptical beam in O3, while Q4 changes it back into a round beam. Now the third-order octupole aberration contribution $C_{3,4}$ has the same scattering angle dependence along two orthogonal directions normal to the axis as does the isotropic spherical aberration C_s we wish to cancel, but with the opposite sign. The idea then is to align the long axis of the ellipse in the first octupole O1 with the direction which produces cancellation, then convert the beam to an ellipse whose long axis is orthogonal to the first ellipse, and pass it though another octupole O3 in order to cancel the aberration in the remaining direction. (This process is also shown in Section 3.3.)

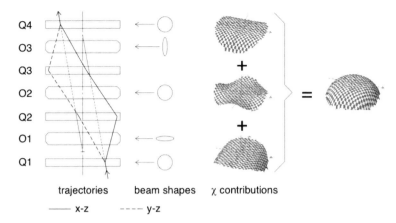

Figure 2.18 Trajectories and aberration function contributions in a quadrupole–octupole corrector. (From Krivanek *et al.* (2007).)

There are many complications and subtleties, including the way in which aberrations combine (combination aberrations), aberrations due to misalignment, parasitic aberrations, off-axis aberrations, and, most important, the stability of the power supplies in order to achieve accuracy in beam positioning and the angles needed. (These can be as small as a few nanometres and microradians.) Figure 2.19 shows a cross-section of the Nion cold-field-emission aberration-corrected STEM.

The correction of chromatic aberration has proceeded more slowly and is more difficult, but was foreseen in Scherzer's (1947) paper. A detailed design was given by Rose (1971) for an instrument which corrects chromatic, spherical, and off-axial coma aberrations—finally an *achroplanatic* microscope. This complex instrument has been constructed, based on quadrupole and octupole elements, and first results were reported by Kabius *et al.* (2009). Chromatic aberration (including the significant amount introduced by the corrector itself) is corrected using electric–magnetic quadrupole fields.

A chromatically corrected lens with $C_c = 0$ brings all rays to the same focus at all energies (eqn 2.38). Both the energy spread in the beam due to the source and due to energy losses in the sample must be considered. For a STEM without post-specimen lenses, the second effect is absent. It is found that the effect of spherical-aberration correctors is to increase chromatic aberration slightly, so that, via the damping function given in Chapter 4, the information limit may actually be slightly degraded, with C_c increased by perhaps 50%. The use of a monochromator may reduce this effect, if sample energy losses are not important for the thinnest samples. Chromatic-aberration correctors are expected to be important in lower-voltage machines (where the beam energy spread is otherwise a larger fraction of the beam energy), which may be important for biological samples, if they produce greater contrast with reduced radiation damage at lower beam energy—this remains to be determined. In thicker samples where sample energy losses dominate, the advantages of having all the electrons which pass through the sample contribute to the in-focus image are clear, since image background and radiation dose would be reduced.

42   *Electron optics*

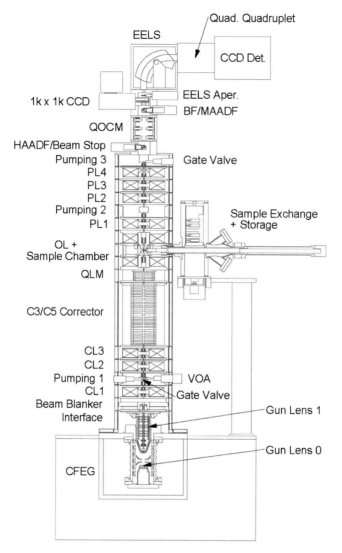

Figure 2.19 Cross-section of the microscope column of the Nion UltraSTEM™ 200 cold field-emission aberration-corrected STEM. (From Delby *et al.* (2011).)

An optimum strategy for imaging 'thick' biological samples (e.g. a whole cell) may be to use an imaging energy filter (Senoussi *et al.* 1971, Rose 1974) tuned to the maximum in the energy-loss spectrum, together with a chromatic-aberration corrector. The optimum sample thickness may be about equal to the elastic mean free path.

In Section 7.4 we discuss the optimum set of aberration coefficients required for aberration-corrected phase-contrast imaging (where no contrast is seen at Gaussian focus in an unaberrated image). The optimum conditions for corrected STEM are discussed in Crewe and Salzman (1982). Section 10.6 describes a Ronchigram analysis for aberration measurement. In Section 3.3 we give the full expression for the wavefront aberration phase-shifts to fifth order. Applications of aberration-corrected microscopy are treated in Section 5.15.

References

Archard, G. D. (1958). An unconventional electron lens. *Proc. Phys. Soc.* **72**, 135.
von Ardenne, M. (1941). Zur Prufung von kurzbrennweitigen Elektronenlinsen. *Z. Phys.* **117**, 602.
Beck, V. D. (1979). A hexapole spherical aberration corrector. *Optik* **53**, 241.
Busch, H. (1926). Berechnung der Bahn von Kathodenstrahlen im axia symmetrischen elektromagnetischen Felde. *Ann. Phys.* **81**, 974.
Carey, D. C. (2006). Third order TRANSPORT with MAD input. <http://www.slac.stanford.edu/pubs/slacreports/slac-r-530.html>
Cosslett, V. E. (1946). *Introduction to electron optics.* Oxford University Press, London.
Crewe, A. V. (1984). The sextupole corrector. 1: Algebraic calculations. *Optik* **69**, 24.
Crewe, A. and Salzman, D. B. (1982). On the optimum resolution for a corrected STEM. *Ultramicroscopy* **9**, 373.
Delby, N., Bacon, N., Hrncirik, P., Murfitt, M., Skone, G., Szilagyi, Z., and Krivanek O. (2011). Dedicated STEM for 200 to 40 keV operation. *Eur. Phys. J. Appl. Phys.* **54**, 33505.
Deltrap, J. H. M. (1964). Aberration correction. *Proc. 3rd Eur. Electron Microsc. Congress (Prague)* p. 45.
Dietrich, I. (1977). *Superconducting electron optic devices.* Plenum, New York.
Dugas, J., Durandeau, P., and Fert, C. (1961). Lentilles electroniques magnetiques symetriques et dyssymetriques. *Rev. Opt.* **40**, 277.
Erni, R. (2010) *Aberration-corrected imaging in transmission electron microscopy.* Imperial College Press, London.
Glaser, W. (1952). *Grundlagen der Elektronenoptik.* Springer-Verlag, Berlin.
Glaser, W. (1956). In *Handbuch der Physik*, Vol. 33. Springer-Verlag, Berlin.
Grivet, P. (1965). *Electron optics.* Pergamon, London.
Hall, C. E. (1966). *Introduction to electron microscopy.* Second edition. McGraw-Hill, New York.
Hardy, D. F. (1967). Aberration correction in electron optics. PhD Thesis, University of Cambridge.
Hawkes, P. W. (1965). Geometrical aberrations of general optical systems. *Phil. Trans. R. Soc. Lond.* **A 257**, 479.
Hawkes, P. W. (1970). The addition of round lens aberrations. *Optik* **31**, 592.
Hawkes, P. W. (1972). *Electron optics and electron microscopy.* Taylor and Francis, London.
Hawkes, P. W. (2009) Aberration correction past and present. *Phil. Trans. A, Math. Phys. Eng. Sci.* **367**, 3637.
Hawkes, P. W. and Kasper, E. (1996). *Principles of electron optics.* Academic Press, New York.
Haider, M., Braunshausen, G., and Schwan, E. (1995). Correction of the spherical aberration of a 200 kV TEM by means of a hexapole corrector. *Optik* **99**, 167.
Haider, M., Uhlemann, S., Schwan, E., Rose, H., Kabius, B., and Urban, K. (1998). Electron microscopy image enhanced. *Nature* **392**, 768.
Haider, M., Muller, H., Uhleman, S., Zach, E., Loebach, U., and Hoeschen, R. (2008). Pre-requisites for a C_s/C_c corrected ultra-high resolution TEM. *Ultramicroscopy* **108**, 167.
Ichinose, H., Sawada, H., Takuma, E., and Osaki, M. (1999). Atomic resolution HVEM and environmental noise. *J. Electron Microsc.* **48**, 887.
Jap, B. and Glaeser, R. (1978). The scattering of high energy electrons. I. Feynman path integral formulation. *Acta Crystallogr.* **A34**, 94.

Kabius, B., Hartel, P., Haider, M., Muller, H., Uhlemann, S., Loebau, U., Zach, J., and Rose, H. (2009). First applications of a C_c corrected imaging for high-resolution and energy-filtered TEM. *J. Electron Microsc.* **58**, 147.

Krivanek, O. L. (2010). Atom-by-atom structural and chemical analysis by annular dark-field electron microscopy. *Nature* **464**, 571.

Krivanek, O., Delby, N., Spence, A. J. H., Camps, R. A., and Brown, L. M. (1997). STEM aberration correction. *Inst. Phys. Conf. Ser. 153 (Proc. EMAG)*, ed. J. Rodenburg, p. 35. Bristol, Institute of Physics, Bristol.

Krivanek, O. L., Delby, N., and Murfitt, M. (2007). Aberration correction in electron microscopy. In *Handbook of charged particle optics*, ed. J. Orloff, pp. 601–640. CRC Press, Boca Raton, FL.

Liebmann, G. (1949). An improved method of numerical ray tracing through electron lenses. *Proc. Phys. Soc.* **B62**, 753.

Liebmann, G. (1955a). A unified representation of magnetic electron lens properties. *Proc. Phys. Soc.* **B68**, 737.

Liebmann, G. (1955b). The field distribution in asymmetrical magnetic electron lenses. *Proc. Phys. Soc.* **B68**, 679.

Mollenstedt, G. (1956). Elektronenmikroskopische Bilder mit einem nach O. Scherzer sphärisch korrigierten Objektiv. *Optik* **13**, 2009.

Mulvey, T. and Wallington, M. J. (1969). The focal properties and aberrations of magnetic electron lenses. *J. Phys. E* **2**, 446.

Mulvey, T. and Wallington, M. J. (1972). Electron lenses. *Rep. Prog. Phys.* **36**, 348.

Nellist, P., Nellist, P. D., Chisholm, M. F., Dellby, N., Krivanek, O. L., Murfitt, M. F., Szilagyi, Z. S., Lupini, A. R., Borisevich, A., and Sides Jr, W. H. (2004). Direct sub-angstrom imaging of a crystal structure. *Science* **305**, 1741.

Nussbaum, A. (1968). *Geometric optics: an introduction*. Addison-Wesley, Reading, MA.

Orloff, J. (ed.) (2007). *Handbook of charged particle optics*. CRC Press, Boca Raton, FL.

Overwijk, M. H. F., Bleeker, A. J., and Thust, A. (1997). Correction of three-fold astigmatism in ultra-high resolution TEM. *Ultramicroscopy* **67**, 163.

Reimer, L. (1984). *Transmission electron microscopy*. Springer-Verlag, Berlin.

Rose, H. (1971). Electronenoptische Aplanate. *Optik* **34**, 285.

Rose, H. (1974). Phase contrast in STEM. *Optik* **39**, 416.

Rose, H. (1990). Outline of a spherically corrected semiplanatic medium-voltage transmission electron microscope. *Optik* **85**, 19.

Rose, H. (2005). Prospects for aberration-free electron microscopy. *Ultramicroscopy* **103**, 1.

Rose, H. (2008). History of direct aberration correction. *Adv. Imaging Electron Phys.* **153**, 1.

Rose, H. and Plies, E. (1974). Design of a magnetic energy analyzer with small aberrations. *Optik*. **40** 336.

Scherzer, O. (1936). Uber einige Fehler von Elektronenlinsen. *Z. Phys.* **101**, 593.

Scherzer, O. (1947). Spharische and chromatishche Korrektur von Electronen-Linsen. *Optik* **2**, 114.

Seeliger, R. (1953). Uber die Justierung Spharisch Korrigierter Elecktonenoptischer systeme. *Optik* **10**, 29.

Senoussi, S., Henry, L., and Castaing, R. (1971). Aberration correction. *J. Microscopie* **11**, 19.

Septier, A. (1967). *Focusing of charged particles*, Vol. 1. Academic Press, New York.

Shao, Z. (1988). On the fifth-order aberration in a sextupole corrected probe forming system. *Rev. Sci. Instrum.* **59**, 2429.

Shao, Z., Beck, V. and Crewe, A. V. (1988). A study of octupoles as correctors. *J. Appl. Phys.* **64**, 1646.

Smith, D. (2008). Development of aberration-corrected electron microscopy. *Microsc. Microanal.* **14**, 2.

Spence, J. C. H. and Zuo, J. M. (1988). Large dynamic range CCD area detector for electron microscopy. *Rev. Sci. Instrum.* **59**, 2102.

Thomson, M. (1968). Electron-optical aberration correction. PhD Thesis, University of Cambridge.

Uchida, Y. and Fujimoto, F. (1986). Effect of higher order spherical aberration term on transfer function in electron microscopy. *Jpn J. Appl. Phys.* **25**, 644.

Zach, J. and Haider, M. (1995). Correction of spherical and chromatic aberration in a low-voltage SEM. *Optik* **99**, 112.

Zworykin, V. K., Morton, G. A., Ramberg, E. G., Hillier, J., and Vance, A. W. (1945). *Electron optics and the electron microscope*. Wiley, New York.

Bibliography

Useful introductory accounts of electron optics which stress the practical considerations involved in the design of an instrument can be found in the books by M. E. Haine (*The electron microscope*; Spon, London, 1961), and by Hall (referenced above). The books by Hawkes, Septier, Orloff, and Grivet also referenced above provide more advanced treatments.

3
Wave optics

Improvements in instrumental stability and the development of brighter electron sources, such as the field-emission gun, have enabled high-resolution images to be formed under highly coherent conditions. The microscope then becomes an electron interferometer, and the out-of-focus image must be interpreted as a coherent interference pattern. The particle model of the fast electron, useful for the classical ray-tracing techniques described in Chapter 2 to determine simple lens properties, must then be replaced by a wave-optical theory taking account of the finite electron wavelength. A quantum-mechanical theory is required to do this correctly. However, this shows (Komrska and Lenc 1972) that, since the spin of the beam-electron can usually be neglected, the simple wave-optical methods give accurate results for electron interference and imaging experiments with a suitable change of wavelength and allowance for image rotation. Consequently, in this chapter, for the purposes of discussing interference effects, the electron lens is simply replaced by an equivalent ideal thin lens whose wave-optical properties are discussed. The squared modulus of the complex wave amplitude $\psi(x,y)$ is then interpreted as a probability density (with dimensions Length$^{-3/2}$), rather than giving the time average of the electric field, as in optics. The image current density in an electron microscope, formed from non-interacting fast electrons, is then proportional to $\psi(x,y)\psi^*(x,y)$, the electron 'intensity'.

For complicated optical systems it is convenient to use the concepts of entrance and exit pupil to account for the diffraction limit of the system. Aberrations are introduced as a phase shift across the exit pupil added to the spherical wavefront which would converge to an ideal image point. In electron microscopy the important aberrations for high resolution are confined to the objective lens; consequently the following sections treat a single lens only. Then the objective aperture forms the exit pupil and the intensity distribution across the back-focal plane of the objective lens is the intensity recorded in an electron diffraction pattern. In this chapter we are interested in finding both the amplitude of the back-focal plane and the image amplitude and to relate these to the wave leaving the object. The relationship between this complex amplitude and the specimen structure is a more difficult problem. Most of the work on image analysis has concentrated on retrieving an improved version of the object exit-face wavefunction, rather than relating this to the specimen structure. This second problem of structure analysis is discussed in Chapter 5, and in this chapter we are mainly concerned with understanding the way in which the microscope aberrations contribute to the image complex amplitude. The emphasis in this chapter is on interference effects, so that perfectly coherent illumination is assumed throughout.

The diffraction limit, an important 'aberration' for high-resolution images, is also discussed. The integral solutions of the wave equation in the scalar theories of Kirchhoff and Helmholtz, together with Sommerfeld's work in the 1890s showed that the earlier physical models based on 'secondary sources' for diffraction due to Huygens and Young were a useful basis for understanding diffraction. That is, that diffraction can be thought of as arising from interference between the wave transmitted through the aperture unobstructed and waves generated at fictitious secondary sources, including Sommerfeld's 'edge wave' which originates as a line source around the aperture rim. This model is may also be applied to Fresnel diffraction at an edge, producing the bright 'edge wave' which Thomas Young first observed. Huygens saw diffraction and wave propagation arising from the interference between an infinite number of fictitious secondary spherical waves constructed along a wavefront. The mathematical description of one of these spherical waves is the Green function for the integral solution of the wave equation. A comprehensive treatment and bibliography for the fascinating and complex history of early diffraction theory is given in Born and Wolf (1999), including, in this seventh edition, the relationship of these theories to the Born approximation (reconciling their apparently different wavelength dependence) which we will use later.

A more modern view of the diffraction limit is expressed by the Fourier optics approach and is based on the Abbe theory and concepts taken from communications engineering in the early years after World War II. In the jargon of electrical engineering, image formation under coherent illumination is band-limited; that is, the image is synthesized from a truncated Fourier series. A scattering angle θ is associated with each object periodicity (or 'spatial frequency') d through Bragg's law and the image is synthesized from a Fourier summation of those beams which are included within the objective aperture. Only for the very thinnest specimens can these Fourier coefficients be used to represent a simple object property such as its projected electric potential or charge density. On the Fourier optics interpretation, the image formation consists of two stages—a Fourier transform from the specimen exit-face wavefunction giving the amplitude of the back-focal plane, followed by a second transform from this plane to reconstruct the image. An accessible account of Fourier electron optics can be found in the book by Hawkes (1972).

The subject matter of this chapter has been treated from slightly different points of view by many authors, including Lenz (1965), Hanszen (1971), and Misell (1973). The properties and wider applications of Fourier transforms and convolutions are also discussed in many books, including those by Bracewell (1999) and Goodman (2004). For a general account of optical aberration theory see Welford (1974).

3.1. Propagation and Fresnel diffraction

At high resolution the best images are not usually those obtained at exact Gaussian focus. Figure 3.1 shows an out-of-focus plane conjugate to the viewing phosphor which is imaged by an electron lens—the detector will record a copy of the wavefield across the plane containing P'. In order to interpret out-of-focus images it is important to understand the propagation of the electron wave across the defocus distance Δf. The mathematical development of Huygen's principle, due to Fresnel and Kirchhoff, gives a good approximation to the wave amplitude at P' as

48 Wave optics

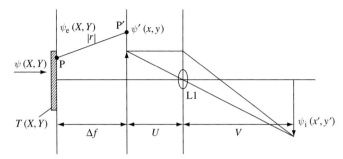

Figure 3.1 The out-of-focus image ψ_i, of an object with transmission function $T(X, Y)$. The wave amplitude conjugate to the image plane is $\psi'(x, y)$ and the incident wave $\psi(X, Y)$. For coherent imaging this is taken to have unit amplitude with the phase origin at the object.

$$\begin{aligned}\psi(x,y) &= \frac{i}{\lambda \Delta f} \int_{-\infty}^{\infty}\!\!\int \psi_e(X,Y) \exp\!\left(-\frac{2\pi i |r|}{\lambda}\right) dX\, dY \\ &\approx \frac{i \exp(-2\pi i \Delta f/\lambda)}{\lambda \Delta f} \int_{-\infty}^{\infty}\!\!\int \psi_e(X,Y) \\ &\quad \times \exp\!\left\{\frac{i\pi}{\lambda \Delta f}[(x-X)^2 + (y-Y)^2]\right\} dX\, dY \end{aligned} \qquad (3.1)$$

The essential features of this important expression can be understood from Fig. 3.2, and a more detailed account of its derivation can be found in many optics texts (see, e.g., Goodman (2004)). Equation (3.1) expresses the amplitude at P as the sum of the amplitudes due to an infinite number of fictitious secondary sources along $\psi_e(X, Y)$, each emitting a coherent spherical wave. The justification for using this expression in electron microscopy and the approximations involved (e.g. use of a scalar theory, neglect of backscattering) are discussed elsewhere (Cowley 1995). The scattering angles are all much smaller for electrons than for light. In eqn (3.1) $\psi_e(X, Y)$ is the two-dimensional complex amplitude across the exit face of the specimen. It is related to the incident wave $\psi(X, Y)$ by a transmission function $T(X, Y)$, where

$$\psi_e(X, Y) = T(X, Y)\,\psi(X, Y) \qquad (3.2)$$

Equation (3.2) can be used to compute the out-of-focus image of a specimen if the in-focus wave amplitude $\psi_e(X, Y)$ is known. For example, the Fresnel fringes seen at the edge of a partially absorbing specimen with finite refractive index have been calculated using this equation by Fukushima et al. (1974). The anomalous difference in contrast between the over-focus and under-focus fringes, is investigated in this way. For an edge, $\psi_e(X, Y) = \psi_e(X)$ and so, performing the Y integration,

$$\psi'(x) = \frac{1+i}{\sqrt{2}} \frac{\exp(-2\pi i \Delta f/\lambda)}{\sqrt{\lambda \Delta f}} \int_{-\infty}^{\infty} \psi_e(X) \exp\!\left\{-\frac{i\pi}{\lambda \Delta f}(x-X)^2\right\} dX \qquad (3.3)$$

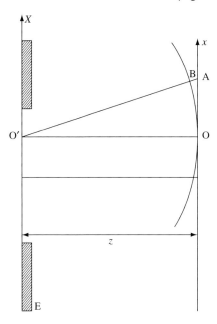

Figure 3.2 A spherical wave (a Huygens 'secondary source') emitted by a single point O' in the plane of a diffracting aperture E. The phase is constant along OB and the path difference $AB = [z^2 + (x - X)^2]^{1/2} - z \approx (x - X)^2/2z$ where x is the coordinate of A and X that of O'. The phase variation across Ox is thus $\psi(x) = \exp(-2\pi i AB/\lambda) = \exp[-i\pi(x - X)^2/\lambda z]$ due to a single point source at O'. Equation (3.1) expresses the total wavefunction across Ox as the sum of many such contributions from an object such as a transparency in the plane O'X, each weighted by the wavefunction $\psi_e(X, Y)$ leaving the object.

It is the real and imaginary parts of this integral, with the integration limits taken from 0 to a, say, which are represented by the Cornu spiral. While eqn (3.3) has been used to predict the form of Fresnel fringes in electron microscopy, more accurate work clearly shows that at high resolution both the properties of the edge-scattering material and the effect of spherical aberration must be considered. Dynamical calculations which include all these effects can be found in Wilson et al. (1982). Similar problems arise in the simulation of HREM 'profile' images, and in the simulation of images of planar defects such as nitrogen platelets in diamond.

A convenient notation results if eqn (3.1) is written as a convolution. Then

$$\psi'(x, y) = A\psi_e(x, y) * \mathcal{P}_z(x, y) = A \int_{-\infty}^{\infty} \psi_e(X, Y)\mathcal{P}_z(x - X, y - Y) \mathrm{d}X\, \mathrm{d}Y \tag{3.4}$$

where A is a complex constant and

$$\mathcal{P}_z(x, y) = \exp\left(-\frac{i\pi(x^2 + y^2)}{\lambda \Delta f}\right) \tag{3.5}$$

50 Wave optics

is called the Fresnel propagator.

Thus the evolution of the electron wave amplitude in the near-field region, important for the out-of-focus image, is described qualitatively by Huygens' principle, and quantitatively, within approximations accurate for electrons, by eqn (3.4). A list of the properties of the propagation factor $\mathcal{P}_Z(x)$ is given in Table 3.1. Note that two successive applications of the propagator leave the wavefield unchanged, in support of Huygens' original physical intuition.

Section 3.2 discusses the image of a point formed by a lens of finite aperture. The aperture limitation results in a blurred image amplitude, known as the lens impulse response. From eqn (3.4) the amplitude at the entrance surface of a lens distance U from a point object is, in one dimension,

$$\psi'(x) = A\delta(x) * \mathcal{P}_z(x) = A\exp\left(-\frac{i\pi x^2}{\lambda U}\right) \tag{3.6}$$

which is a quadratic approximation to the wavefield of a spherical wave diverging from the object point. This is shown in Fig. 3.2, taking $X = 0$. For a converging spherical wave, the sign in the exponential is reversed.

3.2. Lens action and the diffraction limit

The analogy between electron and light optics suggests that the action of a magnetic lens is equivalent to the introduction of a 'focusing' phase shift $(2\pi/\lambda)(x^2/2f)$, where f is the focal length of the lens and x the radial coordinate in the plane of the lens. The varying thickness of an ideal thin glass lens of suitable refractive index introduces such a phase shift. Thus, for a plane-wave incident on such a lens (with the phase origin taken at the entrance to the lens) the emerging wave would be proportional to

$$\psi_e = \exp\left(\frac{\pi i x^2}{\lambda f}\right) \tag{3.7}$$

From Fig. 3.2 we see that this is a quadratic approximation to a spherical wave converging to the optic axis at $z = f$. Thus a lens which introduces the phase shift of eqn (3.7) produces a focusing action. For the diverging spherical wave given by eqn (3.6), the wave immediately behind the lens will be, in one dimension (for a lens of infinite aperture),

$$\psi_e = A\exp\left(\frac{-i\pi x^2}{\lambda U}\right)\exp\left(\frac{i\pi x^2}{\lambda f}\right) = A\exp\left(\frac{i\pi x^2}{\lambda}\left[\frac{1}{f} - \frac{1}{U}\right]\right)$$

Now from the lens law, $1/f - 1/U = 1/V$, so that the wave emerging from this lens is a spherical wave converging to the conjugate image point P' at $z = V$, that is,

$$\psi_e = A\exp\left(\frac{i\pi x^2}{\lambda V}\right)$$

Lens action and the diffraction limit

Table 3.1 Properties of the Fresnel propagator $\mathcal{P}_z = (i/\lambda z) \exp(-i\pi (x^2 + y^2)/\lambda z)$.

1.	$\mathcal{P}_{z_1} * \mathcal{P}_{z_1} = \mathcal{P}_{z_0}$	where $z_0 = z_1 + z_2$
2.	$1 * \mathcal{P}_{z_1} = 1$	
3.	$\delta(x,y) * \mathcal{P}_{z_1} = \mathcal{P}_{z_1}$	
4.	$\mathcal{P}_{z_1} \mathcal{P}_{z_2} \propto \mathcal{P}_{z_3}$	where $1/z_3 = 1/z_1 + 1/z_2$
5.	$\mathcal{P}_{z_1} \mathcal{P}^*_{z_2} \propto \mathcal{P}_{z_4}$	where $1/z_4 = 1/z_1 - 1/z_2$

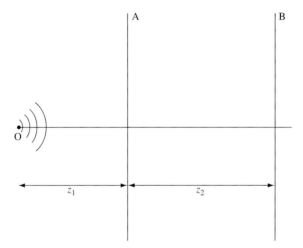

The wave amplitude a distance z from a plane on which it is known is obtained by convoluting the wave across that plane with $\mathcal{P}_z$. Notice that transmission through a thin lens is given by multiplication with $\mathcal{P}_{-f} = \mathcal{P}^*_f$. Property 1 shows that the effect of choosing an arbitrary intermediate plane A for the construction of Huygens' sources produces the same wave amplitude at B as would be obtained by a single propagation from 0 to B. Property 2 shows the effect of Fresnel propagation on a plane wave, while the third property expresses the spherical wave emitted by a point source. The final properties 4 and 5 can be used to show how the focal lengths of thin lenses in contact combine. These results are all obtained using standard integrals (a scaling constant has been omitted in 4 and 5).

The effect of a limiting aperture on such a wave is now considered, as shown in Fig. 3.3. This aperture may be due to the lens itself or, as in electron optics, to the objective aperture situated near the back-focal plane. In electron microscopy the objective aperture is rarely exactly in the back-focal plane since the aperture position is fixed, whereas the focal length of the objective lens is variable. Applying the Fresnel propagator from the plane of the aperture, where the wave is converging over a distance d to P', to the image gives

$$\psi(X) = AP(x) \exp\left(\frac{i\pi x^2}{\lambda d}\right) * \exp\left(-\frac{i\pi x^2}{\lambda d}\right)$$

$$= A' \int_{-\infty}^{\infty} P(x) \exp(2\pi i u x) \mathrm{d}x \qquad (3.8)$$

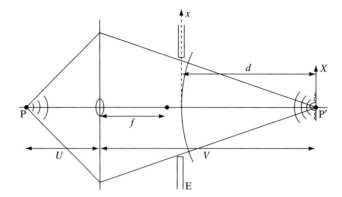

Figure 3.3 The conjugate image P' of a point P. The image is broadened by the effect of the diffraction limit set by the aperture E. The amplitude at P' is proportional to the Fourier transform of the pupil function $P(x)$ describing the aperture.

where A' is a quadratic phase factor, $u = X/\lambda d$, and $P(x)$ is the exit pupil function, equal to unity within the aperture and zero elsewhere. Thus, regardless of the position of the aperture, the image of a point formed by an ideal lens is given by the Fourier transform of the limiting aperture pupil function. Fourier transforms are also used to describe far-field or Fraunhofer scattering. In this special case of an aperture illuminated by a coherent converging spherical wave, the intensity of the Fresnel diffraction pattern equals that of the Fraunhofer pattern. The image intensity for a point object using a circular aperture is shown in most optics texts. The phase contrast impulse response (point image) for an electron lens is further discussed in Section 3.4 and shown in Fig. 3.8. In passing, it is interesting to note that, if a lens were not used, the condition for Fraunhofer diffraction ($z > d^2/\lambda$) with an object of size d would require a viewing screen to be placed more than 27 m from a 10-µm objective aperture in order to observe the Fraunhofer pattern with 100 kV electron illumination.

The imaging properties of simple lenses can also be understood using the Fresnel propagator to trace the progress of a wavefunction through the lens system. With unit amplitude coherent illumination on an object with transmission function $T(x, y)$, the wavefunction incident on the lens becomes

$$\psi_1(x_1, y_1) = \frac{i}{\lambda U} \psi_0(x, y) * \mathcal{P}_U(x, y) \quad \text{where } \psi_0(x, y) = T(x, y)$$

while the wavefunction immediately beyond the lens is

$$\psi_2(x_2, y_2) = \psi_1(x_2, y_2) \exp\left\{ \frac{i\pi (x_2^2 + y_2^2)}{\lambda f} \right\}$$

and the complex amplitude in the back-focal plane is given by

$$\psi_d(X, Y) = \frac{i}{\lambda f} \psi_2(x_2, y_2) * \mathcal{P}_f(x_2, y_2)$$

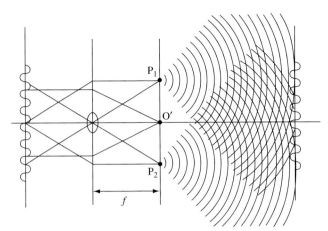

Figure 3.4 The Abbe interpretation of imaging. An object with sinusoidal amplitude transmittance produces two diffracted orders in directions given by the Bragg law, together with the unscattered beam. These are focused to points in the back-focal plane. The image is formed from the interference between spherical waves from each of the three 'sources' in the back-focal plane.

Evaluating these integrals and using eqn (3.5) gives

$$\psi_d(X, Y) = \frac{i}{\lambda f} \exp\left(-\frac{i\pi (X^2 + Y^2)}{\lambda} \left\{\frac{1}{d} - \frac{U}{f^2}\right\}\right)$$
$$\times \iint \psi_0(x, y) \exp\left(2\pi i \left\{\left(\frac{X}{f\lambda}\right) x + \left(\frac{Y}{f\lambda}\right) y\right\}\right) dx dy \quad (3.9)$$

The quadratic phase factor disappears if the object is placed at $U = f$, as is approximately the case in electron microscopy. This important result shows that a function proportional to the Fourier transform of the object exit-face wave amplitude (the transverse wavefield leaving the object) is found in the back-focal plane of the lens, in accordance with the Abbe interpretation of coherent imaging. All parallel rays leaving the object are gathered to a focus in the back-focal plane at a point distance X from the optic axis where $X = f\theta$, where θ is the angle at which rays leave the object. Figure 3.4 illustrates the Abbe interpretation for a periodic specimen where the diffraction pattern consists of a set of point amplitudes, if the illumination is coherent. These points can be thought of as a set of secondary sources like the holes in Young's pin-hole experiment which produce waves which interfere in the image plane to form fringes. The image is a Fourier synthesis of all these fringe systems. To confirm this mathematically, the back-focal plane amplitude of eqn (3.9) is written

$$\psi_d(X, Y) = \frac{Ai}{\lambda f} F(u, v)$$

where

$$F(u,v) = \iint \psi_o(x,y)\exp\{2\pi i(ux+vy)\}\,dxdy = \Im\{\psi_o(x,y)\} \quad (3.10)$$

In order to use published tables of Fourier transforms it is convenient to introduce the variables $u = X/\lambda f = \theta/\lambda$ and $v = Y/\lambda f = \beta/\lambda$, where θ and β are angles made with the optic axis. An imperfect lens with an aberration phase shift $\chi(u,v)$ can be incorporated into this analysis by defining the modified back-focal plane amplitude

$$\psi'_d(X,Y) = \frac{Ai}{\lambda f}F(u,v)P(u,v)\exp(i\chi(u,v)) \quad (3.11)$$

where $P(u,v)$ is the objective aperture pupil function. Then the image amplitude is

$$\psi_i(x',y') = \frac{i}{\lambda z_0}\psi'_d(X,Y) * \mathcal{P}_{z_0}(X,Y) \quad (3.12)$$

where $z_0 = V - f$ is the propagation distance between the back-focal plane and the image. From the lens law for conjugate planes this is $z_0 = Mf$, where M is the lateral magnification. Evaluating eqn (3.12) gives the image amplitude as

$$\psi_i(x',y') = -\frac{1}{M}\exp\left(-\frac{i\pi(x'^2+y'^2)}{\lambda M f}\right)\int_{-\infty}^{\infty}\!\!\int F(u,v)P(u,v)\exp(i\chi(u,v))$$
$$\times \exp\left(2\pi i\left\{u\frac{x'}{M}+v\frac{y'}{M}\right\}\right)du\,dv \quad (3.13)$$

It is convenient to assume unit magnification in most image calculations. Then the image amplitude, aside from a quadratic phase factor, is seen to be given by the Fourier transform of the product of the back-focal plane amplitude $F(u,v)$ and the transfer function $P(u,v)\exp(i\chi(u,v))$. The back-focal plane amplitude is proportional to the Fourier transform of the object exit-face wave amplitude. These two Fourier transforms produce an inverted image, the sign of M being taken as positive.

Inserting the expression for $F(u,v)$ in eqn (3.13) gives finally for the image amplitude in an ideal lens with $P(u,v)\exp(i\chi(u,v)) = 1$:

$$\psi_i(x',y') = \frac{1}{M}\exp\left(-\frac{i\pi\left(x'^2+y'^2\right)}{\lambda f M}\right)\psi_0\left(-\frac{x'}{M},-\frac{y'}{M}\right)$$

where the properties of the Dirac delta function have been used. Equation (3.13) can be written symbolically at unit magnification

$$\psi_i(x',y') = \mathcal{F}\{F(u,v)A(u,v)\} \quad (3.14a)$$

where
$$A(u, v) = P(u, v)\exp(i\chi(u, v)) \tag{3.14b}$$
is the microscope transfer function. Using the Fourier transform convolution theorem this becomes
$$\psi_i(x', y') = \mathcal{F}\{F(u, v)\} * \mathcal{F}\{A(u, v)\} = \psi_0(-x', -y') * \mathcal{F}\{A(u, v)\} \tag{3.15}$$
showing that each point of the ideal image is smeared or broadened by the impulse response function $\mathcal{F}(A(u,v))$, shown in figure 3.8, where $\mathcal{F}$ denotes the Fourier transform. This result applies only in a small isoplanatic patch near the optic axis where the form of $\mathcal{F}(A(u,v))$ does not depend on image position. The imaging is seen also to be linear in complex amplitude from the distributive properties of convolution (convolution is also commutative and associative). More generally $\psi_0(x', y')$ can be taken as the image predicted by geometrical optics, taking into account image inversions, magnification, and rotation.

3.3. Wave and ray aberrations (to fifth order)

In this section the aberration phase shift $\chi(u, v)$ is investigated, commencing with the term arising from a small error in focusing, from spherical aberration and from chromatic aberration. The important relationship between ray and wave aberration functions is given. We then give general results for higher-order aberrations.

Aside from a constant term, the wavefield imaged in Fig. 3.1 is, using eqn (3.1),
$$\psi'_{\Delta f}(x, y) = \psi_e(x, y) * \mathcal{P}_{\Delta f}(x, y)$$
with $\mathcal{P}_{\Delta f}(x, y)$ given by eqn (3.5). Using the convolution theorem for Fourier transforms the back-focal plane amplitude can be written as
$$\psi_d(u, v) = \mathcal{F}\{\psi_e(x, y)\} \cdot \mathcal{F}\{\mathcal{P}_{\Delta f}(x, y)\}$$
A table of Fourier transforms then gives the transform of the Fresnel propagator as
$$\mathcal{F}\{\mathcal{P}_{\Delta f}(u, v)\} = \frac{i}{\lambda \Delta f} \iint \exp\left(-\frac{i\pi x^2}{\lambda \Delta f}\right) \exp\left(-\frac{i\pi y^2}{\lambda \Delta f}\right)$$
$$\times \exp(2\pi i(ux + vy)) \, dx \, dy = \exp\left(i\pi \Delta f \lambda (u^2 + v^2)\right) \tag{3.16}$$
which is a radially symmetric function of the scattering angle $\theta = (u^2 + v^2)^{1/2}\lambda$ and can be written as $\exp(i\chi_1(\theta))$, where
$$\chi_1(\theta) = \frac{2\pi}{\lambda} (\Delta f \theta^2/2) \tag{3.17}$$

Spherical aberration can also be incorporated into the complex transfer function as follows. Figure 3.5 shows an aberrated ray whose deviation r_i in the Gaussian image plane

56 Wave optics

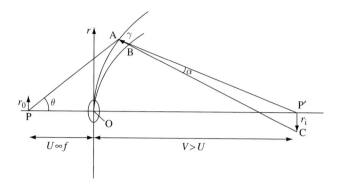

Figure 3.5 Determination of the phase shift due to spherical aberration (see text). The curvature of the wavefronts has been exaggerated. The rays AP' and BC run normal to the surfaces tangential to the wavefronts and containing OA and OB, respectively.

has been obtained as a function of θ by, say, numerical solution of the full ray equation (see Section 2.8). Only the case of high magnification is of interest and we ignore any dependence of C_s on M. Then, from Section 2.8.1, the ray aberration is found to be proportional to the cube of the scattering angle (hence $C_s = C_{3,0}$ as discussed below)

$$r_i = MC_s\theta^3 \qquad (3.18)$$

The rays drawn run normal to the surfaces of constant phase OA and OB. The surface OA is spherical, corresponding to the spherical wave converging from the exit pupil to an ideal image point P'. We require an expression for

$$\chi_2(\theta) = \frac{2\pi}{\lambda}\gamma$$

Now

$$r_i = \left(\frac{V}{f}\right) C_s\theta^3 = \frac{VC_sr^3}{f^4}$$

and

$$\alpha = \frac{d\gamma}{dr} \quad \text{with } \alpha = \frac{r_i}{V} \qquad (3.19)$$

so that, using $r = f\theta$ (for small angles)

$$\gamma = \frac{1}{V}\int r_i dr = C_s\theta^4/4$$

giving $\chi_2(\theta) = (2\pi/\lambda)(C_s\theta^4/4)$ for the spherical aberration phase shift, whose dependence on scattering angle is one power higher than the ray aberration. Combining this result with that obtained for defocus gives

$$A(\theta) = P(\theta)\exp\left(\frac{2\pi}{\lambda}\left(C_s\theta^4/4 + \Delta f\theta^2/2\right)\right) = P(\theta)\exp(i\chi(\theta)) \quad (3.20)$$

where a positive value of Δf implies an over-focused lens.

The focusing error Δf due to image formation by radiation of energy $V_0 - \Delta V_0$ must be considered with respect to the Gaussian image formed by radiation of energy V_0. The focusing action of a magnetic lens is greater for electrons of lower energy and longer wavelength. At high magnification $\Delta U \approx \Delta f$, so that eqns (2.34) and (3.17) give

$$\chi_3(\theta) = \frac{2\pi}{\lambda}\left[C_c\left(\frac{\Delta V_0}{V_0}\right)\theta^2/2\right] \quad (3.21)$$

where C_c is the chromatic aberration constant. To a good approximation the intensities of the various chromatically aberrated images can be added to form the final image, weighted by the normalized distribution function of incident electron energies (Misell 1973), as described in Section 4.2 on damping envelopes. Only for recording times less than $\Delta\tau = h/\Delta V_0$ should a coherent addition of image amplitudes be considered.

The azimuthally dependent focusing effect of astigmatism is similarly found to contribute a term (Fig. 2.20)

$$\chi_4(\theta) = \frac{2\pi}{\lambda}\left(\frac{1}{4}C_a\theta^2\sin 2\phi\right) \quad (3.22)$$

where ϕ is the azimuthal scattering angle. Collecting all these terms together gives a transfer function

$$A(\theta) = P(\theta)\exp\left(\frac{2\pi i}{\lambda}\left(\Delta f\theta^2/2 + C_s\theta^4/4 + \frac{1}{4}C_a\theta^2\sin 2\phi\right)\right) \quad (3.23)$$

Thus the effects of aberrations and instabilities in the microscope can be understood by imagining the scattering (diffraction pattern) in the object lens back-focal plane to be multiplied by this function.

In terms of the back-focal plane coordinates X and Y with $u = X/\lambda f$ and $v = Y/\lambda f$ the transfer function including defocus and spherical aberration effects only becomes

$$A(u,v) = P(u,v)\exp\left(2\pi i\left[\Delta f\lambda\left(u^2+v^2\right)^2/2 + C_s\lambda^3\left(u^2+v^2\right)^2/4\right]\right)$$
$$= P(u,v)\exp(i\chi(u,v)) \quad (3.24)$$

With recent improvements in resolution to the sub-angstrom level, the three-fold astigmatism aberration has been exposed. This introduces a phase shift

$$\chi_5(\theta) = \frac{2\pi}{3\lambda}|A_2|\theta^3\cos 3(\theta-\theta_2)$$

where A_2 is the coefficient of three-fold astigmatism and $\cos 3(\theta - \theta_2)$ represents the 120° azimuthal periodicity. The recent installation of a corrector for this aberration (Overwijk et al. 1997) has reduced A_2 from 2.46 to 0.05 μm or less (O'Keefe et al. 2001), allowing a resolution of 0.89 Å to be synthesized from through-focal series.

In general the higher-order aberrations discussed in Chapter 2 will be present, in which case the wavefront aberration phase shift, describing the deviation of the wavefield from a spherical wave, can be expanded in the isoplanatic approximation using complex coefficients $\mathbf{C}_{n,m} = C_{n,m,a} + iC_{n,m,b}$ (order n, symmetry m, units of length) as (Krivanek et al. 2007):

$$\chi(\theta) = (2\pi/\lambda)\Sigma_n\Sigma_m \chi_{n,m}(\theta) = (2\pi/\lambda)\Sigma\Sigma \, \mathrm{Re}\left[C_{n,m}\theta^{(n+1)}e^{-im\phi}\right]/(n+1) \qquad (3.25)$$

$$= (2\pi/\lambda)\Sigma_n\Sigma_m \, \mathrm{Re}\,\{[C_{n,m,a}\,(\theta_x - i\theta_y)^m]$$
$$+ iC_{n,m,b}\{(\theta_x - i\theta_y)^m]\}\,(\theta_x^2 + \theta_y^2)^{[n-(m+1)]/2}/(n+1) \qquad (3.26)$$

where we have used the complex scattering angle $\theta = \theta_x + i\theta_y = \theta\,e^{i\phi}$. Real coefficients describe aberrations with mirror symmetry, imaginary ones are anti-symmetric. Here $|\theta|/\lambda = (u^2 + v^2)^{1/2} = \theta/\lambda$ in the small-angle approximation, where θ is the scattering angle for HREM or the convergence angle for the focused beam in STEM. Writing this out in full to fifth order, we have

$$\chi(\theta) = (2\pi/\lambda)\,\Big\{C_{0,1,a}\theta_x + C_{0,1,b}\theta_y$$
$$+ C_{1,0}\left(\theta_x^2 + \theta_y^2\right)/2 + C_{1,2,a}\left(\theta_x^2 - \theta_y^2\right)/2 + C_{1,2,b}\theta_x\theta_y$$
$$+ C_{2,1,a}\left(\theta_x^3 + \theta_x\theta_y^2\right)/3 + C_{2,1,b}\left(\theta_x^2\theta_y + \theta_y^3\right)/3$$
$$+ C_{2,3,a}\left(\theta_x^3 - 3\theta_x\theta_y^2\right)/3 + C_{2,3,b}\left(3\theta_x^2\theta_y - \theta_y^3\right)/3$$
$$+ C_{3,0}\left(\theta_x^4 + 2\theta_x^2\theta_y^2 + \theta_y^4\right)/4 + C_{3,2,a}\left(\theta_x^4 - \theta_y^4\right)/4 + C_{3,2,b}\left(\theta_x^3\theta_y + \theta_x\theta_y^3\right)/2$$
$$+ C_{3,4,a}\left(\theta_x^4 - 6\theta_x^2\theta_y^2 + \theta_y^4\right)/4 + C_{3,4,b}\left(\theta_x^3\theta_y - \theta_x\theta_y^3\right)$$
$$+ C_{4,1,a}\left(\theta_x^5 + 2\theta_x^3\theta_y^2 + \theta_x\theta_y^4\right)/5 + C_{4,1,b}\left(\theta_x^4\theta_y + 2\theta_x^2\theta_y^3 + \theta_y^5\right)/5$$
$$+ C_{4,3,a}\left(\theta_x^5 - 2\theta_x^3\theta_y^2 - 3\theta_x\theta_y^4\right)/5 + C_{4,3,b}\left(3\theta_x^4\theta_y + 2\theta_x^2\theta_y^3 - \theta_y^5\right)/5$$
$$+ C_{4,5,a}\left(\theta_x^5 - 10\theta_x^3\theta_y^2 + 5\theta_x\theta_y^4\right)/5 + C_{4,5,b}\left(5\theta_x^4\theta_y - 10\theta_x^2\theta_y^3 + \theta_y^5\right)/5$$
$$+ C_{5,0}\left(\theta_x^6 + 3\theta_x^4\theta_y^2 + 3\theta_x^2\theta_y^4 + \theta_y^6\right)/6$$
$$+ C_{5,2,a}\left(\theta_x^6 + \theta_x^4\theta_y^2 - \theta_x^2\theta_y^4 - \theta_y^6\right)/6 + C_{5,2,b}\left(2\theta_x^5\theta_y + 4\theta_x^3\theta_y^3 + 2\theta_x\theta_y^5\right)/6$$
$$+ C_{5,4,a}\left(\theta_x^6 - 5\theta_x^4\theta_y^2 - 5\theta_x^2\theta_y^4 + \theta_y^6\right)/6 + C_{5,4,b}\left(2\theta_x^5\theta_y - 2\theta_x\theta_y^5\right)/3$$
$$+ C_{5,6,a}\left(\theta_x^6 - 15\theta_x^4\theta_y^2 + 15\theta_x^2\theta_y^4 - \theta_y^6\right)/6$$
$$+ C_{5,6,b}\left(3\theta_x^5\theta_y - 10\theta_x^3\theta_y^3 + 3\theta_x\theta_y^4\right)/3\Big\} \qquad (3.27)$$

These wave aberrations are related to the ray aberrations of Chapter 2 (e.g. eqn 2.30; also eqn 3.19 and Fig. 3.5) as follows. For an aberrated wavefield converging to a Gaussian focus

Wave and ray aberrations 59

on a plane normal to the optic axis at P, rays travelling at an angle θ will intersect this plane a vector distance

$$\Delta \mathbf{r}_0 = \lambda \left(-d\chi(\boldsymbol{\theta})/d\boldsymbol{\theta}_x, -d\chi(\boldsymbol{\theta})/d\boldsymbol{\theta}_y \right) / 2\pi \qquad (3.28)$$

from P. These rays give the direction of energy flow, normal to surfaces of constant phase. Using the wave aberration phase shift of eqn (3.20), to include only defocus and spherical aberration, we obtain for the ray aberration

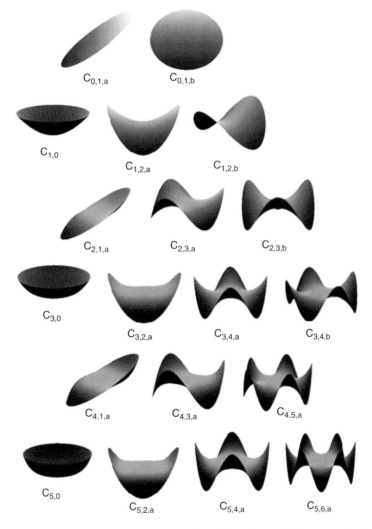

Figure 3.6 Wave aberration functions from eqn (3.27) for the axial (aperture) aberrations of a round lens. (From Krivanek *et al.* (2007).)

$$\Delta r_0 = C_s \theta^3 + \Delta f_\theta \qquad (3.29)$$

which is the familiar expression for the broadening of a focused beam due to these aberrations (within numerical factors determined by definitions of the intensity threshold).

Among these aberration coefficients, probe position (in STEM) or image shift (in HREM) is $C_{0,1}$, defocus $\Delta f = C_{1,0}$ (or C_1 in the notation of Rose and Haider), spherical aberration (third-order) $C_s = C_{3,0}$ (abbreviated C_3), while two-fold ($m = 2$) astigmatism of first order ($n = 1$), as shown in Fig. 2.15, is $C_{1,2}$. The shapes of these wave aberration functions are shown in Fig. 3.6. For a round lens, the third-order aberrations are the Siedel aberrations, and these are all zero except for C_s in the isoplanatic approximation.

The effect of adding a multipole element of $2m$ poles to an optical column is mainly to introduce an aberration $C_{m-1,m}$. One then has a multiparameter optimization problem—find the strength and directions of these multipole fields which minimize the total aberrations of the system to the largest scattering angle possible. The modern auto-tuning strategy is to measure the aberration coefficients of the whole system using either a Ronchigram (Fig. 7.5) or diffractogram tableaux (Fig. 12.8), then feed back corrections to the multipoles which will produce the desired compensating aberrations. The process of measurement, and the extraction of aberration coefficients from diffractograms and Ronchigrams, is described in Sections 10.6 and 10.10.

In Section 2.10 we briefly outlined the operation of the quadrupole–octupole corrector. Figure 3.7 shows this using the wave-aberration functions. The round beam is contracted to a thin ellipse by a quadrupole before entering the first octupole O1 in Fig. 2.18. This cancels the angular fourth-power C_s aberration along one of the white-line directions in Fig. 3.7 if the strength of the octupole is adjusted correctly. The beam is then expanded and contracted to a thin ellipse running along the orthogonal direction of the second white line in Fig. 3.7 as it enters another octupole O3, thereby cancelling spherical aberration in

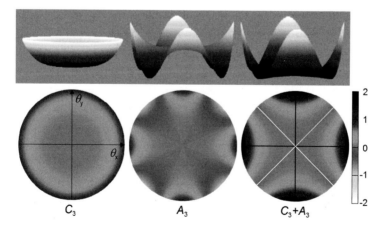

Fig 3.7 Addition of four-fold astigmatism from an octupole ($A_3 = C_{3,4}$) to a spherically aberrated round lens ($C_3 = C_{3,0}$). The sum of the wave-aberrations can be cancelled along the two crossed white lines, but doubled in the orthogonal directions of the black lines. Doing this twice with a rotation between eliminates the aberration everywhere. (From Erni (2010).)

most directions. The remaining warping of the wavefield can be corrected by an additional octupole. It should be noted that a similar design is capable also of correcting chromatic aberration (Haider et al. 2008).

3.4. Strong-phase and weak-phase objects

A simplified theory based on the weak-phase object and useful for low- and medium-weight molecules and ultra-thin unstained biological specimens ($t < 10$ nm) has become popular owing to its convenience for a posteriori image analysis. The neglect of Fresnel diffraction (focus variation) within the specimen (but not of multiple scattering) allows the exit-face complex wave amplitude to be written as

$$\psi_e(x, y) = \exp(-i\sigma\phi_p(x, y)) \tag{3.30}$$

as suggested by Fig. 2.2. Here $\sigma = 2\pi me\lambda_r/h^2$ (a positive quantity, with relativistically corrected values of λ_r and m) and $\phi_p = \int_{-t/2}^{t/2} \phi(x, y, z)\, dz$ is the projected specimen potential in volt-nanometres. At 100 kV, $\sigma = 0.009244$. This strong-phase object takes the Ewald sphere to be a plane normal to the incident beam direction defined by the z-axis. It represents the complete N-beam solution to the dynamical scattering problem in the limit of infinite voltage (Moodie 1972), while the terms in its expansion correspond to those of the Born series for the case where the Ewald sphere is flat—in this sense it sums the Born series. (In high-energy physics it is known as the Molière high-energy approximation (Wu and Ohmura 1962).)

The further approximation

$$\psi_e(x, y) \approx 1 - i\sigma\phi_p(x, y) \tag{3.31}$$

is known as the weak-phase object approximation and assumes single scattering within the specimen. This requires that the central 'unscattered' beam be much stronger than the diffracted intensity. Like the strong-phase object, this approximation is a projection approximation—atoms could be moved vertically within the specimen without affecting the computed image. The potential $\phi_p(x, y)$ may be made complex with an imaginary component used to represent either depletion of the elastic wave-field by inelastic scattering or the presence of an objective aperture, as described in Appendix 3.

Leaving aside constants, the complex amplitude in the back-focal plane of the objective lens is given by the Fourier transform of eqn (3.32), that is, with $\phi_p(x, y)$ a real function,

$$\psi_d(u, v) = \delta(u, v) - i\sigma\mathcal{F}\{\phi_p(x, y)\} \tag{3.32}$$

Introducing the microscope transfer function for a bright-field image formed with a central objective aperture then gives

$$\psi'_d(u, v) = \delta(u, v) - i\sigma\mathcal{F}\{\phi_p(x, y)\}P(u, v)\exp(i\chi(u, v)) \tag{3.33}$$

62 Wave optics

A further Fourier transform then gives the image complex amplitude as

$$\psi_i(x, y) = 1 - i\sigma\phi_p(x, y) * \mathcal{F}\{P(u, v)\exp(i\chi(u, v))\} \tag{3.34}$$

Because the sine and cosine of $\chi(u, v)$ are even functions their transforms are both real, so that the image intensity can be found to first order as

$$I(x, y) = \psi_i(x, y)\psi_i^*(x, y) \approx 1 + 2\sigma\phi_p(-x, -y) * \mathcal{F}\{\sin\chi(u, v)P(u, v)\} \tag{3.35}$$

The function $\mathcal{F}\{\sin\chi(u, v)P(u, v))\}$ is negative and sharply peaked as shown in Fig. 3.8, so the bright-field image of a weak-phase object consists of dark detail on a bright background at the optimum value of Δf and correctly chosen $P(u, v)$, determined by the objective aperture size (see Section 6.2). On this simplest theory of image contrast, the contrast is proportional to the projected specimen potential, convoluted with the impulse response of the instrument. Where this is radially symmetric (no astigmatism), it can be written

$$\sigma\mathcal{F}\{\sin\chi(u, v)P(u, v)\} = \frac{2\pi}{\lambda^2} \int_0^{\theta_{ap}} \sin\chi(\theta) J_0\left(\frac{2\pi\theta r}{\lambda}\right) \theta d\theta \tag{3.36}$$

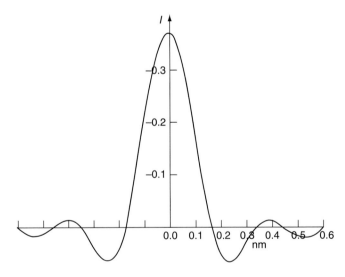

Figure 3.8 The impulse response of an electron microscope for phase contrast. The parameters are $C_s = 0.7$ mm, 100 kV, $\theta_{ap} = 12.8$ mrad, $\Delta f = -61$ mm. This negative function, added to the constant bright-field background, gives the image of an ideal point object under the optimum imaging conditions for phase contrast. The full width at half maximum height of this function is 0.204 nm, which gives an indication of the resolution. A more accurate estimate of resolution would be based on the coherent superposition of two adjacent functions and take into account the scattering properties of the specimen and instrumental instabilities.

as shown in Fig. 3.8 for a modern high-resolution instrument. This function is laid down at every point in the phase-contrast image, and so limits the instrumental resolution. The parameters used for Fig. 3.8 are the optimum values of θ_{ap} and Δf for phase contrast on an instrument with $C_s = 0.7$ mm (Scherzer conditions; see Section 6.2).

Higher-order terms in the expansion of eqn (3.30) correspond to the multiple-scattering terms of the Born series. Their importance has been investigated by Erickson (1974) and detailed numerical calculations comparing these two approximations with the correct N-beam dynamical solution have also been given by Lynch et al. (1975). These workers found serious discrepancies for middle-weight atoms between the amplitudes and phases of diffracted beams calculated using the kinematic eqn (3.31) and those calculated from the complete N-beam solution for specimens as thin as 1 nm. For a lighter element such as carbon, eqn (3.31) may hold for perhaps 6 mm at 100 kV. As has been pointed out, the validity of the weak-phase object approximation depends on the specimen structure and orientation. For crystalline specimens seen in projection with all atoms aligned in the beam direction, the phase change between an image point beyond a column of heavy atoms (see Fig. 2.2) and a neighbouring empty tunnel will increase rapidly with thickness. For a light amorphous specimen such as an unstained biological specimen in ice, the local deviation of phase from the mean phase shift will increase less rapidly with thickness. This important fact explains much of the success of quantitative cryo-EM. This field also benefits from the fact that, for typical macromolecules embedded in a slab of vitreous ice about 50 nm thick, application of eqn (3.30) shows that only differences between the mean phase shift due to the ice and that of the molecule will appear in the image intensity.

3.5. Diffractograms for aberration analysis

Coherent optical diffraction patterns of electron micrographs, obtained on an optical bench, were originally used to measure the defocus, spherical aberration, and astigmatism present in a high-resolution image; however, these have been replaced by digital Fourier transforms of either a digitized film recording (for cryo-EM) or the image directly acquired into a computer, using on-line image acquisition software. Since there are some dangers in the interpretation of these 'diffractograms' due, for example, to non-linearities in the recording device, the relationship between the electron microscope specimen and a diffractogram of its electron image is outlined in this section.

We are concerned only with low-contrast TEM specimens imaged in bright field. Then the optical density of the electron image recorded on film or an image plate is

$$D(x, y) = D_0 + D_0 f(x, y) \tag{3.37}$$

where D_0 represents the optical density of the bright-field background and $f(x, y)$ is the contrast term given either by

$$f(x, y) = 2\sigma\phi_{\text{p}} * \mathcal{F}\{\sin \chi(u, v) P(u, v)\} \tag{3.38}$$

or

$$f(x, y) = \frac{\Delta f \lambda \sigma}{2\pi} \nabla^2 \phi_{\text{p}}(x, y) \tag{3.39}$$

depending on the precise experimental conditions, and provided the specimen is sufficiently thin (see Chapter 5). For read-out, the intensity of light transmitted by the micrograph is, from the definition of optical density,

$$I_e(x,y) = I_0 \exp(-D(x,y))$$

so that the amplitude immediately beyond the micrograph is

$$\psi_e(x,y) = \sqrt{I_0} \exp\left(-\frac{1}{2}D(x,y)\right)$$
$$= \sqrt{I_0} \exp(-D_0/2) \exp(-f(x,y)D_0/2)$$
$$= C \exp(-f(x,y)D_0/2)$$

where C is a constant.

For a dark-field micrograph $D_0 \approx 2$, and if the contrast is low, $f(x) < 1$ so that

$$\psi_e(x,y) \approx C - \frac{1}{2}D_0 C f(x,y)$$

The complex amplitude in the optical lens back-focal plane is given, aside from phase factors, by the Fourier transform of this,

$$\psi_d(u,v) = C\delta(u,v) - \frac{1}{2}D_0 C \mathcal{F}\{f(x,y)\} \tag{3.40}$$

with

$$u = \theta/\lambda_e, \quad v = \phi/\lambda_e$$

where λ_e is the laser light wavelength. The intensity recorded (the diffractogram) is

$$I(u,v) = \psi_d \psi_d^* = B\delta(u,v) + \frac{1}{4}D_0^2 C^2 |\mathcal{F}\{f(x,y)\}|^2 \tag{3.41}$$

where B is a constant. For the weak-phase object (eqn 3.31)

$$|\mathcal{F}\{f(x,y)\}|^2 = 4\sigma^2 |\mathcal{F}\{\phi_p(x,y)\}|^2 \sin^2 \chi(u,v) P(u,v) \tag{3.42}$$

If $|\mathcal{F}\{\phi_p(x,y)\}|^2$ is sufficiently slowly varying, the important structure seen will be the term $\sin^2 \chi(u,v)$, that is, the square of the transfer function shown in Fig. 6.1. This is the basis of the interpretation given to Thon's diffractograms (Section 10.6). The interpretation depends on two important conditions—low bright-field contrast and a monotonic form for $|\mathcal{F}\{\phi_p(x,y)\}|^2$. An example of a diffractogram is given in Fig. 10.4.

Notice that the intensity of the electron diffraction pattern, given by the squared modulus of eqn (3.42), does not reveal the transfer function $\sin \chi(u,v)$. It does, however, reveal the form of $|\mathcal{F}\{\phi_p(x,y)\}|^2$, which for thin carbon films shows appreciable structure indicating the degree of order in these 'amorphous' specimens.

For thin, low-atomic-number periodic specimens (e.g. unstained catalase) we see from eqn (3.42) that the optical diffraction spots will appear in the same places (though with incorrectly weighted intensities) as the corresponding electron diffraction spots. This allows the possibility of optical filtering of electron micrographs under these conditions. Masks are used to exclude particular diffraction orders or the diffuse scattering between orders, thereby enhancing the contrast of the optically reconstructed image. A bright-field micrograph can be used to form a high-contrast dark-field optical image by excluding the central optical diffraction spot. The crucial requirement is that the specimen is thin enough for the central electron diffraction spot to be much more intense than any diffracted orders (kinematic conditions). In particular it is incorrect to interpret the optical diffraction spot intensities as if they were proportional to the corresponding electron diffraction pattern spot intensities. However, by carrying the analysis through to the optical image plane it can be shown that this image intensity is proportional to the electron micrograph density. Thus the image intensity recorded with an optical mask in place will be similar to the micrograph, apart from the effects of the mask. The use of a mask corresponds to the introduction of a priori information, and results should be treated accordingly. For example note that the periodicity of the optical image is determined by the positions of the holes in the mask. For a particular mask, an image will be formed whose periodicity is completely independent of the micrograph used, so long as it provides some scattering to illuminate the mask.

If an optical image of an 'almost periodic' micrograph (such as a damaged biological crystal) is formed using a back-focal plane mask which allows only the sharp Bragg spots to contribute to the image, a 'periodically averaged' image of the micrograph will be formed. This periodically averaged image is perfectly periodic, and each unit cell in this image is proportional to the sum of all the individual damaged unit cell images in the original micrograph, added into a single unit cell, and then periodically continued. A fuller discussion of the use of optical diffractograms is given in Beeston *et al.* (1972).

For dark-field electron micrographs, a simple interpretation of the optical diffraction pattern is not generally possible, since the density of the micrograph is proportional to the square of the deviation of the potential from the mean potential (on a single-scattering model) (see Section 6.5).

The use of diffractograms for electron microscope testing is described in Section 10.6. These are now provided on-line in digital form by the image acquisition software, which performs a Fourier transform of HREM images. This provides a kind of 'microdiffraction' capability, when a small region of a HREM image is transformed. But, as we see from the above, the intensities in this transform are not proportional to electron microdiffraction pattern intensities. The use of diffractograms for auto-tuning is described in Section 12.5.

References

Beeston, B. E. P., Horne, R. W., and Markham, R. (1972). Electron diffraction and optical diffraction techniques. *Practical methods in electron microscopy* (ed. A. M. Glauert), Vol. 1. North-Holland, Amsterdam.

Born, M. and Wolf, E. (1999). *Principles of optics*. Seventh edition. Cambridge University Press, Cambridge.

Bracewell, R. (1999). *The Fourier transform and its applications*. Third edition. McGraw-Hill, New York.

Cowley, J. M. (1995). *Diffraction physics*. Third edition. North-Holland, Amsterdam.
Erickson, H. P. (1974). The Fourier transform of an electron micrograph—first order and second order theory of image formation. *Adv. Opt. Electron Microsc.* **5**, 163.
Erni, R. (2010). *Aberration-corrected imaging in transmission electron microscopy*. Imperial College Press, London.
Fejes, P. L. (1973). Approximations for the calculation of high resolution electron microscope images of thin films. *Acta Crystallogr.* **A 33**, 109.
Fukushima, K., Kawakatsu, H., and Fukami, A. (1974). Fresnel fringes in electron microscope images. *J. Phys. D.* **7**, 257.
Goodman, J. W. (2004). *Introduction to Fourier optics*. Third edition. Roberts and Company, Englewood, CO.
Hanszen, K. J. (1971). The optical transfer theory of the electron microscope: fundamental principles and applications. *Adv. Opt. Electron Microsc.* **4**, 1.
Hawkes, P. (1972). *Electron optics and electron microscopy*. Taylor and Francis, London.
Haider, M., Muller, H., Uhlemann, S., Zach, J., Loebau, U., and Hoeschen, R. (2008). Pre-requisites for a C_s/C_c corrected ultra-high resolution TEM. *Ultramicroscopy* **108**, 167.
Komrska, J. and Lenc, M. (1972). Wave mechanical approach to magnetic lenses. In *Proc. 5th Eur. Congr. Electron Microsc.*, p. 78. Institute of Physics, Bristol.
Krivanek, O. L., Delby, N., and Murfitt, M. (2007). Aberration correction in electron microscopy. In *Handbook of charged particle optics*, ed. J. Orloff, pp. 601–640. CRC Press, Boca Raton, FL.
Lenz, F. (1965). The influence of lens imperfections on image formation. *Lab. Invest.* **14**, 70.
Lynch, D. F., Moodie, A. F., and O'Keefe, M. (1975). N-beam lattice images. V. The use of the charge density approximation in the interpretation of lattice images. *Acta Crystallogr.* **A31**, 300.
Misell, D. L. (1973). Image formation in the electron microscope with particular reference to the defects in electron-optical images. *Adv. Electron. Electron Phys.* **32**, 63.
Moodie, A. F. (1972). Reciprocity and shape functions in multiple scattering diagrams. *Z. Naturforsch.* **27a**, 437.
O'Keefe, M. A., Hetherington, C. J. D., Wang, Y. C., Nelson, E. C., Turner, J. H., Kiselowski, C., Malm, J.-O., Mueller, R., Ringnalda, J., Pan, M., and Thust, A. (2001). Sub-angstrom HREM at 300 keV. *Ultramicroscopy* **89**, 215.
Overwijk, M. H. F., Bleeker, A. J., and Thust, A. (1997). Three-fold astigmatism in HREM. *Ultramicroscopy* **67**, 163.
Spence, J. C. H., O'Keefe, M. A., and Kolar, H. (1977). High resolution image interpretation in crystalline germanium. *Optik* **49**, 307.
Wade, R. H. and Frank, J. (1977). Electron microscope transfer functions for partially coherent axial illumination. *Optik* **49**, 81.
Welford, W. T. (1974). *Aberrations of the symmetrical optical system*. Academic Press, London.
Wilson, A. R., Spargo, A. E. C., and Smith, D. J. (1982). The characterisation of instrumental parameters in the high resolution electron microscope. *Optik* **61**, 63.
Wu, T. and Ohmura, T. (1962). *Quantum theory of scattering*. Prentice Hall, Englewood Cliffs, NJ.

Bibliography

Two excellent texts on imaging theory which emphasize ideas that are important for high–resolution electron microscopy are:

Goodman, J. W. (2004). *Introduction to Fourier optics*. Third edition. McGraw-Hill, New York.
Martin, L. C. (1966). *The theory of the microscope*. American Elsevier/Blackie, London.

4
Coherence and Fourier optics

The coherence of a wavefield refers to its ability to produce interference effects. As described in Section 4.3, this ability is commonly measured by picking off two points along a wavefield (using the two pin-holes of Young's pin-hole experiment, for example) and measuring the contrast of the interference fringes which result. The high-resolution detail in an electron micrograph arises from coherent interference. In a bright-field image, for example, it is the interference between the central beam and the various waves scattered by the specimen which forms the image. So long as the resolution of the electron microscope was limited by electronic and mechanical instabilities to distances much larger than that coherently illuminated, the question of coherence remained unimportant. Wave-optical interference controls the fine detail in modern HREM images, whether it be the probe-formation process in STEM or the interference between the scattered and direct beams in TEM.

Some of the important ideas of optical coherence theory are described in this chapter. This theory was developed in optics, but most of it also applies to electron optics, and the validity of a fundamental optical coherence theorem (the van Cittert–Zernike theorem) has been tested experimentally for electrons (Burge *et al.* 1975). In optics, the waves emanating from different atomic oscillators in a light source are treated as incoherent. For a heated filament, each atom will emit a wavetrain of finite duration (inversely related to the energy spread of the light), with random phase jumps at different times for different atoms. The total intensity in the interference patterns due to each atomic oscillator must be added together, and such a source is therefore treated as ideally incoherent. Similarly, the wavefields of successive fast electrons emitted from the filament in the electron microscope are incoherent, with no fixed, time-independent phase relationship between them. In the words of Paul Dirac 'each electron interferes only with itself'. Now the image theory outlined in Chapter 3 was developed for a specimen illuminated by an idealized infinite electron wavetrain, originating from a point. Where many electrons, arriving from slightly different directions, are used to illuminate the specimen the image intensities due to each fast electron must be added together.

Some further qualitative ideas, described in more mathematical detail in the sections which follow, are set out below.

1. An *effective source* can be defined for an electron microscope. It lies in the exit pupil of the final condenser lens, which is usually taken as coincident with the illuminating aperture. The effective source is an imaginary electron emitter filling the illuminating aperture. A mathematical definition is given in Hopkins (1957). Each point within the aperture is supposed to represent a point source of electrons. The emerging spherical wave is approximately plane at the specimen and this is focused to a point in the lens back-focal plane

68 Coherence and Fourier optics

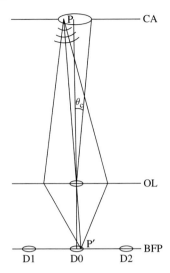

Figure 4.1 Formation of the central 'unscattered' diffraction spot in an electron microscope. Each point P in the illuminating aperture CA is focused to a point P' in the central diffraction spot D0 in the back-focal plane BFP of the objective lens OL. D1 and D2 are two other Bragg reflections. The beam divergence θ_c is also shown.

(Fig. 4.1). At the specimen, each electron can be specified by the direction of an incident plane wave. Increasing the size of the illuminating aperture increases the size of the central diffraction spot accordingly. The conditions under which the illuminating aperture may not be incoherently filled are discussed in Section 4.5, in which case this model does not apply.

2. A second important concept is that of *spatial coherence width*, also known as *lateral or transverse coherence width*. The term coherence length should be reserved for temporal or longitudinal coherence. The coherence width is the distance at the object over which the illuminating radiation may be treated as perfectly coherent. Thus the scattered waves from a specimen consisting of two atoms separated by less than this distance, X_c, will interfere and it is the complex amplitudes of these waves which must be added, in this case to produce a cosine-modulated atomic scattering factor. Atoms separated by distances much greater than X_c scatter incoherently, and the intensities of their scattered radiation must be added. The intermediate range is described by the theory of partial coherence (see Fig. 4.2).

Under normal operating conditions there is a simple relationship between the coherence width X_c and the semi-angle θ_c subtended by the illuminating aperture at the specimen (the beam divergence). This result is given in Section 4.3 as

$$X_c = \lambda/2\pi\theta_c$$

The objective lens pre-field must also be considered (see Section 2.9).

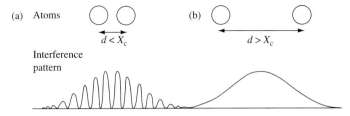

Figure 4.2 The intensity of scattering recorded at a large distance from two atoms separated by (a) less than the coherence width X_c and (b) a distance much greater than X_c. There is no interference in the second case. The unscattered beam is not shown.

3. The effect of coherence width or beam divergence on the *contrast* of phase-contrast images is discussed in Section 4.2. This is an important question since it enables the experimentalist to make the best choice of illuminating aperture size for a particular experiment. Fresnel diffraction, lattice fringes, and single-atom images are three examples of high-resolution phase-contrast detail. To obtain this type of contrast a choice of θ_c must be made which keeps X_c larger than the coarsest detail of interest. Only a range of specimen detail smaller than X_c will produce phase contrast of the type described in Chapter 3 which can be enhanced by the Scherzer optimum focus technique, since this relies on Fresnel interference.

4.1. Independent electrons and computed images

The elastic scattering of a fast electron by a thin specimen is generally treated as a two-body problem in which the specimen is described by a suitable complex optical potential and a solution is obtained for the wavefunction $\psi_0(\mathbf{r}_0, \mathbf{k}_i)$ of the fast electron (incident wavevector $\mathbf{k}_i$) on the specimen exit face. Successive fast electrons are assumed to be independent and any interaction between them (such as the Boersch effect) is neglected. We assign a separate wavevector and direction to each incident electron. Two electrons with the same wavevector would arrive at the specimen at different times. For an extended source, the intensity at a point in the final image $I(\mathbf{r}_i)$ can be obtained by summing the intensities of the images due to each fast electron. Thus

$$I(\mathbf{r}_i) = \int_{-\infty}^{\infty} \int_{-\infty}^{\infty} |\psi_i(\mathbf{r}_i, \Delta F, \mathbf{K}_0)|^2 \, F(\mathbf{K}_0) B(\Delta f) \mathrm{d}\mathbf{K}_0 \, \mathrm{d}\Delta f \tag{4.1}$$

where $\mathbf{r}$ and $\mathbf{K}_0$ are two-dimensional vectors ($\mathbf{K}_0 = u\hat{\mathbf{i}} + v\hat{\mathbf{j}}$) and $F(\mathbf{K}_0)$ describes the normalized distribution of electron wavevectors. Thus $F(\mathbf{K}_0)\mathrm{d}\mathbf{K}_0$ is the probability that the incident electron has a wavevector with $\hat{\mathbf{i}}$ and $\hat{\mathbf{j}}$ components in the range $\mathbf{K}_0$ to $\mathbf{K}_0 + \mathrm{d}\mathbf{K}_0$. Here $\mathbf{K}_i = \mathbf{K}_0 + w\hat{\mathbf{k}}$ and $|\mathbf{k}_i| = 1/\lambda_r$ with $\hat{\mathbf{i}}$, $\hat{\mathbf{j}}$, and $\hat{\mathbf{k}}$ orthogonal unit vectors. $B(\Delta f)$ describes the distribution of energy present in the electron beam, and may also include effects due to fluctuations in the objective lens current. All these effects can be represented as time-dependent variations in the focus setting Δf.

We will show in Section 4.5 (and this can be demonstrated experimentally using an electron biprism) that to a good approximation the filled final illuminating aperture for a TEM fitted with a thermionic electron source can be treated as a perfectly incoherent source of electrons. It is thus not necessary to trace each electron back to its source at the filament in using eqn (4.1). For practical computations, the illuminating cone of radiation under focused illumination may be taken as uniformly filled, corresponding to the choice of a 'top-hat' function for $F(\mathbf{K}_0)$ with $|\mathbf{K}_0|_{max} = |\mathbf{k}_i| \sin \theta_c$, where θ_c is the beam divergence. The exact profile of $F(\mathbf{K}_0)$ can be measured from the intensity distribution across the central diffraction spot.

There are three possible approaches to the problem of understanding and simulating partial coherence effects in HREM:

1. The images may be computed exactly, using eqn (4.1) and the result of a multiple-scattering computer calculation for $|\psi_i(\mathbf{r}_i, \mathbf{K})|^2$ (see Chapter 5). This method makes no approximation but requires a separate mulitple-scattering calculation for each component wavevector $\mathbf{K}_0$ in the incident cone of illumination. Early results of such calculations, exploring the effect of variations in coherence, can be found in O'Keefe and Sanders (1975) (see Fig. 4.9).
2. To avoid the need for many such calculations, an approximation valid for small beam divergence may be adopted. This requires a single dynamical calculation for $\psi_i(\mathbf{r}_i, \mathbf{K}_0)$, and is described in Section 5.8.
3. In addition to assuming small θ_c, we may make the further weak-phase object approximation (eqn 3.31), in order to obtain a result in the form of a multiplicative transfer function (eqn 3.24).

When using a field-emission source (see Section 4.5) it may happen that the illuminating aperture is coherently filled. Then the complex image amplitudes for each incident direction must be summed, rather than their intensities. The appropriate transfer function for this case has been derived by Humphreys and Spence (1981) and O'Keefe and Saxton (1983) and compared with the incoherent illumination case.

4.2. Coherent and incoherent images and the damping envelopes

The labour of detailed image simulation can be avoided and a simple understanding of the effect of coherence on image contrast can be obtained for specimens sufficiently thin that the approximation

$$\psi_0(\mathbf{r}_0, \mathbf{K}_0) = \psi_0(\mathbf{r}_0, 0) \exp(2\pi i \mathbf{K}_0, \mathbf{r}_0) \tag{4.2}$$

can be made. This approximation is satisfied for both strong- and weak-phase objects. Here $\psi_0(\mathbf{r}_0, 0)$ is the specimen exit-face wave for normally incident illumination. The approximation neglects the orientation dependence of scattering within the specimen due to the rotation of the Ewald sphere with respect to the crystal lattice. The neglect of excitation error effects is equivalent to the neglect of Fresnel diffraction (describing Huygens wavelets) within the specimen if the refractive index and propagation effects are taken as separable (see Section 6.3).

For specimens sufficiently thin that eqn (4.2) applies ($t < 5$ nm for low-atomic-number amorphous specimens), we now consider the two extremes of spatially coherent and incoherent illumination. For the present we ignore chromatic aberration effects and take $B(\Delta f) = \delta(\Delta f)$. The transfer equation for imaging is from equation 3.15, with $A'(r)$ the Fourier Transform of $A(u,v)$,

$$\psi_i(\mathbf{r}_i, \mathbf{K}_0) = \int \psi_0(\mathbf{r}_0, \mathbf{K}_0)\tilde{A}(\mathbf{r}_i - \mathbf{r}_0)\mathrm{d}\mathbf{r}_0 \qquad (4.3)$$

Using eqns (4.1), (4.2), and (4.3) gives

$$I(\mathbf{r}_i) = \int\int \psi_0(\mathbf{r}_0,0)\psi_0^*(\mathbf{r}_0',0)\tilde{A}(\mathbf{r}_i - \mathbf{r}_0)\tilde{A}^*(\mathbf{r}_i - \mathbf{r}_0')\gamma(\mathbf{r}_0' + \mathbf{r}_0)\mathrm{d}\mathbf{r}_0\mathrm{d}\mathbf{r}_0' \qquad (4.4)$$

where

$$\gamma(\mathbf{r}_0) = \int F(\mathbf{K}_0)\exp(-\pi i \mathbf{K}_0 \cdot \mathbf{r}_0)\,\mathrm{d}\mathbf{k} \qquad (4.5)$$

The function $\gamma(\mathbf{r}_0)$, if normalized, is known as the complex degree of coherence and will be discussed in Section 4.3. (This is a different gamma from that in figure 3.5 or eqns 5.45 or 6.18)

For coherent illumination, $F(\mathbf{K}_0) = \delta(\mathbf{K}_0)$ and $\gamma(\mathbf{r}_0) = 1$, so that eqn (4.4) becomes

$$I(\mathbf{r}_i) = \left|\int \psi_0(\mathbf{r}_0,0)\tilde{A}(\mathbf{r}_i - \mathbf{r}_0)\,\mathrm{d}\mathbf{r}_0\right|^2 \qquad (4.6)$$

in agreement with the image intensity given by eqn (4.3) for normal plane-wave illumination. For perfectly incoherent illumination, $F(\mathbf{K}_0)$ is constant and $\gamma(\mathbf{r}_0) = \delta(\mathbf{r}_0)$, and we obtain

$$I(\mathbf{r}_i) = \int |\psi_0(\mathbf{r}_0,0)|^2|\tilde{A}(\mathbf{r}_i - \mathbf{r}_0)|^2\,\mathrm{d}\mathbf{r}_0 \qquad (4.7)$$

For a pure-phase object, as discussed in Section 1.1 and used as a model for ultrathin biological specimens, we have

$$\psi_0(\mathbf{r}_0,0) = \exp(-i\sigma\phi_\mathrm{p}(\mathbf{r}_0))$$

where $\phi_\mathrm{p}(\mathbf{r})$ is real. Since the squared modulus of this function is unity, eqn (4.7) indicates that no contrast is possible from such a specimen using perfectly incoherent illumination (see also Section 1.2). In practice one is normally dealing with partially coherent illumination. A perfectly incoherent imaging system would require an illumination aperture of infinite diameter, supplying statistically independent plane-waves arriving from all directions.

Loosely speaking, then, using a very small condenser aperture in the electron microscope is rather like using a laser source in optics, while a large aperture corresponds to a tungsten lamp source.

For specimens satisfying eqn (4.2) the transfer of information in the electron microscope is thus linear in complex amplitude under coherent illumination and linear in intensity under incoherent illumination.

We now consider the important intermediate case of partial coherence with which microscopists are chiefly concerned in practice. The incorporation of the effects of partial spatial and temporal coherence into the transfer function described earlier (eqn 3.24) has been described by many workers. We will rely mainly on the works of Frank (1973) and Fejes (1977) which give an estimate of the effects of partial coherence on images for the simple case of a Gaussian distribution of intensity across the effective source. For small effective source widths and a central zero-order diffracted beam much stronger than any other, these papers show that the combined effects of partial coherence and electronic instabilities leads to a transfer function of the form

$$A(\mathbf{K}) = P(\mathbf{K}) \exp\{i\chi(\mathbf{K})\} \exp\left\{-\pi^2 \Delta^2 \lambda^2 \mathbf{K}^4/2\right\} \gamma(\nabla\chi/2\pi)$$
$$= P(\mathbf{K}) \exp\{i\chi(\mathbf{K})\} \exp\left\{-\pi^2 \Delta^2 \lambda^2 \mathbf{K}^4/2\right\} \exp\{-\pi^2 u_0^2 \mathbf{q}\} \quad (4.8a)$$

in the absence of astigmatism. Here gamma is defined by equation 4.5. This important result is obtained by considering a Taylor expansion of the wavefront aberration function $\chi(\mathbf{K})$ across the illumination aperture. As discussed in Sections 3.3 and 2.8, the quantities $\chi(\mathbf{K})$ and Δ are defined by

$$\chi(\mathbf{K}) = \pi \Delta f \lambda \mathbf{K}^2 + \pi C_s \lambda^3 \mathbf{K}^4/2 \quad (4.8b)$$

with $\mathbf{K}$ the vector $u\hat{\mathbf{i}} + v\hat{\mathbf{j}}$ where $(u^2 + v_2)^{1/2} = |\mathbf{K}| = \theta/\lambda$, and

$$\Delta = C_c Q = C_c \left[\frac{\sigma^2(V_0)}{V_0^2} + \frac{4\sigma^2(I)}{I_0^2} + \frac{\sigma^2(E_0)}{E_0^2}\right] \quad (4.9)$$

where $\sigma^2(V_0)$ and $\sigma^2(I)$ are the variances in the statistically independent fluctuations of accelerating voltage V_0 and objective lens current I_0, respectively. (Correlated fluctuations may also occur.) The root-mean-square value of the high-voltage fluctuation is thus equal to the standard deviation $\sigma(V_0) = [\sigma^2(V_0)]^{1/2}$. A term has also been added to account for the energy distribution of electrons leaving the filament. The full width at half the maximum height of the energy distribution of electrons leaving the filament is

$$\Delta E = 2.345 \sigma(E_0) = 2.345 \left[\sigma^2(E_0)\right]^{1/2}$$

Typical values of Δ lie between 1 and 5 nm. The normalized Gaussian distribution of intensity assumed for the incoherent effective electron source has the form

$$F(\mathbf{K}_0) = \left(\frac{1}{\pi u_0^2}\right) \exp\left(-\frac{\mathbf{K}_0^2}{u_0^2}\right)$$

If the beam divergence is chosen as the angular half-width θ_c for which this distribution falls to half its maximum value, then u_0 is defined by

$$\theta_c = \lambda u_0 \left(\ln 2\right)^{1/2}$$

The quantity **q** in eqn (4.8a) is

$$\mathbf{q} = \left(C_s \lambda^3 \mathbf{K}^3 + \Delta f \lambda \mathbf{K}\right)^2 + \left(\pi^2 \lambda^4 \Delta^4 \mathbf{K}^6 - 2\pi^4 i \lambda^3 \Delta^2 \mathbf{K}^3\right) \tag{4.10}$$

Equation (4.8) expresses all the resolution-limiting factors of practical importance, except specimen movement, for HREM in the absence of aberration correction. The relative importance of these factors is discussed in more detail in Appendix 3. For the present, we note several important features of eqn (4.8a).

A crucial approximation here is the assumption of a strong zero-order diffracted beam. This condition is satisfied both in very thin crystals and in thicker areas of wedge-shaped crystals showing strong Pendellösung fringes, as described in Section 5.6. Note that terms such as those containing $\phi_h \phi_{-h}$ in eqn (5.9) which lead to the appearance of 'half-period fringes' are neglected in this analysis. These fringes have been used to give a misleading impression of high-resolution detail.

The last bracketed complex term in eqn (4.10) expresses a coupling between the effects of using a finite incident beam divergence angle θ_c (partial spatial coherence) and the consequences of using a non-monochromatic electron beam (partial temporal coherence, $\Delta \neq 0$). The magnitude of this coupling term has been investigated in detail by Wade and Frank (1977), who found that, under high-resolution conditions (e.g. $\theta_c < 0.001$ rad, $\Delta < 20$ nm at 100 kV) this term can frequently be neglected. Then eqn (4.8a) contains three multiplicative factors, each of which imposes a resolution limit by attenuating high-order spatial frequencies. The first term $P(\mathbf{K})$ expresses the diffraction limit imposed by the objective aperture. The third term in $\mathbf{K}^4$ describes a damping envelope more severe than Gaussian attenuation with a width

$$u_0(\Delta) = [2/(\pi \lambda \Delta)]^{1/2} \tag{4.11}$$

which will *always* be present even if the objective aperture is removed. This resolution limit $d \approx 1/u_0(\Delta)$ is independent of the illumination conditions used and depends on the existence of instabilities in the objective lens and high-voltage supplies and on the thermal spread of electron energies. The last term γ in eqn (4.8a) shows an apparently complicated dependence on illumination semi-angle θ_c, focus Δf, spherical aberration constant C_s, and wavelength λ. Its behaviour can be given a simple interpretation, however, since the function $\gamma(\nabla \chi / 2\pi)$ is just the Fourier transform of the source intensity distribution evaluated with the function's argument equal to the local slope of the aberration function $\chi(\mathbf{K})$. For a Gaussian source, $\gamma(\mathbf{r})$ is also Gaussian and has a width which is inversely proportional to the width of the source. Thus $\gamma(\nabla \chi / 2\pi)$ is small in regions where the slope of $\chi(\mathbf{K})$ is large, resulting in severe attenuation of these spatial frequencies. Conversely, in the neighbourhood of regions where the slope of $\chi(\mathbf{K})$ is small, all spatial frequencies are well transmitted by the microscope with high contrast.

Extended regions over which the slope of $\chi(\mathbf{K})$ is small are called passbands or contrast transfer intervals, and these can be found for many focus settings, given by

$$\Delta f_n = [C_s \lambda (8n+3)/2]^{1/2} \tag{4.12a}$$

This result may be obtained as follows. By differentiation, it is easily shown that the slope of $\chi(\mathbf{K})$ is zero at K_1 for the corresponding 'stationary phase' focus $\Delta f_0 = -C_s \lambda^2 K_1^2$ (see Fig. A3.1). We require K_1 to lie at the centre of the pass-band, in order to minimize the damping effects of limited spatial coherence. As a separate condition, however, we also require $\chi = n\pi/2$, with $n = -1, -5, -9, -13$, etc., for good phase contrast (see Section 3.4). Then both the scattering phase shift $\exp(-i\pi/2) = -i$ in eqn (3.30) and the lens phase shift (the last term in eqn (3.33), which ideally would have the value $\exp(-i\pi/2)$) have the same sign, as needed to obtain a high-contrast image which is darker in regions of high potential. We might therefore impose the additional condition that

$$\chi(\Delta f_0) = -\frac{\pi}{2}(1, 5, 9, 13, \ldots)$$

in order to select only negative maxima in $\sin \chi$ for the centre of the passband. However, the passband can be made broader if the value of $\sin \chi(K_1)$ is allowed to decrease slightly, as shown by the dip in the passband of Fig. 4.3(b). This is achieved by taking

$$\chi(\Delta f_0) = -\frac{\pi}{2}\left(\frac{8n+3}{2}\right) = -\pi C_s \lambda^3 K^4/2$$

Solving this for K and using this value for K_1 in the stationary phase focus expression gives eqn (4.12a). This procedure guarantees both that the slope of $\chi(\mathbf{K})$ is zero (as sketched in Fig. A3.1) and that $\sin \chi = -1$ in the middle of the passband. The zero-order passband ($n = 0$) is commonly known as the 'Scherzer focus' and is the optimum choice of focus for images of defects or single molecules for which a straightforward interpretation in terms of object structure is required (see Section 6.2). Some examples of these passbands are shown in Fig. 4.3. They are seen to move out toward higher spatial frequencies with increasing n and, as discussed in more detail in Appendix 3, to become narrower with increasing n and C_s. Once the slope of $\chi(\mathbf{K})$ exceeds a certain value beyond these passbands, all spatial frequencies are severely attenuated. This attenuation is the major consequence of using a cone of illumination to illuminate the specimen (in order to produce a brighter final image) from an extended incoherent source. By collecting several images at, say, the $n = 0, 1, 2, 3$ focus values specified by eqn (4.12a) and processing these by computer, a composite image can be built up using only the well-transmitted spatial frequencies within the passband from each image, and this idea is the basis of image-processing schemes discussed in Chapter 7. These passbands cannot, however, be moved out beyond the resolution limit set by electronic instabilities (eqn 4.11). Figure 4.3 shows the transfer function drawn out for a typical instrument with $C_s = 2.2$ mm, $\Delta = 120$ Å, and $\theta_c = 0.9$ mrad at an operating voltage of 100 kV. To a good approximation the functions shown can be taken to be the last two terms of eqn (4.8a) multiplied by $\sin \chi(|\mathbf{K}|)$ with the second bracketed term in eqn (4.10) set equal to zero (see Appendix 3).

There are, therefore, two resolution limits which can be quoted for an electron microscope. The first, generally called the point-resolution of the instrument, is set by the first zero crossing of the transfer function at the Scherzer focus ($n = 0$ in eqn 4.12a). This is the useful resolution limit of the instrument for the analysis of defects and other non-periodic specimens. On high-voltage machines, the stability-resolution limit (eqn 4.11) may occur at

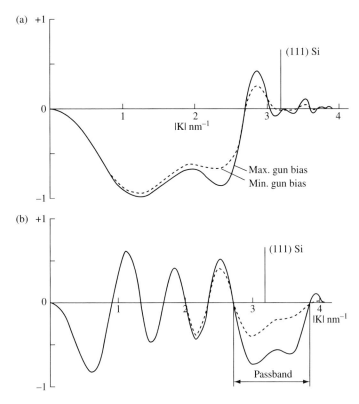

Figure 4.3 Transfer functions for a 100 kV electron microscope without aberration correction, with $C_s = 2.2$ mm and beam divergence $\theta_c = 0.9$ mrad. The cases $n = 0$ (Scherzer focus, $\Delta f = -110.4$ nm) and $n = 3$ ($\Delta f = -331.5$ nm) of eqn (4.12a) are shown in (a) and (b) respectively. In (a) the 'passband' extends from $u = 0$ out to the point-resolution limit of the instrument; in (b) the passband has moved out to the position indicated. The solid curves are drawn for maximum gun-bias setting ($\Delta = 5.4$ nm) and the broken curves show the effect of using the minimum gun-bias setting (maximum beam current $\Delta = 12.0$ nm) resulting in increased attenuation of the higher spatial frequencies. Note that the value of n is equal to the number of minima which precede the passband. The position of the (111) Bragg reflection is indicated and is seen to fall beyond the point-resolution limit (see Section 5.8). The imaginary part of eqn (4.8a) has been plotted.

a lower spatial frequency than the Scherzer cut-off (see eqn 6.17), in which case eqn (4.11) would determine the point-resolution of the machine (see Appendix 3).

The second resolution specification for an instrument might be called the information-resolution limit. It is set by electronic instabilities and is given by eqn (4.11). For uncorrected 100 kV machines this resolution limit generously exceeds the point-resolution, which is chiefly limited by spherical aberration. The information-resolution limit expresses the highest-resolution detail which could be extracted from a micrograph by the methods of image processing (leaving aside problems of electron noise) and can be measured by the Young's fringe diffractogram technique of Frank (see Section 10.6) or by finding the finest three-beam lattice fringes from a perfect crystal which the instrument is capable of recording under axial, *kinematic* conditions (but see Section 5.8). If there is no diffuse scattering between the Bragg reflections, a focus setting can then be found which places one of the passbands of eqn (4.12a) across the Bragg reflection of interest. Defects and non-periodic detail in such an image cannot usually be simply interpreted.

A similar damping envelope has been derived for a uniformly filled illumination disc rather than a Gaussian source. For such a source, Anstis and O'Keefe (unpublished) have obtained a damping function of the form

$$\frac{2J_1\left|2\pi\theta_c\left(\Delta f\mathbf{K}+\lambda\left(\lambda C_s-i\pi\Delta^2\right)\mathbf{K}^3\right)\right|}{\left|2\pi\theta_c\left(\Delta f\mathbf{K}+\lambda(\lambda C_s-i\pi\Delta^2)\mathbf{K}^3\right)\right|} = \frac{2J_1(q)}{q} \tag{4.12b}$$

where $J_1(x)$ is a Bessel function of the first order and kind and $\mathbf{K}$ is a vector in the back-focal plane. Again this function can only be used if the central beam is stronger than any other. Note that there is an interaction between the effects of the focus instability due to the incident electron energy spread and the illumination angle θ_c. This effect is discussed in a simpler way in Section 5.2, and a fuller discussion of damping envelopes can be found in Appendix 3.

The availability of gun monochromators for HREM machines now allows a reduction of ΔE, at the cost of image intensity. If the other electronic instabilities in eqn (4.9) allow it, this may improve the image quality, as described in den Dekker et al. (2001).

4.3. The characterization of coherence

The extent to which the wavefield at neighbouring points on the object vibrates in unison is expressed naturally by the correlation between wave amplitudes at points $\mathbf{r}_1$ and $\mathbf{r}_2$, and is given by the cross-correlation function

$$\Gamma(|\mathbf{r}_1-\mathbf{r}_2|, T) = \lim_{\tau\to\infty}\int_{-\tau}^{\tau}\psi^*(\mathbf{r}_1,t)\psi(\mathbf{r}_2,t+T)dt \tag{4.13}$$

A spatially stationary field has been assumed. When normalized, this function is called the complex degree of coherence $\gamma(x_{1,2}, T)$. Here $x_{1,2} = |\mathbf{r}_1-\mathbf{r}_2|$. The function contains a spatial dependence expressing lateral or transverse coherence and a time dependence expressing temporal or longitudinal coherence. In electron microscopy, the temporal coherence is large and we are chiefly concerned with $\gamma(x_{1,2}, 0) = \gamma(x_{1,2})$. In order to obtain strong interference effects such as Bragg scattering from adjacent scattering centres we require the wavefield at these points to be well correlated. That is, that $\gamma(x_{1,2})$ is large for this value of $x_{1,2}$.

The van Cittert–Zernike theorem relates $\gamma(x_{1,2})$ through a Fourier transform to the function $F(k)$ used in Sections 4.1 and 4.2. Despite differences in the nature of the particles (photons are bosons, electrons are fermions) and differing interpretations of the wavefunction, the results of electron interference experiments suggest that this important optical theorem may be taken over into electron optics. It will be seen that the range of object spacings which can be considered coherently illuminated is proportional to the width of

$\gamma(x_{1,2})$, so that a narrow source (for which $\gamma(x_{1,2})$ is a broad function) produces more coherent radiation than does a larger source. The theorem only applies to perfectly incoherent sources.

While the effects of partial coherence are important for images, they are seen most dramatically in interference experiments. A familiar example of a near-field interference experiment is the observation of Fresnel fringes at an edge. Note that questions of partial coherence only arise when more than one idealized point source of radiation is used. An interference experiment which may be used to measure $\gamma(x_{1,2})$ is Young's slit experiment. This experiment gives an important physical interpretation to $\gamma(x_{1,2})$—it is the contrast of the interference fringes (if the pin-holes are sufficiently small). Figure 4.4 shows the experimental arrangement used in optics. The angular period of the fringe intensity (seen from B) is $\lambda/x_{1,2}$, with $x_{1,2}$ the separation of the pin-holes I_1 and I_2. The relationship between the fringe contrast and source size is described in optics texts (e.g. Born and Wolf 1999). Sharp fringes are obtained from a single point-source P_1. Moving this source to P_2

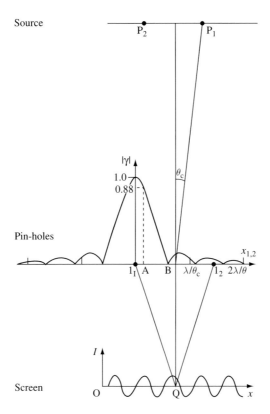

Figure 4.4 Young's slit experiment in optics. Interference fringes are formed with a single point source P_1 because of the path difference between each of the pin-holes I_1 and I_2 and an image point Q. The visibility of these fringes when an extended source P_1P_2 is used is equal to the ordinate of $|\gamma(x_{1,2})|$ evaluated at the pin-hole I_2. The abscissa has the value $\lambda/2\pi\theta_c$ at A and $0.61\lambda/\theta_c$ at B.

78 Coherence and Fourier optics

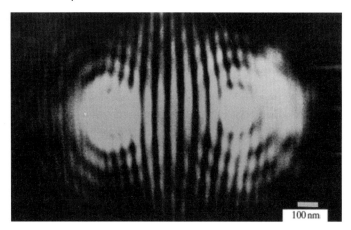

Figure 4.5 Young's slit experiment performed with electron waves. The 'pin-holes' in this case are two small holes in a thin carbon film. Fringes build up even if the electron arrivals are very widely spaced in time. Electrons travel as a wave and arrive as a particle. (Courtesy of E. Zeitler. Similar images have been obtained by A. Tonomura and H. Lichte).

translates the fringes in the opposite direction. The incoherent superposition of many sets of fringes, slightly out of register and arising from a set of sources along $P_1 P_2$, results in a fringe pattern of reduced contrast. A small increase in the width of the source has more effect on the contrast of fine fringes than on coarse fringes. A similar result holds for the effect of source size (condenser aperture) on high-resolution phase-contrast images.

Figure 4.5 shows the result of performing Young's slit experiment with electrons. If the pin-holes are sufficiently small in the optical case, the Michelson visibility (contrast) of these fringes is given by

$$V = \frac{I_\mathrm{max} - I_\mathrm{min}}{I_\mathrm{max} + I_\mathrm{min}} = |\gamma(x_{1,2})| \tag{4.14}$$

for a symmetrical source whose temporal coherence is large. The use of larger pin-holes (as in the electron case) introduces an additional modulating envelope.

The function $|\gamma(x_{1,2})|$ for a uniformly filled incoherent disc source is sketched in Fig. 4.4, centred about the pin-hole I_1. The contrast of the fringes formed in this experiment is then equal to the value of the function evaluated at the second pin-hole I_2. The form of $\gamma(x_{1,2})$ is given by

$$\gamma(x_{1,2}) = \frac{2 J_1(u)}{u}; \quad u = \frac{2\pi \theta_c x_{1,2}}{\lambda} \tag{4.15}$$

for a circular disc source. Here θ_c is the semi-angle subtended by the source at the pin-holes. By convention, distances smaller than that for which the ordinate has fallen by 12% are said to by coherently illuminated. This occurs when $u = 1$, and so allows the coherence width X_c to be defined for

$$x_{1,2} = \lambda/2\pi \theta_c = X_c \tag{4.16}$$

A reasonable criterion for incoherent illumination of two points is that $u = 2\pi$ or larger. Thus we may define an 'incoherence' width

$$X_i = \lambda/\theta_c \qquad (4.17)$$

Points separated by distances greater than X_i are incoherently illuminated. The range of spacings between X_c and X_i is described by the theory of partial coherence.

To summarize, coherence is characterized by the function $\gamma(x_{1,2})$ (the Fourier transform of the source intensity function) whose width gives a measure of the maximum separation between points which can be considered coherently illuminated. The contrast of a pure-phase object decreases as this function become narrower. Physically, the function gives the contrast of fringes in a certain interference experiment. A fuller discussion is given in Barnett (1974) for light and in Lichte and Lehmann (2007) for electrons.

4.4. Spatial coherence using hollow-cone illumination

As a further example of the application of the van Cittert–Zernike theorem to electron microscopy, we consider the use of a condenser aperture containing an annular gap. This is taken to be uniformly filled with quasi-monochromatic illumination. Practical aspects of the use of these apertures are discussed at the end of this section. This imaging mode also has an important parallel in scanning microscopy, through the reciprocity theorem (see Chapter 8).

Take the outer radius of the annulus to be r_0 and the inner radius βr_0, as shown in Fig. 4.6. If a function $g(a, x)$ is defined such that

$$g(a, x) = 1 \quad \text{for } x < a$$
$$= 0 \quad \text{elsewhere}$$

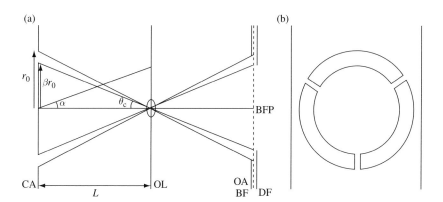

Figure 4.6 (a) Ray diagram for illumination using an annular condenser aperture. The annular gap in the second condenser aperture CA is filled with incoherent radiation and this is focused to a bright ring in the objective back-focal plane (BFP). Part (b) shows the shape of a practical aperture whose dimensions must match a particular instrument.

then the intensity across an incoherently filled annular gap can be described by

$$f(x) = g(r_0, x) - g(\beta r_0, x)$$

where x is the radial coordinate in the condenser aperture plane. The normalized Fourier transform of $f(x)$ with argument $u = \sin\alpha/\lambda$ gives the complex degree of coherence whose behaviour can be studied as a function of β, the ratio of the inner to outer aperture radii. The normalized transform of $f(x)$ is

$$F(X) = \frac{2}{(1-\beta^2)}\left[\frac{J_1(X)}{X} - \beta\frac{J_1(\beta X)}{X}\right] \qquad (4.18)$$

where $J_1(X)$ is a Bessel function of the first order and kind and

$$X = 2\pi r_0 u \approx 2\pi r_0 \alpha/\lambda = 2\pi x_{1,2}\theta_c/\lambda \qquad (4.19)$$

For $\beta = 0$ (the normal circular aperture) the transverse coherence width X_c is defined as the specimen spacing $x_{1,2}$ corresponding to the abscissa value for which $F(X) = 0.88$. Two points in the specimen plane are considered 'coherently' illuminated if their spacing is less than X_c. For $F(X) = 0.88$, $X = 1$ and the commonly used coherence criterion is obtained. From eqn (4.19), for $X = 1$,

$$x_{1,2} = X_c = \lambda/2\pi\theta_c$$

Note that $x_{1,2}$ refers to spacings in the object so than any object point can be taken as the origin. As β increases, $F(X)$ becomes narrower (the coherence deteriorates) but retains a form somewhat similar to the Airy's disc function until

$$\lim_{\beta \to 1} F(X) = J_0(X)$$

corresponding to the case of a circular line source. Figure 4.7 shows that the coherence under conical illumination is always less than that obtained using a conventional circular aperture equal in size to the outer diameter of the annular aperture ($\beta = 0$).

The illuminating hollow cone of rays incident on the specimen forms a bright ring of unscattered intensity in the objective lens back-focal plane (Fig. 4.6). By using a matching objective aperture of optimum size for phase contrast which either excludes or includes this ring, a dark- or bright-field image may be formed. For matching apertures $\theta_{ap} = \theta_c$, so that $X_c = \lambda/2\pi\theta_{ap}$, where θ_{ap} is the semi-angle subtended by the objective aperture. Thus for a 10 mrad objective aperture, about the size commonly used for phase contrast, we find $X_c = 0.05$ nm. This would appear to restrict the use of conical illumination to very high-resolution detail for phase contrast. Non-uniform illumination of the condenser aperture and the partially coherent image contribution may account for the limited success of this method at high resolution.

To construct an annular aperture, a diagram similar to Fig. 4.6 must be drawn for the particular instrument which will give the approximate dimensions of the annulus. Depending on the strength of the objective lens, some deviation from this simple ray diagram must be

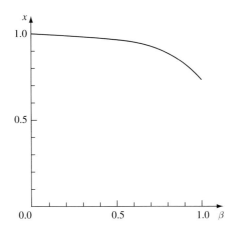

Figure 4.7 The width of the modulus of the complex degree of coherence (for $F(X) = 0.88$) for an annular illumination aperture as a function of β, the ratio of the inner to outer annular radii. For a particular value of β the coherence width is $X_c = \lambda/2\pi\theta_c$.

expected and a certain amount of trial and error is required to obtain matching apertures. Since it is reasonably large (r_0 is typically about 2.5 mm), the annular aperture can be made by the photoetching methods used for electronic circuit boards. Some manufacturers supply these apertures ready made for their instruments. Because L is much larger than the focal length of the objective lens, the scale of a modified condenser aperture is much larger than that of an objective aperture, making interventions in the illuminating system easier than in the objective back-focal plane. This is an important advantage of the conical illumination technique over other dark-field methods, such as the use of a wire beam stop in the objective lens back-focal plane, where the minute scale of the aperture and contamination of the beam stop are important practical problems. Another advantage is that as for high-resolution tilted-illumination bright-field imaging, the range of spatial frequencies included within the aperture is about twice that obtained with untilted illumination and a central aperture. The asymmetrical 'Schlieren' distortion (see Section 6.5) which accompanies tilted-illumination dark-field images is also not present using conical illumination. A wave-optical analysis of hollow-cone illumination is given by Niehrs (1973) and Rose (1977), and applications to amorphous materials are discussed in Saxton et al. (1978) and Gibson and Howie (1978/9).

4.5. The effect of source size on coherence

The preceding discussion has been based on the assumption that the illuminating system can be replaced by an effective source filling the final condenser aperture. This makes the degree of coherence at the specimen dependent only on the size of the condenser aperture used (eqn 4.16), and not on the excitation of the condenser lenses or the physical source size. In this section the conditions under which source size may be important are investigated.

Equation (4.16) applies only if the illuminating aperture is incoherently filled. This will be so to a good approximation if the coherence width X_a in the plane of the illuminating aperture is small compared with the size of that aperture ($2R_a$). Using the van Cittert–Zernike theorem an expression can be obtained for X_a as

$$X_a = \lambda / 2\pi \theta_s$$

where θ_s is now the semi-angle subtended by the focused spot at the aperture (see Fig. 4.8). This angle depends both on the excitation of the lenses and on the source size.

Consider first a hair-pin filament with a tip radius of 15 µm, using the illuminating system shown in Fig. 9.1. For this microscope (a JEM-100C), leaving aside the effect of the objective pre-field, the total demagnification of the two lenses C1 and C2 varies between 0.54 and 0.03 for the minimum and maximum settings of C1 ('spot size'). This would give maximum and minimum focused spot sizes of $r_s = 8.1$ and 0.45 µm. With $D3 = 251$ mm we then have $X_a = 0.018$ and 0.327 µm, respectively, both of which are small compared with a 100 µm condenser aperture. Thus we are justified in treating the illuminating aperture as incoherently filled in this case.

For a sharp tungsten filament with tip radius of 1 µm, a similar argument gives $X_a = 0.26$ and 5.0 µm for the minimum and maximum C1 settings at 100 kV using eqn (4.15). This second value is no longer small compared with, say, a 40 µm illuminating aperture of the kind used in some minimum-exposure techniques for which eqn (4.16) would therefore underestimate the coherence width at the specimen. Note that if the illuminating aperture is coherently filled ($X_a > 2R_a$), then the focused spot at the specimen is also perfectly coherent, as is all subsequent wavefield propagation.

We can conclude that the effect of source size on image coherence will only become important for sources smaller than about 1 µm, such as a field-emission source, the finest pointed filament, or the smaller LaB$_6$ sources (see Table 9.1). Further, a comparison of the coherence conditions in two different experiments can only be made if both θ_s and θ_c are known. At higher voltages an even smaller source is required to produce partially coherent illumination of the illuminating aperture. A method for measuring the degree of coherence across the illuminating aperture is described by Dowell and Goodman (1973).

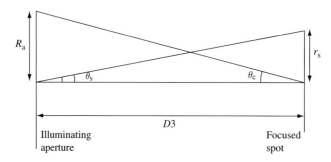

Figure 4.8 Formation of a focused spot of radius r_s beyond a condenser aperture of radius R_a. The quantities defined enable the degree of coherence in the illuminating aperture plane to be found. In high-resolution scanning electron microscopy where a field-emission source is used the illuminating aperture may be coherently filled.

4.6. Coherence requirements in practice

Phase-contrast effects such as Fourier images, lattice images, and images of small molecules all show increased contrast as the coherence of the illumination is increased. The rapid change in image appearance with focus predicted by 'coherent' calculations of the type discussed in Chapters 5 and 6 will also only be seen if highly coherent illumination is used. In practice, this means a beam divergence of less than about 0.8 mrad. The fine structure of a high-resolution image, controlled by wave-optical interference, can be washed-out either through the use of large beam divergence or too large an energy spread in the incident beam (see Section 5.6). This attenuation of high spatial frequencies is represented the damping envelope which multiplies the coherent transfer function, described in Sections 3.3 and 4.2. The result may be that important structural information about the specimen is lost. For periodic specimens, the safest procedure is to compute the image of interest and assess the effect of increasing beam divergence using eqn (4.1). Figure 4.9 shows the effect of increasing beam divergence on the computed image of a large-unit-cell crystal. Much of this loss of fine detail is due to the phase shift introduced by spherical aberration, which becomes appreciable across the angular width of each diffraction spot, particularly those at large scattering angles where the transfer function oscillates rapidly. This degradation in image quality with increasing beam divergence is borne out by experiment (Fig. 4.9). The severity of the effect depends on the particular specimen structure. The excellent match

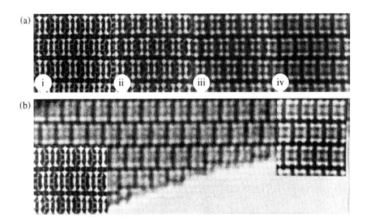

Figure 4.9 The effect of increasing beam divergence (illuminating semi-angle θ_c) on the quality of high-resolution images. Here computer-simulated images of $Nb_{12}O_{29}$ are shown for beam divergences of (i) 0.0 mrad, (ii) 0.6 mrad, (iii) 1.0 mrad, and (iv) 1.4 mrad. The loss of fine detail with increasing condenser aperture size is evident. This effect can be understood physically by extending the analysis of Section 5.2 to the three-beam case with near-axial illumination, or from Section 3.3. The simulated experimental conditions are $C_s = 1.8$ mm, 79 beams included within the objective aperture, $\Delta f = -60$ nm, 100 kV, crystal thickness $t = 5$ nm. Images provided by M. O'Keefe and J. Sanders. (b) An experimental image of $Nb_{12}O_{29}$ taken by S. Iijima under the experimental conditions given above. The two inserts show computed images for $\theta_c = 0.0$ (left) and $\theta_c = 1.4$ mrad (right) as used experimentally.

84 Coherence and Fourier optics

shown in this figure when the correct beam divergence is used does not take account of overall contrast, which has been scaled to fit—this 'Stobbs factor' effect is discussed in Section 5.13.

If the final image obtained at a beam divergence of, say, 0.8 mrad and high magnification (about 300 K) is insufficiently intense to allow accurate focusing, a reduced specimen height (shorter objective lens focal length) should be tried which will increase the objective prefield focusing effect. This will increase the image current density and produce a more intense final image. In doing so, the illuminating spot size is decreased and the beam divergence is increased (the product of these is constant—see Section 2.2). A trade-off between coherence width X_c and specimen intensity j_0 is expected since $j_0 = \pi \beta \theta^2$ (eqn 9.1) while $X_c = \lambda/2\pi\theta_c$. However, it appears that the reduction in spherical aberration accompanying this shorter focal length more than compensates for the loss of coherence, since the highest-quality images have generally been obtained at the shortest possible objective focal length, subject to the limitations of lens current and space within the pole-piece. The optimum choice of specimen position is further discussed in Appendix 3.

The relationship between coherence width and object current density has been studied experimentally by Harada et al. (1974). Figure 4.10 shows their results for a hair-pin filament and compares this with the result obtained with a field-emission source. The ordinate in this figure is

$$X_c = \frac{\lambda}{2\theta_c}$$

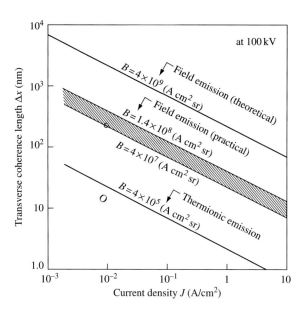

Figure 4.10 Theoretical and experimental (circles) values for coherence width as a function of illumination current density at the specimen. The field-emission gun is compared with a hair-pin filament.

which differs by a factor of π from that used by Born and Wolf (1999). The theoretical curves of this figure assume an incoherently filled illuminating aperture.

The range of focal settings over which sharp 'Fourier images' (Cowley and Moodie 1960) can be seen has been analysed by Fujiwara (1974), showing explicitly the dependence of this range on the source coherence.

With non-periodic specimens, an impression of the coherence conditions can be obtained from a glance at the diffraction pattern. Figure 4.11 shows a sketch of the diffraction pattern of an amorphous specimen formed with a large condenser aperture. Incoherent instabilities such as vibration and electronic fluctuations may limit the resolution to, say, 0.3 nm. This corresponds to the imposition of an 'aperture' (outer circle) of angular radius $\theta_1 = 0.82\lambda/0.3$, from the resolution limit of eqn (4.11). In the absence of a physical aperture, scattering outside this outer circle will contribute to the diffuse image background, resulting in a loss of image contrast. At the other extreme, eqn (4.17) gives the spacing of incoherently illuminated points as $X_i = \lambda/\theta_c$. Specimen spacings larger than this lie inside the inner continuous circle at $\theta_2 = \theta_c$ and are imaged under incoherent conditions. Their contrast cannot therefore be enhanced by the optimum-focus phase-contrast technique.

Notice that the coherence requirement becomes more severe as one attempts to form phase-contrast images of larger spacings. For some biological crystals with very large unit cells, a very large coherence width would be required for coherent imaging. Most published images of these specimens are formed under partially coherent imaging conditions.

When using a LaB$_6$ source, or a field-emission source, the degrading effect of finite-beam divergence on image quality can usually be made negligible, since these sources generally provide adequate intensity for focusing with a small beam divergence (about 0.6 mrad). It is the increased brightness of these sources which makes them valuable, since this enables the image intensity obtainable with a hair-pin filament to be attained using a smaller beam divergence and therefore more coherent illumination (Hibi and Takahashi 1971). Certainly one would expect a change in experimental conditions which increases the number of Fresnel fringes observed to provide enhanced image contrast. A count of the number of fringes seen is probably the simplest qualitative guide to the coherence in the specimen plane (see

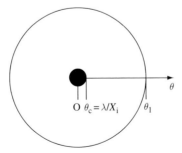

Figure 4.11 The appearance of the diffraction pattern of an amorphous specimen. The central disc represents the 'unscattered' central beam, and the outer circle is drawn at the resolution limit of the microscope set by incoherent instabilities. Specimen spacings corresponding to scattering angles within the central disc cannot be enhanced by phase-contrast microscopy.

Section 10.7 and Fig. 10.7) and can be used to compare sources if other experimental conditions remain unchanged.

Since the only focus setting which allows a simple image interpretation not requiring computer analysis is the Scherzer focus (eqns (6.16) and (4.12), with $n = 0$), a check on coherence conditions can be made by obtaining diffractograms from thin carbon films recorded at the Scherzer focus using several different condenser aperture sizes but otherwise identical conditions. The largest condenser aperture size is then selected for which the corresponding diffractogram shows good contrast across the entire inner ring out to the point-resolution limit of the instrument (eqn 6.16). The theoretical basis for this procedure is described in Section 4.2 and Appendix 3.

References

Barnett, M. E. (1974). Image formation in optical and electron transmission microscopy. *J. Microsc.* **102**, 1.
Born, M. and Wolf, E. (1999). *Principles of optics*. Seventh edition. Cambridge University Press. Cambridge.
Burge, R. E., Dainty, J. C., and Thorn, J. (1975). The spatial coherence of electron beams. In *Proc. EMAG 1975, Bristol*, ed. J. Venables, p. 221. Academic Press, London.
Cowley, J. M. and Moodie, A. F. (1960). Fourier images IV: The phase grating. *Proc. Phys. Soc.* **76**, 378.
den Dekker, A. J., Van Aert, S., Van Dyck, D., van den Bos, A., and Geuen, P. (2001). Does a monochromator improve the precision in quantitative HRTEM? *Ultramicroscopy* **89**, 275.
Dowell, W. C. T. and Goodman, P. (1973). Image formation and contrast from the convergent electron beam. *Phil. Mag.* **28**, 471.
Fejes, P. L. (1977). Approximations for the calculation of high-resolution electron microscope images of thin films. *Acta Crystallogr.* **A33**, 109.
Frank, J. (1973). The envelope of electron microscope transfer functions for partially coherent illumination. *Optik* **38**, 519.
Fujiwara, H. (1974). Effects of spatial coherence on Fourier imaging of a periodic object. *Opt. Acta* **21**, 861.
Gibson, J. M. and Howie, A. (1978/9). Investigation of local structure and composition in amorphous solids by high resolution electron microscopy. *Chem. Scripta* **14**, 109.
Harada, Y., Gota, T., and Someya, T. (1974). Coherence of field emission electron beam. In *Proc. 8th Int. Congr. Electron Microsc., Canberra*, p. 110.
Hibi, T. and Takahashi, S. (1971). Relation between coherence of electron beam and contrast of electron image of biological substance. *J. Electron Microsc.* **20**, 17.
Hopkins, H. H. (1957). Applications of coherence theory in microscopy and interferometry. *J. Opt. Soc. Am.* **47**, 508.
Humphreys, C. J. and Spence, J. C. H. (1981). Resolution and illumination coherence in electron microscopy. *Optik* **58**, 125.
Lichte, H. and Lehmann, M. (2007). Electron holography – basics and applications. *Rep. Prog. Phys.* **70**, 1.
Niehrs, H. (1973). Zur Formulierung der Bildintensitat bei ringformiger Objektbestrahlung in der Electronen-Mikroskopie. *Optik* **38**, 44.
O'Keefe, M. A. and Sanders, J. V. (1975). n-Beam lattice images. VI. Degradation of image resolution by a combination of incident beam divergence and spherical aberration. *Acta Crystallogr.* **A31**, 307.
O'Keefe, M. A. and Saxton, O. (1983). *Proc. 41st EMSA Meeting*, p. 288. Claitors, New York.

Rose, H. (1977). On hollow-cone illumination. *Ultramicroscopy* **2**, 251.

Saxton, W. O., Jenkins, W. K., Freeman, L. A., and Smith, D. J. (1978). TEM observations using bright field hollow cone illumination. *Optik* **49**, 505.

Wade, R. H. and Frank, J. (1977). Electron microscope transfer functions for partially coherent axial illumination. *Optik* **49**, 81.

Bibliography

Clear accounts of partial coherence in imaging are given in the references by Barnett, and especially by Lichte and Lehmann, above. More advanced treatments can be found in:

Beran, M. J. and Parrent, G. B. (1964). *Theory of partial coherence*. Prentice-Hall, Englewood Cliffs, NJ.

Perina, J. (1971). *Coherence of light*. Van Nostrand, London.

Thompson, B. J. (1969). Image formation with partially coherent light. In *Progress in optics*, Vol. 7, ed. E. Wolf. North-Holland, Amsterdam.

5
TEM imaging of thin crystals and their defects

A long-term aim of electron microscopy has been the direct imaging of atoms in solids in three dimensions. To achieve this, a point-resolution of about 0.15 or better is required for most crystalline materials. While this can now be obtained on some TEM instruments without aberration correction (by the methods described in this chapter) much sharper and clearer images of even higher resolution can be obtained with aberration correction. Image-processing techniques, which can take resolution below 0.1 nm on uncorrected machines, are described in Chapter 7. At a resolution limit of 0.20 nm the image of a single atom changes little over a range of focus of, say, 5 nm, so that for specimens of this thickness or less, containing atoms which are well separated laterally, we can at best hope to see a projection of the specimen structure in the direction of the incident beam when using an uncorrected instrument. Optical sectioning, of the kind used with light microscopes, is only possible in TEM on aberration-corrected instruments.

We commence with some notes on three early imaging methods for simple lattice fringes, because of their pedagogic value. After a period of early optimism (Menter 1958) it became clear that only very limited information on the sample is given by these few-beam 'lattice images', as they were called. However, the 'line' resolution records achieved by early researchers, e.g. 0.088 nm by Yada and Hibi (1969) and 0.062 nm by Matsuda et al. (1978), were remarkable and have only very recently been improved upon. The most important use for these images now is for demonstrating the stability of electron microscopes. Nevertheless a sound grasp of the theory of these few-beam lattice images is essential for an understanding of the many-beam structure images described later. The observation of two-dimensional many-beam images from large-unit-cell crystals in 1972 (Cowley and Iijima 1972), building on earlier work by J. Sanders, L. Bursill, L. Hewatt, K. Yada, H. Hashimoto, T. Komoda, and others, marked an important turning point in the subject by showing that useful structural information could be obtained by the HREM method, most importantly, that non-stoichiometry in complex oxides can be accounted for by shear-plane defects, in addition to point defects (as previously assumed), with clear implications for thermodynamics (for reviews see Eyring (1989) and Spence (2003)).

The term 'lattice imaging' has been widely accepted to describe electron microscope images of crystals. This is unfortunate usage, since a lattice is a mathematical abstraction, and one is normally interested in the contents of the unit cell, not the lattice. 'Structure image' (defined in Appendix 4) seems a better term, by analogy with structural biology. Early reviews by Allpress and Sanders (1973) and Cowley (1976, 1985) complement the material of this chapter.

5.1. The effect of lens aberrations on simple lattice fringes

In this section we investigate the effects of changes in lens parameters and illumination conditions on simple lattice fringes, formed from two or three diffracted beams.

The three common imaging methods used for two- and three-beam lattice images are shown in Fig. 5.1. If the complex amplitudes of the Bragg beams are $\Phi_{hk0} = \phi_{hk0} e^{i\varepsilon}$ (ϕ_{hk0} real), the image amplitude in general will be

$$\psi_i(\mathbf{r}_i, \Delta f, K_0) = \sum_\mathbf{g} \Phi_\mathbf{g} \exp(2\pi i \mathbf{g} \cdot \mathbf{r}) \exp\{i\chi(u_\mathbf{g})\} \tag{5.1a}$$

For the simple case of an orthorhombic crystal with the beam parallel to the c-axis this becomes

$$\psi(x,y) = \sum \Phi_{hk0} \exp\{+2\pi i(hx/a + ky/b)\} \exp\{i\chi(u_{h,k})\} \tag{5.1b}$$

where the sum is taken over all beams within the objective aperture and a and b are the orthogonal unit-cell dimensions normal to the electron beam direction. Here $\chi(u_{h,k})$ is the phase shift introduced by instrumental aberrations described in Section 3.3 and given, in the absence of astigmatism, by

$$\chi(u_{h,k}) = 2\pi/\lambda \left\{ \frac{1}{2}\Delta f \lambda^2 u_{h,k}^2 + \frac{1}{4} C_s \lambda^4 u_{h,k}^4 \right\} \tag{5.2}$$

where

$$u_{h,k}^2 = \frac{h^2}{a^2} + \frac{k^2}{b^2} = \mathbf{g}^2$$

The diffracted amplitudes Φ_{hk0} are functions of the specimen orientation and thickness. We now consider the one-dimensional cases shown in Figs 5.1(a)–(c), and choose the Miller indices $\mathbf{g}$ of the reflection included within the aperture to be $(h, 0, 0)$. Then $u_h = h/a = \theta_0/\lambda$ where a is the lattice spacing and θ_0 is twice the Bragg angle. The image intensity in the kinematic case where the scattering phase $\varepsilon = -\pi/2$ is then, for the two beams shown in Fig. 5.1(a),

$$I(x) = \psi_i(x)\psi_i^*(x) \tag{5.3}$$

$$I(x) = \phi_0^2 + \phi_h^2 + 2\phi_0\phi_h \sin(2\pi hx/a - \chi(u_h)) \tag{5.4}$$

The form of this expression indicates that any change in $\chi(u_h)$ will produce a sideways movement of the fringe image. Fluctuations in objective lens current or high voltage cause Δf to vary (see Section 3.3) and so result in a small fluctuating fringe displacement during an exposure. The image resolution is therefore limited by these fluctuations. Lattice fringes formed under these conditions will be seen to move sideways (normal to their length) as

90 TEM imaging of thin crystals and defects

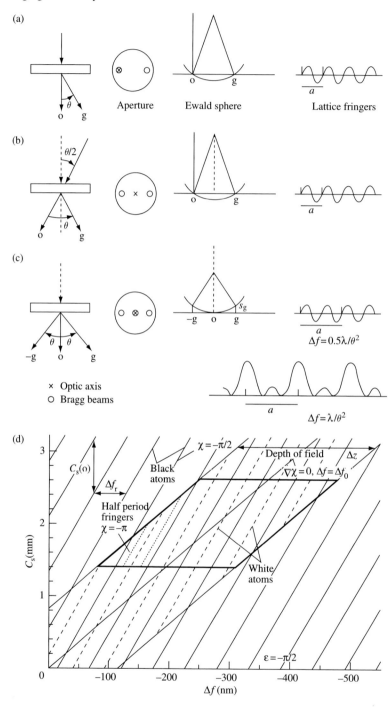

Figure 5.1 (*continued*)

the focus control is altered, often well beyond the edge of the specimen. Leaving aside the effect of spherical aberration, eqn (5.4) gives the displacement of the fringes due to defocus as

$$I(x) = \phi_0^2 + \phi_h^2 + 2\phi_0\phi_h \sin\{2\pi h(x - \Delta f\theta_0/2)/a\} \tag{5.5}$$

In Fig. 5.1(b) the illumination has been tilted with respect to the optic axis, resulting in an improvement in the incoherent resolution limit imposed by electronic instabilities (Dowell 1963, Komoda 1964). The image amplitude is then

$$\psi_i(x) = \Phi_0 \exp(i\chi(-u_h/2)) + \Phi_h \exp(-2\pi ihx/a + i\chi(u_h/2)) \tag{5.6}$$

Since the function $\chi(u_h)$ is even, the same instrumental phase shift now applies to each beam. The image intensity is

$$I(x) = \phi_0^2 + \phi_h^2 + 2\phi_0\phi_h \cos((2\pi hx/a) - \varepsilon) \tag{5.7}$$

which is independent of the focus condition (for perfectly coherent illumination). It is also therefore independent of fluctuations in the high voltage and lens currents within the accuracy of the illumination alignment. Chromatic aberration arising from the thermal energy spread of the incident electrons (Section 2.8.2) is similarly reduced by this technique. The highest-'resolution' two-beam lattice fringes are invariably formed using this imaging method, since it confers a limited immunity to these important image defects; however, it is extremely difficult to extract crystal structure information from these images. This mode has a very important relationship through the reciprocity relation, with annular dark-field (ADF)-STEM (Chapter 8).

Figure 5.1 Three imaging methods for simple lattice fringes. The first figure in each row suggests the directions of the incident and diffracted beams with respect to the crystal; the second shows the appearance of the objective aperture and the positions of the diffracted beams (o) and the optic axis (×); the third gives the corresponding Ewald sphere orientation (see Fig. 5.2), while the last figure shows the form of the lattice fringes produced. The cases of untilted illumination (a), tilted illumination (b), and three-beam fringes (c) are shown. In case (c) fringes are shown for two values of defocus Δf, to illustrate the occurrence of half-period fringes ($C_s = 0$). (d) Axial three-beam lattice imaging conditions summarized. Continuous lines are the locus of full-period images which coincide with the lattice. Broken lines show images of reversed contrast, while the dotted line shows one of the many half-period images. For an instrument with $1.4 < C_s < 2.6$ mm, images are only seen within the heavy parallelogram. Thus, if a high-contrast lattice image were recorded under these conditions the possible pairs of values of Δf and C_s would be greatly restricted. The depth of field Δz, the periodicity in focus Δf_f and the period $C_s(0)$ in spherical aberration are all shown. The figure is drawn for a lattice period $a = 0.313$ nm (silicon (111)) with $\theta_c = 0.0014$ rad at 100 kV and $\varepsilon = -\pi/2$.

Figure 5.1(c) shows three-beam fringes formed with axial illumination. The image amplitude is

$$\psi_i(x) = \Phi_0 + \Phi_h \exp(2\pi i h x/a + i\chi(u_h)) + \phi_{-h} \exp(-2\pi i h x/a + i\chi(-u_h)) \qquad (5.8)$$

and the image intensity becomes

$$I(x) = \phi_0^2 + \phi_h^2 + \phi_{-h}^2 + 2\phi_h \phi_{-h} \cos(4\pi h x/a)$$
$$+ 4\phi_0 \phi_h \cos(2\pi h x/a) \cos(\chi(u_h) + \varepsilon) \qquad (5.9)$$

In the simplified case where $\phi_h = \phi_{-h}$ and $\varepsilon = -\pi/2$ the image is seen to consist of constant terms plus cosine fringes with the lattice spacing, together with weaker cosine fringes of half the lattice spacing. These 'half-spacing' fringes are often seen on micrographs taken under these conditions. The focus condition for observing the half-spacing if $\varepsilon = -\pi/2$ is obtained by setting $\chi(u_h) = n\pi$, giving

$$\Delta f = \frac{n\lambda}{\theta_0^2} - \frac{C_s \theta_0^2}{2}, \qquad n = 0, \pm 1, \pm 2, \ldots. \qquad (5.10)$$

Some examples are sketched in Fig. 5.1(c). It has been suggested that since the half-period fringes are insensitive to focus, while the lattice-period fringe contrast may reverse with changes in focus, the effect of electronic instabilities (leading to a variation in $\chi(u_h)$) should be to allow the half-period fringes to accumulate during an exposure at the expense of the lattice-period fringes.

It is clear that any change of 2π in $\chi(u_h)$ occurring in eqn (5.8) leaves these axial three-beam fringes unaffected, while a change of π reverses their contrast. The change in focus needed to alter $\chi(u_h)$ by 2π is $\Delta f_f = 2n/(\lambda u_h^2)$. The change in C_s needed to change $\chi(u_h)$ by 2π is $C_s(0) = 4n/(\lambda^3 u_h^4)$, and the lattice fringes are thus periodic in both Δf and C_s. All these features of three-beam images are summarized in Fig. 5.1(d). The equation for the straight lines shown on the figure is obtained by setting $\chi(u_h) = (2n - 1/2)\pi$ and solving for C_s. From eqn (3.24) this gives

$$C_s = \frac{(4n-1)}{(\lambda^3 u_h^4)} - \frac{2\Delta f}{(\lambda^2 u_h^2)}$$

as the condition for identical three-beam axial images. The phase angles $(2n - 1/2)\pi$ have been chosen since these result in an electron image which 'coincides' with the lattice in the sense that atom positions appear dark. This can be shown by using the kinematic amplitudes of eqn (5.15) in eqn (5.8) with $S_g = 0$. The 90° phase shift ε due to scattering, represented by the factor $-i = \exp(-i\pi/2)$ in eqn (5.15), and the 90° lens phase shift $\chi(u_h)$ (for $n = 0$, say) are then superimposed to give a total phase shift of $-180°$. (This factor is responsible for the minus sign in eqn (6.6), giving dark atom images.) Diffracted beams whose phase differs from that of the central beam by an amount not equal to 90° are common under dynamical scattering conditions, and may result in severely distorted images.

In practice fringes are not seen for all focus settings, only those near the 'stationary-phase' focus $\Delta f_0 = -C_s \lambda^2 u_h^2$ (see discussion following eqns (4.12a) and (5.66)). This limited depth of field, resulting from the use of an extended electron source, is discussed in Section 5.2 and drawn in Fig. 5.1(d) as the envelope within which clear lattice fringes can be seen. The periodicity in focus in this case is actually a special simplified example of Fourier or Talbot self-imaging discussed in Section 5.8. For a general one-dimensional many-beam image, a focus change of $\Delta f_f = 2a^2/\lambda$ changes the phase of all beams by $2n\pi$, and so leaves even many-beam images unaltered. For many crystals, this result can also be extended to cover two-dimensional lattice images (see Section 5.8). Then a focus change of $\Delta f_f/2$ results in an image shifted by half a unit cell along the cell diagonal (for cubic projections), which cannot usually be distinguished from the 'true' image through the viewing binocular. Half-period images can also be obtained at certain focus settings in two-dimensional images. Similar half-period fringes occur at certain specimen thicknesses (see Fig. 12.4a) and the spurious impression of high-resolution detail which they may give when used as an instrumental resolution test is discussed in Section 10.9.

In larger-unit-cell crystals, the Fourier image period Δf_f frequently becomes larger than the depth of field Δz, so that only a single lattice fringe image is seen. This may also occur for a certain choice of illumination semi-angle θ_c for any lattice spacing a. The general condition of restricting the depth of field to a single Fourier image period is that $a/\theta_c < 2a^2/\lambda$, or $\theta_c > \lambda/2a$. On combining this with the Bragg law, we see that this is just the condition that adjacent diffraction discs overlap.

5.2. The effect of beam divergence on depth of field

It is possible to estimate the range of focus Δz over which lattice fringes of a particular periodicity should be visible. A limit is set to this range by the finite beam divergence θ_c. The depth of field obtained applies to lattice images formed with partially coherent illumination and should be distinguished from the depth of field given in Section 2.2, which applies to the incoherent imaging of non-periodic specimens.

Taking each point within the effective source as an independent source of electrons (neighbouring points assumed incoherent), the total intensity can be obtained by summing the intensities of images formed by each source point. For the lattice fringes of Fig. 5.1(a) the methods of Chapter 3 give the image intensity when the illumination direction makes a small angle α with the optic axis as

$$I(x, \alpha) = \Phi_0^2 + \Phi_h^2 + 2\Phi_0 \Phi_h \cos\{\chi(-u_h - u') - \chi(u') + 2\pi u_h x\}$$
$$= \Phi_0^2 + \Phi_h^2 + 2\Phi_0 \Phi_h \cos\{2\pi h(x + \Delta f \alpha)/a + \pi \Delta f \theta_0^2/\lambda\} \quad (5.11)$$

with $u' = \alpha/\lambda$.

Physically, this means that fringes out of focus by an amount Δf suffer a translation normal to their length proportional both to the focus defect and to the illumination tilt. Assuming a uniform illumination disc, the total image intensity is then

$$I_T(x) = \frac{1}{\theta_c} \int_{-\theta_c}^{\theta_c} I(x, \alpha) d\alpha \tag{5.12}$$

for an illuminating cone of semi-angle θ_c. Evaluating this shows that the fringe contrast $C' = (I_{\max} - I_{\min})/(I_{\max} + I_{\min})$ is proportional to

$$C = \sin(\beta)/\beta \tag{5.13}$$

where $\beta = 2\pi \Delta f \theta_c / a$. The contrast falls to zero for $\beta = \pi$, so that the range of focus Δz over which fringes are expected is

$$\Delta z = a/\theta_c \tag{5.14}$$

With $\theta_c = 1.5$ mrad and $a = 0.34$ nm, we have $\Delta z = 227$ nm. There would thus be little point in seeking lattice fringes under these conditions on an instrument whose smallest focal increment was much greater than about 200 nm.

It is instructive to derive a similar result using the damping envelope construction of Section 4.2. We consider axial three-beam lattice fringes as shown in Fig. 5.1(c). It is shown in Section 5.15 that, if the central beam is stronger than the diffracted beams, three-beam lattice fringes of highest contrast occur near the focus setting $\Delta f_0 = -C_s \lambda^2 u_h^2$ which makes the slope of $\chi(u_h)$ zero. (Neglecting the second bracketed term in eqn (4.10), this condition can also be obtained by setting $\mathbf{q} = 0$ in eqn (4.10) and solving for Δf.) If an arbitrary focus setting $\Delta f'$ is measured by its deviation from Δf_0, then the focus defect Δf measured from Gaussian focus is

$$\Delta f = \Delta f_0 + \Delta f'$$

The damping envelope due to beam divergence alone is (eqn 4.8a)

$$A(u_h) = \exp(-\pi^2 u_c^2 q)$$

where

$$q = (C_s \lambda^3 u_h^3 + \Delta f \lambda u_h)^2 \quad \text{and} \quad \theta_c = \lambda u_c (\ln 2)^{1/2}$$

Combining these expressions gives, for the focus dependence of the fringe contrast,

$$A(u_h, \Delta f') = \exp(-B^2 \Delta f'^2) \quad \text{where} \quad B = \pi u_c \lambda u_h$$

This is a Gaussian of width $1/B$, so that the depth of field Δz is

$$\Delta z = \frac{1}{B} = \frac{(\ln 2)^{1/2}}{\pi} \frac{a}{\theta_c}$$

for a (100)-type reflection ($u_h = 1/a$). This differs only by a numerical factor (ln $2)^{1/2}/\pi = 0.26$ from the result of eqn (5.14).

The contrast of these axial three-beam fringes will also be affected by the term $\exp(i\chi(|\mathbf{K}| = u_h))$ of eqn (4.8a), which has been neglected in the above discussion. In general, then, these fringes will show a rapid variation of contrast with focus (due to the terms in $\cos\chi(u_h)$ of eqn (5.9)) modulated by the more slowly varying Gaussian envelope $A(u_h, \Delta f')$ described above and centred about the focus setting Δf_0. Note, also, that the term $\exp(i\chi(u_h))$ has the same value for all the 'Fourier image' focus settings (see Section 5.8) Δf_n which differ from the reference focus Δf_0 by integral multiples of $2n/(\lambda u_h^2)$ (n being an integer). Thus, if a strong image occurs at the reference focus Δf_0 (this will be so, if, from eqns (5.9) and (5.66), $C_s = 2m/(\lambda^3 u_h^4)$ with m an integer) then similar strong images will occur at focus settings

$$\Delta f = \Delta f_0 \pm 2n/(\lambda u_h^2)$$

The contrast of these images will fall off smoothly as the focus is increased (or decreased) beyond Δf_0 due to the Gaussian damping of $A(u_h, \Delta f')$. This behaviour is also seen in the calculated three-beam axial fringe contrast shown in Fig. 5.2 for silicon. This figure is the result of a dynamical calculation including both chromatic and spatial coherence effects, with $\theta_c = 1.4$ mrad, $C_s = 2.2$ mm, and $\Delta = 5$ nm at 100 keV.

We may conclude that the finer the image detail, the smaller is the focal range over which it may be observed, for a fixed illumination aperture.

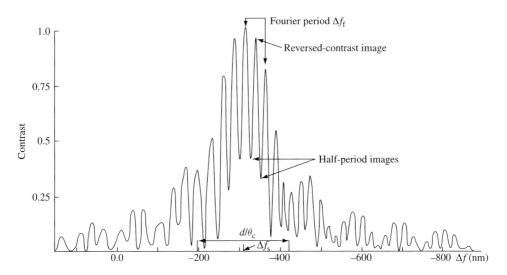

Figure 5.2 Contrast of three-beam (111) lattice fringes in silicon as a function of focus. The depth of field Δz is limited by finite spatial coherence (beam divergence) and temporal coherence (chromatic effects). Contrast is a maximum at the stationary-phase focus Δf_s and high-contrast images would be seen in a through-focus series for a range $\Delta z \simeq d/\theta_c$ around Δf_s. (From Olsen and Spence (1981).)

5.3. Approximations for the diffracted amplitudes

It is important to understand how the amplitudes of a set of Bragg reflections depend on the specimen structure, thickness, and orientation. In Section 5.2 we saw how these amplitudes form the Fourier coefficients in a series whose sum gives the image amplitude. The dimensionless diffracted amplitudes $\Phi_\mathbf{g}$ express the total scattering by the crystal in a particular direction relative to a unit-amplitude incident wave. For all but the thinnest specimens, the possible directions are confined to a narrow range of angles around each reciprocal lattice point (see Fig. 5.3). This scattering is focused to a small spot by the objective lens in the lens back-focal plane. The intensities $|\Phi_\mathbf{g}|^2$ are therefore proportional to

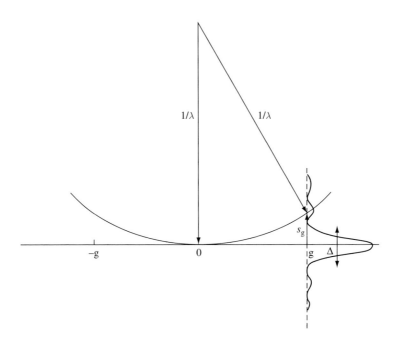

Figure 5.3 The Ewald sphere construction. A vector of length $1/\lambda$ is drawn in the crystal reciprocal lattice in the direction of the incident beam. In the zone-axis orientation as shown, this is normal to a plane of the reciprocal lattice. The kinematic amplitude diffracted into the beam $\mathbf{g}$ is proportional to the amplitude of the 'shape transform' $t \sin(\pi s_\mathbf{g} t)/(\pi s_\mathbf{g} t)$ at the point where the sphere crosses this function. The square of the function is suggested in the figure. The width Δ between first minimum of this function is $2/t$, so that the angular range of incident beam directions which will produce scattering into the beam $\mathbf{g}$ increases as the specimen thickness decreases. A crystal of large transverse dimensions has been assumed. The excitation error $s_\mathbf{g}$ is drawn normal to the reciprocal lattice plane and is negative if the lattice point is outside the sphere. At 100 kV, $1/\lambda = 270$ nm^{-1} and with $|\mathbf{g}| = h/a \approx 2.5$ nm^{-1} the sphere is well approximated by a plane in the neighbourhood of the origin (i.e. to some limited 'resolution') for a particular crystal thickness (which determines the width of the 'shape transform'). At higher voltage, the sphere becomes larger and the flat sphere approximation holds to larger crystal thickness—that is, with a narrower shape transform.

the intensities of the diffraction spots seen when the microscope is focused for the specimen diffraction pattern.

In the following sections, the kinematic and two-beam expressions for the diffracted amplitudes are given for the information they supply about the effect of changes in experimental conditions on simple 'lattice fringes'. In particular, we are interested in the best choice of specimen thickness and orientation for two-beam fringes.

5.3.1   The kinematic theory

For those ultrathin, low-atomic-weight specimens for which the central beam is very much stronger than any diffracted beam, the kinematic or single-scattering theory of image formation may be applied. This includes the case of bent organic monolayers, as discussed in Section 7.5. The results of this theory give a useful qualitative guide to the effects of changes in experimental conditions. Measurable dynamical effects (multiple scattering of the beam inside the sample) leading to failure of this theory have been observed in graphite flakes just 4 nm thick. Nevertheless, the simplicity of the kinematic model has ensured its popularity in discussions on electron imaging.

The weak-phase object approximation (Section 3.4) can be extended to include the effect of Fresnel diffraction through the thickness of the specimen (and so take account of the curvature of the Ewald sphere) by integrating the contribution to scattering in a particular direction (Howie 1971). This gives the kinematic result for electron diffraction from a crystal with Fourier coefficients of potential $v_\mathbf{g}$ and thickness t as

$$\Phi_\mathbf{g} = -i\sigma v_\mathbf{g} t \lfloor \sin \pi s_\mathbf{g} t / \pi s_\mathbf{g} t \rfloor \exp(-i\pi s_\mathbf{g} t) \tag{5.15}$$

with

$$\Phi_0 = 1$$

and the sign convention for phases of Chapter 3. The orientation of the crystal with respect to the incident beam is specified by $s_\mathbf{g}$, the excitation error, shown in Fig. 5.3. Practical examples of the way in which $s_\mathbf{g}$ can be measured from a diffraction pattern are given in Spence and Zuo (1992). Phase factors which depend on the lens focal length and object distance have been omitted. The exponential phase factor in eqn (5.15) depends on the choice of origin in the crystal—it can be eliminated by choosing an origin midway though the thickness of the specimen. Whatever origin is chosen, it is important that this same origin is used when reckoning the image focus defect. Equation (5.15) assumes that any focus defect is measured from the top surface of the specimen. Inserting this expression into eqn (5.6) gives the fringe intensity for tilted illumination as

$$I(x) = 1 - 2\sigma v_\mathbf{g} t [\sin \pi s_\mathbf{g} t / \pi s_\mathbf{g} t] \sin(-\pi s_\mathbf{g} t - 2\pi hx/a) \tag{5.16}$$

This expression suggests the following:

1. Sinusoidal fringes are expected which, for a parallel-sided crystal, have the periodicity of the lattice. The positions of the fringe maxima will not in general coincide with the projected position of atoms in the crystal, but will depend on the specimen thickness and orientation.

2. For a centrosymmetric crystal the $v_\mathbf{g}$ are real if absorption is negligible. Otherwise the $v_\mathbf{g}$ are complex and contain a phase factor which appears in the argument of the second sin function in eqn (5.16). The phases of the diffracted beams thus determine the relative displacements of the various fringes which make up an image. In this sense, image formation in the electron microscope 'solves' the phase problem of X-ray diffraction.
3. For fringes running parallel to the edge of a wedge-shaped specimen, we have a situation akin to phase modulation in electronic engineering. If the thickness varies with x (taken normal to the wedge edge) then the fringe spacing for a particular value of x will depend on the local variation of t with x. Only for a perfect wedge, with t proportional to x, would the spacing be constant (see Section 5.4).
4. A similar variation in fringe spacing is expected for bent crystals, where $s_\mathbf{g}$ varies with x.
5. Fringes running perpendicular to the wedge edge are likely to be curved, since a different value of t applies to each chosen value of x (taken parallel to the wedge edge).

5.3.2 The two-beam dynamical theory

The highly restrictive assumption of an undepleted central beam in the kinematic theory is improved upon in the two-beam dynamical theory which provides a practical guide to the best specimen thickness for two-beam fringes. In this theory the real crystal is replaced by a fictitious crystal supporting only two diffracted beams. The accuracy of the theory can be judged from a glance at the diffraction pattern, which will indicate how much energy is scattered into reflections other than the central beam and the one other beam used for image formation. In this theory the sum of the intensities of the two beams is constant, and the contrast of two-beam lattice fringes at the exact Bragg angle is a maximum when the beams are of equal strength. This occurs when the specimen thickness is

$$t_\mathbf{g} = \frac{\xi_\mathbf{g}}{4} = \frac{\pi}{4\sigma v_\mathbf{g}} \tag{5.17}$$

where $\xi_\mathbf{g}$ is the extinction distance for the reflection $\mathbf{g}$. Typical values at 100 kV are $t_{111} = 4$ nm (gold), $t_{111} = 14$ nm (aluminium), and $t_{111} = 15$ nm (silicon). Values of $t_\mathbf{g}$ for other materials can be obtained from published scattering factors (e.g. Doyle and Turner 1968). Scattering factor calculations are also discussed in Burge (1973), and Fourier coefficients for many elements can be found in Radi (1970). The most complete tabulation of high-energy electron scattering factors is that contained in the International Tables for Crystallography.

If the scattering factors $f_j(\theta)$ are calculated from the first Born approximation, then

$$v_\mathbf{g} = 0.04787 F_\mathbf{g}/\Omega \tag{5.18}$$

where Ω is the unit-cell volume in cubic nanometres and $F_\mathbf{g}$ is the kinematic structure factor obtained from the atomic scattering amplitudes according to

$$F_\mathbf{g} = F_{h,k,l} = \sum_j f_i(\theta_\mathbf{g}) \exp(2\pi i(hx_j/a + ky_j/b + lz_j/c)) \tag{5.19}$$

where (x_j, y_j, z_j) are the coordinates of the atom j in the unit cell, and $\theta_\mathbf{g}$ is the Bragg angle for the reflection with Miller indices h, k, l. Here $v_\mathbf{g}$ is in volts and $F_\mathbf{g}$ in angstroms, as usually tabulated (e.g. in the International Tables for Crystallography).

In practice the best specimen thickness for two-beam lattice fringes is generally found by trial and error; however, an indication of thickness is given by Pendellösung thickness fringes. These broad dark and bright bands are seen parallel to the edge of wedge-shaped specimens at the Bragg angle if a small aperture is placed around the central beam. The thickness increment between successive fringe minima is approximately equal to the extinction distance $\xi_\mathbf{g}$, where $\mathbf{g}$ refers to the reflection satisfied. The clearest lattice fringes are expected between the edge of the wedge and the first bright-field Pendellösung minimum.

The two-beam theory can also be used to predict the variation of lattice fringe contrast with orientational changes of the specimen. For reference, the complex amplitudes of the two diffracted beams are

$$\Phi_0 = \left\{\cos\left(\pi t(1+w^2)^{1/2}/\xi_\mathbf{g}\right) - iw(1+w^2)^{-1/2}\sin\left(\pi t(1+w^2)^{1/2}/\xi_\mathbf{g}\right)\right\}\exp(-i\pi s_\mathbf{g} t) \tag{5.20}$$

and

$$\Phi_\mathbf{g} = i(1+w^2)^{-1/2}\sin\left(\pi t(1+w^2)^{1/2}/\xi_\mathbf{g}\right)\exp(-\pi i s_\mathbf{g} t)$$

where $w = s_\mathbf{g}\xi_\mathbf{g}$. These expressions can be used with eqns (5.6) and (5.3) to give complicated expressions for the two-beam lattice fringe intensity. The resulting expression gives a much better indication of the variation of lattice fringe contrast with specimen orientation and thickness, the only important approximation being the neglect of other diffracted beams. These neglected diffracted beams become increasingly important in crystals such as minerals with large unit cells. Whereas the width in reciprocal space of the central maximum of the kinematical 'shape transform' shown in Fig. 5.3 is $1/t$, the width of the corresponding two-beam dynamical function tends to $1/\xi_\mathbf{g}$. The kinematic theory does, however, remain accurate to larger thickness for weak beams with large excitation errors or small values of $v_\mathbf{g}$.

The theory of lattice imaging in the two-beam approximation is given in Cowley (1959). This and similar work (Hashimoto et al. 1961) using the two-beam expressions indicates that a reversal of fringe contrast is expected for every increase of half an extinction distance in thickness. In addition, the two-beam lattice image spacing is found to agree with the specimen lattice spacing only at the exact Bragg angle.

5.3.3 The thick phase grating

We have seen that the commonly used phase object approximation (eqn 3.30), while including multiple-scattering effects, is a projection approximation and does not take account of the curvature of the Ewald sphere, which is approximated by a plane. The development of the multislice theory of dynamical scattering suggests that this curvature corresponds physically to Fresnel diffraction within the specimen, to the extent that refractive index and propagation effects can be separated. Thus eqn (6.20), the condition for the neglect of Fresnel broadening of the electron wave, also represents the condition under which the Ewald sphere can be approximated by a plane (all $s_\mathbf{g}$ set to zero).

An approximation which takes account of multiple scattering and, in a limited way, the Ewald sphere curvature has been proposed by Cowley and Moodie (1962). This 'thick-phase grating' gives the object exit-face wave function as

$$\psi(x,y) = \exp\left[-i\sigma t \sum_{h,k} v_{h,k,0}(\sin \pi s_{h,k,0} t / \pi s_{h,k,0} t) \exp(2\pi i(hx/a + ky/b))\right] \quad (5.21)$$

with a and b cubic unit-cell dimensions and the incident beam approximately in the direction of the c-axis. The Fourier coefficients of $\psi(x, y)$ then give the diffracted amplitudes $\Phi_\mathbf{g}$. We note that the first-order expansion of eqn (5.21) gives the kinematic result (eqn 5.15) and that eqn (5.21) incorporates propagation effects if the $s_{h,k,0}$ are measured onto the Ewald sphere, as in Fig. 5.3. Since a plane in reciprocal space represents a projection in real space, the summed expression in eqn (5.21) above would represent simply a re-projection of the thin crystal potential if the $s_{h,k,0}$ were measured from the $(h, k, 0)$ zone onto a plane through the origin and normal to the electron beam. The resulting expression is then closely similar to the 'high-energy approximation' of Molière described in texts on the quantum theory of scattering. This expression, unlike the Born approximation, satisfies the optical theorem. The phase-grating or 'high-voltage limit' approximation of eqn (3.25) can also be derived by the method of partial wave analysis, or by a term-by-term comparison of the Born series (with $K_z = 0$) with the Fourier transform of the series expansion of eqn (3.25). The validity domain of eqn (5.21) has been investigated in unpublished work by A. F. Moodie.

5.3.4 The projected charge density approximation

Unlike the weak-phase object approximation, the projected charge density (PCD) approximation incorporates the effects of multiple scattering in a limited way, and is most useful for cases where the effects of spherical aberration are small, such as images formed in an aberration-corrected instrument.

To expose the principle of the PCD approximation, consider an image formed with no objective aperture and no spherical aberration. The exit-face wave amplitude for a phase object (including some multiple-scattering effects) is, from eqn (3.25),

$$\psi_\mathrm{e}(x, y) = \exp(-i\sigma\phi_\mathrm{p}(x, y)) \quad (5.22)$$

with $\phi_\mathrm{p}(x, y)$ the projected specimen potential in volt-nanometres. The back-focal plane amplitude is obtained by Fourier transformation as

$$\begin{aligned}\psi_\mathrm{d}(u,v) &= \mathcal{F}\{\exp(-i\sigma\phi_\mathrm{p}(x,y))\}\exp\left(i\pi\Delta f \lambda(u^2 + v^2)\right) \\ &\approx \Phi(u,v)\left[1 + i\pi\Delta f \lambda(u^2 + v^2)\right]\end{aligned} \quad (5.23)$$

if a small focusing error Δf is allowed and $\Phi(u, v)$ is the Fourier transform of $\exp(-i\sigma\phi_\mathrm{p}(x, y))$. The image amplitude at unit magnification and without rotation is given by inverse transformation as

$$\psi_\mathrm{i}(x,y) = \exp(i\sigma\phi_\mathrm{p}(x,y)) + i\pi\Delta f \lambda \mathcal{F}^{-1}\{(u^2 + v^2)\,\Phi(u,v)\} \quad (5.24)$$

A theorem from Fourier analysis can now be used (Bracewell 1999) which shows that if $f(x, y)$ and $\Phi(u, v)$ are a Fourier transform pair according to eqn (3.10), then

$$\mathcal{F}^{-1}\{(u^2 + v^2)\Phi(u, v)\} = \frac{-1}{4}\pi^2 \nabla^2 f(x, y)$$

Using this result gives the image amplitude as

$$\psi_i(x, y) = \exp(-i\sigma\phi_p(x, y)) - (i\Delta f\lambda/4\pi)\nabla^2\{\exp(-i\sigma\phi_p(x, y))\}$$
$$= \exp(-i\sigma\phi_p(x, y)) + (i\Delta f\lambda\sigma/4\pi)\exp(-i\sigma\phi_p(x, y))$$
$$\times \{\sigma\nabla\phi_p(x, y) + i\nabla^2\phi_p(x, y)\} \tag{5.25}$$

so that the image intensity, to first order, is

$$I(x, y) = 1 - (\Delta f\lambda\sigma/2\pi)\nabla^2\phi_p(x, y) \tag{5.26}$$

Using Poisson's equation, $\nabla^2\phi_p(x, y) = -\rho_p(x, y)/\varepsilon_0\varepsilon$, we finally obtain

$$I(x, y) \approx 1 + (\Delta f\lambda\sigma/2\pi\varepsilon_0\varepsilon)\rho_p(x, y) \tag{5.27}$$

Notice that a first-order (kinematic) expansion of the phase object expression has not been made so that, through the inclusion of multiple-scattering effects, this approximation can be expected to hold to greater thickness than the kinematic approximation. The PCD approximation does, however, assume a flat Ewald sphere. The quantity $\rho_p(x, y)$ is the projected total charge density including the nuclear contribution and not the electron charge density as measured in X-ray diffraction.

Using the sign convention for phases from Chapter 3, we see that the under-focused electron image (Δf negative) is deficient in regions of high specimen charge density. Here a photographic print of a micrograph would appear dark in regions of high specimen charge density, such as around groups of heavy atoms. Also, for small defocus, the contrast is proportional to defocus. Experimental confirmation of eqn (5.27) is shown in Fig. 5.4.

The extension of this theory to include the effects of spherical aberration and a limiting objective aperture are discussed by Lynch et al. (1975). Their work shows that within a certain range of focus, specimen thickness, and resolution, eqn (5.26) can still be used as a basis for the interpretation of spherically aberrated images of limited resolution. That is, *within certain experimental conditions, the high-resolution image obtained represents the specimen projected total charge density as if seen through a spherically aberrated lens of limited resolution*. Note that PCD images are found in the neighbourhood of every Fourier image plane for periodic specimens (see Section 5.8). The PCD approximation would seem to be ideally suited to the interpretation of aberration-corrected HREM images, where C_s is small. The PCD analysis bears a close similarity to the transport of intensity method (Section 7.6).

102 TEM imaging of thin crystals and defects

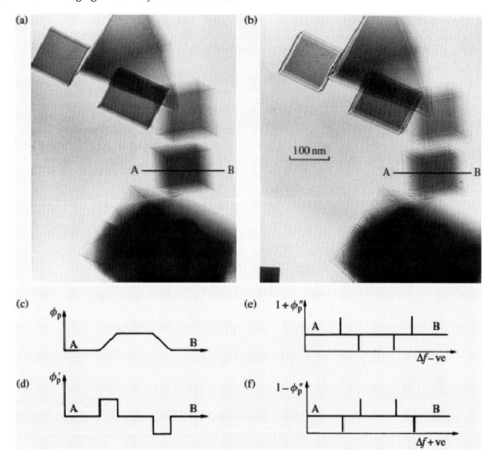

Figure 5.4 Experimental verification of the PCD interpretation of electron microscope images at low resolution where the effect of spherical aberration can be completely neglected. Magnesium oxide cubes are shown imaged at slight under-focus (a) and over-focus (b). The diagrams suggest the form of the projected specimen potential (c), its first derivative (d), and a constant plus its second derivative (e) along the line AB; (e) then gives a rough indication of the contrast predicted by eqn (5.26). Part (f) shows the expected image with the sign of the focus reversed (Δf positive). That (e) and (f) correctly predict the contrast seen at the cube edges can be seen by comparison with (a) and (b). Notice that the cubes do not lie flat on the substrate and are balanced on a corner or edge. A clear plastic model cube or an ice block viewed in a direction close to one of its body diagonals will clarify the form of the projected potential in (c). The cubes appear as a cage of bright or dark wires.

5.4. Images of crystals with variable spacing—spinodal decomposition and modulated structures

Image artefacts may arise in crystals in which either the lattice spacing or the chemical composition (or both) varies periodically, if resolution is limited so that individual atom columns are not resolved. This is relevant to attempts to use Vegard's law to relate local composition to local strains in alloys. For example, Hall et al. (1983) were able to relate lattice spacings to the carbon content of retained austenite in a dual-phase steel. Although the tilted-illumination super-resolution mode we will discuss (and its important STEM analogue) is now rarely used, it is instructive to understand in detail how this unconventional imaging mode (in which aberrations cancel) can nevertheless introduce artefacts. We first consider artefacts due to the lens alone, when using inclined illumination in order to obtain the highest 'resolution'. Following eqn (5.15) we saw that in wedge-shaped crystals the fringe spacing parallel to the edge will depend on the local variation of the phase ε of Φ_g with t. For the tilted illumination two-beam geometry of Fig. 5.1(b), the local periodicity observed in the image will be $1/u'_h(x)$, where

$$u'_h(x) = u_h + (1/2\pi)\mathrm{d}\varepsilon(x)/\mathrm{d}x \qquad (5.28)$$

This result is obtained by analogy with the theory of phase modulation in electrical engineering. Equation (5.7) gives the variation of image intensity. However, it is clear that any other spatially dependent contribution to the phase $\varepsilon(x)$ in eqn (5.7) will similarly influence the local image period. For crystals in which the lattice spacing varies, the objective lens phase factor $\chi(u)$ will also provide such a contribution, as follows.

Following Spence and Cowley (1979), we consider an idealized thin crystal containing a single atomic species for which the modulated crystal potential is proportional to

$$\phi(x) = \cos[2\pi u_0 x + A\sin(2\pi u_\mathrm{L} x)]$$
$$= \cos[2\pi u_0 x + \phi'(x)]$$

Here A is the amplitude of the modulation and $1/u_\mathrm{L}$ is its period. The true local period of this potential is $1/u'_x$ where

$$u'_x(x) = u_0 + (1/2\pi)\mathrm{d}\phi'/\mathrm{d}x$$
$$= u_0 + u_\mathrm{L} A\cos(2\pi u_\mathrm{L} x) \qquad (5.29)$$

We wish to compare this true period with that observed in the image. If $u_\mathrm{L} \ll u_0$, the image intensity for tilted illumination is

$$I(x) = |\Phi_0|^2 + 2|\Phi_0||\Phi'_u|\cos[2\pi u'_x x + \chi(u_x) - \chi(u_0/2) + \varepsilon]$$
$$= |\Phi_0|^2 + 2|\Phi_0||\Phi'_u|\cos[2\pi u'_x x + \phi(x)] \qquad (5.30)$$

where

$$u_x = u'_x - u_0/2 \qquad (5.31)$$

We assume that the optic axis bisects the angle $2\theta_\mathrm{B} = u_0 \lambda$ between the direct 'beam' and that corresponding to the average period u_0. Here we have implicitly adopted the column approximation (which may fail unless $u_\mathrm{L} \ll u_0$), and we further assume that $\chi(u)$ varies slowly in the neighbourhood of u_0 (see Cockayne and Gronsky (1981) for a discussion of these points). The inverse local period observed in the image will be

$$u''_x = u' + (1/2\pi)\mathrm{d}\chi/\mathrm{d}x \tag{5.32}$$

The distorting effect of the lens may now be characterized by defining $F(x)$ = (local image spatial frequency – local object spatial frequency)/(local object spatial frequency). This gives a measure of the extent to which the lens 'amplifies' variations in object periodicity. Thus we define

$$F(x) = (u''_x - u'_x)/u'_x = [\mathrm{d}\chi/\mathrm{d}x]/2\pi u'_x$$

using eqn (5.32). Calculations of the maximum value of $F(x)$ (maximized over x) as a function of focus setting show that F is minimized at the stationary phase focus $\Delta f = -C_s\lambda^2 u_0^2/4$. Under these conditions $\chi(u)$ varies least across the spectrum of satellite reflections around u_0 which result from the modulated structure.

This amplification effect may be as great as 400% under readily obtainable conditions, but may be only 90% at the optimum condition. Cockayne and Gronsky (1981) have studied the effect of periodic segregation of the atomic species on these images, using the artificial superlattice method (see Section 5.11). In the axial geometry structure, however, images may be obtained for which a straightforward interpretation is possible provided that sufficiently thin samples are used and that the point-resolution exceeds the lateral interatomic spacing.

The effects of elastic relaxation must also be considered in these very thin films. Shear stresses have been shown (Gibson and Treacy 1984) to distort the local unit-cell dimensions appreciably. This effect is present for all imaging conditions, including axial structure imaging at atomic resolution. In particular, the elastic response of a thin modulated film will cause a bending of the lattice planes about the Y-axis (normal to the beam and x). Calculations for this important effect (Gibson and Treacy 1984) show that it may lead to differences of up to 33% between the bulk lattice spacing and that present in very thin films. A comprehensive analysis of TEM images of spinodally decomposed $In_xGa_{1-x}As_yP_{1-y}$ films may be found in Treacy et al. (1985), who also discuss in detail the elastic relaxation expected in these very thin films.

A method of mapping strains using digital transforms of a portion of the diffraction pattern of axial bright-field HREM images is described in Hytch and Plamann (2001), and the dramatic effects of using aberration correction to avoid all the artefacts discussed above for strain mapping are demonstrated in Hue et al. (2008) for mapping the strain-field within a transistor. (An important problem, since induced strain is used to increase carrier mobility.) Section 13.3 describes strain mapping in transistors using the convergent-beam method, and a method based on electron holography is described in Section 7.3.

5.5. Are the atom images black or white? A simple symmetry argument

Since a change of focus of $\Delta f_f/2$ (see Section 5.1 and eqn (5.53b)) reverses the contrast of lattice images, the question arises as to how one can tell whether dark or light regions

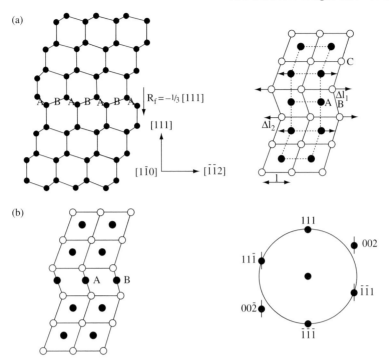

Figure 5.5 (a) The structure of an intrinsic stacking fault projected along $[1\bar{1}0]$ in the diamond structure. (b) The simplified projection seen at limited resolution. The open circles are tunnel sites and filled circles are pairs of atomic columns. The screw axis at A falls on a filled circle, B is a two-fold axis. Arrows indicate the layers which change position with focus. (From Olsen and Spence (1981).)

in an image coincide with the atomic column positions. If a diffractogram is obtainable, and the thickness is known, a determination can be made by image-matching techniques (see Section 5.11). It is sometimes possible, however, to use the point group symmetry of a defect of known structure to resolve this question (see Olsen and Spence (1981) for detailed simulations). As an example, Fig. 5.5 shows the structure of an intrinsic stacking fault in silicon viewed along $[1\bar{1}0]$. At limited resolution, the image appears as shown in Fig. 5.5(b). The stacking fault contains two two-fold symmetry axes (in projection) as shown at A and B. Since both are known a priori to fall on a layer of atoms (rather than tunnels), they may be used to determine whether a given image shows white tunnels or white atoms. (A 'layer' is a line of dots running across the figure.) This is not possible in a perfect crystal. The contrast along a horizontal line across these images will be found to be that of the atoms, not the tunnels. This powerful argument holds for any thickness of sample and focus setting if the image is correctly aligned and stigmated. A similar argument can be used to identify the position of a sub-nanometre-diameter electron probe in coherent microdiffraction (see Section 13.4).

5.6. The multislice method and the polynomial solution

The multislice theory of dynamical electron diffraction was developed before computers became available commercially (Cowley and Moodie 1959). It was not, therefore, originally intended as a numerical algorithm but as an attempt to obtain a closed-form expression for the dynamical Bragg beam amplitudes $\Phi_\mathbf{g}$ by considering the limiting case of vanishing slice thickness. The result is (Moodie 1972)

$$\Phi_\mathbf{g} = \sum_{n=1}^{\infty} E_n(\mathbf{g}) Z_n(\mathbf{g}) \tag{5.33}$$

where the terms E_n depend only on structure factors, while the functions $Z_n(\mathbf{g})$ involve polynomials which are functions of the geometric factors only (excitation errors, wavelength, lattice constant). This polynomial result has subsequently been obtained by other authors using different methods, and can be shown to reduce to the phase-grating approximation, to a power series in thickness, or to the two-beam approximation under suitable approximations. Equation (5.33) has been used to explain the occurrence of dynamically forbidden reflections (see Section 5.10). With the advent of electronic computers, a form of the multislice theory more convenient for computational work was applied in calculations, first by Goodman and others (Goodman and Moodie 1974). This recursion relation, which has become the basis of all modern multislice computations, is

$$\Phi_\mathbf{g}(n) = \sum_\mathbf{h} \Phi_\mathbf{h}(n-1) P_{\Delta Z}(\mathbf{h}) Q(\mathbf{g} - \mathbf{h}) \tag{5.34}$$

where $P_{\Delta Z}(\mathbf{h})$ is given by eqn (3.16) and $Q(\mathbf{g})$, the 'phase grating amplitudes', are given by the Fourier coefficients of eqn (5.22). Here the crystal has been divided into slices normal to the beam (not necessarily equal in thickness to the period of the lattice in that direction) and $\Phi_\mathbf{g}(n)$ is the set of Bragg diffracted beams emerging from slice n. Thus $\phi_p(x, y)$ is the specimen potential projected through a single slice. Computational aspects of eqn (5.34) are discussed in Section 5.11. $\Phi_\mathbf{g}(1)$ are taken to be the Fourier coefficients of the incident plane-wave. The derivation of the time-independent Schrödinger equation from eqn (5.34) is outlined in Goodman and Moodie (1974), where the relationship of this approach to other dynamical theories is fully discussed (see also Section 5.11). Researchers wishing to write a computer program based on the multislice algorithm are referred to Section 5.11, and, in particular, to the books by Kirkland (2010) and Spence and Zuo (1992) (which contains a FORTRAN source-code listing) and articles by Self et al. (1983) and Ishizuka (1985). The propagation function $P_{\Delta Z}(\mathbf{g})$ can be written

$$P_{\Delta Z}(\mathbf{g}) = \exp(-2\pi i S_\mathbf{g} \Delta Z) \tag{5.35}$$

where $S_\mathbf{g}$, the excitation error for beam $\mathbf{g}$, is taken negative for reflections outside the Ewald sphere (see Section 5.12). A review of 'real space' multislice methods may be found in Van Dyck and Coene (1984).

The effects of steeply inclined boundary conditions on the multislice formulation have been discussed by Ishizuka (1982). Anstis (1977) has shown that a unity sum of beam intensities is not a sufficient condition to ensure that the calculation includes enough beams. The calculation will be normalized for any number of beams in the limit of sufficiently small slice thickness. The method may be extended to include three-dimensional diffraction effects by choosing the slice thickness to be much less than the periodicity in the beam direction. An early example of such a calculation, used to analyse the occurrence of 'forbidden' termination reflections, can be found in Lynch (1971).

5.7. Bloch wave methods, bound states, and 'symmetry reduction' of the dispersion matrix

A solution to the N-beam dynamical problem of kilovolt electron diffraction was first given by Bethe (1928) for both the Bragg and Laue geometries. Pedagogical treatments of this theory can be found in Howie (1971), Dederichs (1972), Metherall (1975), Hirsch *et al.* (1977), Humphreys (1979), and Reimer (1984), and a FORTRAN source listing of the Bloch wave code in Spence and Zuo (1992). We consider here aspects of dynamical theory which are relevant to lattice imaging, such as symmetry reduction of the dispersion matrix, the theory of dynamically forbidden reflections, and the direct observation of separate Bloch waves in real space. The 'quantum-mechanical' sign convention used in this section (and in the above reviews) is the opposite of the 'crystallographic convention' adopted elsewhere throughout this book (see Section 5.12).

It is conventional to define quantities

$$U_\mathbf{g} = \frac{\sigma v_\mathbf{g}}{\pi \lambda} = \frac{1}{\lambda \xi_\mathbf{g}} = \frac{2m|e|v_\mathbf{g}}{h^2} \tag{5.36}$$

where $v_\mathbf{g}$ are the Fourier coefficients of crystal potential (in volts) (see also eqn (5.18)). Note that σ here involves the relativistically corrected electron mass m and wavelength λ (see eqn (2.4) and following). It has been shown by K. Fujiwara and A. Howie (see Gevers and David (1962) and Humphreys (1979) for reviews) that electron spin effects are negligible in transmission electron diffraction. Thus, rather than solve the Dirac equation, we may derive a relativistically corrected Schrödinger-like equation as follows. The relativistic energy–momentum conservation equation for an electron accelerated to a potential $e(V_0 + \phi)$ is

$$p^2 c^2 + m_0^2 c^4 = [|e|(V_0 + \phi) + m_0 c^2]^2 = W^2 \tag{5.37a}$$

Here V_0 is the accelerating voltage and $\phi = \phi(\mathbf{r})$ is the crystal potential, whose Fourier coefficients are given by eqn (5.18). The total relativistic energy is

$$W = mc^2 = |e|(V_0 + \phi) + m_0 c^2 = \gamma m_0 c^2$$

with γ given by eqn (6.18).

108 TEM imaging of thin crystals and defects

Neglecting a term $|e|^2\phi^2$, these equations give

$$p^2c^2 - 2mc^2|e|\phi = |e|V_0\left(2m_0c^2 + |e|V_0\right)$$

Replacing p by the operator $-i\hbar\nabla$, and operating on Ψ gives the Schrödinger-like equation

$$(\hbar^2/2m_0)\nabla^2\Psi(\mathbf{r}) + |e|\gamma\phi(\mathbf{r})\Psi(\mathbf{r}) + |e|V_r\Psi(\mathbf{r}) = 0 \qquad (5.37\text{b})$$

where $\gamma = m/m_0$ and V_r, the relativistically corrected accelerating voltage, is given following eqn (2.4).

The solution to eqn (5.37a) for a free electron, with $p = \hbar k_i$ and $\phi = 0$, gives the vacuum electron wavevector as

$$|\mathbf{k}_i| = \frac{W\beta}{hc} = \frac{1}{\lambda} = \frac{\gamma\beta m_0 c^2}{12.4}$$

Here if $m_0c^2 = 511$ kV then $|\mathbf{k}_i|$ is given in Å^{-1}. This result is consistent with eqn (2.4).
The solution to eqn (5.37b) is taken to be a linear sum of Bloch waves of the form

$$\Psi(\mathbf{r}) = \sum_j \alpha_j b(\mathbf{k}^{(j)}, \mathbf{r}) \qquad (5.37\text{c})$$

where

$$b(\mathbf{k}^{(j)}, \mathbf{r}) = \sum_\mathbf{g} C_\mathbf{g}^{(j)} \exp((2\pi i \mathbf{k}^{(j)} + \mathbf{g}) \cdot \mathbf{r}) \qquad (5.37\text{d})$$

which describes the total electron wavefunction inside the crystal. Inside the crystal the beam electron wavevector corrected for the effect of the mean inner potential is

$$\mathbf{k}_0^2 = (1/\lambda)^2 + U_0 = \mathbf{k}_i^2 + U_0 \qquad (5.37\text{e})$$

where U_0 has dimensions length^{-2} and λ is given by eqn (2.5). One Bloch wave is associated with each branch j of the dispersion surface, and values of $C_\mathbf{g}^{(j)}$ and $\mathbf{k}^{(j)}$, the Bloch wave coefficients and wavevectors, are obtained by numerical solution of the eigenvalue equation (5.38a). This results from substituting eqn (5.37d) into the Schrödinger equation (5.37b). The Fourier coefficients of $\phi(\mathbf{r})$ are $v_\mathbf{g}$. If backscattered waves are neglected and the small-angle approximation

$$\mathbf{k}_0^2 - (\mathbf{k}^{(j)} + \mathbf{g})^2 \approx 2\mathbf{k}_0(S_\mathbf{g} - \gamma^{(j)}) \qquad (5.37\text{f})$$

is made, where $\gamma^{(j)} = k_z^{(j)} - |\mathbf{k}_0|$, and we define $U_{\mathbf{g}-\mathbf{g}} = 2k_0 S_\mathbf{g}$, we then obtain

$$\sum_\mathbf{h} U_{\mathbf{g}-\mathbf{h}} C_\mathbf{h}^{(j)} = 2|\mathbf{k}_0|\gamma^{(j)} C_\mathbf{g}^{(j)} \qquad (5.38\text{a})$$

which can be written

$$\begin{bmatrix} 2|\mathbf{k}_0|S_\mathbf{h} & \cdots & U_{\mathbf{g-h}} & \cdots & U_{\mathbf{1-h}} \\ \vdots & & \vdots & & \vdots \\ U_{\mathbf{h-g}} & \cdots & 2|\mathbf{k}_0|S_\mathbf{g} & \cdots & U_{\mathbf{1-g}} \\ \vdots & & \vdots & & \vdots \\ U_{\mathbf{h-1}} & \cdots & U_{\mathbf{g-1}} & \cdots & 2|\mathbf{k}_0|S_\mathbf{1} \end{bmatrix} \begin{bmatrix} C_\mathbf{h}^{(j)} \\ \vdots \\ C_\mathbf{g}^{(j)} \\ \vdots \\ C_\mathbf{1}^{(j)} \end{bmatrix} = 2|\mathbf{k}_0|\gamma^{(j)} \begin{bmatrix} C_\mathbf{h}^{(j)} \\ \vdots \\ C_\mathbf{g}^{(j)} \\ \vdots \\ C_\mathbf{1}^{(j)} \end{bmatrix} \qquad (5.38b)$$

or

$$\mathbf{AC} = 2|\mathbf{k}_0|\gamma^{(j)}\mathbf{C} \qquad (5.39)$$

Here the eigenvalues $\gamma^{(j)}$ measure the deviation of the dynamical dispersion surfaces from spheres of radius $|\mathbf{k}_0|$ centred on each reciprocal lattice point. We take $\mathbf{A}$ to be an $n \times n$ matrix. The spheres correspond to the dispersion surfaces of plane waves in the 'empty lattice' approximation, in which all $U_\mathbf{g} = 0$. Here $\mathbf{k}^{(j)2} = k_z^2 + k_t^2$. The excitation errors $S_\mathbf{g}$ (see Fig. 5.3) are related to the two-dimensional vector $\mathbf{K} = -\mathbf{k}_t^{(j)}$ drawn from the origin of reciprocal space to the centre of the Laue circle by

$$2|\mathbf{k}_0|S_\mathbf{g} = 2\mathbf{K} \cdot \mathbf{g} - \mathbf{g}^2 \qquad (5.40)$$

For most HREM work in axial orientations $K = 0$ so that

$$S_\mathbf{g} = -\lambda \mathbf{g}^2/2 \qquad (5.41a)$$

in accordance with the sign convention of Fig. 5.3. In deriving eqn (5.38a) the Laue (transmission) geometry has been assumed for a parallel-sided slab of crystal; back-scattering has been neglected, and a type of projection approximation assumed where only reflections in the zero-order Laue zone are considered, but curvature of the Ewald sphere is allowed for.

The effects of steeply inclined boundary conditions on lattice images are discussed in Ishizuka (1982) from the multislice viewpoint. A discussion of this problem in the Bloch wave formulation can be found in Metherall (1975) and Gjønnes and Gjønnes (1985). For the three-dimensional case, where $g_z \neq 0$ and reciprocal lattice vectors are included which do not lie in a plane normal to the beam, it is necessary to renormalize the eigenvectors $\mathbf{C}$ (see Lewis et al. (1978)), and eqns (5.38a,b) cannot be used. For the most general case of a parallel-sided slab of crystal traversed by an electron beam inclined to the surface normal $\hat{\mathbf{n}}$ the renormalized equation to be solved is (Spence and Zuo 1992)

$$\frac{B_g(2\mathbf{k}^{(j)} \cdot \mathbf{g} + \mathbf{g}^2)}{1 + g_n/k_n} - \sum_h \frac{B_h U_{g-h}}{(1 + g_n/k_n)^{1/2}(1 + h_n/k_n)^{1/2}} = -2k_n\gamma B_g \qquad (5.41b)$$

Here $g_n = \mathbf{g} \cdot \hat{\mathbf{n}}$, $k_n = \mathbf{k}_0 \cdot \hat{\mathbf{n}}$ and the result includes the effects of multiple scattering along the beam direction (three-dimensional dynamical diffraction).

In the absence of absorption effects, the dispersion or structure matrix $\mathbf{A}$ is Hermitian (therefore $\gamma^{(j)}$ are real) and, for centrosymmetric crystals, it is real and symmetric

($U_\mathbf{g} = U_{-\mathbf{g}}$). It reflects the point symmetry about the centre of the Laue circle in the Brillouin zone. The following relations can be also shown to hold amongst the eigenvectors and eigenvalues:

$$C_\mathbf{g}^{(j)}(\mathbf{K} + \mathbf{h}) = C_{\mathbf{g}-\mathbf{h}}^{(j)}(\mathbf{K}) \tag{5.42}$$

$$\gamma^{(j)}(\mathbf{K} + \mathbf{h}) = \gamma^{(j)}(\mathbf{K}) + S_\mathbf{h} \tag{5.43}$$

$$\mu^{(j)}(\mathbf{K} + \mathbf{h}) = \mu^{(j)}(\mathbf{K}) \tag{5.44}$$

$$\gamma^{(j)}(\mathbf{K}) = \gamma^{(j)}(-\mathbf{K}) \tag{5.45}$$

$$C_\mathbf{g}^{(j)}(\mathbf{K}) = C_{-\mathbf{g}}^{(j)*}(-\mathbf{K}) \tag{5.46a}$$

$$\sum_\mathbf{g} C_\mathbf{g}^{(i)} C_\mathbf{g}^{(j)*} = \delta_{ij} \tag{5.46b}$$

$$\sum_j C_\mathbf{g}^{(j)} C_\mathbf{h}^{(j)*} = \delta_{\mathbf{gh}} \tag{5.46c}$$

Thus, the eigenvectors form a complete orthogonal and normalized set. The matrix $\mathbf{C}$ whose columns are the complex eigenvectors $\mathbf{C}^{(j)}$ is unitary. Here $\mathbf{K}$, the vector to the centre of the Laue circle, is assumed to be 'reduced', to lie within the first Brillouin zone, and $\mu^{(j)}$ is the absorption parameter for the jth Bloch wave. In the axial orientation, the point group of the dispersion matrix is equal to that of the projected crystal structure at the point of highest symmetry. The symmetries of the Bloch wave coefficients can be found from group character tables. From eqns (5.42)–(5.44) we see that it is only necessary to solve for $C_\mathbf{g}^{(j)}, \gamma^{(j)}$, and $\mu^{(j)}$ within the two-dimensional reciprocal-space unit cell in order to obtain solutions for all incident beam directions. Note that while the Bloch waves given by eqn (5.37d) are periodic in reciprocal space, the total wavefunction given by eqn (5.37c) (which incorporates the effects of boundary conditions) is not. Bloch waves are numbered in order of decreasing $k_z^{(j)}$, so that the wave with largest $k_z^{(j)}$ has index 1. All Bloch waves correspond to the same total energy. Whereas the dispersion matrix expresses the point group symmetry of the crystal, the translational symmetry is expressed by the total wavefunction (eqn 5.37c), which depends on boundary conditions. The excitation amplitudes α_j are obtained by matching the incident wavefunction and its gradient to the crystal wavefunction at the top surface of the crystal. We then find that $\alpha^{(j)} = C_0^{(j)*}$. The result for the total HREM image wavefunction, including the effects of lens aberrations, is

$$I(\mathbf{r}) = \sum_{\mathbf{g},\mathbf{h}} \sum_{i,j} C_0^{(i)} C_0^{(j)} C_\mathbf{g}^{(i)} C_\mathbf{h}^{(j)} \exp(-2\pi(\gamma_{\text{Im}}^{(i)} + \gamma_{\text{Im}}^{(j)})t)$$
$$\times \exp\{i[2\pi(\gamma_R^{(i)} - \gamma_R^{(j)})t + 2\pi(\mathbf{g} - \mathbf{h}) \cdot \mathbf{r} - \chi(\mathbf{g}, \Delta f) + \chi(\mathbf{h}, \Delta f)]\} \tag{5.47}$$

$$= |\psi_i(\mathbf{r}, \Delta f, \mathbf{K}_0)|^2 \tag{5.48}$$

The image is seen to contain, in general, all possible 'false' periodicities $(\mathbf{g} - \mathbf{h})$, including the half-spacings with $\mathbf{h} = -\mathbf{g}$. Only under linear imaging conditions (small t) is this intensity simply related to the crystal potential (Pirouz 1981). Here $\chi(\mathbf{h}, \Delta f)$ is defined by eqn (3.24) (Δf positive for overfocus), and $C_{\mathbf{g}}^{(i)} = C_{\mathbf{g}}^{(i)}(K)$, etc. The quantities $\gamma_{\text{Im}}^{(i)}$ are the imaginary parts of the eigenvectors $\gamma^{(i)} = \gamma_{\text{R}}^{(i)} + i\gamma_{\text{Im}}^{(i)}$, introduced to account for absorption. This is described by the use of an optical potential $U(\mathbf{r}) = U_{\text{R}}(\mathbf{r}) + iU_{\text{Im}}(\mathbf{r})$ so that the Fourier coefficients $v_{\mathbf{g}}$ are now replaced by complex coefficients $(v_{\mathbf{g}} + iv'_{\mathbf{g}})$. Values of $\gamma_{\text{Im}}^{(i)}$ may be found (for given $U'_{\mathbf{g}}$) by matrix diagonalization of the complex matrix $\mathbf{A}$. A good approximation, however, based on perturbation theory, expresses $\gamma_{\text{Im}}^{(i)}$ in terms of $U'_{\mathbf{g}}$ and $C_{\mathbf{g}}^{(j)}$ (Humphreys 1979). Some difficulties in accurately accounting for absorption effects in HREM are further discussed in Section 5.9.

For thin crystals showing a centre of symmetry in projection, we may take $U'_{\mathbf{g}} = \gamma_{\text{Im}}^{(i)} = 0$; eqn (5.47) then becomes the fundamental equation for lattice imaging in the Bloch wave formulation. Partial coherence effects are discussed in the next section. For real coefficients $U_{\mathbf{g}}$ the solution of the eigenvalue problem of eqns (5.38) may be accomplished by standard computational methods. Programs are also available for complex $U_{\mathbf{g}}$, although slower. A perturbation expression for $\gamma_{\text{Im}}^{(j)}$ has also been given (Hirsch et al. 1977) which avoids the need to work with a complex dispersion matrix in centrosymmetric projections.

Further treatments of lattice imaging from the point of view of the Bloch wave can be found in the work of Desseaux et al. (1977), Fujimoto (1978), Pirouz (1981), Kambe (1982), and Marks (1984). Pirouz (1981) has related the projected charge density approximation (see Section 5.3.4) and the projected potential approximation (see Section 3.4) to the Bloch wave formulation through a second-order expansion of the term $\exp(i\gamma^{(j)}t)$ (see eqn 5.47). In the theory of electron channelling and channelling radiation (see Sections 13.1 and 8.4), it is customary to use a modified Bloch wave theory in which the total wavefunction is written as a product of a z-dependent plane wave of very high kinetic energy (a 'free' state) and a transverse wavefunction. The eigenvalue problem for the transverse states is solved, and these may be either free or bound depending on whether the corresponding transverse energy $h^2 k_0 \gamma^{(j)}/m$ is either larger or smaller than the maxima in the interatomic potential (see Kambe et al. (1974) and Section 8.4 for more detail). This approach has been applied to the problem of lattice imaging by several authors (e.g. Van Dyck and Op de Beeck 1996, Buxton et al. 1978). Fujimoto (1978) has used it to analyse the apparent periodicity in thickness of some lattice images (see Section 5.14), while Marks (1984) has combined this approach with the 'k.p' method of band structure calculation to assess the effects of small misalignments on lattice images (including instrumental aberration effects). Kambe (1982) has shown that, for Ge [110] images, a similar three-Bloch wave analysis indicates that at certain thicknesses two of these interfere destructively, so that one may obtain directly by HREM real-space images of individual Bloch waves at certain thicknesses. (This analysis does not make the 'independent Bloch wave' approximation.) It is a feature of these transverse-wavefunction treatments that the role of the excitation errors is disguised. They have the advantage, however, of exposing the rather small number of bound states normally excited, even in cases where the number of Bragg beams is very large (Vergasov and Chuklovskii 1985).

The computation of HREM images in axial orientations by direct application of eqn (5.39) involves a large amount of redundant calculation if the crystal structure possesses some symmetry. There have been two (related) approaches to the problem of reducing the dimension of the dispersion matrix in axial orientations of high symmetry, and these are now discussed. In non-axial orientations some symmetry may also be preserved if, for example, **K** lies on a symmetry element. These methods have variously been known as 'beam reduction', 'Niehrs reduction', or 'symmetry reduction'.

General group-theoretical approaches to this problem have been given by Tinnappel (1975) and Kogiso and Takahashi (1977). Here the known point group symmetry of the crystal projection is used to find a matrix **T** which, by a similarity transformation, gives **A** in block-diagonal form. Eigenvalues and eigenvectors can then be found separately for each of these blocks which are of lower dimension than that of **A**. The columns of **T** belong to the irreducible representation of the group of the incident wavevector, and may be obtained from group character tables (Cracknell 1968).

A second approach relies on more physical arguments, and inspection of the possible symmetries of the Bloch waves which are consistent with the known symmetry of the crystal projection. Simple worked examples can be found in Hirsch et al. (1977). The most extensive applications of this method have been made by Fukahara (1966), Howie (1971), and Taftø (1979). Other examples can be found in Serneels and Gevers (1969) for the systematics case and Desseaux et al. (1977) for [011] projections of the diamond structure. In this manner cases involving as many as 12 beams (related by symmetry) can be reduced to a 2×2 matrix eigenvalue problem—for a worked example, see the second edition of this book (Spence 1988).

When account is taken of the incident boundary conditions for a parallel-sided slab of crystal, the complete solution may be written

$$\Phi(\mathbf{g}) = \mathbf{S}\Phi(0) \tag{5.49}$$

with

$$\mathbf{S} = \mathbf{C}\{\mathbf{D}\}\mathbf{C}^{-1} = \exp(2\pi i \mathbf{A} t) \tag{5.50}$$

Here $\Phi(\mathbf{g})$ is a column vector containing the complex amplitudes of the diffracted beams, $\Phi(0)$ is a similar vector describing the incident beam, and **D** is a diagonal matrix whose diagonal elements are $\exp(2\pi i \gamma^{(i)} t)$, with t the crystal thickness. Since **C** is unitary, its inverse is equal to the transposed conjugate form of **C**.

The matrix **S** is known as the scattering matrix, and was first introduced by Sturkey (1950). Matrix **A** is known as the structure matrix, or the dispersion matrix if the quantities $2|\mathbf{k}_0|\gamma^{(j)}$ are included on the diagonal. Since **S** is unitary, we see that dynamical electron diffraction in the Laue geometry consists of a unitary transformation (Sturkey 1962).

It is clear that the development of a completely general computer program which will reduce and diagonalize the dispersion matrix of any crystal of given symmetry is a large undertaking, and this fact probably accounts in part for the popularity of the multislice method for HREM computations. Symmetry reduction is also possible in that method. An important advantage of the Bloch wave method is that a single diagonalization gives results for all thicknesses.

5.8. Partial coherence effects in dynamical computations—beyond the product representation. Fourier images

All partial temporal and spatial coherence effects may be included exactly using eqns (4.1) and (5.47) (for the Bloch wave method) or, for the multislice method, using eqn (4.1) and the results of the recursion relation (5.34) combined with eqn (5.1a). The results of such a calculation are shown in Fig. 4.9. In this way, variations in diffraction conditions (i.e. rotation of the Ewald sphere) within the angular range θ_c of the beam divergence are correctly incorporated, and no approximations are made to the aberration function $\chi(u_g)$ (O'Keefe and Sanders 1975).

Calculations of this type are time-consuming and, for the small values of θ_c and thin samples normally used for HREM work, unnecessarily accurate. If variations in diffraction conditions over the angular range θ_c are neglected (a good approximation for thin crystals—see Fig. 5.3 and Section 5.3.2) and a first-order Taylor expansion is made for $\chi(u_g)$ about u_g over the angular range θ_c, then the expression for a single Fourier coefficient of the image *intensity* becomes (Fejes 1977, O'Keefe and Saxton 1983),

$$I_g = \sum_h \Phi_h \exp(i\chi(\mathbf{h}))\gamma\{(\nabla\chi(\mathbf{h}) - \nabla\chi(\mathbf{h}-\mathbf{g}))/2\pi\}\Phi^*_{\mathbf{h}-\mathbf{g}}$$
$$\times \exp\{-i\chi(\mathbf{h}-\mathbf{g})\}E\left\{\frac{1}{2}(\mathbf{h}^2 - |\mathbf{h}-\mathbf{g}|^2)^2\right\} \quad (5.51)$$

Here $\gamma(\mathbf{r})$, the transform of the angular distribution of source intensity, has been defined in eqn (4.5), while

$$E(q) = \exp(-2\pi^2\Delta^2\lambda^2 q^2) \quad (5.52)$$

is the Fourier transform of $B(\Delta f)$, the distribution of focus settings (or, equivalently, incident beam energies) given in eqn (4.1). The linear imaging approximation of eqns (4.8) and (A3.1) involves the term $E(\mathbf{K}^2/2)$. The quantity $\Delta = C_c Q$ describes the variances in the statistically independent fluctuations in accelerating voltage, lens current, and source energy spread (see eqn 4.9).

Equation (5.51) should be used for the computer simulation of dynamical HREM images where θ_c is small under the 'crystallographic' sign convention (see Section 5.12). It assumes that the effective illumination aperture is perfectly incoherently filled, and so is appropriate for LaB$_6$ and tungsten hair-pin filaments under normal conditions (see Section 4.5 for discussion). For the other sign convention, the sign in front of $\chi(\mathbf{g})$ should be reversed everywhere. Since the longitudinal (or temporal) coherence time $\Delta\tau = L/V = h/\Delta E$ is much shorter than typical film exposure times, eqn (5.51) performs an incoherent addition of intensities for electrons of different energies. (Here L is the coherence length, V is the electron velocity, and ΔE is the source energy spread.)

For field-emission sources, the illumination aperture may be coherently filled if $X_a > 2R_a$ (see Section 4.5). Transfer functions for linear imaging in this case are derived in Humphreys and Spence (1981), where the different plane-wave contributions of the illumination are added coherently at the image plane on the detector. The expression corresponding to eqn (5.51) (not restricted to linear imaging) then becomes (O'Keefe and Saxton 1983)

$$Ig = \sum_{\mathbf{h}} \Phi_{\mathbf{h}} \exp(i\chi(\mathbf{h}))\gamma\{(\nabla\chi(\mathbf{h})/2\pi)\}\Phi_{\mathbf{h-g}}^*$$
$$\times \exp\{-i\chi(\mathbf{h-g})\}\gamma\{\nabla\chi(\mathbf{h-g}/2\pi)\}E\left\{\tfrac{1}{2}(\mathbf{h}^2 - |\mathbf{h-g}|^2)^2\right\} \quad (5.53a)$$

Here again wavefunction components of different energies have been added incoherently. Equations (5.53a) and (5.51) predict identical images for a disc-shaped source in the weak-phase object approximation (Humphreys and Spence 1981).

For $\theta_c = 0$ and $\Delta = 0$, the image intensities given by eqns (5.51) and (5.53a) are periodic functions of both focus setting Δf and spherical aberration constant C_s, as first pointed out by Cowley and Moodie (1960) (see also the discussion of particular cases in Sections 5.1 and 5.15). Here we consider the general n-beam case for a projection which allows orthogonal unit-cell axes (possibly non-primitive) to be chosen in directions normal to the electron beam. Let these cell dimensions be a and b. First we consider the case $a = b$, so that $g^2 = (h^2 + k^2)/a^2$ where h and k are integers. Then it is easily shown, from eqn (4.8b), that

$$\chi(\mathbf{g}, \Delta f + 2na^2/\lambda) = \chi(\mathbf{g}, \Delta f) + 2n\pi(h^2 + k^2)$$

where n is an integer. For ideal monochromatic illumination $\gamma(\mathbf{r}_0) = E(q) = 1$ and eqns (5.51) and (5.53a) involve only $\exp(i\chi(\mathbf{G}))$, where $\mathbf{G}$ is any reciprocal lattice vector. Then $\exp(i\chi(\mathbf{G}, \Delta f)) = \exp(i\chi(\mathbf{G}, \Delta f + 2na^2/\lambda))$ for all $\mathbf{G}$, since n, h, and k are all integers. Hence, changes in focus by

$$\Delta f_{\mathrm{f}}' = 2na^2/\lambda \quad (5.53b)$$

leave the two-dimensional image intensity distribution unchanged. This result holds for any number of beams. In practice the effects of finite coherence mean that these 'Fourier images' can be seen only over a limited range of focus. (The effect is often referred to as 'Talbot self-imaging'). As pointed out in Section 5.1, the general condition for restricting the depth of field to a single Fourier image half-period is

$$\theta_c > \lambda/(2a)$$

From Bragg's law, this is just the condition that adjacent diffraction discs overlap.

Equation (5.53b) also limits the number of dynamical images which need to be computed in image simulations—these need only cover the range Δf_{f}. By a similar argument it can be shown that images are periodic in C_s (see Section 5.1).

For $a \neq b$ it will be found that the images are also periodic if $a^2/b^2 = p$, where p is an integer. This condition holds for many projections of simple structures. For non-orthogonal axes with angle γ, periodic images occur if $a^2/b^2 = p$ and $\cos\gamma = (b/a)m$, where m is an integer. For smaller changes in focus, such as $\Delta f_{\mathrm{f}}/2$, an image of reversed contrast is formed (see Fig. 5.2 for the three-beam case). It is instructive to plot out the transfer function $\sin\chi(K)$ for two focus settings differing by Δf_{f} in order to confirm that these functions differ everywhere except at reciprocal lattice points.

5.9. Absorption effects

'Absorption' of the elastic electron wavefunction in HREM may refer either to the exclusion of elastic scattering from the image by the objective aperture, as discussed in Section 6.1 and Appendix 2, or to the depletion of the elastic wavefield by inelastic scattering events in the specimen, which we consider here. The agreement between computer-simulated images and experimental images for the thicker crystals ($t > 40$ nm) for which the effects of absorption are not negligible has generally been poor. This problem therefore remains as an important challenge for the future. Magnesium oxide smoke forms in perfect cubes of sub-micron dimensions, providing 90° wedges when viewed along [110] on which the thickness is accurately known from the geometry at each image pixel. In addition, there have been several measurements of the electron absorption coefficients for this material. It would therefore appear to provide an excellent opportunity for testing HREM theory quantitatively in the presence of absorption (see Appendix 5). Equation (5.47) gives the lattice image in terms of $\gamma_R^{(j)}$ and $\gamma_{\text{Im}}^{(j)}$ which can be determined if $U_\mathbf{g} + iU'_\mathbf{g}$ is known. In the multislice approach (using the crystallographic sign convention), the Fourier coefficients $v_\mathbf{g}$ are replaced by complex coefficients $v_\mathbf{g} - iv'_\mathbf{g}$ (see eqn 5.36). Here $v_\mathbf{g}$ are the Fourier coefficients of the 'real' potential $\phi_R(x, y)$, while $v'_\mathbf{g}$ are the coefficients of the 'imaginary' potential $\phi_i(x, y)$, given in eqn (6.2). These real-space potentials may not possess inversion symmetry, in which case both $v_\mathbf{g}$ and $v'_\mathbf{g}$ will be complex. Values of $v'_\mathbf{g}$ (or, equivalently, $U'_\mathbf{g}$) are spread throughout the literature (see Reimer (1984) and Spence and Zuo (1992) for a summary), and measurements exist for silicon, magnesium oxide, germanium, and a few other materials. Theoretical calculations showing the strong effects of absorption on lattice images in 'thick' crystals have been published by Pirouz (1979). However, there are some important difficulties with the direct use of these coefficients for HREM image simulation, which we now discuss.

Microdiffraction patterns from regions of thicker crystal used to form high-resolution images show a considerable amount of scattering in non-Bragg directions, even from perfect crystals. In the absence of defects, this scattering arises from inelastic scattering of the fast electrons by the various crystal elementary excitations, chiefly plasmons, phonons, inner-shell excitations, and valence-band 'single-electron' excitations (see Section 6.6). The question arises of what contribution, if any, these scattered electrons make to a high-resolution image.

The usual assumption has been that electrons that lose energy ΔE in traversing the specimen will, if the objective lens is correctly focused for the elastic or 'zero loss' electrons, be out of focus by an amount $C_c(\Delta E/E_0)$ due to the chromatic aberration of the lens, and so contribute only a slowly varying blurred background to the HREM image, as indicated by eqn (2.34). In addition to contributing a low-resolution background to HREM images, a second effect of inelastic scattering is to deplete the elastic wavefield, and so modify the amplitudes and phases of the beams used for image formation. Several historically distinct bodies of literature have a bearing on the problem—these include the theory of the Debye–Waller factor, the justification for the use of an 'absorption' or 'optical' potential in electron diffraction, and work on the effects of absorption on diffraction contrast images.

The Debye–Waller factor expresses the attenuation of the elastic Bragg beams due to the excitation of phonons in a crystal according to the kinematic theory. When used in the dynamical theory, it does not affect the normalization of intensities or cause 'absorption'.

The use of a complex optical potential to describe depletion of the elastic wavefield by inelastic processes in the dynamical theory (absorption) was justified by Yoshioka (1957). It is not inconsistent to include both a Debye–Waller factor and a contribution to v'_g from phonon excitations in dynamical calculations for the elastic wavefield (Ohtsuki 1967). In the quantum picture, the Debye–Waller factor describes the modification to the real parts V_g due to virtual inelastic scattering which results from the creation and annihilation of phonons in a single interaction (at the same time). The imaginary part of the potential v'_g however, describes the redistribution of elastic scattering which results from 'real' inelastic phonon scattering through non-Bragg angles. In the classical picture, the Debye–Waller factor describes the time-averaged periodic crystal potential responsible for the purely elastic Bragg scattering in Bragg directions. For a review of the electronic contribution to the absorption potential, see Ritchie and Howie (1977).

For isotropic materials, the Debye–Waller factor M is incorporated by replacing the F_g of eqn (5.19) by

$$F'_g = F_g \exp(-2\pi^2 B g^2) \tag{5.54}$$

where the constant $B = <u^2> = M/(8\pi^2)$ describes the temperature-dependent vibration of atoms with displacement amplitudes u. Tabulated values of B can be found in the International Tables for X-ray Crystallography. Since B increases with temperature, hot atoms will appear larger in an atomic-resolution image.

In principle, the 'classical' method of treating phonon absorption effects in HREM is straightforward. The interaction time of a fast electron traversing a thin crystal is much shorter than that of any of the inelastic crystal excitations. Therefore a time-average is required of the intensity of the dynamical, many-beam, aperture-limited image for every instantaneous configuration of the crystal potential. This requires a complete description of all the atomic displacements due to the thermal motion of the atoms. The time-averaging must be performed on the dynamical image, rather than on diffracted beams or the crystal potential. A second aspect of such a calculation is the estimation of the distribution of diffuse inelastic scattering due to electronic excitations, from which the background due to those inelastic electrons which pass through the objective aperture could be estimated. The contribution of these electronic processes to the imaginary part of the optical potential must also be determined. The result would give the high-resolution image observed without the use of an energy filter, that is, the image due to both elastically and inelastically scattered electrons. This is the approach described in Chapter 8 for STEM, and given in detail in Kirkland (2010).

Since, due to the presence of chromatic aberration, only electrons within a small range of energies around the accelerating potential will contribute to the in-focus HREM image, we may think of lattice imaging as a kind of energy filtering for elastically scattered electrons. Then two effects must be considered—the modification to the amplitudes and phases of the elastic beams due to inelastic scattering, and the contribution to the low-resolution background from inelastically scattered electrons which pass through the objective aperture. The existence of virtual inelastic scattering and exchange effects between the fast electron and crystal electrons also makes a small contribution to V_g (the elastic potential); however, Rez (1977) has shown that these effects are very small. We note in passing that these exchange effects between the fast electron and crystal electrons are fundamentally different from the exchange effects included in band-structure calculations, which take account of exchange amongst crystal electrons. Thus, the potential measured by electron diffraction, while sensitive to bonding effects, is not the same as that used for band-structure calculations. Diffraction-contrast imaging (where images are formed using a small aperture around a single Bragg beam) is concerned with the effect of inelastic scattering on a particular Bragg beam, which may then be related to a diffraction contrast image formed from this beam alone through the column approximation—these measured absorption coefficients, which depend on aperture size, are not relevant to HREM imaging.

In summary, it may be said that the few reliable values of $V'_\mathbf{g}$ useful for HREM calculations are those which have been measured experimentally by convergent-beam or similar diffraction methods (see Spence and Zuo (1992) for a summary). Empirical relationships, such as

$$V'_\mathbf{g} = V_\mathbf{g}(A|\mathbf{g}| + Bg^2) \tag{5.55}$$

have also been used, where A and B are fitting parameters, and $\mathbf{g}$ is a reciprocal lattice vector. Calculated values (for phonons) can be found in Bird and King (1990).

The second step in correcting HREM image calculations for the effects of inelastic scattering is to estimate the low-resolution background contribution to the images from inelastically scattered electrons which pass through the objective aperture. This problem has received little attention, no doubt owing to the difficulty of determining the detailed angular redistribution of multiple inelastic scattering within the objective aperture. In practice a simpler approach would be to include as background an appropriate fraction of the energy lost from the elastic wavefield as a consequence of the use of an absorption potential.

The situation is less clear for phonon scattering. Transmitted electrons which lose energy owing to the creation of phonons in a crystalline sample give up very small amounts of energy (less than 1 eV even for multiple losses) and so remain 'in focus', and, in addition, the angular distribution of phonon-loss electrons shows a broad maximum for $u \sim 2$ Å^{-1}, in addition to peaks at the Bragg positions. Most of this broad maximum falls within the resolution limit of modern HREM instruments. The question therefore arises of what contribution this scattering makes to HREM images. It has been shown (Cowley 1988) that phonon-loss electrons (if they could be filtered out separately) would produce an atomic-resolution image. This may be relevant to the 'Stobbs factor' problem (Section 5.13).

Lattice images formed from inner-shell excitations are discussed in Spence and Lynch (1982); the important point is that elastic scattering of energy-loss electrons cannot be suppressed. This means that if a crystal containing two species is imaged using an energy filter tuned to the inner-shell peak of one species, the resulting image may also show the second species, due to elastic scattering from the second species of beam-electrons which are inelastically scattered by the first species.

5.10. Dynamical forbidden reflections

The observation of intense Bragg reflections whose kinematic structure factors are zero will be familiar to most electron microscopists. For example, in the diamond structure the (200) reflection, for which $F_{200} = 0$ (see eqn 5.19), is commonly observed as a result of double scattering, involving the $(11\bar{1})$ and $((1\bar{1}1)$ reflections. All these reflections lie in the [011] zone axis pattern, and we note that $(11\bar{1}) + (1\bar{1}1) = (200)$.

It has been shown, however, that for certain incident beam directions, certain reflections remain absent, despite the effects of multiple scattering, *for all thicknesses and accelerating voltages* (Cowley and Moodie 1959, Gjønnes and Moodie 1965). The reflections which remain absent are those whose absence results from the existence of screw or glide translational symmetry elements, and are not 'accidental' absences such as those due to atoms which lie on special positions in the lattice, or arise from a non-primitive choice of unit cell. The incident beam directions which cause the continued absence of these space-group-forbidden reflections are given in this section, and include the case, important for lattice imaging, in which the beam runs normal to a screw axis or glide plane. This powerful theoretical prediction was first confirmed experimentally by Goodman and Lehmpfuhl (1964) and Fujime *et al.* (1964). It forms the basis of a general method of crystal space-group determination by convergent-beam electron diffraction (CBED; see Section 13.3). In this brief summary, we consider only the results in the projection approximation for which only reflections in the zero-order Laue zone (ZOLZ) are considered. For the general three-dimensional result, the reader is referred to the classic paper of Gjønnes and Moodie (1965).

The importance of this result for HREM lies in the fact that only if these space group-forbidden reflections do not contribute to the image will the projected potential seen in the image reveal the true crystal structure. Yet the angular range of incident beam directions for which these reflections remain absent is rather small (about 0.1°, depending on crystal thickness and structure), so that small errors of alignment may result in misleading images of the crystal structure.

In the projection approximation, the only plane groups corresponding to a two-fold screw axis or glide plane normal to the beam are pg, pmg, or pgg in the rectangular system or $p4g$ in the square system. Screw axes other than two-fold may be thought of as giving rise to accidental forbidden reflections in projection. To expose the principle of the effect, we consider the simple case of a crystal with projected symmetry pmg, of which CdS forms an example if the electron beam is normal to the c-axis. CdS is hexagonal, with a 6_3 screw axis along **c**. In projection, this screw axis becomes a glide line. Because of this translational symmetry element, the structure factors have the signs shown in Fig. 5.6(a), with

$$F_{hk} = (-1)^k F_{\bar{h}k} \tag{5.56}$$

and

$$F_{0k} = 0 \quad \text{for} \quad k = 2n+1 \tag{5.57}$$

(see Stout and Jensen (1968)). Here the index k is parallel to **c**, and the beam direction is normal to the page. When these conditions are imposed on eqn (5.33), it will be found that the resulting series of terms can be arranged in pairs with products of F_{hk} of equal magnitude and opposite sign. This occurs only if the excitation errors also preserve certain symmetry, in particular we require that K (the projection of the incident electron wavevector in the ZOLZ) lie on the bold cross shown in Fig. 5.6(a). Here, either the (010) reflection is at the Bragg condition or the beam lies in the plane containing the $(0k0)$ systematics and the

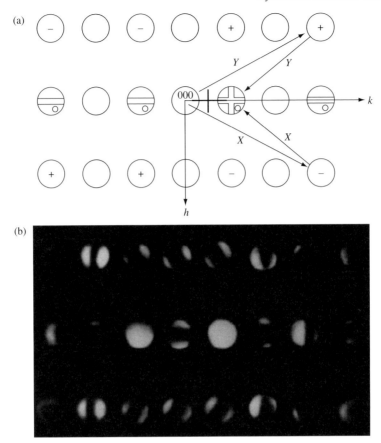

Figure 5.6 The principle of dynamical extinction. For a screw axis along k normal to the beam the structure factors have the signs indicated. Those containing O are forbidden. Unmarked discs have equal structure factors. Double scattering along X into the (010) reflection is exactly cancelled by that along Y. The bold cross indicates the locus of K along which extinction is expected in the projection approximation. (b) Experimental convergent-beam electron diffraction pattern from CdS showing dynamical extinctions. The beam is approximately normal to the screw axis (which runs across the page) along which every second reflection remains absent, despite multiple scattering.

zone axis. Physically, the extinction arises because of the exact cancellation (by destructive interference) of all pairs of multiple-scattering paths (such as X and Y in Fig. 5.6(a)) which are related by crystal symmetry. Thus, near the axial orientation (and within the projection approximation) the experimental convergent beam discs F_{0k} appear as shown in Fig. 5.6(b), for a two-fold screw axis normal to the beam (Goodman and Lehmpfuhl 1964).

For a glide plane normal to the beam, again in the projection approximation, the period of the projected potential will be halved, leading to the extinction of alternate reflections in the ZOLZ plane.

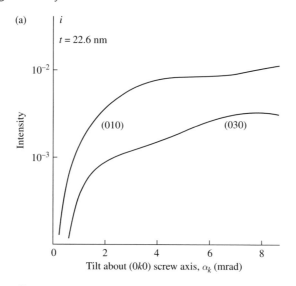
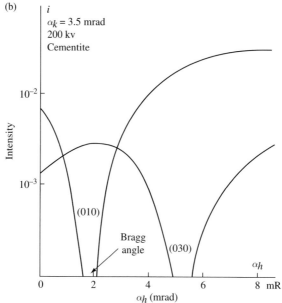

Figure 5.7 Computed Bragg beam intensities for two dynamically forbidden reflections due to a screw axis along $[0k0]$ in cementite at 200 kV as a function of α_k (in mrad) at a thickness of 22.6 nm (Fig. 5.7a) and as a function of α_h (Fig. 5.7b). Here α_k is the angular misalignment from the exact zone axis orientation and describes a rotation about the k-axis in Fig. 5.6(a) (similarly for α_h). The angular width of the extinction band decreases with increasing crystal thickness. (From Nagakura and Nakamura (1983).)

For a glide plane which contains the beam direction, the situation (after projection of the crystal structure in the beam direction) is the same as that for the two-fold screw axis.

The effects of dynamically forbidden reflections on lattice images have been studied by several researchers. Nagakura and Nakamura (1983), for example, observed image artefacts in cementite and studied their dependence on beam misalignment. Figures 5.7(a) and (b) show the intensity of the first-order dynamically forbidden reflection in this material as a function of incident beam misalignment in the directions K_x and K_y in Fig. 5.6(b). For this structure, the incident beam must be aligned to within less than 1 mrad of the zone axis in order to produce a faithful representation of the projected crystal structure in the HREM image. A review of the effects of dynamically forbidden reflections in several materials, including TiO_2, SiC, Mo_5O_{14}, CuAgSe, and SnO_2 can be found in Smith *et al.* (1985). Figure 5.8 (taken from that work) shows clearly the artefactual 4.6 Å fringes which appear in thicker regions of this sample unless the alignment is within about 0.2 mrad of the exact zone axis orientation. Note that it is important to distinguish the linear image fringes due to the (0, 2) interaction (see Fig. 5.6(a), with (0, 1) dynamically forbidden) from the (0, 1) (0, $\bar{1}$) interaction (producing fringes of the same period) in thicker crystals due to non-linear imaging effects.

The more subtle effects of translational symmetry elements whose translational component lies parallel to the beam direction are described in Ishizuka (1982). Many experimental examples, including three-dimensional effects, can be found in the excellent compilation of convergent-beam electron diffraction (CBED) patterns produced by Tanaka and Terauchi (1985).

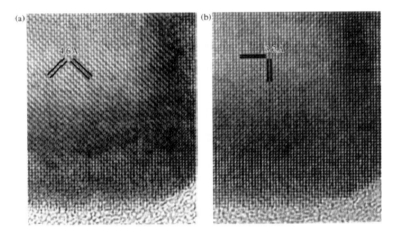

Figure 5.8 A lattice image of a wedge-shaped thin crystal of rutile recorded at 500 kV in the [001] projection. Artefactual 4.6 Å fringes can be seen in thicker crystal, resulting from a slight error in alignment. The contribution of these dynamically forbidden reflections has been eliminated in (b), following an alignment correction of less than 0.12 mrad. (From Smith *et al.* (1985).)

5.11. Relationship between algorithms. Supercells, patching

Computational algorithms for solving the dynamical electron diffraction problem have mainly been based on the Bloch wave method (see Section 5.7) or the multislice method (Section 5.6). Other methods include the Feynman path-integral formulation (Jap and Glaeser 1978, Van Dyck 1978) and numerical evaluation of the polynomial solution (eqn 5.33). The method of Sturkey (1962) is based on the expansion (from eqn 5.50)

$$S = \exp(2\pi i \mathbf{A} t) = [\exp(2\pi i \mathbf{A} t/n)]^n = [\exp(2\pi i \mathbf{A} \Delta z)]^n \tag{5.58}$$

where n is chosen sufficiently large to ensure that the series expansion of $\exp(i\mathbf{A}\Delta z)$ converges rapidly. The differential equation method of Howie and Whelan (see Hirsch *et al.* (1977)) can be obtained by differentiating the fundamental equation given above, while the multislice iterative technique is obtained from this equation by writing

$$\mathbf{A} = \mathbf{V} + \mathbf{T} \tag{5.59}$$

where $\mathbf{T}$ is a diagonal matrix containing only the excitation errors. Since $\mathbf{V}$ and $\mathbf{T}$ do not commute, a theorem due to Zassenhaus must be used to obtain the multislice iterative formula from this expansion as described by Goodman and Moodie (1974) in more detail. The formal relationship between all these algorithms is summarized in Fig. 5.9.

It is often instructive to consider two limiting cases of the fundamental eqn (5.50) when attempting to understand dynamical scattering effects. For $\mathbf{T} = 0$ (all excitation errors zero) a term-by-term comparison of the series expansion of $\mathbf{S}$ with that of the Fourier transform of the phase-grating expression (eqn 3.30) shows that the observed column of $\mathbf{S}$ contains just the Fourier coefficients of $\psi_e(x, y)$. Thus, the simple phase-grating approximation is recovered from the fundamental equation as the accelerating voltage tends to infinity ($S_\mathbf{g} \to 0$), in which limit $\sigma \to 2\pi/(hc)$, the Compton interaction parameter (c is the speed of light, h is Planck's constant). A term-by-term comparison of this same series with the Born series for electron scattering also confirms that the phase-grating approximation (known elsewhere as the Molière high-energy approximation) provides an exact summation for the series in the limiting case where the component of the scattering vector in the beam direction is zero. (The Born series does not contain the experimentally verified 90° phase shift between scattered and unscattered waves, due to differences in boundary conditions.)

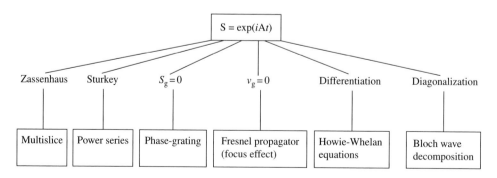

Figure 5.9 Algorithms and approximations in dynamical theory.

The phase-grating approximation (eqn 5.22) also gives an approximation for Pendellösung effects, since the Fourier coefficients of $\psi_e(x, y)$ are periodic in t, unlike the kinematic expression (eqn 5.16), which predicts intensities proportional to t^2 for $S_g = 0$. By varying the direction of projection in eqn (3.30), the phase-grating approximation can also be seen to give an estimate of the angular dependence of diffracted beams (their 'rocking curve').

The second limiting case of the fundamental equation is obtained when $\mathbf{V} \to 0$, in which case, since $\mathbf{A}$ is diagonal, the observed column of $\mathbf{S}$ contains entries

$$\phi_g = \exp(2\pi i S_g t) = \exp(-2\pi i \lambda u_g^2 t / 2) \tag{5.60}$$

since

$$S_g = -\lambda u_g^2 / 2 \tag{5.61}$$

in the zone axis orientation. Comparison with eqn (3.24) shows that this phase shift corresponds to free-space Fresnel propagation over a distance equal to the crystal thickness, as expected in the 'empty lattice' approximation of zero crystal potential.

For perfect crystals, it should be noted that the images will be periodic functions of the focus setting (see Section 5.8). Therefore, it is only necessary to compute images through a single 'Fourier image' period in order to obtain all possible images for a particular thickness. In addition it can be shown that dynamical solutions are functions only of the product $\lambda t = X$ at non-relativistic energies ($V_0 < 100$ kV). All (non-relativistic) diffraction patterns for the same value of X are identical for a given structure, so that halving the specimen thickness is equivalent to halving the electron wavelength. Specimens thus 'look' thinner to the fast electron as its energy is increased. This property has been used to directly invert soft X-ray scattering patterns (which also suffer from multiple-scattering perturbations) to a sample charge-density map, by recording patterns at two different beam energies, which mimic the effect of a small change in thickness (Spence 2009).

The articles by Goodman and Moodie (1974) and Self et al. (1983), and the books by Head et al. (1973), Cowley (1995), Hirsch et al. (1977), Stadelmann (1987), Spence and Zuo (1992), Chen and Van Dyck (1997), and Kirkland (2010) contain ample information to assist programmers. The text by Kirkland (2010) refers to web support, the text by Spence and Zuo (1992) contains listings of both Bloch wave and multislice codes, while the EMS program described by Stadelmann (1991) has become very popular. A web-based program is available for simulations hosted by J. M. Zuo at <http://emaps.mrl.uiuc.edu/> where experimental parameters can be entered directly on the web and a simulation program will run remotely, with output returned as a pdf. This is an excellent teaching aid, whereby many students can learn simultaneously during lectures from the same program run on the web.

Computing methods based on the multislice technique are most efficient for the simulation of large unit-cell crystals and thin specimens. Matrix methods, which depend on the diagonalization of the structure matrix $\mathbf{A}$, are most efficient for thick, perfect specimens of the small-unit-cell crystals. All these numerical methods provide solutions to the same fundamental equation, and have close similarities to the numerical techniques used in solid-state electron band-structure calculations. Several workers have demonstrated the equivalence of the various computing techniques. In the multislice technique, most of the computing time is occupied performing a numerical convolution using fast Fourier transforms.

For the multislice method the computing time per slice for n beams is proportional to $n \log_2 n$, whereas for the Bloch wave method, matrix diagonalization times are proportional to n^2. Thus, the multislice method is faster for $n > 16$ (Self et al. 1983). Computing space increases roughly as n^2 for the diagonalization method, and as n for the multislice method.

An important question which arises in image simulations concerns the lateral spread of the electron wavefunction. For the simulation of images of defects it is common to use the method of periodic continuation (see Section 5.11), as shown in Fig. 5.10. We wish to know how large a cell is needed for a given defect. Thus we wish to know the extent to which the dynamical wavefunction at A in Fig. 5.10 (immediately below a defect) depends on the crystal potential at B. Within the weak phase object approximation (eqn 3.31) there is clearly no spreading; however, in thicker crystals this may not be so. This problem has been studied theoretically in two main ways. Firstly, dynamical calculations for a STEM probe wavefunction incident at E have been made for various thicknesses. If the width of this wave packet at E is small, then its width at A gives a measure of the lateral spreading of the wavefunction. The results of these calculations (Spence 1978, Humphreys and Spence 1979, Marks 1985, Etheridge et al. 2011) show that, in the axial orientation for a specimen whose thickness is less than $t = 50$ nm, this broadening is very small and given very approximately by $\Delta x = 2t \tan \theta_c$. A more realistic estimate for imaging calculations (based on the 'Takagi triangle' construction; see Hirsch et al. (1977)) might be $\Delta x = 2t \tan \theta$, where θ is twice the largest Bragg angle which contributes to the dynamical diffraction process. Alternatively, we might argue that the contribution at A from a spherical wave source at B will be negligible if the lateral distance between A and B exceeds the width of the first Fresnel fringe formed at 'defocus' Δz. (This gives an estimate of the effects of free-space propagation alone.) From eqn (10.7) we have, approximately

$$\Delta x = (\lambda \Delta z)^{1/2} \tag{5.62}$$

for the propagation effect. In practice this equation is found to overestimate Δx. A more straightforward method of analysing the spreading of the wavefunction is to perform dynamical image simulations for the same defect within 'superlattices' of varying dimension CD. The dependence of the wavefunction at the point A can then be studied as function of the supercell size CD and the thickness t. The results of extensive computational trials of this

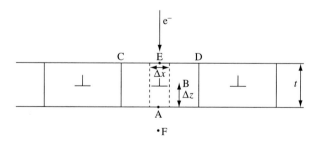

Figure 5.10 The principle of periodic continuation. Each identical 'supercell' contains the defect of interest. Because the wavefunction at A is almost independent of the potential at B, dynamical images can be computed in 'patches'.

type (Spence et al. 1978, Matsuhata et al. 1984) show that this broadening is typically less than 0.4 nm for thicknesses of less than 50 nm. However, it will depend, weakly, on the structure and atomic number of the crystal, on the accelerating voltage, and on the order of the image Fourier coefficients considered.

In summary, this important qualitative result indicates that the axial, dynamical image wavefunction on the exit face of a crystal is locally determined by the crystal potential within a small cylinder whose axis forms the beam direction and whose diameter is always less than a few angstroms for typical HREM conditions. This is essentially a consequence of the forward-scattering nature of high-energy electron diffraction. Thus, for the purposes of finding a few particular atomic column positions, a rather small artificial supercell can be used, with CD typically about 2.5 nm if $t < 20$ nm. This argument only applies to the axial orientation.

The importance of this result is that for the purposes of the image simulation of defects at interior points of a supercell such as A, there is no need to obtain perfectly smooth periodic continuation of the crystal potential at the boundaries C and D. An abrupt discontinuity of the potential at C will only influence the dynamical image within a few angstroms (laterally) of this point. *It follows that images of large defects can be simulated in 'patches' using separate computations.* The image patches may subsequently be joined together, if sufficient allowance has been made at the borders for discontinuities in the potential. For example, Fig. 8 of Olsen and Spence (1981) actually consists of three panels, resulting from separate computations, which have been joined at overlapping regions. But these techniques of patching and lack of smooth periodic continuation cannot be used for computation of electron diffraction patterns.

At large focus defects, the most important broadening effect will occur beyond the exit face of the crystal, between A and F in Fig. 5.10. It has therefore been suggested (Matsuhata et al. 1984), that, since most of the computer time is devoted to the multislice, it may be possible to use a small supercell for the multislice calculation within the crystal, followed by an image-synthesis calculation referred to a much larger supercell to account for propagation between A and F. A knowledge of the crystal structure surrounding the defect is assumed.

5.12. Sign conventions

Since high-energy electron diffraction is described by the Dirac equation (or the relativistically corrected Schrödinger equation), all the signs in the scattering and image formation process are fixed by this equation if the time dependence is included. Most modern quantum mechanics texts write the time-dependent Schrödinger equation as

$$(h^2/2m)\nabla^2\psi = -ih\frac{\mathrm{d}\psi}{\mathrm{d}t} \tag{5.63}$$

with the time-dependent solution

$$\psi_\mathrm{t}(\mathbf{r}) = A\exp(+2\pi i \mathbf{k}\cdot\mathbf{r} - i\omega t) \tag{5.64}$$

Note that $\psi_\mathrm{t}^*(\mathbf{r})$ is *not* a solution of eqn (5.63), since this would require an imaginary wavevector. If the minus sign on the right-hand side of eqn (5.63) is replaced by a plus sign, $\psi_t^*(\mathbf{r})$ becomes a solution. This form is used in some texts. For the time-independent Schrödinger equation, either

$$\psi_s(\mathbf{r}) = A \exp(2\pi i \mathbf{k} \cdot \mathbf{r}) \qquad (5.65)$$

or $\psi_s^*(\mathbf{r})$ are suitable solutions. While most textbooks give the Dirac equation and the Schrödinger equation in forms which require $\psi_t(\mathbf{r})$ as the solution, others (e.g. Jauch and Rohrlich 1955) adopt forms for these equations which require $\psi_t^*(\mathbf{r})$ for the positive-energy, electron plane-wave states. In this book (with the exception of Section 5.7 on Bloch wave methods), ψ_s^* is used throughout. Section 5.7 uses $\psi_s(\mathbf{r})$, for conformity with most texts on quantum mechanics. The theory of inelastic electron scattering (for which a time-dependent Schrödinger equation may be used) is not discussed in detail in this book (see Section 6.6).

The situation is complicated by the need for conformity with the established conventions of X-ray crystallography, as codified in the International Tables for Crystallography. This is desirable since most modern HREM work involves crystallographic problems and one often wishes to compare the results of X-ray and electron work, to use existing X-ray computer software, or to make use of the extensive tabulations of structure factors given in the International Tables.

The X-ray tradition has been based, like this book, on the use of $\psi_s^*(\mathbf{r})$ (von Laue 1960; International Tables for Crystallography, Vol. 1, p. 353), and indeed this form appears to have been the most common in the early days of electron and X-ray diffraction, out of which the new quantum mechanics grew. Either sign convention, if used consistently, will produce identical images. These sign conventions are summarized in Table 5.1 (from Saxton et al. (1983), with corrections), and it is therefore proposed that these be known as the 'quantum-mechanical convention' (for $\psi_s(\mathbf{r})$) and the 'crystallographic convention' (for $\psi_s^*(\mathbf{r})$). Note that in *both* conventions in this book it is also assumed that a weakened lens (underfocus, first Fresnel fringe bright, toward Scherzer focus) is represented by a *negative* value of Δf. (The opposite convention has been used in some papers.)

An instructive case concerns the comparison of stacking faults imaged by X-ray topography and by electron microscopy. The various sign conventions used in both fields have been reviewed by Lang (1983). Here the distinction between extrinsic and intrinsic faults depends on the contrast of the first thickness fringe, which may be reversed by a sign error in calculations.

5.13. Image simulation, quantification, and the Stobbs factor

Most image-simulation software programs consist of three sections: a program to compute the Fourier coefficients of crystal potential, a dynamical scattering program, and an image-simulation program. The first of these (based on eqn 5.18) may be checked by hand (note the sign convention required in Table 5.1 and eqn 5.19). The second program must reproduce experimentally observed thickness fringes and convergent beam patterns (see Section 13.3). An approximate check can be made by running the program for a small-unit-cell crystal under 'two-beam' conditions (Bragg condition satisfied for beam g). The approximately sinusoidal variation of the intensity of beam g with thickness should have a 'period' within about 15% of that given by $\xi_\mathbf{g}$ in eqn (5.17). If no resolution limits are imposed, the imaging

Table 5.1 Sign conventions used in HREM.

	Quantum-mechanical convention	Crystallographic convention (used in this book)
Free-space wave	$\exp\{+2\pi i(\mathbf{k}\cdot\mathbf{r}-\omega t)\}$	$\exp\{-2\pi i(\mathbf{k}\cdot\mathbf{r}-\omega t)\}$
Transmission function	$\exp\{+i\phi_p(x)\cdot\Delta z\}$	$\exp(-i\sigma\phi_p(x)\cdot\Delta z)$
Phenomenological absorption	$\phi_p(x)\to\phi_R(x)+i\phi_i(x)$	$\phi_p(x)\to\phi_R(x)-i\phi_i(x)$
Propagation function	$\exp\{2\pi i S_g(u)\cdot\Delta z\}$	$\exp\{-2\pi i S_g(u)\cdot\Delta z\}$
Wave aberration function (modification of diffracted wave by objective lens)	$\psi'(u)=\psi(u)\exp\{-\pi\lambda u^2 \times(\Delta f+\tfrac{1}{2}\lambda^2 C_s u^2)\}$	$\psi'(u)=\psi(u)\exp\{+i\pi\lambda u^2 \times(\Delta f+\tfrac{1}{2}\lambda^2 C_s u^2)\}$
Object to diffraction space	$\int \psi(\mathbf{r})\exp\{-2\pi(\mathbf{u}\cdot\mathbf{r})\}\,d\mathbf{r}$	$\int \psi(\mathbf{r})\exp\{+2\pi i(\mathbf{u}\cdot\mathbf{r})\}\,d\mathbf{r}$
Diffraction to object space/reciprocal to real	$\int \psi(\mathbf{u})\exp\{+2\pi i(\mathbf{u}\cdot\mathbf{r})\}\,d\mathbf{u}$	$\int \psi(\mathbf{u})\exp\{-2\pi(\mathbf{u}\cdot\mathbf{r})\}\,d\mathbf{u}$
Structure factors	$\Sigma_j f_j \exp\{-2\pi i(\mathbf{u}\cdot\mathbf{r}_j)\}$	$\Sigma_j f_j \exp\{+2\pi i(\mathbf{u}\cdot\mathbf{r}_j)\}$

σ, electron interaction constant = $2\pi me\lambda/h^2$; m, (relativistic) electron mass, λ, electron wavelength; e, (magnitude of) electron charge; h, Planck's constant; Φ_p, crystal potential averaged along beam direction (positive); Δ_z, slice thickness; ϕ_i, absorption potential (positive); S_g, excitation error (negative for reflections lying outside the Ewald sphere); Δf, defocus (negative for underfocus); C_s, spherical aberration coefficient; f_j, electron scattering factor for jth atom; $\mathbf{r}_j$, position of jth atom.

program should give images in very thin crystal ($t < 20$ Å) which agree exactly (apart from a uniform background) with a map of the projected crystal potential as given by the first program. It is worth checking that the phase change of 90° introduced by the lens (for moderate spatial frequencies) at the Scherzer focus has the *same* sign (i.e. adds to) that which arises from scattering (the factor i in eqn 3.33) Thus a total phase shift of 180° (not 0°) is required for phase contrast. Finally, the images at all thicknesses should have the same symmetry as the projected potential, unless a resolution limit has been imposed, in which case the image symmetry may be higher.

The ultimate test of an image-simulation procedure is its ability to reproduce experimental results. The most convincing image simulations are those in which computed and experimental images are matched as a function of some parameter, usually thickness or defocus. It is desirable to determine these parameters independently. The focus setting may be determined by the methods described in Section 10.1. Specimen thickness may be determined by using the image width of an inclined planar fault known to lie on a particular crystallographic plane, or from the external crystal morphology, or by using the CBED method for a known structure. For high-angle annular dark-field (HAADF)-STEM images, where coherent orders overlap, unwanted coherent interference effects may be eliminated by scanning the probe during the exposure to produce a conventional CBED pattern

(Le Beau et al. 2009). Perhaps the simplest case for testing a program is that of MgO smoke crystals, which form in perfect cubes. Images taken with the beam parallel to [110] therefore allow an exact thickness calibration, so that the 'turning points' (at which the fringe contrast reverses) can be compared with the results of calculation over a range of thickness (Hashimoto 1985, O'Keefe et al. 1985) (see Appendix 5). These may depend sensitively on the absorption potential used in a thicker crystal. Ionic scattering factors for MgO have been measured by CBED (Zuo et al. 1997).

As shown in Fig. 5.16, the image blobs may not coincide with true atom positions. An analysis of the silicon [110] case has been given by Krivanek and Rez (1980). For sufficiently thin crystals, we have seen (eqns 5.27 and 3.35) that peaks of crystal potential must coincide with corresponding peaks in the image intensity if no resolution limits are imposed. For dynamical images perturbed by multiple scattering, there is, in general, no one-to-one correspondence between the crystal exit-face wavefunction and the projected crystal potential (see the discussion of resolution in Section 10.9). A detailed analysis of this problem (Saxton and Smith 1985) shows that even under kinematic scattering conditions, lateral shifts between the position of a maximum in the projected crystal potential and that of the corresponding image 'peak' of about 0.03 nm are common at some focus settings. These results are somewhat model-sensitive; however, it is also found that the choice of focus which maximizes image contrast is not that which minimizes the lateral image shift. Dynamical calculations for model structures give broadly similar results, and we may summarize these by saying that image peak separations may vary by up to 20% as a function of thickness and focus. These calculations were all performed for the case in which the instrumental point-resolution (eqn 6.17) is comparable to the interatomic spacing. For spacings much larger than the point-resolution in very thin crystals, these image shifts will become negligible.

We note that the errors found in this study are a small fraction (about 10%) of the point-resolution of the instrument (about 0.2 nm). Just as the position of the maximum in a sine wave can be located to within a small fraction of its wavelength, so it is common in X-ray diffraction to locate atom positions to within an accuracy far smaller than the d-spacing of the highest-order reflection used in the refinement. Similarly, if it is known a priori that an image feature consists of a *single* atomic column, then the centre of this image 'peak' may be located to within a small fraction of the point-resolution. This is possible because of the interferometric nature of electron images, for which a difference in optical path length of one electron wavelength between adjacent image points may cause a contrast reversal. The problem of *distinguishing* two adjacent or overlapping columns in a crystal of unknown structure is discussed in Section 10.9.

The 'Stobbs factor' refers to a discrepancy between experimental and simulated structure images, in which simulated images are found to show a contrast level about three times greater than experimental images (Hytch and Stobbs 1994). A review of possible origins of this discrepancy can be found in Howie (2004), where attention is drawn to the importance of flexure vibrational modes in thin films, and their associated quasi-elastic scattering. In an early survey of the causes of systematic error in image matching (Boothroyd 1998), the influence of the following factors has been evaluated: scattering factors, phonon-loss electrons, contaminating layers, fringing fields, beam damage, errors in measurement of experimental parameters, backscattered X-rays generated beyond the film, secondary electrons, and the detector response functions. None of the above factors, acting alone, could account for the effect. The excellent agreement between computed and elastically filtered experimental CBED patterns (Zuo et al. 1997) suggests that the error does not lie in the diffraction computations or scattering factors, but rather in the imaging process. Thermal scattering seems a likely cause in thicker samples—at the zone axis orientation there is a strong maximum in total phonon scattering, so that the background should decline with temperature; however, the mean-free path for phonon scattering of at least 100 nm is much greater than most sample thicknesses. It is significant that this thermal diffuse scattering in rather thick samples can be fitted to theoretical simulations rather accurately in the coherent CBED diffraction patterns (Muller et al. 2001).

Additional understanding comes from off-axis HREM holography, which acts as an almost perfect elastic energy filter, so that, for recording time T, losses exceeding $\hbar/T$ $\sim 10^{-15}$ eV are excluded from images reconstructed from the side-band diffraction patterns (Van Dyck et al. 1999). Only the on-axis image (not filtered) is found to suffer from high background, so that the much higher contrast of the side-band holographic images suggests that the Stobbs factor is due to phonon scattering, which has been removed by the holographic filtering (Verbeek et al. 2011).

Since no Stobbs factor discrepancy is found for bright-field STEM imaging, which is related by reciprocity to bright-field HREM imaging, it was suggested that the CCD detector (used in HREM but not in STEM) rather than thermal diffuse scattering was responsible (LeBeau et al. 2009a). (The electron source size in STEM is reciprocally related to detector pixel size in HREM, and so must be included in STEM simulations, as discussed in Chapter 8.) This possibility was investigated in detail by Thust (2009) (however an independent measurement of sample thickness was not used). Using the new direct-injection detectors, P. Dennes (personal communication, 2012) found that a Stobbs factor does exist when using this new type of detector (see Section 9.10). Others have suggested that the effect arises because experimental HREM images are usually recorded without any objective aperture in place, whereas simulations usually include a hard aperture cut-off, leading to differences in the thermal diffuse contribution (Forbes et al. 2011). It is of interest to note that early comparisons using film detectors did observe a Stobbs factor error, suggesting that the modulation transfer function (MTF) of the CCD camera may not be entirely responsible (Boothroyd and Stobbs 1989). Recent quantitative analysis of image intensities from graphene monolayers using a CCD detector in an aberration-corrected TEM found excellent agreement between simulation and experiment without the need for Stobbs factor correction, provided the simulations were convoluted with the CCD camera response function before being compared with experiment (Meyer et al. 2011). There is an urgent need for more work on this problem, perhaps using MgO cubes viewed along a face diagonal, where sample thickness is accurately known at each point, or from analysis of graphene layers of increasing thickness.

5.14. Image interpretation in germanium—a case study

In this section we give, as a teaching exercise, a detailed study of the imaging conditions in crystalline germanium. Germanium and silicon are particularly convenient specimens for electron microscopy—they are not contaminated rapidly, they are sufficiently good conductors to prevent accumulation of specimen charge, both are brittle and so allow stable images

130   *TEM imaging of thin crystals and defects*

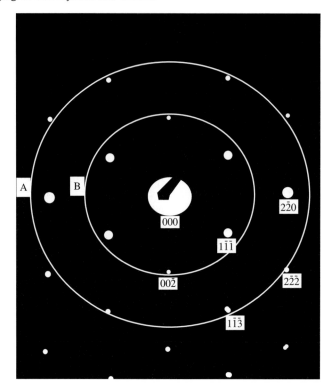

Figure 5.11 The diffraction pattern from a germanium crystal oriented with the [110] direction parallel to the electron beam. An aperture placed at A, from which a 13-beam image can be formed, contains the smallest number of beams necessary to resolve every atom in the structure.

of immobile defects to be obtained, and finally, when viewed in the [110] direction the diamond structure which they share exposes tunnels whose size lies within the resolution limit of most modern uncorrected TEM instruments. For silicon, the atomic column separations are 0.19 nm in projection down [001], 0.136 down [110], and 0.078 nm down [112]. (The point resolution, eqn (6.17), is 0.17 nm for $C_s = 0.6$ mm at 300 kV.) Specimen preparation equipment for silicon and germanium is available commercially, based on the successful early methods of Booker and Stickler (1962) (see also Goodhew (1972)). A wedge-shaped region of specimen is sought and this is oriented to give the diffraction pattern shown in Fig. 5.11. If a through-focus series of images of such a specimen is recorded using the objective aperture A shown at 100 kV, images similar to those shown in Figs 12.4(a) and (b) will be obtained, with the following characteristics.

1. The images are approximately periodic with thickness. In particular, identical images resembling the specimen structure are found at thickness increments of about 12 nm, the first occurring at 7.5 nm thickness. All these images are of 'reversed' contrast in which the atoms appear white in a positive print. Figure 5.12 shows the diamond structure, while Fig. 5.13 shows three identical computed images expected at three different thicknesses, each resembling the crystal structure. Figure 5.13(d) shows a curious image obtainable at

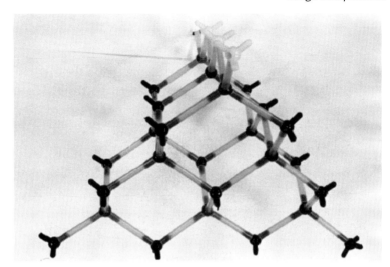

Figure 5.12 A model of the diamond structure viewed approximately in the [110] direction. The white 'atoms' form a line in the [110] direction which runs down the exposed six-sided tunnels. The positions of the columns of atoms seen in this projection coincide with those shown in the computed micrograph in Fig. 5.13.

certain thicknesses, in which the atoms appear to have a hole in their centre. (These annular single-atom images have also been observed by H. Hashimoto in crystals of gold.) The image appearance is a consequence of multiple scattering and does not reveal the internal structure of the atom! Images of conventional contrast from germanium crystals in which the atoms appear dark in a positive print are only possible at thicknesses of less than 3 nm at 100 kV accelerating voltage.

The occurrence of images which repeat with increasing thickness has been observed experimentally in other materials (Fejes *et al.* 1973) and is found to occur only in certain structures illuminated by electrons of a particular wavelength. The effect indicates that in these special cases all beams oscillate with the same periodicity in depth through the crystal. The special 'tuned' accelerating voltages which produce this effect are analysed in Glaisher and Spargo (1985). An explanation has also been given using the methods of electron channelling theory (Fujimoto 1978). In the simple two-beam case (and cases reducible to it), we see that all beams have the same Pendellösung depth period. Thus, we can expect periodic imaging (in thickness) for the many simple crystal structures for which only two Bloch waves are strongly excited, or bound within the potential well.

2. The images are periodic with focus. That is, as predicted by the theory of Fourier images (Section 5.8), identical images are found at focus settings $\Delta f \pm 2nd^2/\lambda$, where n is an integer, Δf is an arbitrary focus setting, and d is the large unit-cell dimension normal to the electron beam. For germanium, $d = 0.5658$ so that at 100 kV identical images of perfect crystal recur at focus increments of 173 nm on either side of the Scherzer (or any other) focus (see Section 6.2). This periodicity with focus greatly reduces the labour of image calculation, since a computed series of images covering a range of 173 nm therefore contains all possible images. A further simplification arises since, over this range, only four distinct

132 TEM imaging of thin crystals and defects

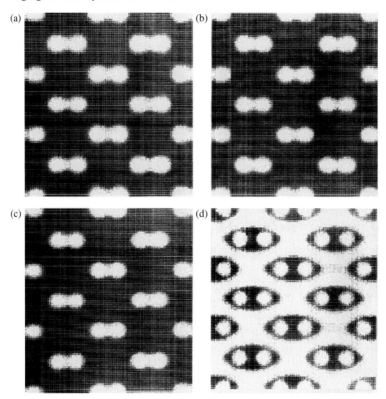

Figure 5.13 Similar images predicted for three different thicknesses of germanium. All are 13-beam images computed for $C_s = 0.7$ mm and $\Delta f = -60$ nm at 100 kV accelerating voltage. The thicknesses are (a) $t = 7.6$ nm, (b) $t = 18$ nm, and (c) $t = 30.8$ nm. (d) The annular type of image found at certain thicknesses ($t = 3.6$ nm in this case). The annular atom image shape is a consequence of multiple scattering in the specimen. A similar effect is seen in the experimental image shown in Fig. 5.14.

images of high contrast appear. These would be likely to be selected by a microscopist in practice and are shown in Fig. 5.14. Two of these resemble the crystal structure; one of them is displaced by a half a unit cell from the true atom positions. These two images recur several times within the Fourier image period—in all, there are seven different focus settings within the range $\Delta f = +26.5$ nm to $\Delta f = -146.5$ nm which give faithful images of the crystal structure similar to either Fig. 5.15(a) or (b).

The theory of Fourier images was worked out in detail by Cowley and Moodie (1960) for electron microscopy. Fourier images have not often been observed in complex oxides because, with d large, the Fourier image defocus period exceeds the range of focus over which partial coherence effects allow sharp imaging (see Section 5.2). In practice, Fourier images are familiar to any microscopist who has noticed that simple axial three-beam lattice fringes fade and recur as the focus setting is adjusted—several focus settings appear to give equally sharp fringes. It is important to emphasize that defects are not be imaged faithfully in Fourier images, as shown in Fig. 5.16.

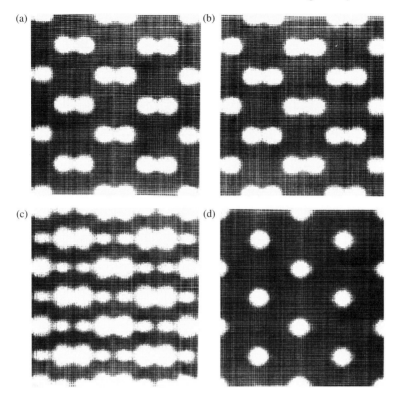

Figure 5.14 The four important image types seen in any through-focus series of crystalline germanium taken under the following conditions: $C_s = 0.7$ mm, $V_0 = 100$ kV, 13 beams included in the objective aperture. The specimen thickness is close to any of the thicknesses shown in Figs 5.13(a)–(c). Only (a) and (b) are a good representation of the crystal structure. The focus values are: (a) -60 nm, (b) $+26.5$ nm, (c) -16.8 nm, and (d) -68.0 nm.

The general result for predicting the occurrence of Fourier images is that identical images of a periodic object are formed at successive focal increments Δf_0 satisfying

$$\Delta f_0 = 2na^2/\lambda \quad \text{and} \quad \Delta f_0 = 2mb^2/\lambda$$

where a and b are the orthogonal two-dimensional unit-cell dimensions and m and n are integers. In many crystallographic projections with non-orthogonal primitive unit-cell vectors, a large orthogonal cell can be found satisfying $a^2/b^2 = m$ and therefore giving periodic images. For the diamond structure viewed in the [110] direction, these equations simplify, since $a = b/\sqrt{2}$, so that the first equation is automatically satisfied if the second is satisfied.

A fuller discussion of imaging conditions in the diamond structure can be found in Spence et al. (1977) and Nishida (1980) using multislice calculations and in Desseaux et al. (1977) for the Bloch wave methods. Figure 5.15 shows an image of crystalline silicon in which every

134 *TEM imaging of thin crystals and defects*

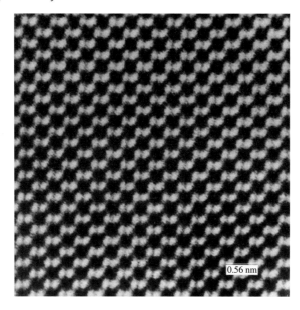

Figure 5.15 Electron micrograph of silicon viewed in the [110] direction (courtesy of K. Izui, *J. Electron Microsc.* **26**, 129 (1977)). This image was recorded using a pole-piece with $C_s = 0.7$ mm fitted to the JEOL 100C electron microscope. The objective aperture used included about 100 beams; however, many of these do not contribute to the image owing to the effects of electronic instability (see Section 3.3). Since the interatomic spacing (0.14 nm) is smaller than the point-resolution (0.28 nm) this is not a true structure image, and gives slightly incorrect atom positions as described in Section 5.13.

atom appears to be resolved; however, this Fourier image in fact shows a false structure. The image has been photoprocessed to reduce noise by superimposing photographic images displaced by one unit cell. This processing method cannot be used for the more challenging problem of defect imaging.

The point-resolution of present-day uncorrected machines is about equal to the primitive unit-cell dimensions of the simple projections of silicon, germanium, and diamond. Therefore, the special considerations of this section apply to these crystals, and the lattice images of highest contrast will occur at rather large under-focus values given by eqn (5.66) and not at the Scherzer focus, as for large-unit-cell crystals. This is a most important practical consideration. In particular, these high-contrast fine-lattice images (such as that shown in Fig. 5.15) do not reveal simply the structure of any defects present.

5.15. Images of defects and nanostructures

Most of the mechanical, electrical, chemical, and thermal properties of crystalline solids depend on the presence of point, line, and planar defects. Thus, resistance to plastic deformation depends on dislocation core and kink structure, first-order phase transformations depend on the atomic processes responsible for diffusion and interface kink and ledge motion, and the resistance of composite materials to fracture depends on the presence of intergranular phases. The nature of flux-pinning centres in superconductors is not understood at the atomic level, and only recently have the role of interfaces in ceramic superconductors and ionic conductors begun to be understood. The fascinating role of charge, spin, and orbital ordering in manganates with colossal magnetoresistance (CMR) has been traced to a complex microstructure strongly dependent on doping and temperature, to the existence of stripes of charge ordering and to mechanisms which have much in common with the high-T_c materials. Finally, the nanoscience boom in general has greatly increased demand for atomic-scale TEM imaging, especially in three dimensions. For nanoscale devices, an atomic-resolution image of relevant interfaces has become almost essential in publications, in order to demonstrate interfacial smoothness at the atomic scale. In this section we review recent attempts to understand materials properties in terms of the atomic mechanisms and defects revealed by HREM, with emphasis on the problems of image interpretation. For general reviews, see Buseck *et al.* (1989), Smith (1997), Van Tenderloo (1998), Spence (1999), and Shindo and Hiraga (1998), which contains a superb collection of HREM images of defects in a variety of materials. Special issues of the journals *Ultramicroscopy*, the *Journal of Microscopy*, *Acta Crystallographica A*, the *Journal of Electron Microscopy*, and *Microscopy and Microanalysis* have been devoted to the electron microscopy of particular materials systems and types of defects (interfaces, surfaces, environmental microscopy, nanoscience, etc.), which may be traced through subject indices on the web.

There has been dramatic progress in this field since the previous editions of this book, due largely to the invention of the aberration corrector discussed in Sections 2.10, 3.3, and 7.4 and the demand for atomic-scale imaging of nanostructures in three dimensions. The image responsible for the discovery of the nanotube shown in Plate 2 demonstrated that materials discovery by HREM is now possible, and the high-density integration of nanotube arrays needed for multitransistor operation now seems feasible (Park *et al.* 2012). More examples of non-crystalline structures and nanocrystals occur throughout this book, and examples of the measurement of strain within gold nanoparticles and imaging of gold nanocrystals can be found described in Section 7.5.

The interpretation of defect images must be confirmed by detailed multislice simulation—for these the method of periodic continuation is used, as described in Section 5.11. The projection problem is being overcome in various ways. For dislocation cores without kinks, straight dislocations in a plane-strain arrangement consist of straight atomic columns, since the displacements are independent of the coordinate along the beam direction, if this is parallel to the core. (Image forces near surfaces and elastic relaxation will tend to rotate partial dislocations in soft materials as they approach surfaces in order to minimize their self-energy.) In silicon, alternating segments of the shuffle and glide structure can be expected to complicate images (Louchet and Thibault-Desseaux 1987). Cores at the intersection of stacking faults lying on different planes can be expected to be straight, as for faulted dipoles and other locks. A new imaging mode based on forbidden reflections has been developed to provide view of dislocation cores and kinks from the side. For interfaces seen in projection along a line in the plane of the interface, several images are therefore desirable in projections rotated about a line normal to the interface. Annular dark-field (ADF)-STEM imaging of interfaces has advantages in reducing projection artefacts. For point defects (now visible by ADF-STEM), defects in all possible equivalent interstitial sites will be seen. The mapping of strain-fields based on HREM images has been discussed

136 *TEM imaging of thin crystals and defects*

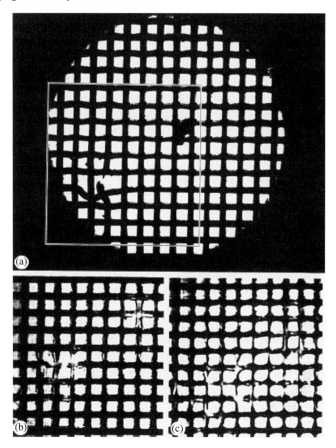

Figure 5.16 Optical images of a wire gauze of period d illuminated with laser light, recorded at three different focus settings. A crystal 'defect' has been simulated by damaging the gauze with a pin, as shown in the in-focus image at (a). (b) The first-order Fourier (Talbot) self-image around the defect at $\Delta f = 2d^2/\lambda$, while (c) shows the second-order Fourier image at twice this defocus ($d = 0.19$ mm, $\lambda = 632.8$ nm, $\Delta f = 114$ mm). The defect is masked in the Fourier images, and the damage appears to be repaired. Similar effects are seen in HREM images. At intermediate focus settings the defect alone is maximized.

in great detail—the state of the art is described in Hytch and Plamann (2001). (Strains may also be mapped to the 10^{-4} level by quantitative convergent-beam electron diffraction (QCBED), as described in Section 13.3.)

We first discuss some practical considerations of defect imaging. Figure 5.16 uses a laser-light simulation to show how the image of a crystal defect can be masked (with the damage 'repaired') if the wrong focus setting is used. In HREM, similarly, focus settings differing by $2d^2/\lambda$ (the Fourier image period) from the Scherzer focus, will perform a kind of local periodic averaging, which preserves perfectly the image of the undamaged lattice around the defect and masks the appearance of the defect. Only in the true Scherzer image is the structure of the defect revealed correctly, resolution permitting.

As beam divergence is increased, phase contrast is lost; however, this occurs much more rapidly if C_s is large. Image simulations readily demonstrate this effect, described by eqn 4.8(a). It follows that, for a given resolution, a much more intense final image can be obtained, using a larger final illumination aperture, if C_s is small. Typical values for uncorrected HREM machines are $C_s = 0.65$ mm and $\theta_c = 0.25$ mrad at 300 kV ($\lambda = 0.00197$ nm). Some experimentation with the sample height may be needed to find the minimum C_s value, which can be done using diffractogram analysis. The lowest position allowed by the maximum focusing lens current will increase both magnification and the intensity of the final image, due to an increase in the pre-field effect. For imaging at the point-resolution of the instrument, the gun-bias setting need only ensure that the resolution limit imposed by chromatic aberration (eqn 4.11) exceed that due to spherical aberration (eqn 6.16). If the image processing schemes of Chapter 7 are to be used, the highest possible information limit is needed, and hence a gun-bias setting is required which reduces the energy spread of the source below the level of other instabilities (eqn 4.9).

Figure 5.17 demonstrates the value of reducing the electron wavelength for uncorrected instruments—the point resolution of these images of the same structure has increased from 0.38 to 0.25 nm in going from 100 kV to 1 MeV. Curves B and C in the contrast transfer function (CTF) plots show the chromatic and spatial coherence damping envelopes, respectively, while D shows the total effect. Note that the spatial coherence (beam divergence) damping is negligible at 1 MeV. The multislice dynamical image simulations

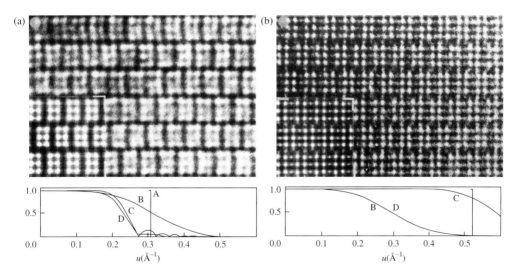

Figure 5.17 Comparison of images of $Nb_{12}O_{29}$ recorded at 100 kV (a) (S. Iijima) and (b) at 1 MeV (S. Horiuchi), showing greatly increased detail at higher voltage due to the reduced electron wavelength. Computer simulated images are shown inset. CTF plots are shown below: A is the physical aperture used, B chromatic damping, C the spatial coherence limit (modulus of eqn (4.12b) with $\theta_c = 1.4$ mrad), D both damping envelopes. Point resolution 0.38 nm at 100 kV and 0.25 nm at 1 MeV. The improvement in resolution to 1.2 Å is seen in Fig. 5.22. Aberration correction achieves a similar effect without the additional radiation damage effects of higher beam energy.

138 *TEM imaging of thin crystals and defects*

(inset) agree well in both cases, indicating a correct choice of thickness. The improvement in going from 0.24 resolution to 0.12 nm at 1.5 MeV is demonstrated in Fig. 5.23. Aberration correction achieves a similar effect without the additional radiation damage effects of higher beam energy.

When imaging a crystal whose first-order reflection u_0 lies beyond the point-resolution of the instrument, there is a tendency for the operator to choose a certain 'stationary phase' focus Δf_s setting instead of the correct Scherzer focus. This important focus setting makes the slope of $\chi(u)$ zero around $u = u_0$. Since the spatial coherence damping is proportional to $\gamma(\nabla\chi/2\pi)$, where γ is the Gaussian coherence function (eqn 4.8a, following), this focus produces the brightest image of spatial frequency u_0, almost unaffected by spatial coherence limitations. From eqn (3.24) we have

$$\Delta f_s = -C_s \lambda^2 u_0^2 \tag{5.66}$$

As a result, if diffractogram analysis is not used, most inexperienced operators will select this focus setting, much larger than the Scherzer value, when focusing images of elemental metals and semiconductors. By contrast, the Scherzer focus may show rather weak contrast.

Some additional experimental considerations for defect imaging are discussed in Chapter 12. We now turn to some examples of the method for line, point, and planar defects.

The mystery of the structure of quasicrystals, which display 10 sharp diffraction spots around a zone-axis ('sharp', suggesting translational symmetry, '10' indicating five-fold symmetry, which is inconsistent with translational symmetry) can be understood using the real-space representation afforded by HREM to directly reveal the tilings. Figure 5.18 shows such an image, viewed along the five-fold axis (Hiraga and Sun 1993).

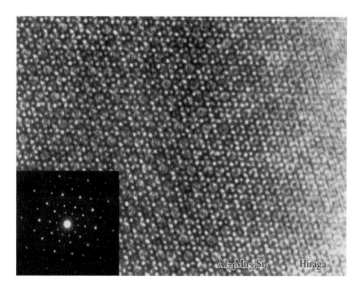

Figure 5.18 HREM image of rapidly quenched $Al_{74}Mn_{20}Si_6$ quasicrystal, viewed along the five-fold axis. The diffraction pattern shows five-fold symmetry (see Hiraga and Sun (1993) and references therein).

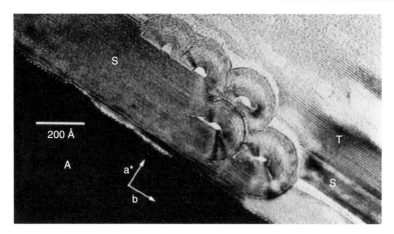

Figure 5.19 Rocks, caught in the act of transformation. Intergrowth of the serpentine structures (S), lizardite (planar layers), and chrysotile (curved layers). Talc (T) can also be seen (Veblen 1985).

HREM has become an indispensable tool for mineralogists, since the images directly reveal the great variety of intergrown microphases which can occur in minerals. These can provide useful information on the thermodynamic history and solid-state reactions which take place. Figure 5.19 (Veblen 1985) shows the fascinating relationship which can occur between chrysotile asbestos, talc, and lizardite. At higher resolution, images of interfaces such as these can give clues to the transformation mechanism.

The imaging of dislocation cores in semiconductors has a long history (Olsen and Spence 1981, Northrup et al. 1981) and must be based on a solid understanding of imaging conditions in the perfect crystal (Glaisher et al. 1989). The field is reviewed by Alexander and Teichler (1991), George (1997), and Jones (2000). In a classic study, the structure of a sigma 5 (130) tilt boundary in silicon and its dislocations has been solved in three dimensions by HREM (Bourret et al. 1988). Structural analysis of screw dislocation cores in Mo is reported at the near-angstrom level of resolution in Sigle (1999), and of a 30° partial dislocation in GaAs at even higher resolution in Beckman et al. (2002). The dramatic improvement in image quality made possible by the aberration corrector is seen in the work of Heuer et al. (2010) on the structure of dislocation cores on the basal planes of sapphire. While the structure of dislocation kinks cannot be determined by HREM, individual kinks can be seen in a projection where the beam is normal to the glide plane, and in this way video images of kink motion in silicon have been obtained, showing kinks pausing at obstacles. The images were also used to determine kink formation ($F_k = 0.727$ eV for kinks on the 90° partial) and migration energies ($W_m = 1.24$ eV) based on measurements of kink velocity under conditions of known temperature and stress (Kolar et al. 1996). Non-equilibrium dissociated dislocations were allowed to relax and move by heating in the microscope, and lattice images recorded before and after partial dislocation motion. To avoid beam-induced motion, motion occurred with the beam switched off. Figure 5.20 shows a typical image formed from the three inner 'forbidden' $[-211]/3$ reflections which are generated by the stacking fault. Since they are absent outside the fault, the lattice image delineates the bounding partial dislocation cores.

140 *TEM imaging of thin crystals and defects*

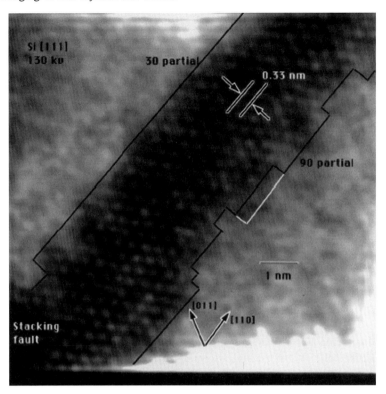

Figure 5.20 A dissociated 60° dislocation in silicon after relaxation. Beam direction [111], normal to the glide plane. The bright diagonal band of regular dots are six-membered rings in the ribbon of the stacking fault layer separating 30° and 90° partial dislocation lines. Black lines run along cores of the two partial dislocations. The fine white line shows a typical alternative boundary used to estimate error in counting kinks. The lattice image was formed using only the inner six 'forbidden' $[-422]/3$ reflections generated by the stacking fault, so that fringes are only seen within the fault, whose boundaries delineate the partial dislocation cores (Kolar *et al.* 1996).

HREM imaging of interfaces in semiconductors, ceramics, and metal–ceramic interfaces has generated a large literature in the attempt to test theories of fracture, of the segregation of dopants at ceramic and ferroelectric interfaces, and theoretical calculations of interface structure. A semiconductor interface is shown in Plate 3—references to prior work can be found in Spence (1999). In ceramics, it has been found, for example, that misfit dislocation cores, as predicted by elasticity theory, lie somewhat displaced back into the softer material rather than at the actual interface on an atomic scale (Kuwabara *et al.* 1989). Figure 5.21 shows an image of the aluminium/spinel (Al/MgAl$_2$O$_4$) interface down [110], recorded on the Stuttgart MPI 1.25 MeV instrument, with a point resolution of 1.2 Å (Schweinfest *et al.* 2001). The sample was prepared by molecular-beam epitaxy, forming a coherent, atomically abrupt (100) interface. These two face-centred cubic structures have cell constants of 0.405 and 0.404 nm, respectively, leading to a very small mismatch of -0.0025 and a coherent interface in which misfit dislocations are not seen. Consequently

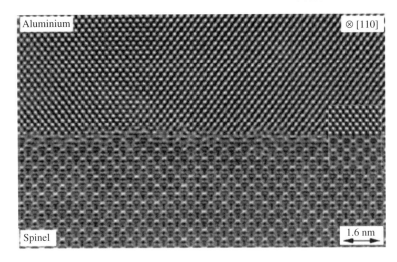

Figure 5.21 Experimental image of the Al/MgAl$_2$O$_4$ interface in [110] projection recorded at 1.25 MeV; underfocus 54 nm (Schweinfest et al. 2001). The inset shows the excellent match with a computer simulated multislice image of the interface. Point-resolution 0.12 nm.

the number of atoms needed in a supercell to represent the structure both in image simulations and electronic structure calculations is small. Of particular interest is the lattice offset or displacement between the two crystals, which can be measured from the HREM images with picometre precision, and the atomic species occurring on the terminating plane. Image analysis was based on the EMS software package (Stadelmann 1991). An automated multiparameter search was used to determine imaging parameters (Chapter 7), using a cross-correlation function to indicate goodness-of-fit (Mobus et al. 1998). Two projections were obtained down [100] and [110] after tilting 45° about an axis normal to the interface; however, not all spacings were accessible in both orientations. The Al atoms of the metal are found to lie above the oxygens of the spinel. *Ab initio* electronic structure calculations based on the local density approximation were completed using pseudopotentials for the atomic structure derived from the images. From these, the work of separation between the interfaces could be obtained as a function of separation for different lateral displacements. The structure observed in the HREM images is found to have lowest energy. A map of charge density, and the density of electronic states (DOS) at the interface are obtained. From this data, electron energy-loss spectra can be predicted, and electrostatic theories of interface adhesion (such as the image charge model) can be tested. A second example of this approach can be found in the study of the Cu/MgO interface by ADF-STEM and electron energy-loss spectroscopy (EELS) by Muller et al. (1998), where metal-induced gap states are detected.

The first convincing images of point defects, in the form of substitutional dopant atoms, have now been obtained using the ADF-STEM method, as shown in Fig. 8.8 (Voyles et al. 2002). With the addition of aberration correctors in STEM, it may be possible to distinguish substitutional from interstitial atoms in these images, in correlation with local energy-loss spectra.

142 *TEM imaging of thin crystals and defects*

The contribution of HREM to understanding the microstructure of CMR materials can be traced through references in Mori *et al.* (1998) and Van Tenderloo *et al.* (2000), in which static charge stripes are directly observed by HREM. The sensitivity of HREM images to ionicity is analysed in detail in Anstis *et al.* (1973), and the sensitivity of structure factors to orbital ordering in Jiang *et al.* (2002). Literature on HREM in the closely related area of oxide superconductors can be found in work by Ourmazd and Spence (1987), and the reviews of Horiuchi (1994) and Shindo and Hiraga (1998). Figure 5.22 (Lebedev *et al.* 2002) shows cross-sectional HREM images of an epitaxial $(La_{0.67}Ca_{0.33}MnO_3)_{1-x}$:$(MgO)_x$ (LCMO:MgO) composite film on MgO(100) with two different MgO concentrations (a, $x = 0$; b, $x = 0.5$) which straddle the metal–insulator phase transition at $x = 0.3$. We note the absence of any secondary phase at the boundary for

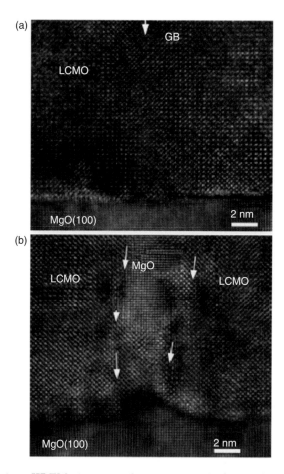

Figure 5.22 Cross-section HREM images of an epitaxial $(La_{0.67}Ca_{0.33}MnO_3)_{1-x}$:$(MgO)_x$ (LCMO:MgO) composite film on MgO(100) with different MgO concentrations: (a) $x = 0$, (b) $x = 0.5$. These straddle the metal–insulator transition. The domain boundary is indicated by a white arrow (Lebedev *et al.* 2002).

$x = 0$ and the heteroepitaxial growth of MgO columns along the domain boundary in the direction normal to the interface for the film with $x = 0.5$. The LCMO films for $x = 0$ were found from HREM and electron diffraction patterns to be orthorhombic ($Pnma$), whereas for $x > 0.3$ the structure becomes rhombohedral ($R3c$). The phase boundary at $x_c \approx 0.3$ coincides with the percolation threshold in conductivity. The largest CMR effect ($\Delta R(5T)/R(5T) \approx 10^5$) is observed for a film with $x \approx x_c$, and is caused by a magnetic field-induced insulator to metal transition. This is explained as a three-dimensional stress accommodation in thick films due to a structural phase transition of $La_{0.67}Ca_{0.33}MnO_3$ from the orthorhombic $Pnma$ structure to a rhombohedral $R\bar{3}c$ structure.

The problem of imaging the very light columns of oxygen atoms in crystals has been a long-standing challenge for HREM. First indications came in the through-focus series work of Coene and colleagues described in Chapter 7, and finally in direct images from an aberration-corrected microscope using the new focus and negative spherical aberration condition discussed in Section 7.4, which produces bright atom images in the bright-field TEM mode (Jia et al. 2003). This has since opened the way to the study of oxygen ordering by direct imaging, which is crucial to understanding the function of many electronic oxides, strongly correlated materials, and oxide superconductors.

5.16. Tomography at atomic resolution—imaging in three dimensions

A long-term goal of HREM has been the reconstruction of *three-dimensional* images of matter at atomic resolution. Biologists have been the leaders in this field, and a brief history of the development of three-dimensional imaging by TEM is given in Chapter 6. In biology, because of the radiation damage induced when multiple images are recorded in different projections of the *same* particle, resolution is normally limited to about 2 nm. Near-atomic resolution is only possible if multiple copies or two-dimensional crystals of a macromolecule are available, in order to spread the radiation dose over many particles. Reconstruction in three dimensions is usually achieved by assembling the various two-dimensional projections in their correct relative orientations in real space. Each two-dimensional density is then extended with constant intensity variation along the direction of projection. The weighted sum of these overlapping intensities gives a three-dimensional reconstruction, as implemented in modern software packages (Mastronarde 1997). The relation between the number of projections needed and resolution is given by eqn (6.33). This weighted back-projection method has also been applied in materials science to bright-field HREM images of MoS_2 fullerene-like particles by Bar Sadan et al. (2008), giving a resolution of about 0.3 nm. For disordered inorganic materials, the problems of image registration after tilting, radiation damage, and tilt calibration have so far proven intractable at atomic resolution, except perhaps for the special case of projections related by rotation about the normal to an interface in a crystal which is insensitive to radiation damage. The methods of compressive sensing have been extensively developed in other imaging fields, and look promising also for electron microscopy—Saghi et al. (2011) describe TEM tomography of iron oxide nanoparticles using this approach.

We consider first the simpler problem of reconstructing the three-dimensional atomic-resolution image of an inorganic nanocrystal lacking any internal defects, in order to determine its external shape at atomic resolution. The first requirement of tomography

144 *TEM imaging of thin crystals and defects*

is to obtain some property of the object which relates sample thickness linearly to the intensity at one pixel in the image. For HREM, this property might be the projected potential of eqn (3.31) in the weak phase approximation. Immediately we see a difficulty with tomographic HREM, since the linear range for light elements then limits the size of our nanocrystal to perhaps 20 nm, unless high-voltage machines are used. A more sophisticated approach, for a nanocrystal of known atomic structure but unknown three-dimensional shape, is to attempt three-dimensional reconstruction of an exit wave as a first step, based perhaps (for heavier atoms such as gold) on the phase-grating (eqn 3.30), two-beam (eqn 5.20) or channelling approximations (eqn 8.16), all of which predict a sinusoidal variation of intensity at one point in the image with changing thickness. This indicates the non-uniqueness of the problem when multiple scattering is allowed. (An Argand diagram for these cases then shows a circular plot, if the real and imaginary parts of the exit-face wavefield at one point are plotted against thickness.).

In the linear range one may then apply the methods of discrete tomography (DM) (Batenburg 2005). Figure 5.23 shows the principle of the method for a simplified two-dimensional nanocrystal consisting of identical atoms on a lattice. We assume that the two-dimensional image intensity at one atomic column in each projection is proportional to the number of atoms in that column. Then, by counting the number of atoms in each column, for several projections, it is possible to determine the external shape of the nanocrystal (Jinschek *et al.* 2008). The crucial simplification is the quantization of intensity levels, so that every atom position has an occupancy of unity or zero. Then a set of linear equation

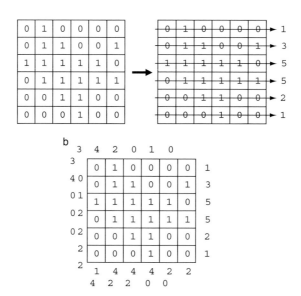

Figure 5.23 Discrete tomography. Digits indicate atom occupancy for an ideal two-dimensional crystal of irregular shape in (a). Occupancy sums are written down the side, as might be obtained from HREM atomic column intensities. (b) The sums for vertical, horizontal, diagonal, and anti-diagonal projections. These linear equations can be inverted to reconstruct a three-dimensional image.

may be written down relating the occupancy of each site to the intensity of each projected column, for every column in each image of a projection, and for every orientation. In Fig. 5.23, with four projections taken along horizontal, vertical, diagonal, and anti-diagonal directions, one obtains 34 equations in 36 unknowns, an insoluble problem. However by restricting the occupancies to integer values, the problem becomes solvable using modern DM algorithms. Errors in the estimation of the number of atoms must be considered, together with the effect of vacancies, defects, and the presence of more than one type of atom—all these may lead to ambiguous reconstructions.

The theory of STEM given in Chapter 8 provides another approach to tomography. Here we detect scattering generated continuously throughout the thickness of the sample, which therefore provides an imaging signal which oscillates less rapidly with sample thickness. ADF-STEM has now been used successfully for tomography at atomic resolution, as described in Section 8.6.

5.17. Imaging bonds between atoms

The increased sensitivity of low-angle electron scattering to ionicity, compared with X-ray scattering, was an important motivation for the earliest researchers in electron diffraction—the electron scattering factor for an ion in a crystal may even change sign (e.g. Cowley 1953a), and the effects of hydrogen bonding have been observed (Cowley 1953b). Equation (8.17b), the so-called Mott–Bethe formula, shows how the electron scattering from an ion increases rapidly with decreasing scattering angle, unlike the X-ray scattering factor which converges to the number of atomic electrons per unit cell (see Spence and Zuo (1992) for a full discussion). The value at zero scattering angle, the mean inner electrostatic Coulomb potential, is thus highly sensitive to the very small changes in potential which occur when atoms bond in a crystal (O'Keeffe and Spence 1994). (For MgO, for example, it changes by several electron volts between the bonded crystal and one formed from neutral atoms, despite extremely small changes in the valence electron distribution.) For many-beam structure images, the effect of multiple scattering can 'amplify' the small changes in low-order scattering factors due to bonding, as demonstrated experimentally by Anstis et al. (1973), who showed how sensitive the thickness dependence of structure images recorded from wedge-shaped crystals is to bonding effects. Plate 6 shows a refinement of the bonding electron distribution in cuprite, based on convergent-beam electron diffraction measurements, that is sufficiently accurate to allow measurement of the amount of charge in one covalent bond. Using a submicron electron probe, many millions of atoms in the crystal contributed to this measurement, for which the bond charge constitutes much less than 1% of the valence electron distribution. For light atoms, however, the bonding effect is proportionately much larger, and the question arises as to whether selected-area electron diffraction is sufficiently accurate to reveal bonding effects in a monolayer of proteins (Chang et al. 1999). Here the angular variation across the CBED rocking curve cannot be used, and it is found that, at resolutions worse than 5 Å, differences in Bragg intensities of up to 10% can result from bonding effects. Nevertheless, the attainment of this accuracy in cryo-EM of two-dimensional crystals is a considerable challenge, given both multiple scattering perturbations and sample imperfections. For the case of nitrogen-substituted atoms in graphene, however, the situation is more hopeful if radiation damage can be controlled

146 TEM imaging of thin crystals and defects

Figure 5.24 (a) Simulated image of graphene with single nitrogen substitutional dopant (neutral atom calculation) at optimum focus—the dopants are not seen. (b) A similar image using DFT for the bonded and relaxed structure. (c). Experimental HREM image, with large defocus—the bonding effect is seen. (The large black feature at top left in (c) is a beam-induced hole.) (From Meyer *et al.* (2011), by permission from Macmillan Publishers Ltd., © 2011.)

(very difficult!). Figure 5.24(a) shows simulated HREM images of nitrogen-doped graphene using neutral (gas-phase) atoms, without relaxation due to the formation of chemical bonds. In this Scherzer-focus image at 80 kV using an aberration-corrected TEM at –9 nm defocus with $C_s = 20$ μm, the nitrogen atoms cannot be seen (unlike similar imaging using the HAADF-STEM mode). Figure 5.24(b) shows an image simulation based on the local density approximation (which allows for bond formation between atoms) at a larger defocus of 18 nm, which reveals the bonds. Additional simulations allowing for strain relaxation did not produce the observed contrast. Figure 5.24(c) shows the experimental image, in agreement with the simulation (Meyer *et al.* 2011).

References

Alexander, H. and Teichler, H. (1991). Dislocations. *Electronic structure and properties of semiconductors*, Vol 4, ed. W. Schroter, p. 254. Weinheim, VCH.

Allpress, J. G. and Sanders, J. V. (1973). The direct observation of the structure of real crystals by lattice imaging. *J. Appl. Crystallogr.* **6**, 165.

Anstis, G. R. (1977). The calculation of electron diffraction intensities by the multislice method. *Acta Crystallogr.* **A33**, 844.

Anstis, G. R., Lynch, D. F., Moodie, A. F., and O'Keefe, M. A. (1973). N-beam lattice images. III Upper limits of ionicity in $W_4Nb_{26}O_{77}$. *Acta Crystallogr.* **A29**, 138.

Bar Sadan M., Houben, L., Wolf, S., Enyashin, A., Seifert, G., Tenne, R., and Urban, K. (2008). Toward atomic-scale bright-field electron tomography for the study of fullerene-like nanostructures. *Nano Lett.* **8**, 891.

Batenburg, K. J. (2005). A new algorithm for 3D binary tomography. *Electron. Notes Discr. Math.* **20**, 247.

Beckman, S. P., Xu, X., Specht, P., and Weber, E. R. (2002). *Ab initio* prediction of the structure of glide set dislocation cores in GaAs. *J. Phys. Cond. Matt.* **14**, 12673.

Bethe, H. (1928). Theorie der Beugung von Elektronenan Kristallen. *Ann. Physik (Leipzig)* **87**, 55.

Bird, D. and King, Q. A. (1990). Absorptive form factors for high energy electron diffraction. *Acta Crystallogr.* **A46**, 202.

Booker, G. R. and Stickler, R. (1962). Method of preparing Si and Ge specimens for examination by transmission electron microscopy. *Br. J. Appl. Phys.* **13**, 446.

Boothroyd, C. B. (1998). Why don't high resolution simulations and images match? *J. Microsc.* **190**, 99.

Boothroyd, C. B. and Stobbs, W. M. (1989). The contribution of inelastically scattered electrons to high resolution [110] images of AlAs/GaAs heterostructures. *Ultramicroscopy* **31**, 259.

Bourret, A., Rouviere, J., and Penisson, J. (1988). Structure determination of planar defects in crystals of germanium and molybdenum by HREM. *Acta Crystallogr.* **A44**, 838.

Bracewell, R. (1999). *The Fourier transform and its applications*. Third edition. McGraw-Hill, New York.

Burge, R. E. (1973). Mechanisms of contrast and image formation of biological specimens in the transmission electron microscope. *J. Microsc.* **98**, 251.

Buseck, P., Cowley, J. M., and Eyring, L. (1989) *High resolution transmission electron microscopy and related techniques*. Oxford University Press, New York.

Buxton, B. F., Loveluck, J. E., and Steeds, J. N. (1978). Bloch waves and their corresponding molecular orbitals in HEED. *Phil. Mag.* **A38**, 259.

Chang, S., Head-Gordon, T., Glaeser, R., and Downing, K. H. (1999). Chemical bonding effects in the determination of protein structures by electron crystallography. *Acta Crystallogr.* **A55**, 305.

Chen, J. and Van Dyck, D. (1997). Accurate multislice theory. *Ultramicoscopy* **70**, 29.

Cockayne, D. J. H. and Gronsky, R. (1981). Lattice fringe imaging of modulated structures. *Phil. Mag.* **44**, 159.

Cowley, J. M. (1953a). Electron diffraction study of the hydrogen bonds in boric acid. *Nature*, **171**, 440.

Cowley, J. M. (1953b). Structure analysis of single crystals by electron diffraction. *Acta Crystallogr.* **6**, 516.

Cowley, J. M. (1959). The electron optical imaging of crystal lattices. *Acta Crystallogr.* **12**, 367.

Cowley, J. M. (1975). *Diffraction physics*. North-Holland/American Elsevier, Amsterdam.

Cowley, J. M. (1976). The principles of high-resolution electron microscopy. In *Principles and techniques of electron microscopy; biological applications*, Vol. 6. Van Nostrand Reinhold, New York.

Cowley, J. M. (1985). The future of high resolution electron microscopy. *Ultramicroscopy* **18**, 463.

Cowley, J. M. (1988). Electron microscopy of crystals with time-dependent perturbations. *Acta Crystallogr.* **A44**, 847.

Cowley, J. M. (1995). *Diffraction physics*. Third edition. North-Holland, Amsterdam.

Cowley, J. M. and Iijima, S. (1972). Electron microscope image contrast for thin crystals. *Z. Naturforsch.* **27a**, 445.

Cowley, J. M. and Moodie, A. F. (1959). The scattering of electrons by atoms and crystals, III. Single-crystal diffraction patterns. *Acta Crystallogr.* **12**, 360.

Cowley, J. M. and Moodie, A. F. (1960). Fourier images IV; the phase grating. *Proc. Phys. Soc.* **76**, 378.

Cowley, J. M. and Moodie, A. F. (1962). The scattering of electrons by thin crystals. *J. Phys. Soc. Japan* **17**(Suppl. B II), 86.

Cracknell, A. P. (1968). *Applied group theory*. Pergamon, London.

Dederichs, P. H. (1972). Dynamical diffraction theory by optical potential methods. In *Solid state physics*, Vol. 27, ed. H. Ehrenreich, F. Seitz, and D. Turnbull, p. 135. Academic Press, New York.

Desseaux, J., Renault, A., and Bourret, A. (1977). Multibeam lattice images from germanium in (011). *Phil. Mag.* **35**, 357.

Dowell, W. C. T. (1963). Das elektronenmikroskopische Bild von Hetzebenenscharen und sein Kontrast. *Optik* **20**, 535.

Doyle, P. A. and Turner, P. S. (1968). Relativistic Hartree–Fock X-ray and electron scattering factors. *Acta Crystallogr.* **A24**, 390.

Etheridge, J., Lazar, S., Dwyer, C., and Botton, G. A. (2011). Imaging high energy electrons propagating in a crystal. *Phys. Rev. Lett.* **106**, 160802.

Eyring, L. (1989). *High resolution transmission electron microscopy and related techniques*, ed. P. Buseck, J. M. Cowley, and L. Eyring, Ch. 10, p. 378. Oxford University Press, New York.

Fejes, P. L. (1977). Approximations for the calculation of high-resolution electron-microscope images of thin films. *Acta Crystallogr.* **A33**, 109.

Fejes, P. L., Iijima, S., and Cowley, J. M. (1973). Periodicity in thickness of electron microscope crystal lattice images. *Acta Crystallogr.* **A29**, 710.

Forbes, B. D., d'Alfonso, A. J., Findlay, S. D., Van Dyke, D., LeBeau, J. M., Stemmer, S., and Allen, L. J. (2011). Thermal diffuse scattering in transmission electron microscopy. *Ultramicroscopy* **111**, 1670.

Fujime, S., Watanabe, D., and Ogawa, S. (1964). On forbidden reflection spots and unexpected streaks appearing in electron diffraction patterns from hexagonal cobalt. *J. Phys. Soc. Japan* **19**, 711.

Fujimoto, F. (1978). Periodicity of crystal structure images in electron microscopy with crystal thickness. *Phys. Status Solidi* **45**, 99.

Fukahara, A. (1966). Many ray approximations in dynamical theory. *J. Phys. Soc. Japan* **21**, 2645.

George, A. (1997). Dislocation review. *Mater. Sci. Eng.* **A233**, 88.

Gevers, R. and David, M. (1962). Relativistic theory of electron and position diffraction at high and low energy. *Phys. Status Solidi* **113(b)**, 665.

Gibson, J. M. and Treacy, M. M. J. (1984). The effect of elastic relaxation on the local structure of lattice modulated thin films. *Ultramicroscopy* **14**, 345.

Gjønnes, J. and Gjønnes, K. (1985). Bloch wave symmetries and inclined surfaces. *Ultramicroscopy* **18**, 77.

Gjønnes, J. and Moodie, A. F. (1965). Extinction conditions in the dynamic theory of electron diffraction. *Acta Crystallogr.* **19**, 65.

Glaisher, R. W. and Spargo, A. E. C. (1985). Aspects of the HREM of tetrahedral semiconductors. *Ultramicroscopy* **18**, 323.

Glaisher, R., Spargo, A., and Smith, D. (1989). HREM of semiconductors. *Ultramicroscopy* **27**, 131.

Goodhew, P. J. (1972). Specimen preparation in materials science. In *Practical methods in electron microscopy*, Vol. 1, ed. A. M. Glauert. North-Holland, Amsterdam.

Goodman, P. and Lehmpfuhl, G. (1964). Verbotene Elektronenbeugungareflexe von CdS. *Z. Naturforsch.* **19a**, 818.

Goodman, P. and Moodie, A. F. (1974). Numerical evaluation of N-beam wave functions in electron scattering by the multislice method. *Acta Crystallogr.* **A30**, 280.

Hall, D. J., Self, P., and Stobbs, W. M. (1983). The relative accuracy of axial and non-axial methods for the measurement of lattice spacings. *J. Microsc.* **130**, 215.

Hashimoto, H. (1985). Achievement of ultra-high resolution by 400 kV analytical atom resolution electron microscopy. *Ultramicroscopy* **18**, 19.

Hashimoto, H., Mannimi, M., and Naiki, T. (1961). Theory of lattice images. *Phil. Trans. R. Soc.* **253**, 459, 490.

Head, A. K., Humble, P., Clarebrough, L. M., Morton, A. J., and Forwood, C. T. (1973). *Computed electron micrographs and defect identification*. North-Holland, Amsterdam.

Heuer, A. H., Jia, C. L., and Lagerlof, K. P. D. (2010). The core structure of basal dislocations in deformed sapphire. *Science* **330**, 1227.

Hiraga, K. and Sun, W. (1993). Al–Pd–Mn quasicrystal studied by HREM. *Phil. Mag.* **A67**, 117.

Hirsch, P. B., Howie, A., Nicholson, R. B., Pashley, D. W., and Whelan, M. J. (1977). *Electron microscopy of thin crystals*. Krieger, New York.

Horiuchi, S. (1994). *Fundamentals of HREM*. North Holland, New York.

Howie, A. (1971). The theory of electron diffraction image contrast. In *Electron microscopy in materials science*, ed. U.Valdre and A. Zichichi. Academic Press, New York.

Howie, A. (2004). Hunting the Stobbs factor. *Ultramicroscopy* **98**, 73.

Hue, F., Hytch, M., Bender, H., Houdellier F., and Claverie, A. (2008). Direct mapping of strained silicon transistor by high resolution electron microscopy. *Phys. Rev. Lett.* **100**, 156602.

Humphreys, C. J. (1979). The scattering of fast electrons by crystals. *Rep. Prog. Phys.* **42**, 1825.

Humphreys, C. J. and Spence, J. C. H. (1979). Wavons—a simple concept for high resolution microscopy. *Proc. EMSA, 1979*, ed. G. W. Bailey, pp. 554–555. Claitor's, Baton Rouge, LA.

Humphreys, C. J. and Spence, J. C. H. (1981). Resolution and illumination coherence in electron microscopy. *Optik* **58**, 125.

Humphreys, C. J., Drummond, R. A., Hart-Davis, A., and Butler, E. P. (1977). Additional peaks in the high-resolution imaging of dislocations. *Phil. Mag.* **35**, 1543.

Hytch, M. J. and Plamann, T. (2001). Imaging conditions for reliable measurement of displacement and strain in HREM. *Ultramicroscopy* **87**, 199.

Hytch, M. J. and Stobbs, W. M. (1994). Quantitative comparison of high resolution TEM images with image simulations. *Ultramicroscopy* **53**, 191.

Ishizuka, K. (1982). Multislice formula for inclined illumination. *Acta Crystallogr.* **A38**, 773.

Ishizuka, K. (1985). Comments on the computation of electron wave propagation in the slice methods. *J. Microsc.* **137**, 233.

Jap, B. K. and Glaeser, R. M. (1978). The scattering of high-energy electrons. *Acta Crystallogr.* **A34**, 94.

Jauch, J. M. and Rohrlich, F. (1955). *The theory of photons and electrons*. Addison-Wesley, New York.

Jia, C. L., Lentzen, M., and Urban, K. (2003). Atomic resolution imaging of oxygen in perovskite ceramics. *Science*. **299**, 870.

Jiang, B., Zuo, J. M., Chen, C. H., and Spence, J. C. H. (2002). Orbital ordering in $LaMnO_3$ and its effect on structure factors. *Acta Crystallogr.* **A58**, 4.

Jinschek, J. R., Batenburg, K. J., Claderon, H. A., Kilaas, R., Radmilovic, V., and Kisielowski, C. (2008). 3-D reconstruction of the atomic positions in a simulated gold nanocrystal based on discrete tomography: prospects of atomic resolution electron tomography. *Ultramicroscopy*, **108**, 589.

Jones, R. (2000). Do we really understand dislocations in semiconductors? *Mater. Sci. Eng.* **B71**, 24.

Kambe, K. (1982). Visualisation of Bloch waves of high energy electrons in HREM. *Ultramicroscopy* **10**, 223.

Kambe, K., Lehmpfuhl, G., and Fujimoto, F. (1974). Interpretation of electron channeling by the dynamical theory of electron diffraction. *Z. Naturforsch.* **29a**, 1034.

Kirkland, E. (2010). *Advanced computing in electron microscopy*. Second edition. Springer, New York.

Kogiso, M. and Takahashi, H. (1977). Group theoretical methods in the many-beam theory of electron diffraction. *J. Phys. Soc. Japan* **42**, 223.

Kolar, H., Spence, J., and Alexander, H. (1996). Observation of moving kinks and pinning in silicon. *Phys. Rev. Lett.* **77**, 4031.

Komoda, T. (1964). On the resolution of the lattice imaging in the electron microscope. *Optik* **21**, 94.

Krivanek, O. L. and Rez, P. (1980). Imaging of atomic columns in [110] silicon. *Proc. 38th Ann. EMSA meeting*, ed. G. W. Bailey. Claitor's, Baton Rouge, LA.

Kuwabara, M., Spence, J. C. H., and Ruhle, M. (1989). On the atomic structure of the Nb/Al_2O_3 interface and the growth of Al_2O_3 particles. *J. Mater. Res.* **4**, 972.

Lang, A. R. (1983). The correct rules for determining the sign of fault vectors. *Phys. Status Solidi* **76(a)**, 595.

von Laue, M. (1960). *Rontgenstrahlinterferenzen*. Third edition. Akademische Verlagsgesellschaft, Frankfurt-am-Main.

LeBeau, J. M., D'Alfonso, A. J., Findlay, S. D., Stemmer, S., and Allen, L. J. (2009a). Quantitative comparisons of contrast in experimental and simulated bright-field scanning transmission electron microscopy images. *Phys. Rev.* **B80** 174106.

LeBeau, J. M., Findlay, S. D., Wang, X., Jacobsen, A. J., and Allen, L. J. (2009b). High-angle scattering of fast electrons from crystals containing heavy elements: Simulations and experiment. *Phys. Rev.* **79**, 214110.

Lebedev, O. I., Verbeeck, J., Van Tenderloo, G., Shapoval, O., and Belenchuk, A. (2002). Structural phase transitions and stress accommodation in $(LaCaMnO_3)(MgO)$ composite films. *Phys. Rev.* **B66**, 104421.

Lewis, A. L., Villagrana, R. E., and Metherall, A. J. F. (1978). Description of diffraction from higher-order Laue zones. *Acta Crystallogr.* **A34**, 138.

Louchet, F. and Thibault-Desseaux, J. (1987). Dislocation cores in semiconductors. From the 'shuffle or glide' dispute to the 'glide and shuffle' partnership. *Rev. Phys. Appl.* **22**, 207.

Lynch, D. F. (1971). Out of zone reflections in gold. *Acta Crystallogr.* **A27**, 399.

Lynch, D. F., Moodie, A. F., and O'Keefe, M. A. (1975). N-beam lattice images V. The use of the charge density approximation in the interpretation of lattice images. *Acta Crystallogr.* **A31**, 300.

Marks, L. D. (1984). Bloch wave HREM. *Ultramicroscopy* **14**, 351.

Marks, L. D. (1985). Direct observation of diffractive probe spreading. *Ultramicroscopy* **16**, 261.

Matsuda, T., Tonomura, A., and Komoda, T. (1978). Observation of lattice images with a field emission electron microscope. *Japan J. Appl. Phys.* **17**, 2073.

Mastonarde, D. N. (1997). Dual-axis tomography: an approach with alignment methods that preserve resolution. *J. Struct. Biol.* **120**, 343.

Matsuhata, H., Van Dyck, D., Van Landuyt, J., and Amelinckx, S. (1984). A practical approach to the periodic continuation method. *Ultramicroscopy* **13**, 343.

Menter, J. W. (1958). The electron microscopy of crystal lattices. *Adv. Phys.* **7**, 299.

Metherall, A. J. E. (1975). Diffraction of electrons by perfect crystals. In *Electron microscopy in materials science*, Part II. Commission of the European Communities, Directorate General 'Scientific and Technical Information Management', Luxembourg.

Meyer, J., Kurasch, S., Park, H. J., Skakalova, V., Künzel, D., Gross, A., Chuvilin, A., Algara-Siller, G., Roth, S. Iwasaki, T., Starke, U., Smet, J. H., and Kaiser, U. (2011). Experimental analysis of charge redistribution due to chemical bonding by HREM. *Nature Mater.* **10**, 209.

Midgely, P. and Dunin-Borkowski, R. (2009). Electron tomography and holography in materials science. *Nature Mater.* **8**, 271.

Mobus, G., Schweinfest, R., Gemming, T., Wagner, T., and Ruhle, M. (1998). Interactive structural techniques in HREM. *J. Microsc.* **190**, 109.

Moodie, A. F. (1972). Reciprocity and shape functions in multiple scattering diagrams. *Z. Naturforsch.* **27a**, 437.

Mori, S., Chen, C. H., and Cheong, S.-W. (1998). Paired and unpaired charge stripes in ferromagnetic $LaCaMnO_3$. *Phys. Rev. Lett.* **81**, 3972.

Muller, D., Shashkov, D., Benedek, R., Yang, L., Silcox, J., and Seidman, D. (1998). Atomic scale observations of metal induced gap states at (222) MgO/Cu interfaces. *Phys. Rev. Lett.* **80**, 4741.

Muller, D. A., Edwards, B., Kirkland, E. J., and Silcox, J. (2001). Simulation of thermal diffuse scattering including a detailed phonon dispersion curve. *Ultramicroscopy* **86**, 371.

Nagakura, S. and Nakamura, Y. (1983). Forbidden reflection intensity in electron diffraction and structure image. *Trans. Japan Inst. Metals.* **24**, 329.

Nishida, T. (1980). Electron optical conditions for the formation of structure images of silicon oriented in (110). *Japan J. Appl. Phys.* **19**, 799.

Northrup, J. E., Cohen, M. L., Chelikowsky, J. R., Spence, J. C. H., and Olsen, A. (1981). Electronic structure of the unreconstructed 30° partial dislocation in silicon. *Phys. Rev.* **B24**, 4623.

Ohtsuki, Y. H. (1967). Normal and abnormal absorption coefficients in electron diffraction. *Phys. Lett.* **A24**, 691.

O'Keefe, M. A. and Sanders, J. V. (1975). N-beam lattice images, V. Degradation of image resolution by a combination of incident-beam divergence and spherical aberration. *Acta Crystallogr.* **A31**, 307.

O'Keefe, M. A. and Saxton, W. O. (1983). The well known theory of electron image formation. *Proc. 41st Ann. EMSA Meeting*, p. 288.

O'Keeffe, M. and Spence, J. C. H. (1994). On the average Coulomb potential and constraints on the electron density in crystals. *Acta Crystallogr.* **A50**, 33.

O'Keefe, M., Spence, J. C. H., Hutchinson, J. L., and Waddington, W. G. (1985). *Proc. 43rd EMSA Meeting*, ed. G. Bailey. San Francisco Press, San Francisco.

Olsen, A. and Spence, J. C. H. (1981). Distinguishing dissociated glide and shuffle set dislocations by high resolution electron microscopy. *Phil. Mag.* **43**, 945.

Ourmazd, A. and Spence, J. C. H. (1987). Detection of oxygen ordering in superconducting cuprates. *Nature* **329**, 6138.

Park, H., Afzali, A., Han, S., Tulevski, G., Franklin, A., Tersoff, J., Hannon, J., and Haensch, W. (2012). High density integration of carbon nanotubes via chemical self-assembly. *Nature Nanotechnol.* **7**, 787.

Pirouz, P. (1979). The effect of absorption on lattice images. *Optik* **54**, 69.

Priouz, P. (1981). Thin crystal approximations in structure imaging. *Acta Crystallogr.* **A37**, 465.

Radi, G. (1970). Complex lattice potentials in electron diffraction calculated for a number of crystals. *Acta Crystallogr.* **A26**, 41.

Reimer, L. (1984). *Transmission electron microscopy*. Springer-Verlag, Berlin.

Rez, P. (1977). PhD Thesis. Oxford University, UK.

Ritchie, R. and Howie, A. (1977). Electron excitation and the optical potential in electron microscopy. *Phil. Mag.* **36**, 463.

Saghi, Z., Holland, D. J., Leary, R., Falqui, A., Bertoni, G., Sederman, A. J., Gladden, A. F., and Midgley P. A. (2011). Three-dimensional morphology of iron oxide nanoparticles with reactive concave surfaces. A compressed sensing-electron tomography (CS-ET) approach. *Nano Lett.*, **11**, 4666.

Saxton, W. O. and Smith, D. J. (1985). The determination of atomic positions from high-resolution electron micrographs. *Ultramicroscopy* **18**, 39.

Saxton, W. O., O'Keefe, M. A., Cockayne, D. J., and Wildens, M. (1983). Sign conventions in electron diffraction and imaging. *Ultramicroscopy* **12**, 75.

Schweinfest, R., Kostelmeier, S., Ernst, F., Elsasser, C., Wagner, T., and Finnis, M. W. (2001). Atomistic and electronic structure of the Al/MgAl$_2$O$_3$ and Ag/MgAl$_2$O$_4$ interfaces. *Phil. Mag.* **81**, 927.

Self, P. G., O'Keefe, M. A., Buseck, P. R., and Spargo, A. E. C. (1983). Practical computation of amplitudes and phases in electron diffraction. *Ultramicroscopy* **11**, 35.

Serneels, R. and Gevers, R. (1969). Systematics reflections in transmission electron diffraction. *Phys. Status Solidi* **31**, 681.

Shindo, D. and Hiraga, K. (1998). *High resolution electron microscopy for materials science.* Springer, Tokyo.

Sigle, W. (1999). HREM and molecular dynamics study of the 111a/2 screw dislocation in Mo. *Phil. Mag.* **79**, 1009.

Smith, D. J. (1997). The realisation of atomic resolution with the electron microscope. *Rep. Prog. Phys.* **60**, 1513.

Smith, D. J., Bursill, L. A., and Wood, G. J. (1985). Non-anomalous high-resolution imaging of crystalline materials. *Ultramicroscopy* **16**, 19.

Spence, J. C. M. (1978). Approximations for dynamical calculations of micro-diffraction patterns and images of defects. *Acta Crystallogr.* **A34**, 112.

Spence, J. C. H. (1988). *Experimental high resolution electron microscopy.* Second edition. Oxford University Press, Oxford.

Spence, J. C. H. (1999). The future of atomic resolution electron microscopy for materials science. *Mater. Sci. Eng.* **R26**(1–2), 1.

Spence, J. C. H. (2003). Oxygen in crystals – seeing is believing. *Science* **299**, 839.

Spence, J. C. H. (2009). Two-wavelength inversion of multiply-scattered soft X-ray intensities to charge density. *Acta Crystallogr.* **A65**, 28.

Spence, J. C. H. and Cowley, J. M. (1979). The effect of lens aberrations on lattice images of spinodally decomposed alloys. *Ultramicroscopy* **4**, 429.

Spence, J. C. H. and Lynch, J. (1982). STEM microanalysis by ELS in crystals. *Ultramicroscopy* **9**, 267.

Spence, J. C. H. and Zuo, J. M. (1992). *Electron microdiffraction.* Plenum. New York.

Spence, J. C. H., O'Keefe, M., and Kolar, H. (1977). High-resolution image interpretation in crystalline germanium. *Optik* **49**, 307.

Spence, J. C. H., O'Keefe, M., and Iijima, S. (1978). On the thickness periodicity of atomic resolution dislocation core images. *Phil. Mag.* **38**, 463.

Stadelmann, P. A. (1991). HREM and CBED image simulation. *Micron. Microsc. Acta* **22**, 175.

Stout, G. H. and Jensen, L. H. (1968). *X-ray structure determination*, p. 136. Macmillan, London.

Sturkey, L. (1950). Multiple diffraction of a scalar wave (electrons) in a periodic medium—Laue case. *Proc. A.C.A. Summer Meeting, New Hampton, NH*.

Sturkey, L. (1962). The calculation of electron diffraction intensities. *Proc. Phys. Soc.* **80**(20), 321.

Taftø, J. (1979). *Point symmetry reduction of the dispersion matrix.* Internal report. University of Olso, Norway. (See Spence 1988.)

Tanaka, M. and Terauchi, M. (1985). *Convergent beam electron diffraction.* JEOL Ltd, Tokyo.

Thust, A. (2009). High-resolution transmission electron microscopy on an absolute contrast scale. *Phys Rev Lett.* **102**, 220801.

Tinnappel, A. (1975). PhD Thesis, Technische Universitat, Berlin.

Treacy, M. M. J., Gibson, J. M., and Howie, A. (1985). On elastic relaxation and long wavelength microstructures in spinodally decomposed In$_x$Ga$_{1-x}$As$_y$P$_{1-y}$ epitaxial layers. *Phil. Mag.* **51**, 389.

Van Dyck, D. (1978). The path integral formalism as a new description for the diffraction of high-energy electrons in crystals. *Phys. Status Solidi* **72(b)**, 321.

Van Dyck, D. and Coene, W. (1984). The real space method for dynamical electron diffraction calculations. *Ultramicroscopy* **15**, 29.

Van Dyck, D. and Op de Beeck, M. (1996). Channelling approach to HREM. *Ultramicroscopy* **64**, 99.

Van Dyck, D., Lichte, H., and Spence, J. C. H. (1999). Inelastic scattering and holography. *Ultramicroscopy* **81**, 187.

Van Tenderloo, G. (1998). HREM in materials research. *J. Mater. Chem.* **8**, 797.

Van Tenderloo, G., Lebedev, O. I., and Amelinckx, S. (2000). Atomic and microstructure of CMR materials. *J. Magn. Magn. Mater.* **211**, 73.

Veblen, D. R. (1985). Direct TEM imaging of complex structures and defects in silicates. *Ann. Rev. Earth Planet. Sci.* **13**, 119.

Verbeek, J., Bertoni, G., and Lichte, H. (2011). A holographic biprism as a perfect filter. *Ultramicroscopy* **111**, 887.

Vergasov, V. L. and Chuklovskii, F. N. (1985). Excitation of bound valence waves of energetic electrons and formation of crystal lattice images with atomic resolution. *Phys. Lett.* **110A**, 228.

Voyles, P. M., Muller, D. A., Grazul, J., Citrin, P., and Gossmann, H. (2002). Atomic-scale imaging of individual dopant atoms and clusters in highly n-type bulk silicon. *Nature* **416**, 826.

Yada, K. and Hibi, T. (1969). Fine lattice fringes around 1 Å resolved by the axial illumination. *J. Electron. Microsc.* **18**, 266.

Yoshioka, H. (1957). Effect of inelastic waves on electron diffraction. *J. Phys. Soc. Japan* **12**, 618.

Zuo, J. M., O'Keeffe, M., Rez, P., and Spence, J. (1997). Charge density of MgO. *Phys. Rev. Lett.* **78**, 4777.

6
Imaging molecules: radiation damage

Electron microscope imaging in biology has an entirely different emphasis from that in materials science because of the dominant effect of radiation damage, which limits resolution. Fine detail is destroyed first. This problem is addressed in X-ray crystallography by using the coherent amplification of scattering into Bragg beams, which, for a crystal, spreads the dose over many identical molecules. But even the smallest three-dimensional protein crystals are too thick to avoid multiple electron scattering perturbations, so that two-dimensional crystals, one molecule thick, have provided the most successful approach in cryo-EM, in the sense of giving the highest-resolution data. As we will discuss in Section 6.9, 'single-particle' imaging is also possible, in which images are recorded from many hopefully identical molecules lying in random orientations in a thin (50 nm thick) film of vitreous ice. (If the molecules exist in a small number of discrete conformations, it may also be possible to sort these out during the orientation determination process—one then has the basis for collecting the frames for a movie.) Again, the radiation dose per molecule is less than that which causes significant damage at the resolution of interest, but is also insufficient to produce a statistically significant image of one molecule, until combined with images from other similar molecules whose orientational relationship must first be determined.

We begin with a discussion of the characteristics of images of single atoms and the dependence of single-atom image contrast on wavelength, atomic number, defocus, and objective aperture size. Practical methods of single-atom imaging are discussed in Section 6.2 and Chapter 12. The techniques of minimum-exposure microscopy are then outlined followed by a survey of sub-nanometre electron microscopy for structural biology, and a discussion of radiation damage processes.

6.1. Phase and amplitude contrast

At medium and low resolution, the contrast in electron microscope images is understood to arise from the creation of an intensity deficit in regions of large scattering where the scattered rays are intercepted by the objective aperture (Cosslett 1958). At high resolution, where one seeks to examine distances comparable with the coherence width of the beam (Section 4.3), this incoherent theory is no longer adequate and phase contrast becomes the dominant contrast mechanism. The theory of phase contrast is then important. Fresnel fringes, lattice fringes, and single-atom images are three examples of phase contrast. To discuss the transition from amplitude to phase contrast in a simple way, it is necessary to use a highly simplified theory, which unfortunately is accurate only for the very thinnest specimens ($t < 10$ nm for light elements at 100 kV) or for rather thick specimens imaged at low resolution. There is no simple theory for the interpretation of image detail in the intermediate range; however, it is possible, though time-consuming, to compute such an image.

Aperture contrast (also known as amplitude or absorption contrast) can be incorporated into the transfer theory outlined in Chapter 3 in the following way. Since the amount of scattering excluded by a small aperture is approximately equal to the total atomic scattering, the transmitted intensity I for a specimen of thickness t and density ρ is

$$I = I_0 \exp(-\mu t)$$

with

$$\mu = \sigma_e \rho L/M \qquad (6.1)$$

where M is the local molecular weight, L is Avogadro's number, and σ_e is the local elastic scattering cross-section. Here the phase difference between waves scattered by successive atoms has been ignored, so the specimen exit-face wave amplitude becomes

$$\psi_e = \sqrt{I} = \psi_0 \exp(-\mu t/2)$$

This expression can be combined with the phase-grating (eqn 3.30) to describe both phase and amplitude variations in the exit-face wavefunction (Grinton and Cowley 1971; Appendix 2):

$$\psi_e = \exp(-i\sigma\phi_P(x,y)) = \exp(-i\sigma\phi_R(x,y))\exp(-\sigma\phi_i(x,y)) \qquad (6.2)$$

where $\phi_P = \phi'_R - i\phi_i$ is now the complex projected specimen potential, with ϕ'_R averaged over the resolution distance (Cowley and Pogany 1968). Here, $\phi_i = \mu t/2\sigma$. This imaginary part of the potential may also be used to represent the depletion of the elastic wavefield by inelastic processes (Yoshioka 1957). Working through the transfer theory of Section 3.4 with this complex potential then gives for the bright-field image intensity

$$I = 1 + 2\sigma\phi_R(x,y) * \mathcal{F}\{A(u,v)\sin\chi(u,v)\}$$
$$- 2\sigma\phi_i(x,y) * \mathcal{F}\{A(u,v)\cos\chi(u,v)\} \qquad (6.3)$$

where the * again indicates the convolution or smearing of the specimen potential (the ideal image) with the point spread function of the instrument, shown in Fig. 3.8. This equation contains three terms. The first represents the uniform bright-field background, while the second represents the phase-contrast image. The third term represents the 'absorption' image. The relative importance of these last terms can be judged from Fig. 6.1, which shows the transfer functions at the Scherzer focus. Medium-resolution biological electron microscopy is limited by radiation damage and stain effects to a resolution poorer than 1 nm. With an aperture cut-off at this point, Fig. 6.1 shows that we have, very approximately,

$$\cos\chi = 1 \quad \text{and} \quad \sin\chi \approx 0$$

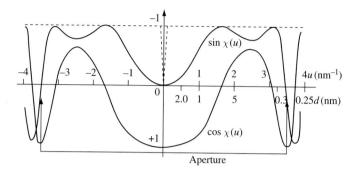

Figure 6.1 The phase-contrast transfer function for bright field at the Scherzer focus $\Delta f = -1.2(C_s\lambda)^{1/2} = -61.0$ nm with $C_s = 0.7$ mm at 100 kV. The limits of a suitable bright-field aperture are indicated. The broken line shows the transfer function required of an ideal optical phase-contrast microscope. In electron microscopy, we approximate this ideal function by a suitable choice of focus and thereby 'convert' a phase object into an amplitude object. The abscissa is marked in units of $|\mathbf{K}| = \theta/\lambda$ and resolution d. The effect of the damping envelope is shown in Fig. A3.2(a).

so that the image intensity is, from eqn (6.3),

$$I \approx 1 - 2\sigma\phi_i(x, y) * \mathcal{F}\{A(u, v)\} \tag{6.4}$$

which predicts an image which is dark in regions of large 'mass-thickness' (ρt). This same image is obtained at the Gaussian focus ($\Delta f = \chi = 0$) where there is no phase-contrast contribution, the effect of spherical aberration being negligible at this resolution. Thus, ultrathin specimens imaged without an objective aperture, so that $\phi_i = 0$, appear transparent at exact focus. This phenomenon is used to obtain a reference focus for high resolution, with some modification for the effects of the transfer function. Note that this analysis assumes perfectly coherent illumination. In practice with $\theta_{ap} = 1$ mrad, the 'incoherence width' is about 3.7 nm at 100 kV (eqn 4.17) and so for resolution much poorer than 3.7 nm the imaging should be considered incoherent. In that case, as is the situation in most lower-resolution biological microscopy, the expression corresponding to eqn (6.3) would be (Section 4.2)

$$I = 1 - 2\sigma\phi_i(x, y) * |\mathcal{F}\{A(u, v)\}|^2 * |\mathcal{F}\{\exp i\chi(u, v)\}|^2 \tag{6.5}$$

where $A(u, v)$ describes the aperture used. For such an incoherent image, multiple-scattering effects will be small for thicknesses much less than the elastic scattering path length. For carbonaceous material at 100 kV this is about 130 nm. To assess the importance of multiple scattering for coherent imaging see Section 6.3.

At high resolution, concentrating on spacings comparable with the coherence length between, say, 0.1 and 1 nm, we see from Fig. 6.1 that, very approximately,

$$\sin\chi \approx -1 \quad \text{and} \quad \cos\chi \approx 0$$

so the image is

$$I = 1 - 2\sigma\phi_R(x,y) * \mathcal{F}\{A(u,v)\} \qquad (6.6)$$

which is a pure phase-contrast image. This accurately reflects the specimen projected potential at this focus setting for sufficiently thin specimens. Like the absorption image, the image is dark in regions of high potential at this focus.

The above is a highly simplified theory and applies only to the very thinnest specimens for phase contrast ($t < 10$ nm for light elements). Nevertheless, eqn (6.3) provides the basis for most computer processing of HREM images in biology (Saxton 1978).

To summarize, images of thick specimens formed with small apertures should be interpreted with the incoherent absorption-contrast model (eqn 6.5), while high-resolution images are formed from phase-contrast effects and require larger (optimized) apertures and the thinnest possible specimens. Physically, absorption contrast arises from the exclusion of electrons from the image by the aperture. Phase contrast arises from the interference between waves within the aperture whose phase must be controlled (by accurate focusing) to produce contrast. Absorption contrast is increased at lower voltages and through the use of smaller apertures. (For details of the sub-angstrom low-voltage electron microscope (SALVE) project, an aberration-corrected low voltage (e.g. 20 kV) TEM instrument for biology, see Lee et al. (2012).) The cut-off angle for absorption contrast can be reduced by increasing the focal length of the lens to provide more low-resolution contrast for stained biological specimens. On the other hand, phase contrast may increase with accelerating voltage (see Fig. 6.8). The maximum specimen thickness which allows a pure phase-contrast interpretation also increases with accelerating voltage. At 100 kV for stained biological specimens about 10 nm thick, both phase and amplitude effects are important if a resolution of 1 nm is expected. In practice, the Fresnel fringe is also commonly used to enhance image contours. Since the width of these fringes is approximately equal to the transverse coherence width (Section 10.7), the approximate rule of thumb can be adopted that the more complicated phase-contrast theory is required to interpret detail smaller than the total width of any Fresnel fringes observed on a micrograph. The contrast for detail larger than this is proportional to the density of scattering matter in the object. The choice of focus under these conditions is also discussed in Agar et al. (1974).

6.2. Single atoms in bright field

In Chapter 3, an expression was given for the two-dimensional complex amplitude across the exit face of a phase object such as a single atom:

$$\psi_e(x,y) = \exp(-i\sigma\phi_p(x,y)) \qquad (6.7)$$

where $\sigma = 2\pi m_0 e \lambda_r \gamma / h^2$ and $\gamma = m/m_0 = (1 - v^2/c^2)^{-1/2}$ incorporate the relativistic correction to the refractive index. A reasonable approximation for atoms of medium and low atomic number is given by the first-order expansion

$$\psi_e = (x,y) = 1 - i\sigma\phi_P(x,y) \qquad (6.8)$$

Imaging molecules: radiation damage

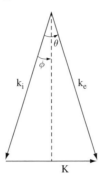

Figure 6.2 Definition of the incident wavevector $\mathbf{k}_i$, the scattered wavevector $\mathbf{k}_e$, and the scattering vector $\mathbf{K}$ for elastic scattering. For elastic scattering, $|\mathbf{k}_e| = |\mathbf{k}_i| = 1/\lambda_r$ and $\mathbf{k}_e - \mathbf{k}_i = \mathbf{K}$. The Bragg angle ϕ is half the scattering angle θ shown.

known as the weak-phase object approximation. An expression for the atomic potential can be obtained as follows. In the first Born approximation, the wave scattered by an atom is given by

$$\psi = \gamma(\exp(-i\mathbf{k}_i r)/r) f(\mathbf{K})$$

where

$$f(\mathbf{K}) = 2\pi m_0 e/h^2 \int_{-\infty}^{\infty} \phi(x,y,z) \exp(2\pi i \mathbf{K} \cdot \mathbf{r}) \mathrm{d}x \mathrm{d}y \mathrm{d}z \tag{6.9}$$

with $\mathbf{K}$ being the scattering vector (see Fig. 6.2). Traditionally, the function $f(\mathbf{K})$ has been tabulated according to eqn (6.9) so that, regardless of the accuracy of the first Born approximation, this publishing convention allows the atomic potential to be synthesized by inverse transform of this equation (Dawson et al. 1974). With the approximation of a flat Ewald sphere, equivalent in this case to the neglect of Fresnel diffraction or propagation within the atom, so that $K_z = 0$, we have, from eqns (6.8) and (6.9),

$$\psi_e(x,y) = 1 - i\lambda_r \gamma \int\int f(\mathbf{K}) \exp(-2\pi i \mathbf{K} \cdot \mathbf{r}) \mathrm{d}K_x \mathrm{d}K_y$$

The methods of Section 3.4 can be used to obtain the image complex amplitude as

$$\psi_i(x,y) = 1 - i\lambda_r \gamma \int\int f(\mathbf{K}) \exp(+i\chi(\mathbf{K})) \exp(-2\pi i \mathbf{K} \cdot \mathbf{r}) \mathrm{d}K_x \mathrm{d}K_y \tag{6.10}$$

where unimportant phase factors of unit modulus have been omitted for clarity. A magnification of unity is assumed and the function

$$\chi(\mathbf{K}) = \pi \Delta f \lambda_r \mathbf{K}^2 + \pi C_s \lambda^3 \mathbf{K}^4/2 = \pi[\Delta f \theta^2 + C_s \theta^4/2]/\lambda_r \tag{6.11}$$

For elastic scattering (Fig. 6.2), we also have

$$|\mathbf{K}| = 2\sin\theta_B/\lambda_r \approx \theta/\lambda = 1/d$$

with θ the angle between the incident beam and the scattered wave—this is twice the X-ray 'Bragg angle' θ_B. Scattering factors are usually tabulated as a function of θ_B/λ. For symmetric atoms

$$f(\mathbf{K}) = f(|\mathbf{K}|)$$

and the image is radially symmetric. Equation (6.10) then becomes

$$\psi_i(r) = 1 - 2\pi i\gamma/\lambda_r \int_0^{\theta_{ap}} f(\theta) J_0(2\pi\theta r/\lambda_r)\exp(+i\chi(\theta))\theta d\theta \tag{6.12}$$

which can be written as

$$\psi_i(\mathbf{r}) = 1 - iz$$

where z is a complex function of $\mathbf{r}$.

The image intensity is then

$$\begin{aligned}I(r) &= \psi_i\psi_i^* = 1 + 2\mathrm{Im}(z) + zz^* \approx 1 + 2\mathrm{Im}(z) \\ &= 1 + 4\pi\gamma/\lambda_r \int_0^{\theta_{ap}} f(\theta) J_0(2\pi\theta r/\lambda_r)\theta\sin\chi(\theta)d\theta\end{aligned} \tag{6.13}$$

for a bright-field image of a single unsupported atom formed with a central objective aperture of semi-angle θ_{ap}. To a good approximation, the integrated image intensity is proportional to the axial value

$$I(0) = 1 + 4\pi\gamma/\lambda_r \int_0^{\theta_{ap}} f(\theta)\sin\chi(\theta)\theta d\theta = 1 + A \tag{6.14}$$

so that the image contrast becomes

$$C = |I(\infty) - I(0)|/I(\infty) = |A| \tag{6.15}$$

The problem of maximizing single-atom contrast then becomes the problem of maximizing the integral in eqn (6.14), for which the experimental variables Δf, θ_{ap}, and $\chi(\theta)$ are available. The integral is dominated by the behaviour of the transfer function $\sin\chi(\theta)$, shown in Fig. 6.1 with its characteristic rapid oscillation at high angles. Several workers have investigated the behaviour of eqn (6.14) and several criteria for choosing Δf and θ_{ap} have been proposed. All these lead to the approximate result that, for maximum contrast,

$$\Delta f = -1.2(C_s\lambda)^{1/2} \tag{6.16a}$$

and
$$\theta_{\mathrm{ap}} = 1.5(\lambda/C_{\mathrm{s}})^{1/4} \qquad (6.16\mathrm{b})$$

a result first obtained by Scherzer (1949). This is the $n = 0$ case in eqn (4.12a). The diffraction limit set by this aperture size corresponds to a point resolution of

$$d_{\mathrm{p}} = 0.66 C_{\mathrm{s}}^{1/4} \lambda^{3/4} \qquad (6.17)$$

in an ideal phase-contrast microscope fitted with the objective aperture of eqn (6.16b). Note, however, that since the imaging is coherent, the resolution cannot strictly be defined in this simple way, since the ability to distinguish neighbouring point objects depends on their scattering properties under coherent illumination (see Chapter 3). The sign of Δf indicates that a slightly under-focused lens is required for optimum contrast (Scherzer focus). Here the lens is slightly weakened from the Gaussian focus condition and is focused above the specimen, while the first Fresnel fringe would appear bright. This value of Δf gives the best approximation to an ideal lens for phase contrast, shown by the broken line in Fig. 6.1 and giving $\chi(\theta) = -\pi/2$ for all θ except at the origin where $\chi = 0$. This would lead to a darkened image on an otherwise bright background. Alternatively, $\chi(\theta) = +\pi/2$ would lead to a bright image. Notice that the low spatial frequencies are severely attenuated so that a pure phase-contrast image would be synthesized from the middle-range spatial frequencies only. In practice, there is a maximum contribution from absorption contrast at low spatial frequencies which may complement this lack of information from the real potential (Section 6.1). All spatial frequencies beyond $1/d = \theta_{\mathrm{ap}}/\lambda$ are excluded by the aperture, corresponding to detail smaller than $d_{\mathrm{c}} = \lambda/\theta_{\mathrm{ap}}$. Resolution for coherent images must be understood in this sense of an allowed band of spatial frequencies with differing weights given by the instrumental transfer function. Focusing the microscope can be thought of as 'tuning' the instrument to the spatial frequencies of interest. Section 7.4 discusses how this analysis must be modified for an aberration-corrected instrument, where the value of C_{s} can be varied at will. (For $C_{\mathrm{s}} = \Delta f = 0$, no phase contrast would be obtained.)

Figure 6.3 shows an idealized bright-field image for a gold atom using the optimum values of Δf and θ_{ap}. The image was computed from eqn (6.13) using the relativistic Hartree–Fock scattering factors of Doyle and Turner (1968). A value of $C_{\mathrm{s}} = 0.7$ mm has been used. An image of lower contrast can be obtained with a small positive value of Δf which produces a bright image more intense than the uniform background. This transition from bright to dark image can be observed experimentally with small clusters of evaporated gold atoms on thin amorphous carbon films as described in Section 1.3, or using the resolution standards referred to in Section 10.9.

Isolated single atoms have been imaged in bright-field transmission microscopy by several workers (e.g. Iijima 1977) and it is important to establish that these are indeed individual atoms. (Since that time, the development of cryomicroscopy and the use of graphene substrates have provided a greatly improved method of supporting heavy atoms and molecules in ice—see Sections 6.9 and 12.6.) Perhaps the most convincing evidence that one is seeing individual atoms comes from experiments in which images of the same substrate area are recorded before and after the addition of a heavy atom. A method for adding atoms inside

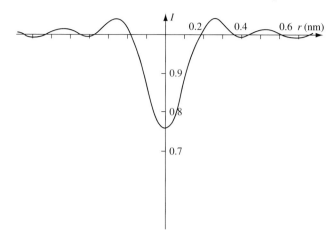

Figure 6.3 The intensity of a bright-field image of a single gold atom plotted as a function of radial distance from the centre of the atom. The contrast is about 25%, which well exceeds the threshold of detectability (5%). The experimental conditions simulated are $C_s = 0.7$ mm, $\Delta f = -61$ nm, accelerating voltage 100 kV. This theoretical calculation based on eqn (6.13) does not take account of the image background due to a substrate-supporting film. The two-dimensional intensity distribution is obtained by rotating the figure about the vertical axis and the intensity distribution for two neighbouring atoms consists of the sum of two displaced functions such as these. For such a pair of atoms the resulting image depends both on the focus condition and the atomic separation—it is possible to obtain a single deep intensity minimum at the point where subsidiary minima from adjacent atoms overlap. This could lead to the false identification of a single atom where two actually exist. The complexities of image interpretation are compounded in dark field where complex amplitudes must be added.

the microscope (*in situ*) was developed by Iijima (1977), who installed a small evaporation furnace in the microscope. Iijima used thin graphite flakes which are etched in the microscope to reduce their thickness still further by the action of the electron beam on tungsten trioxide particles. The thickness of these ultrathin specimen supports has been estimated from the chance observation of graphite crystals whose edges curl up, giving a view down between the graphite crystal layers. Since individual layers are resolved, the thickness of the substrate can be found by counting the atomic layers. Crystals with a thickness of about 2 nm were frequently used as shown in Figs 6.4–6.7. In these figures, taken at the optimum (Scherzer) focus given by eqn (6.16), we note the following: (1) Owing to the use of an extremely thin graphite substrate there is almost no discernible contrast change between the region outside the graphite support film and the region inside the graphite flake. In addition, image noise due to thickness variations of the support film is not seen. (2) The first Fresnel fringe at the edge of the graphite flake appears bright, indicating that the image was recorded slightly under-focus (weakened lens). (3) The atoms and clusters of atoms appear dark, as expected from Fig. 6.3 at the focus setting used.

162   *Imaging molecules: radiation damage*

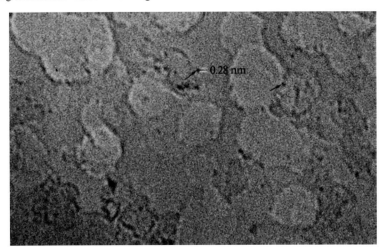

Figure 6.4 Single atoms and clusters of atoms of tungsten on the surface of a few atomic layers of graphite. The black areas are larger tungsten oxide crystals. Fresnel edge fringes can be seen surrounding the region where a single atomic layer of graphite has been removed. Individual atoms are marked by arrows. The experimental conditions are $\Delta f = -61$ nm, $C_s = 0.7$ mm, $\lambda = 0.0037$ nm, illumination semi-angle 1.4 mrad, objective aperture semi-angle 16 mrad, electron optical magnification 500 000. The objective aperture used excludes the Bragg beams diffracted by the graphite substrate from the image. No image processing, background subtraction, or contrast enhancement techniques have been used in printing these micrographs, which were recorded on film.

Three tests can be applied to confirm the interpretation of these images as showing single atoms:

(1) The *size* of an isolated atom image should agree with the width of the function shown in Fig. 6.3 computed from eqn (6.13). This size depends mainly on the objective aperture used, the value of C_s and the focus setting. If no objective aperture is used, an effective aperture must be assumed whose size is determined by the magnitude of the incoherent instabilities (see Sections 3.3 and 4.2 and Appendix 3). Except on aberration-corrected machines, the size of a single-atom image does not depend on the type of atom to any appreciable extent; rather, it depends on the impulse response of the objective lens (see Section 3.4).
(2) The *contrast* of a single-atom image should also agree with calculations. Figure 6.5 shows four optimum focus images of tungsten atoms on a graphite substrate recorded at 2-min intervals. Atoms can be seen to have moved about between exposures, while the contrast of single-atom blobs is between 6 and 9%, in agreement with calculations.
(3) Images showing the same region before and after the addition of a known number of atoms provide the best confirmation of image interpretation.

Figure 6.6(a) and (b) shows such a pair of images, while Fig. 6.6(c) shows the effect of beam-induced atomic mobility. The atoms use some of the energy available from the electron beam to hop about on the surface, finally collecting in the small clusters shown.

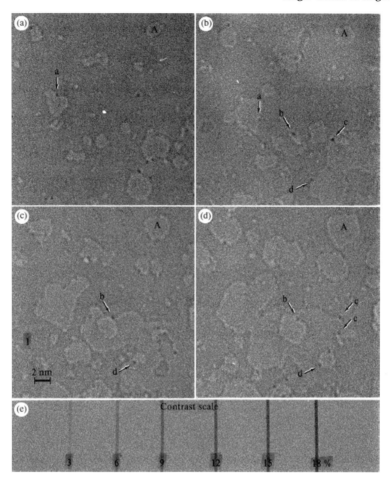

Figure 6.5 These images, recorded under slightly different conditions from those of Fig. 6.4 ($C_s = 1.8$ mm, Scherzer focus), show individual atoms (a, b, d), one of which (atom a) can be seen to move between exposures. All the atoms cling to rough spots on the surface where the electrostatic potential well is deepest. Atom c is believed to be a pair of atoms which separate in the last image. A contrast scale is given below to allow comparison with Fig. 6.4.

An objective lens with low spherical aberration is not essential to image well-separated atoms. A low spherical aberration constant (resulting in reduced width of the atom image) becomes important when one wishes to distinguish closely separated atoms (such as are found in the neighbourhood of defects in solids). However, the reduction in spherical aberration using aberration correction will also increase the height of single-atom peaks above background.

Figure 6.7 beautifully illustrates the 'phase object' nature of single-atom images (see Sections 1.1 and 3.4). In accordance with theoretical predictions, the image recorded at 'exact' focus shown in Fig. 6.7(b) shows little or no contrast. As further predicted by

164 *Imaging molecules: radiation damage*

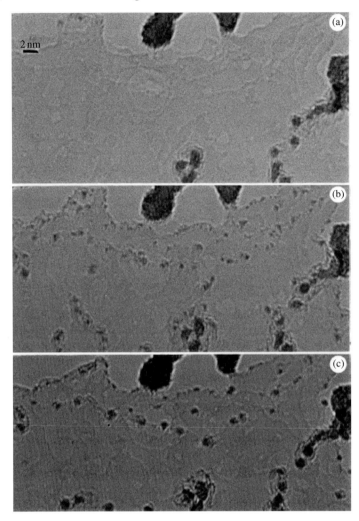

Figure 6.6 Graphite flake shown (a) before and (b) after *in situ* evaporation of tungsten atoms. Again the atoms stick to surface steps. (c) The same region after 5 minutes of beam irradiation when the heavy atoms have migrated together to form metallic tungsten.

eqn (6.13), the over-focus image (Fig. 6.7a) shows bright fluctuations against the bright-field background, changing to dark atom images of high contrast at optimum focus (Fig. 6.7c). The reason why the image of minimum contrast (Fig. 6.7b) occurs at a slight under-focus value rather than for $\Delta f = 0$ is discussed in Section 12.5. It can be seen from a comparison of Fig. 6.7(c) and (d) that for single-atom imaging, a focusing accuracy of a few tens of nanometres is needed. Even greater accuracy (perhaps ± 4 nm) is needed if the atoms are closely spaced so that their images overlap.

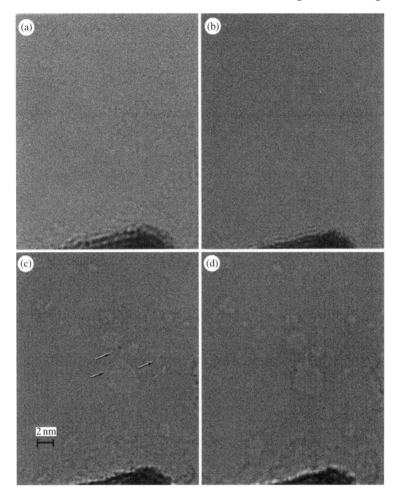

Figure 6.7 Through-focus series showing the phase-object nature of single-atom images. Focus settings are: (a) $\Delta f = +42.5$ nm, (b) $\Delta f = -25.0$ nm, (c) $\Delta f = -92.5$ nm, (d) $\Delta f = -160.0$ nm. Atoms are arrowed. The experimental conditions are the same as in Fig. 6.4. Note the changing appearance of the Fresnel fringes around the clump of tungsten on the lower edge of the images and the absence of contrast in (b) taken under conditions similar to those of Fig. 1.2(b). Owing to the presence of spherical aberration, the focus condition for minimum contrast occurs at a small under-focus value rather than for $\Delta f = 0$. (c) The optimum (Scherzer) focus image.

6.3. The use of a higher accelerating voltage

Prior to the development of aberration correction, the use of a higher accelerating voltage was adopted as a route to improved resolution. This also enables the study of much thicker samples, such as whole cells in biology, but caused much increased 'knock-on' radiation damage (but reduced ionization damage). Since higher voltages also reduce Coulomb interactions between electrons due to relativistic effects, a revival of this approach is being considered for pulsed electron diffraction, in view of the discovery that diffraction with femtosecond pulses can outrun radiation damage (see Spence et al. (2012) for a review). For this reason, and because of the importance of the problem of imaging thicker biological samples at high resolution, we review the literature and underlying concepts in this field with regard to resolution and contrast. Review articles covering the use of high-voltage machines for biological applications include Glauert (1974), Humphreys (1976), Porter (1986), and Renken et al. (1997).

The wavelength reduction resulting from an increase in accelerating voltage has several generally beneficial effects on high-resolution phase-contrast images. Two machines in particular, operating at about 1 MeV in Stuttgart and Tokyo, produced striking images with resolutions close to 1 Å (Ichinose et al. 1999, Schweinfest et al. 2001).

The dependence of single-atom contrast in bright field on wavelength can be obtained from eqns (6.15), (6.14), and (6.16) using the relativistic wavelength and mass correction factor

$$\gamma = m/m_0 = (1 - v^2/c^2)^{1/2} = (1 + eV_0/m_0c^2) \qquad (6.18)$$

The negligible additional effects of electron spin are discussed in Fujiwara (1962). The result of such a calculation is shown in Fig. 6.8 for a gold atom using Turner–Doyle scattering factors. The results of electron-optical calculations and experiment suggest that the product of C_s and λ is approximately constant. The smooth numbered curve in Fig. 6.8

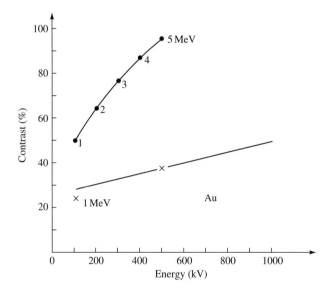

Figure 6.8 Contrast of the bright-field image of a single gold atom as a function of accelerating voltage. Unlike the normal aperture 'absorption' contrast, this phase contrast increases with voltage. The Scherzer focus has been assumed and $C_s = 1.2$ nm at 500 kV, assuming the product of C_s and λ to be independent of wavelength. Note that the assumption that the product of C_s and λ is constant also suggests that the Scherzer 'optimum focus' condition is independent of accelerating voltage (eqn 6.16).

is drawn for $C_s\lambda = 11.3 \times 10^3$ nm^2, while the crosses represent the performance of commercial instruments. With increasing voltage the atomic scattering becomes more narrowly forward-peaked; however, the dependence of the scattering factor on θ differs from that of the transfer function and both effects must be considered. The main cause of the large increase in contrast with accelerating potential is the increase in relativistic electron mass. This increase in phase contrast with increasing accelerating voltage should be distinguished from the conventional (incoherent) aperture absorption contrast of interest at low resolution. In fact, for a fixed objective aperture size, the absorption contrast decreases with accelerating voltage as the electron scattering becomes more strongly forward-peaked.

The resolution of electron images also improves at higher voltage, as shown in Fig. 6.9. Combining the empirical wavelength dependence of spherical aberration with eqn (6.17) suggests that the resolution should improve as the square-root of the wavelength. Experimental work suggests

$$d = 6.8\lambda^{1/2} \tag{6.19}$$

with λ in nanometres. This would give a resolution of 0.1 nm at 4 MeV and 0.07 nm at 10 MeV. At present it appears unlikely that any more machines which operate at voltages

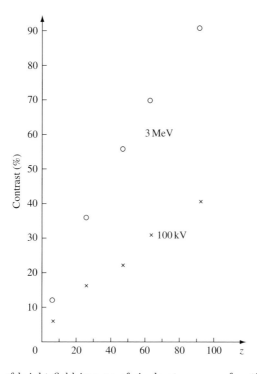

Figure 6.9 The contrast of bright-field images of single atoms as a function of atomic number at two different accelerating voltages. The contrast increases more rapidly with atomic number at higher accelerating voltage. Here the Scherzer conditions are assumed with $C_s = 0.5$ mm at 100 kV and $C_s = 4.7$ mm at 3 MeV.

above 1 MeV will be built, in view of their very high cost. In addition, problems of electronic and mechanical stability have yet to be resolved for instruments operating much above 1 MeV, and the use of aberration correctors provides a much lower-cost alternative, with reduced radiation damage. A more detailed analysis of the dependence of resolution on electron wavelength can be found in Appendix 3.

The effects of chromatic aberration become less severe at high voltage. The incoherent image blurring is (eqn 2.38)

$$\Delta r = C_c \theta_{\mathrm{ap}} \Delta V_0 / V_0$$

which decreases inversely as the accelerating voltage for a fixed energy loss ΔV_0. The focal plane separation is also correspondingly reduced.

We now consider the importance of multiple scattering as a function of wavelength. Tivol et al. (1993) demonstrated the decline of multiple-scattering artefacts in diffraction patterns from organic crystals from 100 to 1200 keV. In the incoherent imaging theory used at low resolution, multiple-scattering effects can be neglected for specimens much thinner than the path length for elastic scattering (see Section 6.1). However, when we consider high-resolution coherent imaging, there are two physical mechanisms which must be considered: the strength of the electron scattering interaction and the effects of Fresnel propagation through the specimen. The strength of the electron-scattering interaction is measured by $\sigma v_{\mathbf{g}}$, where $v_{\mathbf{g}}$ are the crystal potential Fourier coefficients (see Section 5.3.2). Relativistic effects are correctly incorporated if the relativistic mass m, accelerating voltage, and wavelength λ_r (see eqn 2.5) are used in the definition of σ (see Section 2.1). The relativistic extinction distance is then given by eqn (5.17). A rule of thumb which expresses the spreading of a wave from a point source due to Fresnel propagation can be obtained by setting the maximum phase shift allowed by Fresnel propagation equal to π (see Section 3.1), or, equivalently, by restricting the resolution to those reflections for which the Ewald sphere passes through the central maximum of the kinematic shape transform shown in Fig. 5.2. Thus, the curvature of the Ewald sphere expresses physically the process of the propagation of the Huygens wavelets in the crystal. Setting the phase of eqn (3.16) equal to π gives

$$u^{-1} = d = \sqrt{\lambda t} \qquad (6.20)$$

That is,

$$t < d^2/\lambda$$

gives the approximate condition that Fresnel broadening of the diffracted waves may be neglected in a specimen of thickness t if one is interested in detail of size d or larger. Note that eqn (6.20) also gives the approximate width of the first Fresnel edge fringe (see Section 10.7). One approximation which neglects the effect of Fresnel diffraction (or Ewald sphere curvature) is the phase-grating approximation (eqn 6.7), and this belongs to a class of approximations known generally as 'projection approximations'. Using them, it is not possible to distinguish differences in the height of atoms within the specimen. Resolution in the direction of the electron beam is only possible using a theory which takes full account of Fresnel diffraction and the Ewald sphere curvature.

Thus, eqn (6.20) may be used to estimate the maximum thickness for which eqn (6.7) can be used for non-periodic specimens. From eqn (6.20) we see that while this expression accurately describes images of 0.35 nm resolution in 30 nm thick specimens at 100 kV, for the same resolution this thickness is increased to 400 nm at 4 MeV. For crystalline specimens, the orientation is important and detailed calculations must be used to assess the accuracy of the phase-grating approximation.

An important problem in high-voltage microscopy concerns the design of area detectors (which are available commercially), and the reduced sensitivity of viewing phosphors at high accelerating voltages (Iwanaga et al. 1968, Cosslett 1974). Electron source brightness increases with accelerating voltage (see eqn 9.2).

To summarize, with the increasing availability of aberration-corrected machines, it seems unlikely that there will be much further development of MeV instruments, except for specialist applications such as thick-sample imaging and high-speed pulsed imaging. For reasons of cost and performance the 'medium-energy' machines (Cosslett 1974, Kobayashi et al. 1974) have now become the most popular instruments for HREM, with a point-resolution of about 0.17 nm or better. Appendix 3 provides instructive comparisons in more detail on the effects of accelerating voltage on HREM performance. For corrected machines, the wavelength reduction at higher voltage is not needed, and radiation damage can be reduced by using beam energies perhaps as low as 20 kV, well below the threshold for knock-on damage (Lee et al. 2012).

6.4. Contrast and atomic number

Experimental images of heavy atoms show more contrast than those of light atoms (Crewe et al. 1974). This is expected theoretically, as shown in Fig. 6.9 where the bright-field, single-atom contrast is plotted as a function of atomic number. Equation (6.15) has been used with the optimum values of Δf and θ_{ap} at 100 kV and 3 MeV. The steeper slope of the graph at higher voltage suggests that the discrimination of atomic species may be easier at higher voltage.

It is of interest to compare the expression for bright-field single-atom contrast with the total elastic scattering cross-section to clarify the distinction between a scattering experiment and a phase-contrast image. For an ideal phase-contrast image in bright field ($\chi(\theta) = \pi/2$ in eqn 6.10), without an objective aperture, the contrast would be

$$C_{\text{BF}} = \lambda \gamma \int f(\mathbf{K}) d\mathbf{K} \tag{6.21}$$

For comparison the elastic-scattering cross-section is defined as

$$\sigma_e = \int |f(\mathbf{K})|^2 d\Omega \tag{6.22}$$

which can be expressed as an integration over $\mathbf{K}$.

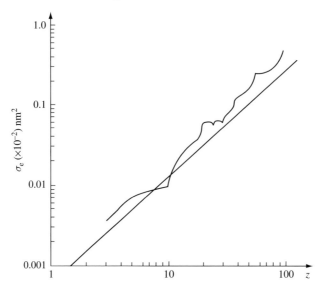

Figure 6.10 The elastic scattering cross-section σ_e plotted as a function of atomic number. In TEM, only a small part of the total scattering shown here is used to form the image. The straight line is a simple expression due to Lenz, the other curve is based on Doyle–Turner scattering factors. Both scales are logarithmic. (Courtesy of C. J. Humphreys.)

Values of σ_e are indicated in Fig. 6.10, obtained from Turner–Doyle scattering factors (Humphreys et al. 1974). On the same plot is shown the expression due to Lenz (1954)

$$\sigma_e = \frac{\lambda^2 Z^{4/3}}{\pi(1 - v^2/c^2)} (\text{Å}^2) \tag{6.23}$$

which provides a rough approximation.

The contrast in a dark-field atom image depends critically on the substrate used; however, the intensity relative to unit incident intensity at the centre of a dark-field image of a single atom formed in an ideal phase-contrast microscope ($\chi(\theta) = 0$ in eqn 6.25) would be

$$I_{\text{DF}} = \lambda^2 \gamma^2 \sigma_e \tag{6.24}$$

if an unsupported central beam stop were used within a large objective aperture.

Accurate calculations must take account of the substrate, the geometry of the beam stop, and the effects of defocus and spherical aberration as in eqn (6.25) (see also Section 6.5).

Through the reciprocity theorem (Cowley 1969), the results given in Figs 6.9 and 6.10 also apply to the case of single atoms imaged in STEM under reciprocal aperture conditions—that is, a small central detector and optimum illuminating aperture and focus condition. The dark-field mode has many advantages in STEM (Langmore and Wall 1973) for elemental discrimination (see Sections 8.3 and 8.5 on STEM imaging).

6.5. Dark-field methods

A large increase in contrast is obtained by eliminating the bright-field background represented by the first term in eqn (6.13). (The human eye, especially, is sensitive to ratios of intensity rather than their absolute magnitude.) The unscattered wave is focused to a small spot in the objective lens back-focal plane and may be eliminated using a suitably shaped aperture, or tilted illumination. Against this advantage must be set the disadvantage that dark-field images are less easily interpreted than bright-field images, since the effect of the aperture geometry on the image Fourier synthesis must be considered.

First consider a single atom imaged using an unsupported central beam stop to exclude the central diffraction maximum. Then, eqn (6.8) gives the image as

$$I(x,y) = \left| \lambda\gamma \int_{-\infty}^{\infty} f'(\mathbf{K}) e^{i\chi(\mathbf{K})} e^{2\pi i \mathbf{K}\cdot\mathbf{r}} d\mathbf{K} \right|^2 \tag{6.25}$$

where $f'(\mathbf{K})$ is the atomic scattering amplitude modified by the aperture. For an ideal central beam stop, $I(x, y)$ is a smooth peak, so that a bright spot will be seen as the dark-field image of an isolated single atom. For more complicated apertures, the function $f'(\mathbf{K})$ must be set to zero in the regions excluded by the beam stop. For groups of atoms and more realistic apertures, the situation is complicated and false image detail is possible in dark field. In practice, the diffraction conditions may not be symmetric, so that a two-dimensional image simulation is required to predict the image of a known structure. Computed images for a variety of dark-field modes (wire beam stop, displaced aperture, tilted illumination) have been published (Chiu and Glaeser 1974, Krakow 1976). Beam stops have been made using fine tungsten wire across the diameter of the objective aperture or a gold-plated spider's web in the same position. Fine quartz fibres have also been used. The effect of electrostatically charged contaminant forming on these wires is an important experimental problem. The effect of the wire thickness on the appearance of the image of an atom is discussed in Chiu and Glaeser's paper. Figure 6.11 shows the image distortion possible under various dark-field conditions for a small molecule.

The optimum defocus and aperture size which must be used in dark field differ from those used in bright field. Unlike the bright-field case, both the real and imaginary parts of eqn (3.23) are equally influential for dark field, as can be seen by comparing eqns (6.25) and (6.13). The optimum focus for dark field is one which makes $\chi(\theta) \approx 0$ to the highest resolution possible. The condition $\sin \chi(\theta_{\mathrm{ap}}) = 0.3$ has been suggested (Cowley 1976), which gives an optimum focus for dark field of

$$\Delta f = -0.44\sqrt{C_s \lambda} \tag{6.26}$$

With $C_s = 0.7$ mm, this gives $\Delta f = -22$ nm (under-focus). Figure 6.12 shows $\sin \chi(u)$ and $\cos \chi(u)$ under these conditions at 100 kV. Since the form of the image is mainly determined by the $\cos \chi(\theta)$ function at small Δf, the image change with defocus is less rapid than for bright field, as is readily seen experimentally. We emphasize that since

172 *Imaging molecules: radiation damage*

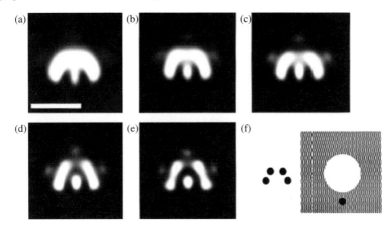

Figure 6.11 Five computed dark-field electron micrographs of the mercury-thiophene molecule whose atoms are arranged in the pattern shown in (f). The simulated conditions are $\Delta f = -75$ nm (under-focus), $C_s = 1.35$ mm, accelerating voltage 100 kV, objective aperture semi-angles: (a) 6.5 mrad, (b) 7 mrad, (c) 7.5 mrad, (d) 8.0 mrad. The optic axis passes through the centre of the objective aperture as shown in (f), while the central 'unscattered' beam falls at the position of the black dot. In practice, any fine structure seen in these images would be lost if a large illuminating aperture were used and the contrast would be further degraded by scattering from the necessary supporting film. These images lead to the identification of five atoms (see (e)) where only four actually exist. (Images courtesy of W. Krakow.)

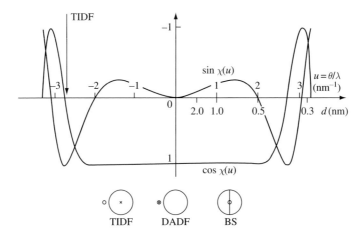

Figure 6.12 The sin and cosine of $\chi(u)$ at the optimum conditions for high-resolution dark-field imaging. The phase shift due to aberrations has been kept close to zero to the highest resolution possible. Here $C_s = 0.7$ mm and $\Delta f = -22$ nm (under-focus). The position of the optic axis (×) and the 'unscattered' beam (○) are shown in relation to the objective aperture for three dark-field imaging modes below. These are tilted-illumination dark-field (TIDF), displaced-aperture dark-field (DADF), and beam stop (BS). The vertical arrow on the left suggests the position of the unscattered beam in the tilted-illumination mode. This is not a 'transfer function' (see text).

the image in dark field is no longer linear in specimen potential, the convenient product representation and envelope functions of the bright-field theory cannot be used (Hanssen and Ade 1978). Thus, Fig. 6.12 should not be thought of as representing a 'transfer function'. Equation (6.26) also gives the minimum-contrast bright-field reference focus (see Section 12.5).

In practice, the best dark-field high-resolution images have been obtained using tilted illumination with the unscattered beam intercepted on the rim of an aperture placed centrally about the optic axis, or using annular illumination (see Section 4.4). The optimum defocus for this mode is the same as that required for the ideal beam stop (BS) and the best aperture size to use can be found from the transfer function. Notice that it is possible to include almost twice the range of spatial frequencies that could be included using axial illumination and a translated aperture (Fig. 6.12). Chapter 7 discusses super-resolution schemes based on this mode. The semi-angle subtended by an aperture is easily measured by taking a double exposure of the objective aperture and the diffraction pattern of a crystal of known structure.

A clue to the kind of false contrast which may appear in dark-field images of non-periodic specimens may be obtained from eqn (6.8) (Cowley 1973). The image intensity formed with an ideal central beam stop is

$$I(x,y) = \sigma^2 \left|\phi_p(x,y) - \bar{\phi}\right|^2 \tag{6.27}$$

where $\bar{\phi}$ is the average value of the projected potential (equal to the excluded zero-order Fourier component) with the average taken over a coherently illuminated region. This equation suggests that both positive and negative deviations from the mean potential could produce the same contrast in dark field. That is, both a void and a heavy-atom inclusion would appear bright. It is also possible to show that a sinusoidal object produces an image whose periodicity is half that of the object, if the dark-field image is formed with a central beam stop. This is similar to the formation of 'half-spacing' fringes when three beams are used for lattice imaging (Section 5.1). Physically, dark-field images are formed by interference between Fourier components across the full width of the aperture. In bright field, the important contribution is from interference between the central beam and a particular Fourier component or spatial frequency. Thus, image detail may appear in dark-field images on a scale which is finer than that seen in a bright-field image taken with the same aperture.

More realistic aperture arrangements, such as tilted-illumination dark field (TIDF), introduce an additional distortion in the form of an elongation in the image in the direction of a line between the unscattered beam and the optic axis. This 'Schlieren' distortion is similar to the effect of astigmatism and can be corrected accordingly. Figure 6.13 shows the individual atoms of the thorium oxide lattice imaged using this method (Hashimoto et al. 1973). Thus, in practice, it does appear to be possible to make small molecules and atoms appear 'round' by careful adjustment of the stigmator. Leaving the distortion aside, we may say that *the intensity variation in a high-resolution dark-field image is proportional to the square of the deviation of the specimen projected potential from the mean projected potential*.

174 Imaging molecules: radiation damage

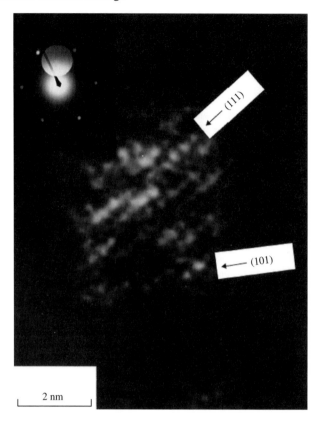

Figure 6.13 Experimental image of thorium oxide obtained by H. Hashimoto using the TIDF method. The Miller indices of two identified planes of atoms are shown. Also shown is the diffraction pattern of the graphite substrate, indicating the position of the objective aperture between the substrate reflections, which is therefore excluded from the image. Inelastic scattering around the central beam is also excluded by this method. The objective aperture size must be matched to the substrate used and the focal length of the objective lens.

6.6. Inelastic scattering

Electrons which lose energy within the specimen suffer a wavelength change and therefore contribute an out-of-focus background to molecular images, generally incoherent with the elastic wavefield. These image intensities can be summed in an image calculation if the energy loss distribution for the specimen is known. If the microscope is focused for electrons of energy V_0, the focus defect for other energies can be described by an additional term in the transfer function, as described in Section 2.8.2. The overall effect of this blurred, incoherent background is then to reduce the elastic image contrast. Note that the expression for image broadening given in Section 2.8.2 deals with the incoherent images of neighbouring

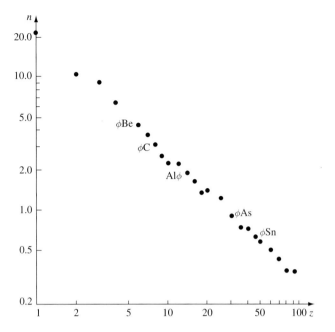

Figure 6.14 The ratio n of total inelastic to total elastic scattering as a function of atomic number for 80 kV electrons. The open circles (with error bars) are experimental points and include scattering angles up to 0.13 rad. Less than 3% scattering occurs outside this range. The experimental points were obtained from amorphous materials similar to biological specimens. The points give the predictions of the theory due to Lenz (1954), which neglects the small plasmon contribution.

point sources. For most thin biological specimens, the only way the microscopist can control chromatic aberration is by choosing a specimen position which minimizes C_c (see Fig. 2.17); however, this can only be done by increasing C_s with a consequent degradation in resolution.

Experimental measurements using energy-selecting microscopes to measure the ratio of inelastic to elastic scattering for amorphous carbon (Egerton 1975) suggest that inelastic scattering dominates by a factor of about 3 if all angles and energies are included (see Fig. 6.14). The objective aperture will increase this ratio to perhaps 10, since the important inelastic processes for amorphous carbon are confined to small angles around the central beam. Practically all the inelastic scattering occurs within a milliradian of the central beam under single-scattering conditions. The mean free path for inelastic scattering in amorphous carbon, which gives an indication of the average distance between scattering events, is about 70 nm at 100 kV. With a substrate much thinner than this, most of the inelastic scattering is therefore excluded in the TIDF technique. A calculation of the inelastic and elastic scattering from amorphous materials, including coherence effects, is difficult, since the atomic coordinates must be known (Howie et al. 1973). Incoherent calculations, accurate for large thickness, have appeared (Misell 1973); however, it is the coherent Fresnel diffraction and possibly any microcrystalline structure in 'amorphous' carbon when used as a substrate which gives rise to the unwanted background 'noise'. The use of graphene as a substrate avoids these issues.

For crystalline substrates where the objective aperture is placed between the sharp Bragg reflections of the substrate, scattering will occur into the aperture owing to the inelastic processes of phonon excitation, plasmon excitation, and single-electron excitations. The intensity of this scattering can be computed if the substrate thickness is known, but an experimental measurement is simpler. An example of an attempt to fit theoretical estimates of diffuse inelastic scattering in a crystal to experimental measurements can be found in Howie and Gai (1975). There will also be an elastic contribution to the image background in these experiments from any non-periodic crystal structure such as surface contamination or crystal defects.

For isolated atoms, the proportion of elastic to inelastic scattering depends on atomic number. At low atomic number inelastic scattering dominates, while at high atomic number the scattering is predominantly elastic. The important early work of Lenz (1954) gives the ratio of inelastic to elastic scattering as inversely proportional to atomic number with a value of unity for copper. Figure 6.14 shows this result compared with experimental measurements (see also Burge 1973). Again the inelastic scattering is narrowly forward-peaked with a half-width of much less than a milliradian. Thus, for a single atom imaged in bright field all the inelastic scattering is included within the aperture, while only a portion of the elastic scattering is included. As an example, the bright-field image of a single aluminium atom is found to include only 3% of the elastic scattering but about 80% of the inelastic scattering, when using an optimum aperture, at 100 kV (Whelan 1972).

The precise form of an image formed only from inelastically scattered electrons depends on the particular inelastic process. Images formed from plasmon-scattered electrons or valence electrons are not expected to contain high-resolution detail, owing to the non-localized plasmon interaction (Craven et al. 1978). However, images formed from electrons which have lost energy owing to inner-shell excitations should show high-resolution detail, thereby allowing the possibility of image formation from a particular atomic species using a selective energy filter 'tuned' to this loss peak, as demonstrated in Section 13.2 using STEM. There are, however, two difficulties with this technique. First, as mentioned above, only large-energy-loss processes are expected to provide high spatial resolution, for reasons based on the uncertainty principle applied to the transverse momentum transfer involved in the inelastic event (Spence 1988b). Secondly, one has a strong mixture of multiple inelastic and elastic scattering present in the specimen, and it is very difficult to devise experimental conditions which truly separate the two. Both these issues are discussed in more detail in Section 13.2. It is notable that in thicker samples the electron diffraction patterns can only be accurately matched to simulations if an elastic energy filter is used (Spence and Zuo 1992).

The contrast improvement which results when the inelastic image contribution is removed from the image by an energy-selecting electron microscope is shown in Fig. 6.15. The image formed by elastically scattered electrons is compared with that formed using all electrons scattered by the specimen. We can conclude that for high-resolution imaging the inelastic background provides a slowly varying low-resolution image which has the effect of reducing the high-resolution elastic image contrast.

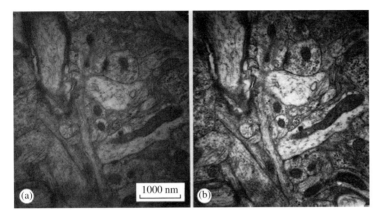

Figure 6.15 Energy-filtered electron images. The image on the left (a) has been formed in the normal way using all electrons. That on the right (b) was formed using only those electrons which are elastically scattered in the specimen and lose little energy. Chromatic aberration due to specimen energy losses is thereby almost eliminated. The specimen is visual cortex tissue (stained). The importance of these inelastic electrons depends strongly on the specimen thickness—an optimum thickness exists since the elastic scattering falls to zero at large thickness but initially builds up with thickness. Here removal of inelastic scattering leads to an improvement in image contrast. (Image courtesy of R. Egerton.)

6.7. Noise, information, and the Rose equation

A measure of the noise in an electron image is given by the standard deviation of the distribution of electrons contributing to each pixel. For Poisson statistics this is $\sqrt{N}$ where N is the mean number of electrons per pixel, so that the signal-to-noise ratio S is $N/\sqrt{N} = \sqrt{N}$ for the rather unrealistic case of a dark-field image recorded with no background noise contribution.

While electron noise is the dominant source of noise in an electron image, there is a further contribution to image noise from the detector itself. Detective quantum efficiency (DQE) is defined as the ratio of the square of the signal-to-noise ratio measured from the output of the detector to the square of this ratio for the electron noise arriving in the electron beam (Hamilton and Marchant 1967). It thus provides a measure of the noise introduced by the detector. A DQE of 0.75 indicates that the signal-to-noise ratio in the recorded image is 87% of that in the electron beam. Most modern detectors have DQEs in the range 0.5–1.

An impression of the severity of electron noise can be obtained by analysing a typical bright-field phase-contrast image. Taking the brightness of a hair-pin filament to be about 3×10^5 A cm^{-2} s^{-1}, the object current density becomes about 1 A cm^{-2} for an illuminating aperture semi-angle of 1 mrad (eqn 9.1). The total charge passing a 0.3 nm^2 image element (referred to object space) after a 2 s exposure is then 1.8×10^{-15} C, corresponding to about 11 250 electrons. With 10% contrast we would expect a reduction of $\Delta N = 1125$ electrons in the vicinity of a heavy-atom image (see Fig. 6.16). The error

178 Imaging molecules: radiation damage

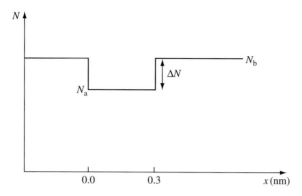

Figure 6.16 Simplified representation of an atom image. A total number N of electrons contribute to the bright-field image within the image area. The background, for the same area, is made up from N_b electrons.

(standard deviation) in an estimate of the peak area would be given by the square root of the variance (eqn 2.40)

$$\sigma^2(\Delta N) = \sigma^2(N_b - N_a) = \sigma^2(N_b) + \sigma^2(N_a) \approx 2N_b \qquad (6.28)$$

since $\sigma^2(N) = N$ and $N_a \approx N_b$. Thus $\sigma(\Delta N) = 2\sqrt{N_b}$ and the electron signal-to-noise ratio for a bright-field single-atom image is

$$S = \Delta N / \sqrt{2 N_b}. \qquad (6.29)$$

Applying this result to the given example gives $S = 7.5$; $S = 5$ is a commonly accepted minimum value. Notice that the signal-to-noise ratio is improved by increasing the image element size, so that a trade-off between resolution and noise is expected. This trade-off between dose and resolution d is expressed by a modified form of the Rose equation (Rose 1948) originally developed for television tubes, as follows. We define contrast $C = \Delta n/n$, where $n = fn_o$ and n_o is the number of electrons incident per unit area, f being the fraction contributing to background (e.g. $f = 0.95$ for bright-field, $f = 0.001$ for dark-field, and f can also be used to represent detector inefficiencies etc.). The idea is that the signal to noise ratio $\Delta N/N^{1/2}$ between adjacent resolution elements for a given contrast should exceed about $k = 5$ (Rose's condition), so that intensity changes due to genuine scattering variations in the sample exceed those due to statistical fluctuations of the number of electrons N in the beam per resolution element. The contrast, a property of the sample, determines how many scattered electrons are available to meet this condition. Then, with $\Delta N = NC = fn_o d^2 C$, we find that the minimum dose needed for contrast C is

$$n_o > k^2 / (fC^2 d^2) \qquad (6.30)$$

giving a lower limit on dose for given contrast and resolution. For example, in bright-field with $f = 1.0$, $C = 0.1$, $k = 5$, and $d = 1$ nm, we have $n_o = n_c = 25$ electrons Å^{-2} as the critical dose. Making resolution d the subject of this equation and setting n_o equal to the critical dose n_c which begins to destroy the sample, we see that, for a given contrast, a minimum value of resolution d is fixed. Resolution evidently goes inversely as the square root of dose, and fine detail is destroyed first during an exposure. It is this equation which is used to justify the claim that a single molecule cannot be imaged by HREM without periodic averaging. (Plate 5 shows an image of a single DNA molecule, which has been averaged five times along its length.) The critical dose is sometimes defined as the dose which causes a reduction of the first-order Bragg intensity to exp(–1) of its initial value.

The choice of magnification M for radiation-sensitive microscopy of isolated objects can now be made by setting $M = d_p/d$, where d_p is the detector pixel size, discussed for various detectors in Chapter 9. The application of the Rose equation to imaging of crystals is discussed in Section 6.7.

The question of electron noise becomes important for radiation-sensitive materials for which the object current density must be limited. The relationship between contrast, resolution, and noise for biological specimens is discussed in more detail in Glaeser (1971) and Glaeser et al. (2007).

We now consider the amount of information in electron images (Saxton 1978, Frank 2006). The approximation of Gaussian, additive noise is frequently made to simplify the analysis of filtering operations. This approximation is justified for bright-field images where N_b is large so that the variation of $\sqrt{N_b}$ with N_b is slow. The Poisson distribution becomes Gaussian when the mean is large. Electron noise is also stationary in the sense that its average value is independent of the choice of time origin, so that many of the techniques of stationary time-series analysis are applicable to electron images.

Statistical fluctuations are responsible for noise contrast, and this has an important consequence for image digitization. As with all analogue-to-digital conversion, the number of discernible discrete levels (in this case the grey levels on the micrograph) are limited by noise superimposed on the analogue signal. A relationship can be obtained for the average number of electrons per image element required for p grey levels at a signal to noise ratio S (Van Dorsten 1971):

$$N = p(p+1)S^2/2 \qquad (6.31)$$

An acceptable image has $p = 5$, $S = 6$. The assumption has been made that there is no correlation between different image elements. The effect of correlations, which always exist between different parts of the image in a partially coherent aberrated optical system, on the information content of an image is discussed in Fellget and Linfoot (1955).

The information contained in a microscope image is measured in bits and is given by the logarithm to the base 2 of the total number of possible pictures. If the electron noise limits the number of grey levels to p per image element with n image elements, the maximum information content is

$$I = \ln_2(p^n) = n \ln p \qquad (6.32)$$

For a micrograph of area A recorded by an instrument with a point spread function of size a in the image space, $n \approx A/a$. The value of p is seen from eqn (6.31) to depend only on the electron exposure. At 0.35 nm resolution an 11 cm × 5 cm micrograph recorded at 200 000× magnification is thus seen to contain about 1.8×10^{12} bits of information.

Equation (6.32) is obtained under the assumption that all images are equally probable, which must be made in the absence of a priori information. The probability that a particular image is recorded is then p^{-n}. The information content of a message or picture is an 'inverse' measure of its probability of occurrence—unlikely messages are supposed to contain a lot of information, while those which are certain to occur (probability of occurrence unity) contain no information or 'surprise'. By taking the negative logarithm of the probability of an event occurring, a suitable additive measure of information is obtained. The quantity so defined depends only on the likelihood of the event (or image) occurring, obtained from past observations. It tells us nothing about the image's usefulness to the observer. The definition of information can, however, be extended to the case where a priori probabilities can be assigned to individual image elements (Brillouin 1962).

The concepts of information theory have become increasingly important in HREM of biological specimens, in which one wishes to extract the maximum amount of structural information from the specimen while delivering the smallest possible radiation dose. The use of entropy and information concepts for the analysis of HREM images of amorphous carbon is described in Fan and Cowley (1988).

6.8. Single-particle cryo-electron microscopy: tomography

Three-dimensional imaging of biological structures by TEM has its origins in three important papers published in 1968. The first (DeRosier and Klug 1968), on helical viruses, showed how these could be reconstructed in three dimensions from a single image, projected normal to the axis. (For these structures, their diffraction pattern intensity is independent of rotation about the axis of the helix.) The second gave general principles for reconstruction of asymmetrical bioparticles from a sufficient number of projects (Hoppe et al. 1968), while the third discussed the improvement in noise when two-dimensional data sets are merged into a three-dimensional reconstruction (Hart 1968). Three forms of three-dimensional molecular imaging emerged from this work—tomography, single-particle imaging, and two-dimensional crystals. Practical experimental details of single-particle and two-dimensional crystal methods, with theoretical background, can be found in the excellent text by Glaeser et al. (2007). Tomography refers to the reconstruction of a three-dimensional image from many projections of the *same* particle, usually stained. Since radiation damage is necessarily severe when multiple exposures of the same particle are used, resolution is limited to at best about 2 nm (Zhang and Ren 2012) (see Frank (2010) and Plitzko (2002) for reviews). But the dose-fractionation theorem tells us that the signal-to-noise ratio required to detect a feature is the same in a single, two-dimensional projection as it is in a full, three-dimensional reconstruction, provided that the same total electron exposure is used (Hoppe and Hegerl 1981, McEwen et al. 1995). Images for tomography might be recorded in 2° increments over a range from −60° to +60° about a single tilt axis, resulting in a missing wedge of data, and reconstructed by, for example, filtered back-projection methods, using gold nanoparticles as fiducial alignment aids. The filter takes account of the more sparse sampling which occurs at high angles in reciprocal space, as the Ewald sphere rotates in fixed angular increments. A dose of perhaps 2 electrons Å^{-2} is common. At 100 kV, a 'dose' of incident illumination of 1 electron nm^{-2} corresponds to a true dose of 3.2×10^4 Gy (energy/mass) for a protein sample. A mathematical solution to the reconstruction problem, the Radon transform, had been given early in the 20th century, but it was the tomographic reconstruction of the helical structure of the T4 phage tail in 1968 by DeRosier and Klug, using images recorded on film, which originated modern electron tomography. (Two Nobel prizes are associated with the development of tomography, one for axial tomography in clinical medicine in 1979, and Klug's 1982 award, in part for his contribution to tomography in electron microscopy.) The later introduction of the CCD camera (Spence and Zuo 1988), allowing immediate electronic processing and reconstruction, unlike film, greatly facilitated this technique. The difficulty of developing these for 1 MeV microscopes may have contributed to the decline of this HVEM approach for tomography of thick objects such as cells. 'Sub-tomogram averaging' has also been developed, in which tomographic reconstructions from identical particles are merged to improve resolution. The second form of three-dimensional molecular imaging, usually referred to as 'single-particle cryo-EM' (not tomography), improves on the tomographic approach by merging phase-contrast images from many copies of the molecule of interest, each one of which receives a dose too small to cause serious damage. (An alternative approach to avoiding radiation damage is to use very brief pulses of radiation from an X-ray laser or fast electron gun, which terminate before damage commences, yet contain enough radiation to give a useful diffraction pattern. See Spence et al. (2012) for a review.) These copies may either lie in random orientations within a film of vitreous ice (as in the single-particle cryo-EM method) or form a two-dimensional crystal, as discussed in the Section 6.9, in which case one gains by the coherent amplification of the Bragg diffraction process and hence obtains higher resolution. It is interesting to estimate the smallest number of copies, for a given critical dose per molecules, from which an image of the average molecular structure can be synthesized at a given resolution (eqn 6.34). The smallest protein crystal, for example, which can produce a high-resolution three-dimensional reconstruction has been estimated as 20 000 Da (Glaeser 1999). Using two-dimensional molecular crystals, at best, a resolution of a few angstroms has been achieved in the three-dimensional reconstructed cryo-EM images, and it should be noted that the alpha helices of the protein fold can be resolved at 6 Å resolution, while about 3 Å is needed to identify and distinguish the individual amino acid residues. Model-fitting, based perhaps on the protein with the most similar sequence in the massive Protein Data Bank archive of solved protein structures (with more than 80 000 entries) can be used to improve resolution. Low-dose methods were subsequently developed (Unwin and Henderson 1975), and an important step forward resulted when methods were developed for embedding unstained molecules in vitreous ice (Adrian et al. 1984). The field-emission gun also had a major impact, providing much sharper phase contrast effects, which, by allowing more accurate alignment of images, improved resolution. Because the refractive index difference for electrons between protein and ice is so small, the single-particle images (with many particles in one image) must be recorded several microns out of focus, so that CTF correction becomes crucial. The phase change for 200 keV electrons in water (density 1 g cm^{-3}) is 36 mrad nm^{-1} (approximately proportional to mass density), while the relative phase shift for protein embedded in vitreous ice is about 14 mrad nm^{-1}. The maximum phase contrast could ideally be therefore be twice this phase modulation (eqn 6.4), or $\Delta I/I = 0.028t$ in a sample of thickness t (nm). Theoretical estimates (Rosenthal and Henderson 2003) suggest that an atomic-resolution three-dimensional reconstruction should be possible by merging no more than 12 000 images of identical particles in random orientations, using proteins as small as 40 kDa. Experience shows that the dose must be kept much below 20 electrons/nm^2 for sub-nanometre resolution. It is clear that a Zernike quarter-wave phase plate for electrons would help greatly by introducing a sharp phase shift of 90° between direct and all scattered beams, giving a high-contrast in-focus image (unlike equation 6.4 for $\Delta f = 0$) and, because these allow more accurate alignment of the images, better resolution. There has been a continuous effort to develop such a device for over several decades, based on three main approaches—carbon films (see Danev et al. (2010) for impressive results), electrostatic minilenses which impart a phase shift, and through which only the central beam in the diffraction pattern passes (Cambie et al. 2007), and a new approach based on use of an infrared laser, focused onto the central beam of the electron diffraction pattern in an optical cavity (Mueller et al. 2010). (This is based on the Kaptiza–Dirac effect, the diffraction of electrons by light.)

Improvements in computing power to facilitate data collection, CTF correction, and determination of molecular orientation proceeded steadily in the decades around the end of the last century. But these bright-field, low-contrast images, consisting of many particles per image, required above all the largest possible number of detector pixels (to provide many pixels per particle), while at the same time only moderate dynamic range was needed in view of their low contrast. Hence only very recently has electronic recording been able to provide enough pixels to become more useful than film (with its archival data storage advantages), especially so with the new direct detection detectors described in Section 9.10.

Many important biomolecules, such as the membrane proteins important for drug delivery, are extremely difficult or impossible to crystallize in three dimensions, and for these this single-particle cryo-EM method has proven invaluable, by providing high resolution images of the molecular shape. The interaction and docking of molecules (such as a small drug molecule binding to a larger protein) can then be understood in terms of their shape, and so related to their function. Ideally one would like to draw conclusions about the binding energies of small drug molecules to different sites on a protein from the three-dimensional density maps. The cells of eukaryotes contain a skeleton of fine microtubules, along which proteins are transported, carrying out the various functions of the cell. As one dramatic example, we cite the fitting of the protein structure of tubulin (in this case determined by electron crystallography) into a model of the microtubule obtained by single-particle cryomicroscopy (Nogales et al. 1999). The location of the anti-cancer drug molecule taxol can be identified in this structure, and its function elucidated. There is also the large number of uncrystallizable 'molecular machines', consisting of many component proteins whose individual structures are known from X-ray crystallography. The shape and organization of these macromolecules might be determined by the single-particle cryo-EM technique described in this section.

The important question of whether high-resolution imaging of a single biomolecule, using scattered radiation, is possible at all, or whether the radiation dose needed to do so would destroy it, has been discussed in many papers (Breedlove and Trammel 1970, Glaeser 1979, Spence et al. 1999). The general conclusion, based on the ratio of the elastic to inelastic cross-sections (Henderson 1995) is that this is not possible. (Plate 5 nevertheless shows a HREM image of a single unsupported, unstained DNA molecule. The image has been averaged five times along its length, and is about as close as we can ever hope to get to this goal.) To improve on this, it is necessary to use some form of averaging, or use femtosecond pulses which outrun damage.

The cryo-EM method should be seen in the context of other techniques which provide similar information. At lower resolution, thin ultramicrotomed sections of whole cells, usually stained, may be imaged and tilted in the high-voltage electron microscope, and the sectional views reconstructed to provide a three-dimensional (tomographic) image at a resolution of perhaps 2 nm (Renken et al. 1977, Porter 1986). No other technique can match this capability for studying the ultrastructural organization of large three-dimensional macromolecular assemblies; however, the method is slow and there are always concerns about the effects of dehydration, radiation damage, and staining. It is generally accepted that below about 2 nm resolution, little significant structural information can be extracted from stained samples. TEM imaging of molecules tagged with different sized nanoscale gold balls is also possible. A comparison of X-rays, neutrons, and electrons for imaging in biology can be found in Henderson (1995). A comparison of magnetic resonance techniques, X-ray crystallography, and electron microscopy is given in Chiu (1993) (membrane proteins are usually too big to be studied by NMR, which is restricted to weights below about 50 kDa). Advances in atomic force microscopy for molecular imaging, mainly in aqueous environments (where real-time imaging in water has recently been achieved) is reviewed in Lindsay (2000), and at low temperatures in Shao (1996). The optical fluorescence microscopies can also provide motion pictures of identified molecules in their native environment, but resolution is limited by the wavelength used. By using semiconductor nanodots or fluorescent protein mutants which fluoresce with different colours as labels for different proteins, the Rayleigh resolution limit may be exceeded in filtered images, while structured illumination, two-photon, multilens, near-field, stimulated-emission depletion, and confocal imaging are currently transforming optical imaging in biology. (Reviews of many of the new super-resolution optical microscopy techniques can be found in Hawkes and Spence (2006) and Diaspro (2010).)

Following earlier work on viruses, the first major success using single-particle unstained methods in cryo-EM was imaging of the ribosome, the molecular machine which makes proteins according to instructions from DNA. The approximately 35 000 genes in the human genome code for a rather larger number of the protein molecules which make up the bulk of all living organisms. Human proteins consist of certain folded sequences of the 20 distinct amino acids, selected according to the genetic code, expressed by the nucleic acid sequence in the DNA of the genes. There are four nucleic acids, arranged in pairs as the rungs of a twisted ladder (the double helix), three of which (reading along one side of the ladder) code for each amino acid. This allows for specification of $4^3 = 64$ (rather than 20) amino acids. The nucleic acids are arranged in complementary pairs in DNA. The proteins are synthesized by the ribosome molecule, a collection of proteins and nucleic acids. In brief, messenger RNA (mRNA) reads off the sequence of three-bit 'words' for one protein from one gene sequence along DNA and transfers this to the ribosome. Transfer RNA (tRNA) acts as an adaptor to bring the required amino acids to the ribosome—there is one tRNA for each amino acid, with connectors at one end for the amino acid and at the other for the triplet code on the mRNA. The ribosome runs along the mRNA chain as the amino acid is built in a series of repetitive steps, in which the ribosome moves to a new triplet on the mRNA, connecting new amino acids via tRNA to the growing chain (see Plate 1). Partly because of its size, the ribosome macromolecule has proven very difficult to crystallize. We now briefly review one cryo-EM project aimed at determining ribosome function from its shape, and provide references to other similar projects. While the resolution in this work was limited to about 0.6 nm, the atomic structure of some of the ribosome subunits which can be crystallized is known from X-ray crystallography, so that a detailed understanding of overall function can now be obtained almost at the atomic level. Subsequent X-ray analysis of crystallized ribosomes, based in part on this electron microscopy work, led to the award of a Noble prize in 2009.

The 'dose' (perhaps 10 electrons $Å^{-2}$ of incident illumination) is chosen so that each molecule receives a dose below that which initiates significant damage, but which will be insufficient to provide a statistically significant image of any one molecule at the desired resolution. The Rose equation (eqn 6.30) links resolution, dose, and contrast for one molecule. In order to sum images from different molecules, so that a statistically significant image can be obtained at better resolution, their orientations must be determined, and the images rotated into coincidence. The accuracy of alignment depends on the size of the molecule, since it depends on the number of features which can be used, and the minimum size has been estimated to be about 50 kDa, slightly greater than the maximum size which can be studied by NMR (Henderson 1995). To determine the three-dimensional angular relationship between molecules, the sample, containing many identical molecules lying in different orientations, may be tilted by a known amount and the new image compared with the first. Orientation relationships can then be established between different particles on the same grid imaged at different tilts. To reduce damage, low-dose methods may be used, or a new sample, containing a fresh selection from a random assortment of identical particles, may be used for each tilt. Initially, molecules are assigned to similarity classes. Projections differing by small tilts are similar, and it is sufficient to determine the sequential relationship between particles along a tilt path. Particles which share a common tilt axis may be rotated into coincidence and added together. Cross-correlation functions are used for this alignment process. The entire process of reference-free alignment and hierarchical classification using Ward's merging criterion is described in detail in Frank (1990). The various summed images, projected from known directions, must now be combined to form a three-dimensional image. A considerable literature, outside the scope of this book, exists on the problem of three-dimensional reconstruction from projections (tomography), which arises in many fields. These methods include real-space methods such as algebraic reconstruction (ART), simultaneous iterative reconstruction (SIRT), iterative least-squares methods (ILST), and filtered back-projections. Fourier methods also exist, including methods for both equal and unequally space points (see Frank (2006) for a review). An important result from this work is that the resolution d depends on the number of projections N and the size of the object D as (Crowther et al. 1970)

$$d = \pi D/N \tag{6.33}$$

Using one of these methods, a three-dimensional image of a single molecule may be built up from the summed projections with acceptable signal-to-noise ratio at the 1 nm resolution level or better. The limited tilting range results in a missing cone of data, which results in an anisotropic point spread function. To avoid staining artefacts, samples are preserved in their frozen hydrated state in a thin (e.g. 50 nm thick) layer of amorphous ice using the method of rapid freezing (Adrian et al. 1984). The image analysis is based in all essentials on eqns (4.8a) and (6.3), and the determination of instrumental parameters by automated refinement is a crucial step. It has been shown that the entire three-dimensional reconstruction process and correction for contrast transfer function can be described by a single matrix representation (Frank 2006), combined with linear regression. The weak-scattering linear bright-field imaging approximation used is often taken to hold for thicknesses of up to about 30 nm at 200 keV for the light elements (H, C, N, O) important in biology—errors due to multiple scattering have been estimated in several papers (e.g. Glaeser and Downing 1993). Images are recorded at several focus settings and the results combined after correction for the transfer function (see Chapter 7). Large focus defects are needed (e.g. 1–2 μm) in order to obtain sufficient phase contrast between the ice and the molecule. The analysis of the images involves many steps, described in detail in Frank (1990, 2006) (for details of the algorithm used to analyse the ribosome see Frank et al. (1996)). Several other large computer algorithms have been described for this purpose (e.g. SPIDER, IMAGIC, EM, MDDP, and SEMPER) which may be found on the web. Figure 6.17(a) shows the as-recorded image of 70S *Escherichia coli* ribosomes in vitreous ice (Frank et al. 1995). The image was recorded below −170 °C using an Oxford Instruments helium-cooled side-entry holder and an in-column, elastically filtering TEM (LEO 912) using Kohler illumination. A dedicated liquid helium-cooled side-entry stage for

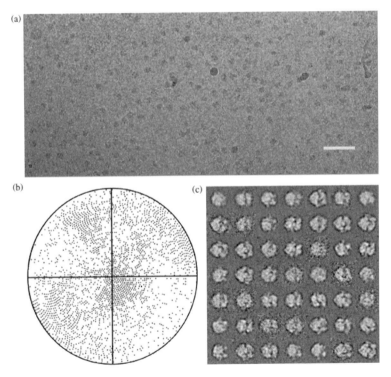

Figure 6.17 (a) Elastically energy-filtered, low-dose, HREM images of ribosomes in random orientations embedded in ice about 50 nm thick (scale bar 100 nm). Large defocus (about 2 μm) must be used to distinguish macromolecule from ice. (b) Orientations of the particles deduced shown on a (θ, ϕ plot). (c) Averages of cluster found by classification (Frank 2006).

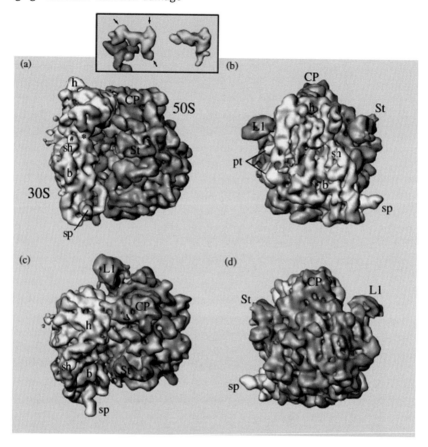

Figure 6.18 Protein manufacture. Cryo-EM map of the *E. coli* 70S ribosome at 1.15 nm resolution (Gabashvili *et al.* 2000): (a) subunit–subunit side view, (b) 30S subunit–solvent side view, (c) top view, (d) 50S subunit–solvent side view. For 30S: b, body; h, head; pt, platform; sp, spur; sh, shoulder; St, stalk. Over 73 000 projections of this ribosome bound to tRNA were used to obtain this reconstruction, on which can be identified RNA helices, peripheral proteins, and inter-unit bridges.

biology is described in Fujiyoshi *et al.* (1991). The benefits of elastic filtering for quantification of electron microscopy data have been clearly demonstrated both for the thin samples used in biology and for the inorganic samples used in materials science, where accurate quantification is impossible without filtering (see Section 13.3). Figure 6.17(b) shows the distribution of the Eulerian angles for the particles from three data sets, one for each focus. Figure 6.17(c) shows the average of 49 classes found after classification from two data sets. Figure 6.18 shows final three-dimensional views of the unstained *E. coli* 70S ribosome molecule at 1.15 nm resolution (Gabashvili *et al.* 2000). A particular choice of charge-density surface must be made to define a suitable boundary for the visual representation of the molecule. Based on these images, the authors were able to identify possible pathways for incoming messenger RNA (see Plate 1) and exiting nascent polypeptides, in addition to suggesting the locations of the anticodon-binding region and peptidyltransferase centre.

Electron tomography also complements X-ray work. For example, it may be used to confirm the positions of the heavy atoms used in the multiple anomalous diffraction (MAD) method used for phase determination in X-ray crystallography. The knowledge of the external shape of a macromolecule which TEM can provide also assists in solving the X-ray phase problem, according to the principles of the solvent-flattening or density modification techniques (see Section 7.9). Specifically, three-dimensional cryo-EM allows studies of the interaction of the ribosome–ligand interactions, which may not be possible if crystallization is needed. (The constraints imposed by packing the molecules into crystals may prevent the interactions with other molecules from being observed.) Unless sorting into a few conformations can be achieved, an assumption of the method is that all molecules are identical—if they exist in many conformations, the resolution of the final reconstruction will be limited. A review of this field can be found in van Heel et al. (2000).

Some experimental requirements for this work include either a large-area CCD cameras (Faruqi et al. 1999, Zuo 2000), an image plate, film-recording, or a direct-injection detector system (see Chapter 9), a field-emission electron source, a spot-scanning system (Downing and Glaeser 1986), perhaps dynamic focusing to refocus the region in highly tilted samples (Downing 1992), and the cryo-sample preparation and transfer systems described in Glaeser et al. (2007). Unbending algorithms have been described to allow for sample bending effects in the reconstruction (Downing 1992).

The quantification of cryomicroscopy images has been more successful than that of images of inorganic crystals, partly because images of proteins in ice are insensitive to the mean inner potential of the ice if the ice slab is parallel sided (only the difference between the mean potential of the ice and the protein matters), allowing it to be rather thick if the atomic structure of the ice is not resolved. By comparison, the aligned columns of the much heavier atoms in an inorganic crystalline sample can generate very large phase shift variations (greater than 2π) across the exit face of the sample, so that mulitple-scattering artefacts rapidly build up with thickness.

The achievable resolution of single-particle cryo-EM reconstructions is limited by many considerations, including the total number of single-particle images, particle conformation uniformity, specimen preservation, detectors, radiation damage, imaging exposure, defocus, spatial and temporal coherence, beam alignment, errors in image reconstruction, and other factors (Rosenthal and Henderson 2003, Glaeser et al. 2007). Perhaps the highest resolution yet attained is 3.3 Å, in single-particle imaging of a virus by Zhang et al. (2010). The single-particle method becomes more difficult for particles without symmetry, and those which support a large number of conformations. Where single macromolecules occur in several conformations within the ice film, these can be sorted both by orientation and conformation class. If these conformations can be correctly sequenced, a form of 'molecular movie' can be obtained, as described by Fischer et al. (2010) for the case of movement of tRNA through the ribosome during translocation. This problem of sorting either diffraction patterns or images according to conformation and orientation has received much attention in connection with 'snap-shot' diffraction from single particles using an X-ray laser (see Spence et al. (2012) for a review). Another important recent development is the direct-injection electron detectors, discussed with results for cryo-EM in Section 9.10.

6.9. Electron crystallography of two-dimensional crystals

Section 6.8 described the general 'single particle' cryo-EM methods than can be used for any biomolecule which cannot be crystallized; however, resolution with that method has been limited to about 0.6 nm except in special cases. For those molecules which can be formed into artificial two-dimensional crystals (or exist naturally so), the method of protein electron crystallography has been developed. Kuhlbrandt (1992) provides a review of methods for forming two-dimensional crystals. This method is now capable of near-atomic resolution (e.g. 0.26 nm, sufficient to resolve individual residues) and was first developed by Henderson and Unwin in 1975 (Unwin and Henderson 1975). Membrane proteins in lipid bilayers are ideal such samples, and the greatest success has been achieved with the bacterium *Halobacterium salinarium* (formerly *Halobacterium halobium*) where three-dimensional images of a membrane protein have been obtained at a resolution of about 0.3 nm. This work produced the first images to reveal the secondary structure of a protein. The membrane protein in question acts as a light-driven proton pump in photosynthesis for the bacterium, which is found in salt marshes, and the molecular mechanism by which the pump operates has now been elucidated in full, largely as a result of the electron crystallography work. Figure 6.19 shows a representation of the final structure based on electron imaging and diffraction data which were obtainable to about 0.3 nm resolution (Mitsuoka *et al.* 1999). The photocycle depends on the isomerization of the chromophore retinal (not resolved in the Fig. 6.19), which is surrounded by the alpha helices shown. The retinal molecule absorbs a photon at 568 nm wavelength, resulting in a series of conformational changes which are thought to pump protons through the cell membrane. These changes in molecular shape result in changes of about 10% in electron structure factors, which have been observed using Fourier difference methods by neutron, X-ray, and electron diffraction. Rapid freezing during the proton pump cycle has thus allowed study of the individual conformations of the molecule during the cycle.

The original Unwin and Henderson work succeeded for several reasons: first glucose embedding was used to substitute for water, moderating the interaction with the supporting carbon grid; second because of the development of the low-dose method (Williams and Fisher 1970); and finally due to the use of the Fourier transform of bright-field images to solve the phase problem (eqn 3.42). The subsequent development of the spot-scanning method (Downing and Glaeser 1986), in which a small beam is scanned over a larger area of sample while this entire area is exposed to the film or CCD camera, further reduces beam-induced movement and charging. A typical acceptable dose is 10 electrons Å^{-2}. The samples are embedded in glucose, trehalose, tannin, or rapidly frozen in amorphous ice (Adrian *et al.* 1984). New preparation techniques are under continuous development, including freeze drying (for high contrast), sugar-embedding and quick freezing, and the use of cubic ice. Vitrification preserves the aqueous environment. The use of liquid helium temperatures has been found to be advantageous for minimizing damage (Kimura *et al.* 1997). An important problem is lack of flatness of the films; this 'cryo-crinkling' may be due to differences between the thermal expansion of the grids and the carbon support film (and their roughness). Charging and damage to the ice can also be important problems—there is an urgent need for doped, conductive ice and this may be provided by coating the ice with graphene, which is conductive. (The use of graphene substrates for cryo-EM is described in Section 12.6.) Both diffraction patterns and HREM images are recorded over a wide range of orientations,

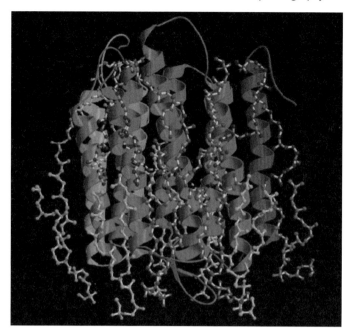

Figure 6.19 Photosynthesis. Representation of bacteriorhodopsin (bR) derived from electron imaging and diffraction data (Mitsuoka et al. 1999). The seven alpha-helices in a single unit cell of the monolayer crystals are seen. Ball-and-spoke structures are lipids. Retinal, responsible for light absorption in the photosynthesis process, lies at the centre of the structure. The absorption of a photon allows the molecule to pump a single proton across the cell membrane.

involving tilts of up to 70° or more. (The tilt limit results in a 'missing cone' of data, making the real-space resolution function anisotropic.) The phase problem is solved by taking phases from the digital Fourier transform of the image, and assuming linear, weak phase-object imaging conditions. However, these phases are only obtainable out to the limited resolution of the electron microscope (in fact more severe limits may apply), whereas diffracted intensities extend to higher resolution. Resolution-limiting factors in this work are discussed in Glaeser (1992). Phase extension methods can be used to supply the missing phases (Gilmore et al. 1992), and iterative methods may be applied which take advantage of the compact support (along the beam direction) offered by a sample in the form of a thin slab (Spence et al. 2003). This iterative phasing (see Section 7.6) reduces the number of images which need to be recorded at high angles. Since the films are typically one molecule thick (perhaps 6 nm) and consist of light elements, errors due to multiple scattering are small (Glaeser and Downing 1993). (The handedness of alpha-helices may therefore be difficult to determine until a three-dimensional high-resolution reconstruction has been obtained.) A large area (many square micrometres) of membrane may be needed to provide a sufficient number of unit cells (molecules) for an acceptable signal-to-noise ratio in the final composite image of one, periodically averaged, unit cell. Since the crystals are highly imperfect, a series of corrections must be applied to the images before analysis, these include 'unbending', background subtraction, local rotations of molecules using cross-correlation functions, and lens transfer function corrections. The sophisticated software developed for applying all these corrections is described in Crowther *et al.* (1996). Field-emission TEMs operating above 200 kV provide improved temporal and spatial coherence, and so have been found to allow finer detail to be extracted from the transfer function correction process. The possibility of extracting information on bonding and ionicity from these diffraction patterns has been investigated (Mitsuoka *et al.* 1999). This analysis must be based on the Mott–Bethe relation between electron and X-ray scattering factors, as discussed in detail in Section 13.3. (The quantity measured by electron diffraction and imaging is the electrostatic potential, not the charge density as measured by X-ray diffraction.) It is important to demonstrate that the effects of charge redistribution between atoms are larger than, for example, the effects of multiple scattering. Bonding effects influence predominantly the low-order reflections accessible to imaging experiments, but attenuated by the $\sin \chi(u)$ lens transfer function at small scattering angles u. Bonding effects may alter structure factor amplitudes and phases by a few per cent in these very large unit cell crystals, and have been imaged directly for dopants in graphene (see Section 5.17).

Apart from bacteriorhodopsin, many other structures have also been studied by this method, providing maps of potential to a resolution of a few angstroms. These include: light-harvesting complex II (to 0.34 nm resolution), bacterial porins, aquaporin-1, holorhodopsin, rhodopsin, gap junctions, the water-channel protein in red blood cell membranes, the nicotinic acetyl choline receptor, plasma membranes from ATPase and microsomal glutathione *S*-transferase. Reviews of the method and references can be found in Walz and Grigorieff (1998) and Fujiyoshi (2011). Recently, artificial periodic three-dimensional lipid structures (lipidic cubic phases; Landau and Rosenbusch 1996) have been created in which the membrane proteins are free to diffuse about until they meet to form three-dimensional crystals, with obvious implications for both X-ray and electron crystallography.

6.10. Organic crystals

Organic crystals were first studied by electron diffraction in the 1930s, and a large body of literature describes the Russian work which followed (for a review in English, see Vainshtein *et al.* (1992)). The first HREM images were obtained in 1970, following the development of epitaxial growth methods for phthalocyanines and aromatic hydrocarbons, and of the low-dose method (see Uyeda *et al.* (1980) and references therein). Figure 6.20 shows a spectacular example of an image of chlorinated copper phthalocyanine, the first to show that HREM can image individual atoms in an organic molecule (Uyeda *et al.* 1980). This material (and its derivatives), which are remarkably radiation resistant, has since become a standard test structure. Thus techniques similar to those used by biologists have been used to study many organic thin films, such as Langmuir–Blodgett films, waxes, and polymers. This field is reviewed in Fryer (1993) and Dorset (1995). Some success has been obtained using the numerical direct methods techniques of X-ray crystallography, applied to selected area electron diffraction patterns in order to determine the phases of the reflections. Here, for lipid bilayers for example, the thickness may be accurately known in terms of the molecular length, and the atoms are light and weakly scattered (Stevens *et al.* 2002). However the effect of the bending of the films on diffraction (which can produce 'secondary scattering';

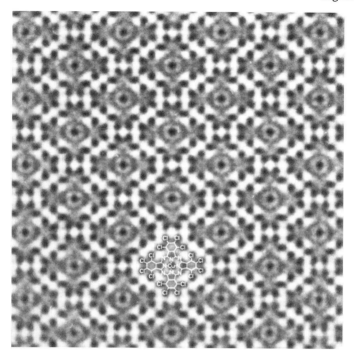

Figure 6.20 Seeing light atoms. Chlorinated copper phthalocyanine. This is the first image to demonstrate that HREM is capable of imaging individual atoms in an organic molecule. No data averaging is used. The molecular structure is shown in the inset (Uyeda *et al.* 1980).

Cowley 1961), and the problem of obtaining structural information along the length of the molecules (normally lying parallel to the beam) creates serious difficulties. (For lipids, these long molecules pack side-by-side in single and double layers. In some cases it is possible, with difficulty, to grow crystals in which the long molecular axis lies in the plane of the film.) Combinations of HREM imaging and diffraction methods have been successful in some cases, and applications to paraffin, alkane derivatives, perchlorocoronene, phthalocyanine, anthracene, and similar radiation-sensitive crystals can be found in these reviews. Lattice images have even been obtained from both monolayers of paraffin (Zemlin *et al.* 1985) and lipids lying on ultrathin amorphous carbon films at low temperature, but this is extremely challenging work in view of the rapid radiation damage—diffraction spots may only be visible for a few seconds. An example of this work, combining imaging with direct methods to phase electron diffraction patterns, can be found in Dorset and Zemlin (1990). The paraffin n-$C_{36}H_{74}$ was grown epitaxially on benzoic acid, and lattice images and diffraction patterns obtained at low temperature. Phases for the diffraction pattern were obtained from the Fourier transform of the HREM image, assuming linear imaging, and lens aberrations corrected. Since few other methods exist for the structural analysis of these films (NMR may be useful, and synchrotron X-ray diffraction patterns are starting to appear from organic monolayers), the combination of HREM imaging and diffraction at liquid helium temperatures seems worth pursuing. This approach is discussed in Chapter 7.

6.11. Radiation damage: organics and low-voltage EM

For X-ray diffraction, most photons do not interact at all with the sample, but pass though it and hit the axial beam dump. A small fraction of the total are annihilated in the sample in the production of photoelectrons. An even smaller fraction undergo elastic scattering to produce the wanted diffraction pattern. The electrons in a TEM, however, can lose any amount of energy on scattering, and do so repeatedly as they move through the sample. Since, in the X-ray case, most photons are annihilated in the production of photoelectrons, they never reach the detector, unlike the electrons, which continue after energy loss to produce background at the detector. However, samples are so thin for HREM that energy-loss spectra show that most electrons are either 'unscattered' (remaining in the central beam), or scatter elastically, thus providing useful imaging information. Inelastic scattering contributes at a different beam energy to the image, and so, because of chromatic aberration, provides mostly a slowly varying out-of-focus background to the high-resolution image if it involves small-angle scattering. Phonon scattering may involve large-angle scattering but very small energy losses, and so may contribute high-resolution detail to HREM images (Cowley 1988).

For the monolayer organic crystals discussed in Sections 6.9 and 6.10 use is made of the redundancy of information in the crystal structure to allow a statistically significant image of the periodically averaged structure to be built up, while applying a sub-critical dose to any one cell. The Rose equation, relating resolution to dose, may be adapted to this situation as follows: let m be the number of unit cells (each containing one molecule) in the image which are added together to form a periodically averaged image. Then eqn (6.30) becomes

$$n_o = k^2/(mfC^2d^2) \qquad (6.34)$$

and we see that the dose n_o needed for contrast C in the averaged image is reduced by the number of molecules, again going inversely as the square of resolution. (A fuller discussion of this issue, including many experimental measurements showing the dependence of resolution on dose for both electron microscopy and X-ray crystallography, can be found in Howells et al. (2009).) If assumptions can be made about molecular symmetry, a further reduction in dose is possible by adding together images after application of these symmetry operations.

Damage has traditionally been measured by the time taken for diffraction spots to fade. The angular dependence of these spots can be modelled using a Debye–Waller factor (eqn 5.54) to give an average lattice displacement arising from the damage, which may then be measured as a function of dose, temperature, and accelerating voltage (Jeng and Chiu 1984). In this way the critical dose which starts to produce damage may be defined. High-order reflections, corresponding to fine detail, fade first. Alternatively, the Patterson function may be studied as a function of dose. The effects of increasing dose from 1 to 14 electrons Å^{-2} on the periodically averaged electrostatic potential of a paraffin layer, reconstructed from transmission electron diffraction (TED) patterns, can be found in Dorset and Zemlin (1985)—the projected hydrocarbon chains lose fine detail and become more rounded. Other critical dose measurements are summarized in Reimer and Kohl (2006)—they vary at 100 kV from about 1 electron Å^{-2} for amino acids, to 8 electrons Å^{-2} for polymers and gelatin, to 30 electrons Å^{-2} for coronene, 240 electrons Å^{-2} for Cu-phthalocyanine, 360 electrons Å^{-2}

for the nucleic acid guanine, and 18 000 electrons Å^{-2} for CuCl_{16}-phthalocyanine. A dose of 1 electron Å^{-2} is about 10^5 times greater than that needed to kill *E. coli* bacteria. Aromatics are found to be more resistant than aliphatics, since the delocalized π bonds in the benzene rings of the aromatics dissipate energy better. The effect is measured by the G-value, or number of chemical events which occur for every 100 eV of energy absorbed. These are typically $G = 3$ (aliphatic) but $G = 0.2$ for aromatic hydrocarbons. The Bethe theory has been compared with measurements for the rate of energy deposition per unit mass-thickness. For a typical organic of unit density (g cm^{-3}), a 100 kV beam loses about 4 eV for every 10 nm travelled, so that with $G = 0.2$, 0.008 irreversible bond scissions will occur from each beam electron in a 10 nm thick HREM crystalline sample. For a resolution of 0.1 nm, eqn (6.30) indicates that 104 electrons nm^{-2} are needed, so that 80 chemical events occur. The area of a phthalocyanine molecule, for example, is about 6.25 nm^2, and there are 26 in a 10 nm thick film. Thus in one column of molecules there will be 19.2 chemical events per molecule during the HREM exposure. It has often been noted that damage starts at corners, defects, and edges. Damage in the DNA bases has been studied extensively by energy-loss spectroscopy (Isaacson *et al.* 1973), and by optical absorption. Infrared absorption spectra provide information on the vibrational changes of the molecules due to damage—a detailed study of damage to amino acids due to TEM irradiation can be found in Baumeister *et al.* (1976). The laser microprobe, electron-spin resonance and nuclear magnetic resonance, and inelastic tunnelling spectroscopy have also been used (Reimer and Kohl 1989). For TEM at 100 keV, the main damage mechanisms in hydrocarbons are ionization, leading possibly to bond breaking, or heating. Loss of hydrogen (4 eV bond energy) and C–H bond breaking are common, and secondary processes such as cross-linking and the diffusion of hydrogen must be considered. Mass loss may occur and has been measured, leading finally to a carbon-rich deposit.

The choice of the correct figure of merit to characterize damage has been controversial. Early workers used the ratio R of inelastic to elastic scattering cross-sections (Breedlove and Trammel 1970) to indicate the ratio of useful image-forming elastic scattering to damaging inelastic scattering. For centric crystals, R is related to the ratio V'_g/V_g of Section 5.9. (Note that detector efficiency also depends on just this inelastic scattering.) The ideal probe for biological imaging would thus consist of particles with a weak inelastic (but strong elastic) interaction at the sample (to minimize damage) which can be ionized or otherwise modified for strong interaction at the detector. For light atoms, R falls to about unity at 100 eV($\lambda = 0.12$ nm, $\sigma_\text{i} = 0.1$ nm^2); below this energy resolution is limited by the de Broglie wavelength. Since at least one scattering event is needed to form an image (see eqn 6.30) and not all the elastic scattering is used to form the image, this suggests that single-molecule imaging using electrons is impossible, a finding consistent with the success of crystal imaging and tomographic methods.

However, complications to this simple picture arise from many effects. These include the effects of secondary electrons liberated by ionization, since their inelastic cross-section increases as they slow down; the differing angular distributions and cut-offs applied to elastic and inelastic scattering; from the fact that coherent bright-field HREM imaging is linear in the scattering potential, so that cross-sections are hardly relevant, and from the fact that not all ionization events result in irreversible nuclear displacements (damage). The critical dose to be used in the Rose eqn (6.30) is rarely known accurately, and depends on temperature, thickness, and other factors. In practice, rather noisy lattice images have been formed from monolayers of several organics (Dorset 1995). The crucial requirement is to make every elastic electron count. Henderson (1995) has used a figure of merit $Q = R\Delta E$, the amount of energy lost by inelastic processes per elastic scattering event, to compare damage in electron, X-ray, and neutron microscopy. Here ΔE is the average energy loss per inelastic event, often taken as 20 eV. With $R = 3$ for TEM, we have $Q = 60$ eV for TEM, $Q = 10^3 \times 400 = 400$ keV for soft X-ray imaging, and $Q = 10 \times 8$ keV $= 80$ keV for hard X-rays. (The poorer values for X-rays result from the larger R values associated with the photoelectric effect and the fact that ΔE is about equal to the beam energy.) These figures strongly favour TEM over X-ray imaging; however soft X-ray imaging using zone-plate lenses at about 550 eV (wavelength 2.5 nm) has achieved a resolution of about 30 nm for samples in water, and cooling is found to produce a much greater reduction in damage than for electrons.

A TEM intended for cryo-EM, corrected for both chromatic and spherical aberration, and designed for operation at 20 and 80 kV, has been constructed at the University of Ulm (the SALVE project; Rose 2009). Equation (2.38) shows that resolution is inversely proportional to beam energy for uncorrected instruments. The possibility of chromatic aberration correction thus makes high-resolution imaging possible at this low energy, where knock-on damage is practically eliminated, and where the phase shift $\theta = \pi\phi_0 t/(\lambda V_0)$, and hence the contrast from very thin samples (for examples molecules on graphene), will be greatly increased. (Ionization damage may, however, be increased somewhat, and penetration is small.) The optimum imaging conditions for this corrected instrument, and experimental graphene images, are given in Lee et al. (2012).

The use of even lower electron energies (a few hundred eV) has also been investigated, since indications are (Isaacson 1976) that it is the carbon K inner-shell event at 285 eV which is most damaging. This is less likely than valence excitation, but dumps much more energy into the system. It was therefore expected (Howie et al. 1985) that spot-fading times should show a threshold increase as beam energies were reduced below 285 eV, and there is some experimental evidence for this in transmission electron diffraction patterns recorded at energies as low as 250 eV (Stevens et al. 2000). As energy is decreased below 100 eV, inelastic scattering reaches a maximum (according to the 'universal curves' used by surface scientists) before decreasing below about 20 eV. Elastic cross-sections continue to increase with decreasing energy (Cartier et al. 1987), so that, despite resolution limitations imposed by the large electron wavelength, the construction of very-low-voltage instruments for biology may have advantages (Schwarzer 1981, Spence 1997). These may be based on the simple point-projection geometry of in-line holography, using a nanotip field-emitter (Fink 1988). Images of unstained tobacco mosaic virus have been obtained at 1 nm resolution and 40 eV energy by this method (Weierstall et al. 1999) and, using atomic-resolution diffractive imaging, of carbon nanotubes at 30 kV (Kamimura et al. 2011). In these instruments, the resolution is to a first approximation about equal to the size of the virtual source size for the field-emitter (Spence et al. 1994). The most important experimental problems at low energy appear to be specimen charging, the poor penetration of the beam, and holographic reconstruction with a non-spherical reference wave. The expected increase in mean-free path for electron energies below the plasmon energy has not been observed experimentally (Spence et al. 1999), perhaps due to sample charging and surface states.

Damage to organic samples may be reduced by several methods, including cooling, use of higher voltages, embedding in a resin or ice, elemental substitution, use of pulses so brief that they outrun the damage processes, or use of an environmental cell. The Bethe energy loss formula shows that energy deposited per unit length is inversely proportional to the square of the beam velocity, so that a higher beam energy reduces the ionization energy deposited. A gain of two in spot fading times is found from 100 to 200 kV, and three at 1 MeV. However, similar ionization processes occur at the detector, so that detector sensitivity decreases with beam energy (see Section 9.8). Instruments are under construction in which the electron velocity is retarded prior to detection. Cooling to 4 K has been reported to increase spot fading times by almost an order of magnitude (Kimura *et al.* 1997)—this is assumed not to affect ionization but to slow down subsequent diffusion processes. Activation energies for the damage processes have been measured (Fryer *et al.* 1992). A description of the design of a superfluid helium stage for biological HREM is given in Fujiyoshi *et al.* (1991).

This subject is reviewed in more detail by Isaacson (1976), Glaeser (1979), Reimer and Kohl (2008), and Zeitler (1982). A special issue devoted entirely to radiation damage can be found in Volume 170 of the *Journal of Electron Spectroscopy and Related Phenomena* (2009).

6.12. Radiation damage: inorganics

Inorganic materials may be classified according to their bond types—metallic, covalent, ionic, or van der Waals. For metallic and covalently bonded materials, because of screening and reversibility, ionization damage is normally not seen in TEM at energies below the beam-energy threshold for ballistic 'knock-on' damage, except perhaps at point defects in semiconductors. (This threshold is about 150 kV for aluminium, increasing with atomic number, with a weak temperature dependence.) The excitations whose decay processes may produce damage are: (1) plasmons, (2) valence excitations, (3) inner shell excitations, and (4) excitons. Beam heating may also be important, as discussed in detail in Reimer and Kohl (2008). Two factors complicate this simple picture—unless an ion trap is fitted, damage may occur due to ions generated in the electron gun. Application of a magnetic field will distinguish ion damage from electron damage by separating the beams. Second, the displacement threshold is reduced for atoms at surfaces and defects. General reviews of damage mechanisms in inorganics in TEM can be found in Hobbs (1979a,b), Urban (1980), and Egerton *et al.* (2006).

Ionic materials, such as the alkali halides are thus highly sensitive to damage through the ionization process and subsequent atomic rearrangement, known as radiolysis, a process extensively studied in the X-ray and UV literature. The Bethe theory (Section 6.6) for single-electron excitation gives a cross-section inversely proportional to the square of the electron velocity, so that ionization damage decreases at higher accelerating voltage. (This cross-section is about 10^{-18} cm^2 at 100 kV for a carbon atom.) Electronic excitations can couple to the nuclear system if their lifetime exceeds vibrational periods (the Debye frequency is about 10^{14} Hz), if they can transfer sufficient energy to break bonds, and if they are localized. Plasmon excitations are delocalized, but the decay processes may involve localized processes. Thus a rough guide in ionic materials has been that for damage to occur, the electron–hole recombination energy should exceed the lattice-binding energy, a condition fulfilled, for example in CaF$_2$, but not in MgO, where radiolysis is not seen. Inner shell excitation may also be important, since, although rarer, it transfers more energy to the system which appears in the form of Auger electrons and X-rays, which may cause damage. Valence excitations generate secondary electrons which can be damaging. Radiolysis may have a threshold in dose, a highly non-linear temperature dependence, and, in the alkali halides may lead to the formation of halogen bubbles and metallic precipitates (Hobbs 1979a,b).

In the ballistic knock-on process, the energy transferred in an elastic collision between the beam electron and the nucleus exceeds the displacement energy. This energy varies from tenths of an electron volt for van der Waals bonding, to tens of electron volts for ionics. A Frenkel pair, consisting of an interstitial and a vacancy, is produced. The rate is proportional to the product of local electron intensity and a cross-section, and so, as in the ALCHEMI (atom location by enhanced channelling microanalysis) technique, it depends on local diffraction (or channelling) conditions. Thus enhanced damage is seen (by TEM diffraction contrast) inside bend contours due to the concentration of electron flux along these lines (Fujimoto and Fujita 1972). Some typical values of the knock-on threshold for bulk polycrystalline material at room temperature are: graphite 54 keV, silicon 145 keV, copper 400 keV, molybdenum 810 keV, gold 1300 keV. In a crystal, in addition to channelling effects, the threshold energy depends on the direction of transferred momentum, being smallest for atoms knocked-on into close-packed directions. Typical displacement cross-sections are 5×10^{-23} cm^2, much smaller than ionization values. A considerable literature exists devoted to the study of this type of damage in materials used in nuclear reactors, since, at very high intensity, the high-voltage EM is very efficient in generating Frenkel pairs, whose temperature-dependent interactions and resulting clusters, dislocations and voids can then be studied in detail. For HREM above the knock-on threshold, these events can sometimes be seen occurring directly, with atomic resolution (Horiuchi 1982). Here changes in lattice images of refractory oxides are studied as a function of dose. Then the HREM method becomes especially powerful since it can reveal, in complex structures, the sites at which displacements occur (in this case preferentially at oxygens in pentagonal tunnel sites). In this way, displacement damage has been studied in quartz, silicon, intermetallic alloys, and other materials (Spence 1988a). Dislocation motion has been observed in CdTe at atomic resolution due to irradiation (Sinclair et al. 1982). Dramatic HREM images of the radiation-induced motion of atoms on the surfaces of small gold particles have been recorded in real-time (Iijima and Ichihashi 1986)—in these and similar experiments, fast video recording has revealed many new aspects of atomic processes (Kolar et al. 1996). When a small probe is used, under clean non-contaminating conditions, the STEM becomes an efficient hole-drilling apparatus with obvious applications to direct-write inorganic lithography (Mochel et al. 1983). (Use of plasma-cleaning apparatus to remove contamination produces sufficiently clean conditions.). A review of the electronic processes involved in this hole-drilling can be found in Berger et al. (1987). The STEM probe is also capable of a kind of direct-write, inorganic lithography, since a focused field-emission beam will 'write' a line of silicon when scanned across silicon dioxide, and many other beam-induced reactions have been studied and used for pattern formation (Jiang et al. 2003). Time-resolved energy-loss spectroscopy of near-edge fine structure using the STEM nanoprobe has proved to be a powerful method of following chemical changes in oxide glasses during irradiation, which can expose the atomic mechanisms of damage (Jiang 2011).

In summary, by a combination of judicious choice of accelerating voltage and exposure conditions, HREM researchers have learnt to live with the problem of radiation damage, which, for most inorganic materials other than the alkali halides, is usually a minor inconvenience. By recording images at high speed, any changes during exposure can be observed directly, and conditions found which minimize damage in most inorganic materials. Aberration correctors are now allowing the use of low accelerating voltages, below the knock-on threshold, without loss of resolution, while pulsed electron diffraction methods are approaching the pulse duration (about 50 fs) needed to outrun damage.

References

Adrian, M., Dubochet, J., Lepault, J., and McDowell, A. W. (1984). Cryoelectron microscopy of viruses. *Nature* **308**, 32.

Agar, A. W., Alderson, R. H., and Chescoe, D. (1974). Principles and practice of electron microscope operation. In *Practical methods in electron microscopy*, ed. A. M. Glauert. North-Holland, Amsterdam.

Baumeister, W. and Typke, D. (1993). Electron crystallography of proteins. *MSA Bull.* **23**, 11.

Baumeister, W., Seredynski, J., Hahn, M., and Herbertz, L. (1976). Radiation damage of proteins in the solid state. *Ultramicroscopy* **1**, 377.

Berger, S. D., Salisbury, I., Milne, R., Imeson, D., and Humphreys, C. J. (1987). Electron energy-loss spectroscopy studies of nanoscale structures in alumina produced by intense electron-beam irradiation. *Phil. Mag.* **B55**, 341.

Breedlove, J. and Trammel, G. (1970). Molecular microscopy: fundamental limitations. *Science* **170**, 1310.

Brillouin, L. (1962). *Science and information theory.* Academic Press, New York.

Burge, R. E. (1973). Mechanisms of contrast and image formation of biological specimens in the transmission electron microscope. *J. Microsc.* **98**, 251.

Cambie, R., Downing, K. H., and Jin, J. (2007). Design of a microfabricated, two-electrode phase-contrast element suitable for electron microscopy. *Ultramicroscopy* **107**, 329.

Cartier, E., Pfluger, P., Pireaux, J., and Vilar, M. (1987). Mean free paths and scattering for 0.1–4500 eV electrons in saturated hydrocarbon films. *Appl. Phys.* **A44**, 43.

Chiu, W. (1993). What does electron cryomicroscopy provide that X-ray crystallography and NMR cannot? *Ann. Rev. Biomol. Struct.* **22**, 233.

Chiu, W. and Glaeser, R. M. (1974). Single atom contrast: conventional dark field and bright field electron microscopy. *J. Microsc.* **103**, 33.

Cosslett, V. E. (1958). Quantitative aspects of electron staining. *J. R. Microsc. Soc.* **78**, 18.

Cosslett, V. E. (1974). Perspectives in high voltage electron microscopy. *Proc. R. Soc. Lond.* **A338**, 1.

Cowley, J. (1961). Diffraction intensities from bent crystals. *Acta Crystallogr.* **14**, 920.

Cowley, J. M. (1969). Image contrast in a transmission scanning electron microscope. *Appl. Phys. Lett.* **15**, 58.

Cowley, J. M. (1973). High resolution dark field electron microscopy I. Useful approximations. *Acta Crystallogr.* **A29**, 529.

Cowley, J. M. (1976). The principles of high resolution electron microscopy. In *Principles and techniques of electron microscopy, biological applications*, Vol. 6, ed. M. P. Hayat. Van Nostrand Reinhold, New York.

Cowley, J. M. (1988). Phonon HREM. *Acta Crystallogr.* **A44**, 847.

Cowley, J. M. and Pogany, A. O. (1968). Diffuse scattering in electron diffraction patterns. I. General theory and computational methods. *Acta Crystallogr.* **A24**, 109.

Craven, A., Gibson, J. M., Howie, A., and Spalding, D. R. (1978). Study of single-electron excitations by electron microscopy I. Image contrast from delocalized excitations. *Phil. Mag.* **38**, 519.

Crewe, A. V., Langmore, J., Isaacson, M., and Retsky, M. (1974). Understanding single atoms in the STEM. In *Proc. 8th Int. Cong. Electron Microsc., Canberra*, Vol. 1. p. 30.

Crowther, R. A., de Rosier, D. J., and Klug, A. (1970). The reconstruction of a three-dimensional structure from projections and its application to electron microscopy. *Proc. R. Soc. Lond.* **A319**, 317.

Crowther, R. A., Henderson, R., and Smith, J. M. (1996). MRC image processing programs. *J. Struct. Biol.* **116**, 9.

Danev, R., Kanamaru, S., and Nagayama, K. (2010). Zernike phase contrast cryo-electron tomography. *J. Struct. Biol.* **171**, 174.

Dawson, B., Goodman, P., Johnson, A. W. S., Lynch, D. F., and Moodie, A. F. (1974). Some definitions and units in electron diffraction. *Acta Crystallogr.* **A30**, 297.

DeRosier, D. J. and Klug, A. (1968). Reconstruction of three-dimensional structures from electron micrographs. *Nature* **217**, 130.

Diaspro, P. (ed.) (2010). *Nanoscopy and multidimensional optical fluorescence microscopy*. Chapman and Hall, Boca Raton, FL.

Dorset, D. L. (1995). *Structural electron crystallography*. Plenum, New York.

Dorset, D. and Zemlin, F. (1985). Structural changes in electron-irradiated paraffin crystals at 15 K and their relevance to lattice imaging experiments. *Ultramicroscopy* **17**, 229.

Dorset, D. and Zemlin, F. (1990). Direct phase determination in electron crystallography: the structure of an *n*-paraffin. *Ultramicroscopy* **33**, 227.

Downing, K. (1992). Automatic focus correction for spot-scan imaging. *Ultramicroscopy* **46**, 199.

Downing, K. and Glaeser, R. (1986). Improvements in high resolution image quality of radiation sensitive specimens achieved with reduced spot-size of the electron beam. *Ultramicroscopy* **20**, 269.

Doyle, P. A. and Turner. P. S. (1968). Relativistic Hartree–Fock X-ray and electron scattering factors. *Acta Crystallogr.* **A 24**, 390.

Egerton, R. F. (1975). Inelastic scattering of 80 kV electrons in amorphous carbon. *Phil. Mag.* **31**, 199.

Egerton, R. F., Wang, F., and Crozier, P. A. (2006). Beam-induced damage to thin specimens in an intense electron probe. *Microsc. Microanal.* **12**, 65.

Fan, G.-Y. and Cowley, J. M. (1988). Assessing the information content of HREM images. *Ultramicroscopy* **24**, 49.

Faruqi, A., Henderson, R., and Subramaniam, S. (1999). Cooled CCD detector for recording electron diffraction patterns. *Ultramicroscopy* **75**, 235.

Fellget, P. B. and Linfoot, E. H. (1955). On the assessment of optical images. *Phil. Trans. R. Soc.* **247**, 369.

Fischer, N., Konevega, A. L., Wintermeyer, W., Rodnina, M. V., and Stark, H. (2010). Ribosome dynamics and tRNA movement by time-resolved electron cryomicroscopy. *Nature* **466**, 329.

Fink, H. (1988). Point sources for ions and electrons. *Phys. Scripta* **38**, 260.

Frank, J. (1990). Classification of macromolecular assemblies studied as single particles. *Q. Rev. Biophys.* **23**, 281.

Frank, J. (2006). *Three-dimensional electron microscopy of macromolecular assemblies*. Oxford University Press, New York.

Frank, J. (2010). *Electron tomography*. Second edition. Springer, New York.

Frank, J., Zhu, J., Penczek, P., Li, Y. H., Srivastava, S., Verschoor, A., Radamacher, M., Grassucci, R., Lata, R., and Agrawal, R. K. (1995). A model of protein synthesis based on cryoelectron microscopy of the *E. coli* ribosome. *Nature*, **376**, 441.

Frank, J., Radermacher, M., Penczek, P., Zhu, J., Li, Y., Ladjadj, M., and Leith, A. (1996). Spider and web: processing and visualisation of images in 3D. *J. Struct. Biol.* **116**, 190.

Fryer, J. R. (1993). Molecular arrays and molecular structure in organic thin films observed by electron microscopy. *J. Phys.* **D26**, B137.

Fryer, J. R., McConnell, C. H., Zemlin, F., and Dorset, D. L. (1992). Effect of temperature on radiation damage to aromatic organic molecules. *Ultramicroscopy* **40**, 163.

Fujimoto, F. and Fujita, H. (1972). Radiation damage induced by channelling. *Phys. Status Solidi* **11a**, K103.

Fujiwara, K. (1962). Relativistic dynamical theory of electron diffraction. *J. Phys. Soc. Japan* **17**, BII, 118.

Fujiyoshi, Y. (2011). Structural physiology based on electron crystallography. *Protein Sci.* **20**, 806.

Fujiyoshi, Y., Mizusaki, T., Morikawa, K., Yamagishi, H., Aoki, Y., Kihara, H., and Harada, Y. (1991). Development of a superfluid helium stage for high resolution electron microscopy. *Ultramicroscopy* **38**, 241.

Gabashvili, I. S., Agrawal, R. K., Spahn, C. M. T., Grassucci, R. A., Svergun, D. I., Frank, J., and Penczek, P. (2000). Solution structure of the *E. coli* 70S ribosome at 1.15 nm resolution. *Cell* **100**, 537.

Gilmore, C. J., Shankland, K., and Fryer, J. (1992). Phase extension for protein monolayers. *Ultramicroscopy* **49**, 510.

Glaeser, R. M. (1971). Limitations to significant information in biological electron microscopy as a result of radiation damage. *J. Ultrastruct. Res.* **36**, 466.

Glaeser, R. (1979). Radiation damage with biological specimens. In *Introduction to analytical electron microscopy*, ed. J. Hren, J. Goldstein, and D. Joy, p. 267. Plenum, New York.

Glaeser, R. M. (1992). Specimen flatness. *Ultramicroscopy* **46**, 33.

Glaeser, R. M. (1999). Electron crystallography: present excitement, a nod to the past, anticipating the future. *J. Struct. Biol.* **128**, 3.

Glaeser, R. and Downing, K. (1993). High resolution electron crystallography of protein molecules. *Ultramicroscopy* **52**, 478.

Glaeser, R., Downing, K., DeRosier, D., Chiu, W., and Frank, F. (2007). *Electron crystallography of biological macromolecules*. Oxford University Press. New York.

Glauert, A. M. (1974). The high voltage electron microscope in geology. *J. Cell. Biol.* **63**, 717.

Grinton, G. R. and Cowley, J. M. (1971). Phase and amplitude contrast in electron micrographs of biological materials. *Optik* **34**, 221.

Hamilton, J. F. and Marchant, J. C. (1967). Image recording in electron microscopy. *J. Opt. Soc. Am.* **57**, 232.

Hanssen, K. J. and Ade, G. (1978). Phase contrast transfer with different imaging modes in electron microscopy. *Optik* **51**, 119 [and references therein].

Hart, R. G. (1968). Electron microscopy of unstained biological material: the polytropic montage. *Science* **159**, 1464.

Hashimoto, H., Kumao, A., Hino, K., Endoh, H., Yotsumoto, H., and Ono, A. (1973). Visualization of single atoms in molecules and crystals by dark field electron microscopy. *J. Electron Microsc.* **22**, 123.

Hawkes, P. and Spence, J. C. H. (eds) (2006) *Science of microscopy*. Springer, New York.

van Heel, M., Gowen, B., Matadeen, R., Orlova, E., Finn, R., Pape, T., Cohen, D., Stark, H., Schmidt, R., Schatz, M., and Patwardhan, A. (2000). Single particle electron cryo-microscopy: towards atomic resolution. *Q. Rev. Biophys.* **33**, 307.

Henderson, R. (1995). The potential and limitations of neutrons, electrons and X-rays for atomic resolution microscopy of unstained biological molecules. *Q. Rev. Biophys.* **28**, 171.

Hobbs, L. W. (1979a). Radiation damage in electron microscopy. In *Introduction to analytical electron microscopy*, ed. J. Hren and D. Joy. Plenum, New York.

Hobbs, L. W. (1979b). Radiation damage in electron microscopy of inorganic solids. *Ultramicroscopy* **3**, 381.

Hoppe, W. and Hegerl, R. (1981). Some remarks concerning the influence of electron noise on 3D reconstruction. *Ultramicroscopy* **6**, 205.

Hoppe, W., Langer, R., Knesch, G., and Poppe, C. (1968). Protein-Kritallstrukturanalyse mit Elecktronenstrahlen. *Naturwissenschaften* **55**, 333.

Horiuchi, S. (1982). Detection of point defects accommodating non-stoichiometry in inorganic compounds. *Ultramicroscopy* **8**, 27.

Howells, M., Beetz, T., Chapman, H., Cui, C., Holton, J., Jacobsen, C., Kirz, J., Lima, E., Marchesini, S., Miao, J., Sayre, D., Shapiro, D., Padmore, H., and Spence, J. C. H. (2009). An assessment of the resolution limitation due to radiation-damage in X-ray diffraction microscopy. *J. Electron. Spectrosc. Relat. Phenom.* **170**, 4.

Howie, A. and Gai, P. (1975). Diffuse scattering in weak beam images. *Phil. Mag.* **31**, 519.

Howie, A., Krivanek, O. L., and Rudee, M. L. (1973). Interpretation of electron micrographs and diffraction patterns of amorphous materials. *Phil. Mag.* **27**, 235.

Howie, A., Rucca, F. J., and Valdre, U. (1985). Electron beam ionization damage processes in p-terphenyl. *Phil. Mag.* **52**, 751.

Humphreys, C. J. (1976). High voltage electron microscopy. In *Principles and techniques of electron microscopy: biological applications*, Vol. 6, ed. M. A. Hayat, p. 1. Van Nostrand Reinhold, New York.

Humphreys, C. J., Hart-Davis, A., and Spencer, J. P. (1974). Optimizing the signal/noise in the dark field imaging of single atoms. *Proc. 8th Int. Congr. Electron. Microsc., Canberra*, p. 248.

Ichinose, H., Sawada, H., Takuma, E., and Osaki, M. (1999). Atomic resolution HVEM and environmental noise. *J. Electron Microsc.* **48**, 887.

Iijima, S. (1977). Observation of single and clusters of atoms. *Optik* **48**, 193.

Iijima, S. and Ichihashi, T. (1986). Motion of surface atoms on small gold particles. *Phys. Rev. Lett.* **56**, 616.

Isaacson, M. (1976). Radiation damage to biological specimens. In *Principles and techniques of electron microscopy: biological applications*, ed. M. A. Hayat, p. 7. Van Nostrand, New York.

Isaacson, M., Johnson, D., and Crewe, A. V. (1973). Electron beam excitation and damage of biological molecules. *Radiat. Res.* **55**, 205.

Iwanaga, M., Ueyanagi, H., Hosoi, K., Iwasa, N., Oba, K., and Shiratsuchi, K. (1968). Energy dependence of photographic emulsion sensitivity and fluorescent screen brightness for 100 kV through 600 kV electrons. *J. Electron Microsc.* **17**, 203.

Jeng, T.-W. and Chiu, W. (1984). Quantitative assessment of radiation damage in a thin protein crystal. *J. Microsc.* **136**, 35.

Jiang, N. (2011). Local ordering of Ca in CaF glasses. *Appl. Phys.* **110**, 013518.

Jiang, N., Hembree, G., Spence, J. C. H., Qiu, J., Garcia de Abajo, F., and Silcox J. (2003). Nanoring formation by direct-write inorganic electron-beam lithography. *Appl. Phys. Lett.* **83**, 551.

Kamimura, O., Maehara, Y., Dobashi, T., Kobayashi, K., Kitaura, R., Shinohara, H., Shioya, H., and Gohara, K. (2011). Low voltage electron diffractive imaging of atomic structure in single-wall carbon nanotubes. *Appl. Phys. Lett.* **98**, 174103.

Kimura, Y., Vassylyev, D. G., Miyazawa, A., Kidera, A., Matsushima, M., Murata, K., Mitsuoka, K., Hirai, T., and Fujiyoshi, Y. (1997). Surface structure of bacteriorhodopsin revealed by high resolution electron crystallography. *Nature* **389**, 206.

Kobayashi, K., Suito, E., Uyeda, N., Watanabe, M., Yanaka, T., Etoh, T., Watanabe, H., and Moriguchi, M. (1974). A new high resolution electron microscope for molecular structure observation. *Proc. 8th Int. Congr. Electron Microsc., Canberra*, p. 30.

Kolar, H., Spence, J., and Alexander, H. (1996). Observation of moving kinks and pinning in silicon. *Phys. Rev. Lett.* **77**, 4031.

Krakow, W. (1976). Computer experiments for tilted beam dark-field imaging. *Ultramicroscopy* **1**, 203.

Kuhlbrandt, W. (1992). Two dimensional crystallisation of membrane proteins. *Q. Rev. Biophys.* **25**, 1.

Landau, E. M. and Rosenbusch, J. P. (1996). Lipidic cubic phases: a novel concept for the crystallization of membrane proteins. *Proc. Natl. Acad. Sci.* **93**, 14532.

Langmore, J. P. and Wall, J. (1973). The collection of scattered electrons in dark field electron microscopy. *Optik* **38**, 335.

Lee, Z., Meyer, J. C., Rose, H., and Kaiser, U. (2012). Optimum HRTEM image contrast at 20 kV and 80 kV – exemplified by graphene. *Ultramicroscopy* **112**, 39.

Lenz, F. (1954). Zur Streung mittleschneller elektronen in klienste Winkel. *Z. Naturforsch.* **9A**, 185.

Lindsay, S. (2000). The scanning probe microscope in biology. In *Scanning probe microscopy, techniques and applications*, ed. D. Bonnel. Wiley, New York.

Liu, J., Bartesaghi, A., Borgnia, M. J., Sapiro, G., and Subramaniam, S. (2008). Molecular architecture of native HIV-1 gp120 trimers. *Nature* **455**, 109.

McEwen, B. F., Downing, K., and Glaeser, R. M. (1995). The relevance of dose-fractionation in tomography of radiation-sensitive specimens. *Ultramicroscopy* **60**, 357.

Misell, D. L. (1973). Image formation in the electron microscope with particular reference to the defects in electron optical images. *Adv. Electron. Phys.* **32**, 63.

Mitsuoka, K., Hirai, T., Murata, K., Miyazawa, A., Kidera, A., Kimura, Y., and Fujiyoshi, Y. (1999). The structure of bacteriorhodopsin at 3.0 angstroms: implications of the charge-density distribution. *J. Mol. Biol.* **286**, 861.

Mochel, M., Humphreys, C., Mochel, J., and Eades, J. (1983). Cutting of 2 nm holes and lines in beta-alumina. *Appl. Phys. Lett.* **32**, 392.

Mueller, H., Jin, J. A., Danev, R., Spence, J. C. H., and Glaeser, R. M. (2010). Design of an electron microscope phase plate using a focused continuous-wave laser. *New J. Phys.* **12**, 073011.

Nogales, E., Whittaker, M., Milligan, R., and Downing, K. (1999). High resolution model of the microtubule. *Cell* **96**, 79.

Plitzko, J., Frangakis, A., Nickell, S., Förster, F., Gross, A., and Baumeister, W. (2002). In vivo veritas: electron cryotomography of cells. *Trends Biotechnol.* **20**, S40.

Porter, K. (1986). High voltage electron microscopy. *J. Electron Microsc. Techn.* **4**, 142.

Reimer, L. and Kohl, H. (2008). *Transmission electron microscopy*. Fifth edition. Springer-Verlag, New York.

Renken, C., Perkins, G., Martone, M., Edelman, V., Deerinck, T., Ellisman, M., and Frey, T. (1997). Three-dimensional reconstruction of mitochondria by electron tomography. *Microsc. Res. Techn.* **36**, 349.

Rose, A. (1948). Television pickup tubes and the problem of noise. *Adv. Electron.* **1**, 131.

Rose, H. H. (2009) Future trends in aberration-corrected electron microscopy. *Phil. Trans. R. Soc.* **A 367**, 3809.

Rosenthal, P. and Henderson, R. (2003). Optimal determination of particle orientation, absolute hand, and contrast loss in single-particle electron cryomicroscopy. *J. Mol. Biol.* **333**, 721.

Saxton, W. O. (1978). Computer techniques for imaging processing in electron microscopy. *Adv. Electron. Phys.* Suppl. 10.

Scherzer, O. (1949). The theoretical resolution limit of the electron microscope. *J. Appl. Phys.* **20**, 20.

Schwarzer, R. A. (1981). Emission electron microscopy. *Microsc. Acta* **84**, 51.

Schweinfest, R., Kostelmeier, S., Erast, F., Elsaskes, C., Wagner, T., and Finnis, M. W. (2001) Atomistic structure of $Al/MgAl_2O_3$ interface. *Phil. Mag.* **81**, 927.

Shao, Z. (1996). Biological atomic force microscopy. *Adv. Phys.* **45**, 1.

Sinclair, R., Smith, D., Erasmus, S., and Ponce, F. (1982). Lattice resolution movie of defect modifications in CdTe. *Nature* **298**, 127.

Spence, J. C. H. (1988a). *Experimental high resolution electron microscopy*. Second edition. Oxford University Press, New York.

Spence, J. C. H. (1988b). Localisation in inelastic scattering. In *High resolution electron microscopy*, ed. P. Buseck, J. M. Cowley, and L. Eyring. Oxford University Press, Oxford.

Spence, J. C. H. (1997). STEM and shadow imaging at 6 V beam energy. *Micron* **28**, 101.

Spence, J. C. H. and Zuo, J. M. (1988). A large dynamic range CCD parallel detection system for electron diffraction and imaging. *Rev. Sci. Instrum.* **59**, 2102.

Spence, J. and Zuo, J. M. (1992) *Electron microdiffraction*. Plenum, New York.

Spence, J. C. H., Qian, W., and Silverman, M. (1994). Electron source information from Fresnel fringes in field emission point projection electron microscopy. *J. Vac. Sci.* **A12**, 542.

Spence, J., Huang, X., Weierstall, U., Taylor, D., and Taylor, K. (1999). Is molecular imaging possible? In *Aspects of electron diffraction and imaging (Festschrift for M. J. Whelan)*, ed. P. B. Hirsch. Institute of Physics, London.

Spence, J. C. H., Weierstall, U., Downing, K., and Glaeser, R. M. (2003). Three-dimensional diffractive imaging for crystalline monolayers with one-dimensional compact support. *J. Struct. Biol.* **144**, 209.

Spence, J. C. H., Weierstall, U., and Chapman, H. (2012). X-ray lasers for structural and dynamic biology. *Rep. Prog. Phys.* **75**, 102601.

Stevens, M. R., Chen, Q., Weierstall, U., and Spence, J. C. H. (2000). TED at 200 eV and damage thresholds below the carbon K edge. *Microsc. Microanal.* **6**, 368.

Stevens, M., Longo, M., Dorset, D., and Spence, J. (2002). Structure analysis of polymerized phospholipid monolayer by TED and direct methods. *Ultramicroscopy* **90**, 265.

Tivol, W. F., Dorset, D. L., McCourt, M. P., and Turnes, J. N. (1993). *MSA Bull.* **23**, 91.

Unwin, N. and Henderson, R. (1975). Molecular structure determination by electron microscopy of unstained crystalline specimens. *J. Mol. Biol.* **94**, 425.

Urban, K. (1980). Radiation damage in inorganic materials in the electron microscope. *Electron microscopy 1980*. EM Congress Foundation, Leiden.

Uyeda, N., Kobayashi, T., Ishizuka, K., and Fujiyoshi, Y. (1980). Crystal structure of Ag·TCNQ. *Nature* **285**, 95.

Vainshtein, B., Zvyagin, B., and Avilov, A. (1992). *Electron diffraction techniques*, ed. J. M. Cowley, Ch. 6. Oxford University Press, New York.

Van Dorsten, A. C. (1971). Contrast phenomena in electron images of amorphous and macromolecular objects. In *Electron microscopy in materials science*, ed. U. Valdre and Z. Zichichi. Academic Press, New York.

Walz, T. and Grigorieff, N. (1998). Electron crystallography of two-dimensional crystals of membrane proteins. *J. Struct. Biol.* **121**, 142.

Weierstall, U., Spence, J. C. H., Stevens, M., and Downing, K. (1999). Imaging of Tobacco Mosaic Virus at 40 V by electron holography. *Micron* **30**, 335.

Whelan, M. J. (1972). Elastic and inelastic scattering. *Proc. 5th Eur. Congr. Electron Microsc.*, p. 430.

Williams, R. C. and Fisher, H. W. (1970). Disintegration of biological molecules under the electron microscope. *Biophys. J.* **10**, 53a.

Yoshioka, H. (1957). Effect of inelastic waves on electron diffraction. *J. Phys. Soc. Japan* **12**, 618.

Zeitler, E. (1982). Cryomicroscopy and radiation damage. *Ultramicroscopy* **10**, 1.

Zemlin, F., Reuber, E., Beckmann, B., Zeitler, E., and Dorset, D. (1985). Molecular resolution electron micrographs of monolamellar paraffin crystals. *Science* **229**, 461.

Zhang, L., and Ren, G. (2012). Tomography of single-molecule structures. *PLOS One* 7, e30249.

Zhang, X., Jin, L., Fang, Q., Hui, W., and Zhou, Z. (2010). 3.3 Angstrom cryo-EM structure of a non-enveloped virus reveals a priming mechanism for cell entry. *Cell* **141**, 472.

Zuo, J. M. (2000). Electron detection characteristics of slow-scan CCD camera, imaging plates and film. *Microsc. Res. Techn.* **49**, 245.

7
Image processing, super-resolution, and diffractive imaging

Prior to the development of the aberration corrector, three main approaches to the attainment of super-resolution were under study—electron holography, image processing, and diffractive (lensless) imaging (including ptychography). By super-resolution we mean the attainment of resolution beyond the Scherzer point-resolution of eqn (6.17). All these methods (and others, including tilt series and combining nanodiffraction patterns with images) continue to be developed for excellent reasons, as we show in this chapter, which is intended to explain the underlying concepts of these innovative approaches, and connect the reader with the relevant literature.

7.1. Through-focus series, coherent detection, optimization, and error metrics

We first review efforts by image processing to improve resolution beyond the point-resolution $d_\text{p} = 0.66\, C_\text{s}^{1/4} \lambda^{3/4}$ (eqn 6.17) out to the information limit $d_\text{i} = (2/(\pi\lambda\Delta))^{1/2}$ (eqn 4.11) set by electronic instabilities. This is possible because, as shown in Figs 4.3 and A3.2, the passband of well-transmitted spatial frequencies moves to higher frequencies with increasing defocus, so that, for *linear* imaging, by adding together members of a focal series a wider total passband is obtained (Schiske 1968). In addition, the zero-crossings of the transfer function move to a different position in each member of the series, facilitating deconvolution. (Division by near-zero quantities in the Fourier domain otherwise amplifies noise, preventing deconvolution.) The optimum choice of focus settings to be used in a through-focus series has been widely discussed—an alternative to the discrete passband settings of eqn (4.12a) is the use of the stationary-phase focus of eqn (5.66) (O'Keefe et al. 2001). However, several problems arise, associated with partial coherence effects, errors in the measurement of the optical parameters needed, and unavoidable zero-crossings at the highest and lowest frequencies. Understanding this method, which dates back to the earliest days of HREM, thus involves many related topics, including coherent detection, the twin image problem (since an out-of-focus HREM image is an in-line hologram), the 'parabola method', and Wiener filters for noise reduction (Schiske 1968, 1973, Hawkes 1974, Coene et al. 1992, Saxton 1994, O'Keefe et al. 2001, and references therein).

The essential idea of coherent detection, on which the through-focus method is based, can be understood as follows. Equation (3.34) gives the image amplitude for a weak phase object. We let

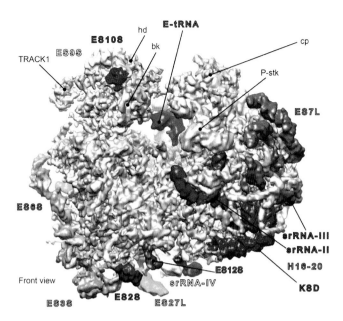

Plate 1 High resolution cryo-em structure of T. brucei" at 0.5 nm resolution. This is the factory which makes proteins in cells according to instructions delivered from the DNA genetic code. (From Hashem et al, *Nature* (2013).)

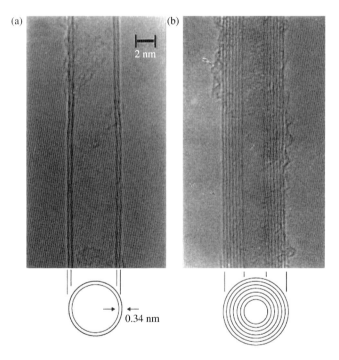

Plate 2 Fullerenes. (a) The discovery of the single-walled nanotube. (b) A multiwalled tube. (c) The first image taken of a buckyball, in 1980. (From S. Iijima, *Nature* **354**, 56 (1991); S. Iijima, *J. Cryst. Growth*, **50**, 675 (1980).)

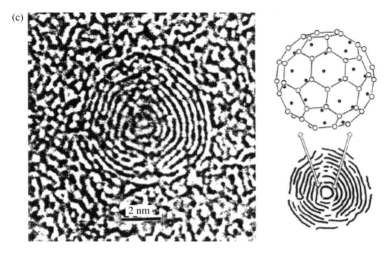

Plate 2 (*continued*)

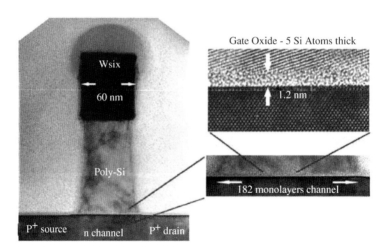

Plate 3 A single transistor. Atomic-resolution (ADF-STEM) image of a single transistor, showing enlargement of the gate oxide, about five atoms thick. Computers contain millions of these nanostructures, which can now only be seen using an atomic-resolution electron microscope (D. Muller, Bell/Lucent, 2002).

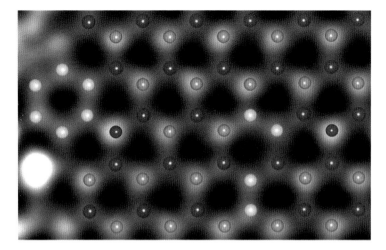

Plate 4 Scanning transmission electron microscope (STEM) image of a single atomic layer of Boron Nitride "graphene", in which carbon (green) and Oxygen (blue) interstitial atoms have been identified. Red are Boron atoms, Blue are Nitrogen. (From Krivanek et al., Nature Letters, **464**, 571 (2010).)

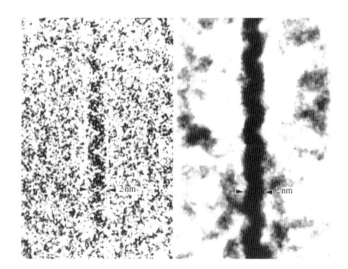

Plate 5 A single unstained DNA molecule imaged by TEM. The molecule is stretched across a hole and dehydrated. Background noise is due to fluctuations in the arrivals of individual beam electrons. The image has been averaged five times along its length. The image on the right has been processed to reduce noise. Whereas cryo-EM allows imaging of molecules in their more useful quasi-hydrated state, this image shows the limits of what is possible for unstained single-molecule imaging with high-energy electrons, given the limitations of noise and radiation damage. (From Y. Fujiyoshi and N. Uyeda, *Ultramicroscopy* **7**, 189 (1981).)

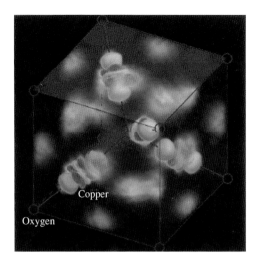

Plate 6 The bond-charge in cuprite (Cu_2O). This map shows the difference between the total charge density measured by CBED, and a collection of spherical ions obtained by calculation. (From J. M. Zuo, M.Kim, K. O'Keefe, and J. C. H. Spence, *Nature*, **401**, 49 (1999).)

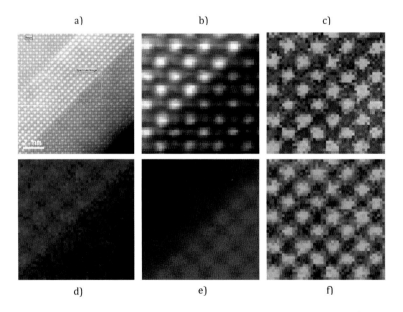

Plate 7 Atomically resolved EELS elemental map of a thin slice of $BaTiO_3/SrTiO_3$ multilayer interface. Sr (M_{45} edge) atom columns are red, Ba blue, and Ti (L_{23} edge) green. (a) An 80 kV ADF-STEM image. An enlarged region around the interface is shown in (b)–(f), recorded with acquisition times of about 30 ms pixel^{-1} (from Botton, G., Lazar, S., and Dwyer, C. (2010) Ultramicroscopy 110, 926). The instrument was an FEI Titan with X-FEG. Energy resolution 1.5 eV.

$$\psi(x, y, \Delta f_n) = -i\sigma\varphi_p(-x, -y) * \Im\{P(u, v)\exp(i\chi(u, v))\}$$
$$= \phi(x, y) * \Im\{P(u, v)\exp(i\chi(u, v, \Delta f_n))\}$$

Then the image intensity is

$$I_i(x, y, \Delta f_n) = |1 + \psi(x, y, \Delta f_n)|^2 = 1 + \psi(x, y, \Delta f_n) + \psi^*(x, y, \Delta f_n) + h \quad (7.1)$$

where h are the non-linear terms (but note that the single-scattering approximation (eqn 3.31) is still retained). We can now pick out the wanted second term and discriminate against both the third 'twin image' and fourth non-linear terms, using a process analogous to holographic reconstruction. (We will see in Section 7.6 that out-of-focus bright-field HREM images are in-line holograms.) We multiply the transform of this intensity $I(u, v, \Delta f_n)$ by the conjugate of the transfer function (to effect deconvolution), and sum over N images recorded at different defoci Δf_n. Thus we form

$$S(u, v) = \sum_n I(u, v, \Delta f_n)\exp(-i\chi(u, v, \Delta f_n))$$
$$= \delta(u, v) + \phi(u, v)\sum_{n=1,N} 1 + \phi^*(-u, -v)\sum_n \exp(-2i\chi(u, v, \Delta f_n))$$
$$+ \sum_n H_n(\Delta f_n)\exp(-i\chi(u, v, \Delta f_n)) \quad (7.2)$$

where $\phi(u, v) = -i\sigma\Im\{\phi_p(x, y)\}$ is the image we seek, and $H_n(\Delta f_n)$ is the transform of the non-linear term. We now see the power of the coherent detection method, since the second term accumulates to N times the wanted image at Gaussian focus, while the third and fourth terms sum to approximately $N^{1/2}$ times their initial value. The last sum behaves as a two-dimensional random walk, while the third term describes images summed at minus twice their recorded foci, and can be shown to behave similarly (Saxton 1994). The effects of spherical aberration appear to have been eliminated, and the only resolution-limiting effects remaining are the pupil function $P(u, v)$, or, in its absence, the damping effects on the transfer function of partial spatial and temporal coherence. The general principle of this powerful method, integration against a kernel to provide a stationary phase condition for a wanted signal, is used throughout physics, including applications to internal-source photoelectron and X-ray holography. (Here a similar procedure, using multiple wavelengths, minimizes both twin image and multiple scattering artefacts.) By extension, parameters other than defocus (such as tilt or wavelength) may be varied with similar effect. The main problems which arise in the implementation of this approach to HREM include the registration (alignment) of the images, the optimum choice of noise filter, errors in the measurement of C_s (which must be independently measured—see Section 10.2), and the correct incorporation of partial coherence effects. The removal of three-fold astigmatism is found to be important (Krivanek and Stadelmann 1995).

A number of papers have extended this analysis to include effects of incomplete coherence and noise. Most introduce a more general restoring filter $r_n(u, v)$ to control noise amplification and deal with partial coherence. Then the experimental image transforms are modified computationally to form

$$S(u,v) = \sum_n I(u,v,\Delta f_n) r_n(u,v,\Delta f_n) \tag{7.3}$$

If a Wiener filter is used to control noise (Schiske 1973), the most general result, which also incorporates the effects of limited space and time coherence is obtained with (Saxton 1994)

$$r_n(u,v) = \frac{[(W+v)w_n^* - (C^*W_n)]}{\left[(W+v)^2 - |C|^2\right]} \tag{7.4}$$

As in Appendix 3 and eqn (4.8a) we use a function $A(\mathbf{K})$ to describe the attenuating effects of partial coherence. Then $w_n(\mathbf{K}) = \exp(i\chi(\mathbf{K}, \Delta f_n))A(\mathbf{K}, \Delta f_n)$, $W(\mathbf{K}) = \sum_n |w_n(\mathbf{K})|^2$ and $C(\mathbf{K}) = \sum_n w_n^2(\mathbf{K})$ (cf. the third term in eqn 7.2), where the vector $\mathbf{K}$ has components (u, v). The term $v(\mathbf{K})$ is an estimate of the noise-to-object spectral power ratio. Modern HREM image restoration software is based on eqn (7.3), taking a variety of forms for $r_n(u, v)$. Figure 7.1 shows an image of diamond projected down [110] using this method. The point-resolution of the modified Philips CM300FEG used was 0.17 nm, whereas the composite reconstructed image shows a resolution of 0.078 nm (O'Keefe et al. 2001). For this instrument, $\Delta = 3.2$ nm.

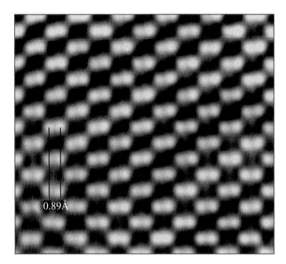

Figure 7.1 HREM image of diamond projected down [110]. This image is a composite of a through-focus series of images, each 'deconvoluted' for the effects of the contrast transfer function (CTF). The point-resolution of the modified Philips CM300 FEG used was 0.17 nm, whereas the composite reconstructed image shows a resolution of 0.078 nm (O'Keefe et al. 2001) ($C_s = 0.65$ mm, 300 kV, $\Delta = 20$ Å, $\theta_c = 0.25$ mrad).

An alternative and enlightening view of this method has also been given by Coene et al. (1992), denoted the parabola method (PAM). This group points out that it is useful to consider the three-dimensional Fourier transform of a through-focus series, the third transformed variable K_z being conjugate to defocus Δf. With $C_s = 0$, the three-dimensional transform of the image intensity (eqn 7.1) then becomes

$$I(\mathbf{K}, \mathbf{K}_z) = \delta(\mathbf{K}, \mathbf{K}_z) + \phi(\mathbf{K})\delta(\mathbf{K}_z - 0.5\lambda \mathbf{K}^2) + \phi^*(-\mathbf{K})\delta(\mathbf{K}_z + 0.5\lambda \mathbf{K}^2)$$
$$+ \int H(\mathbf{K}, \mathbf{K}_z) \exp(-2\pi i \mathbf{K}_z \Delta f) \tag{7.5}$$

This formulation shows that the second term, the linear image we seek, is localized onto a parabolic shell (the Ewald sphere) in this three-dimensional space, and so can be separated easily (except where $\mathbf{K}$ is small) from the unwanted third term, which lies on a reflected sphere. The non-linear final term is smeared out throughout the reciprocal space, and is also readily minimized. Using this approach, experimental image resolution has been improved from a point-resolution of 0.24 nm to the information limit of 0.14 nm (Op de Beeck et al. 1996).

For completeness, we should note that if partial coherence effects are allowed, the transform of the image intensity (eqn 7.1) is

$$I(\mathbf{K}, \Delta fn) = \delta(\mathbf{K}) + \phi(\mathbf{K})w_n(\mathbf{K}) + \phi^*(-\mathbf{K})w_n^*(-\mathbf{K})$$
$$+ \sum_{K'} \phi(\mathbf{K}')\phi^*(\mathbf{K}' - \mathbf{K})m_n(\mathbf{K}', \mathbf{K}' - \mathbf{K}) \tag{7.6}$$

where $m_n(\mathbf{K})$ is the *mutual transfer function*, important in incoherent imaging theory (Section 8.3).

Students may wish to try to develop an expression similar to eqn (7.2) for the cases where the object potential is not centrosymmetric, where it is complex (taking account of absorption as in eqn (6.2)), and where there is double scattering (second-order expansion in eqn (3.30)).

The most sophisticated application of these ideas is encoded in the Philips/Brite–Euram software package (Coene et al. 1996, Thust et al. 1996a,b), which combines the PAM with a non-linear maximum-likelihood (MAL) algorithm derived from Kirkland's maximum a posteriori procedure (Kirkland 1984). The initial linear PAM estimate is improved on by iterative use of the non-linear MAL, based on least-squares fitting and the method of steepest descent, to generate an improved estimate of the exit-face wavefunction. This improved estimate is then used to generate a new simulated through-focus series, and so an improved estimate of the exit-face wavefield. The process is continued to convergence. The success of the through-focus method has been demonstrated in resolution tests where sub-angstrom resolution has been achieved in composite experimental images of diamond (O'Keefe et al. 2001) and of other light elements including the individual nitrogen atom columns in gallium nitride and the oxygen columns in sapphire (Kisielowski et al. 2001). The displacements of atoms near a twin boundary in barium titanate is analysed by this method in Jia and Thust (1999).

For these crystals of *known* structure, excellent matches have been obtained between computed and experimental HREM images, even for strongly non-linear images with multiple scattering (Hytch and Stobbs 1994, King et al. 1998, Mobus et al. 1998), since the beginnings of this effort in the early 1970s (Anstis et al. 1973). The parameters to be varied included electron optical parameters (C_s, Δf, Δ, θ_c), alignment parameters (optic axis, beam direction, crystal axis), and thickness. As suggested in Appendix 5, MgO smoke cubes viewed down [110] form the ideal test sample, since then the sample thickness is accurately known at each point in the image (O'Keefe et al. 1985). (Cleaved wedges of semiconductors may also be used, if the cleavage angle is known.) The most troublesome variable is thereby eliminated from the refinement. This work on known structures revealed, however, that most HREM images show a mean contrast level about three times less than the computed images (the 'Stobbs factor', discussed in Section 5.13).

The extension of this approach to unknown structures has proven difficult. When atomic coordinates and imaging parameters are included, the number of parameters becomes very large and a multiparameter optimization problem results, in which local minima will be encountered in the landscape (Press et al. 1986). The most efficient parameter search method and the choice of goodness-of-fit index become important issues, which we now discuss. A figure-of-merit (FOM) is defined which measures the difference between experimental and computed images for a given structural model. The FOM is then minimized with respect to all parameters. These should be constrained using all a priori information available, such as the crystal space group (perhaps determined by CBED), any known bond lengths, and independently measured electron-optical constants. In view of the large number of parameters, and of the inevitable presence of systematic errors and errors in 'given' parameters, this is a formidable undertaking which has so far achieved only limited success. The underlying theory, based on Bayesian statistics, depends on a priori knowledge of the noise statistics, and no agreement has so far been reached on the best FOM for HREM. Unfortunately the structural models which result from the analysis depend on the choice of FOM. Here we review the most straightforward approach and indicate its limitations. Texts which give more background include Bevington (2002) on the method of maximum likelihood, Prince (1982) on data analysis in crystallography, and Press et al. (1986) on multiparameter optimization. The HREM analysis closely parallels the Reitvelt method used in X-ray and neutron diffraction.

Consider a set of N independent experimental image intensities I_i^e (each with standard deviation σ_i) in the presence of Gaussian, additive, uncorrelated noise. These may be the intensities recorded at each pixel within the image of one projected unit cell of a thin crystal, or some average of many cell images believed to occur at the same crystal thickness. If the symmetry of the crystal is known, the images may be further averaged after application of known plane-group symmetry operations. We assume that these experimental data were generated from a certain theoretical model with the addition of noise, so that $I_i^e = I_i^t(a_i) + e_i$, where e_i is the normally distributed noise and $I_i^t(a_i)$ is the theoretically predicted intensity. This depends on the model parameters a_i (positions of atomic columns, thickness, electron optical coefficients, etc.). The likelihood dP of observing the particular set of experimental intensities (within a small range) that we did actually observe is then just the product of all the Gaussian distributions of noise, centred around each theoretically predicted intensity. The probability distribution function is thus given by

$$P = \prod_i \frac{1}{\sqrt{2\pi}\sigma_i} \exp\{-s^2/2\} \tag{7.7}$$

where s and χ are defined by

$$\chi^2 = \frac{s^2}{N-p} = \frac{1}{N-p} \sum_{i=1}^{N} \frac{1}{\sigma_i^2} [I_i^e - I_i^t(a_i)]^2 \tag{7.8}$$

Here the number of degrees of freedom is $N - p$ (there are p constraints). We then assume that the values of the parameters a_i which maximize P are a good estimate of the true values of the parameters. Then the likelihood $P = P(D|X)$ that the given model X generated the observed data D is a maximum if χ^2 is a minimum (according to this 'method of maximum likelihood'), so that P is a maximum. We imagine a space in which there is one coordinate for each detector pixel, the distance along the coordinate representing the intensity at that pixel. Without constraints, one vector in this space then defines the intensity of every pixel and hence an entire image (or diffraction pattern). Then χ^2 may be interpreted as a metric (Euclidian distance) in this space, which measures the distance between members of two sets, one a member of the set of possible trial-structure images and the other the experimental image. This makes connection with the theory of projection onto convex sets, a powerful tool for image analysis and inversion problems (Spence et al. 1998). This theory shows how to avoid local minima and converge immediately to a global solution if 'convex' constraints can be found (Stark 1987). The search for such constraints would appear to be a profitable approach to the HREM refinement problem. Computer codes (such as the simplex algorithm, and many others) adjust the model parameters a_i (e.g. atom positions, optical constants) until the minimum in χ^2 is found. However, there may be many local (false) minima, and the problem of finding the true global minimum is fundamental, since, in general, apart from an exhaustive search there exists no general method for finding the minimum of a given function. Some algorithms address this by starting again with a different set of starting parameters—a minimum found to be independent of starting conditions is likely to be global. Others (such as simulated annealing) allow for the possibility of jumping out of a shallow local minimum to find a deeper one. The Bayes theorem may be used to obtain the probability $P(X|D)$—the posterior inference—that the model is correct, given the data, as

$$P(X|D) = P(D|X)P(X)/P(D) \tag{7.9}$$

Here the 'prior' $P(X)$ allows for the introduction of a priori information, which limits the probabilities, and $P(D)$ is a constant. The use of maximum entropy analysis constitutes one possible form of prior $P(X)$.

Although widely used, the 'chi-squared' goodness-of-fit index in eqn (7.8) has several disadvantages. Note that, for $\chi^2 = 1$, this equation reduces to the definition of the variance. In particular, a good fit can be obtained (small χ^2) either if the noise is large (large σ^2) but the model poor, or if the model is good and the noise small. Care is therefore needed in comparing χ^2 between different experiments with different noise levels. More importantly, the presence of systematic errors (e.g. wrong value of thickness, absorption,

Debye–Waller factors, point defects, bending of the sample, incorrect values of coherence factors, presence of amorphous layers, thermal diffuse scattering, etc.) can cause serious problems. In summary, the χ^2 optimization analysis works well only under the assumptions of Gaussian, uncorrelated additive noise in the absence of systematic errors. Often, however, the noise distribution is not known, and a different variance may be required for each point. (For the simplest case of normally distributed noise, the variance is equal to the mean.) Optimization will always be subject in practice to the problem of stagnation in local minima, and only the fastest computers can handle minimization with many parameters using HREM data. Independent measurement of thickness, defocus, and spherical aberration coefficients is therefore highly desirable. Saxton (1998) has pointed out that if differences are taken in eqn (7.8) between the Fourier transforms of the image intensities, then the noise on this data is uncorrelated if the image contrast is not too high—cross-talk between CCD pixels due to light-scattering in single-crystal YAG scintillators is otherwise a serious problem.

A procedure is said to be 'robust' if it gives parameter estimates close to the minimum for a wide range of error distributions—for HREM the form of this distribution may not be known. The preceding method is not robust in this sense. A method is 'resistant' if it is insensitive to 'outlier' data points, due, for example, to systematic error. A variety of shaping functions can be introduced into eqn (7.8) to make it more robust and resistant. Six alternative definitions of goodness-of-fit indicies for HREM are compared in Mobus et al. (1998). Different definitions of the Euclidian distance, the cross-correlation function between computed and experimental images, the weighted least squares, mean relative distance, and the traditional X-ray R factor have all been tried. The situation would improve if one could work with uncontaminated samples with atomically smooth surfaces containing defects consisting of columns of atoms perfectly aligned with the beam. The modern trend toward tomographic imaging at near-atomic resolution, by providing much more redundant data, combined with more powerful computers, may eventually solve this problem.

7.2. Tilt series, aperture synthesis

From the earliest days of HREM, it was realized that resolution could be approximately doubled if the incident beam direction (and the sample) were inclined to the optic axis by the Bragg angle θ_B (Dowell 1963). Then the optic axis bisects the angle $\theta = 2\theta_B$ between the incident beam and a first-order Bragg beam. An instrument whose resolution limit d_i was equal to only half this Bragg spacing $d = \lambda/\theta = \lambda/2\theta_B$ would then nevertheless be capable of imaging this spacing, since the two beams would fall at opposites sides of the transfer function, as shown in Fig. 5.1(b). Note that dark-field STEM (Section 8.3), by the principle of reciprocity, always operates in this mode. The two beams then define the diameter of an achromatic circle, on which all even-order aberrations cancel, since the aberration function and its conjugate are then multiplied together in the final image intensity expression (eqn 5.7). Obviously only one side of the diffraction pattern contributes to such an image, and so, for a non-periodic sample, it becomes necessary to record at least four images, each with the incident beam tilted into a different direction. Under the linear imaging approximation (eqn 3.31) these images can be added together. This is the basis of super-resolution schemes based on aperture synthesis. The modern implementation depends on the computer control of microscopes for beam tilting and image acquisition, and is described, for example, in Kirkland *et al.* (1995). In that work, images of gold particles on amorphous germanium were recorded in five orientations (symmetric, and four orthogonal tilts surrounding). Tilt angles of about $(\lambda/C_s)^{1/4}$ are useful. By Fourier analysis and deconvolution of the aberration function for the tilted images, these images could be reassembled into a composite image in which the resolution had improved from 0.23 to 0.14 nm. To align the images, a modified form of the cross-correlation method was used, and a Wiener noise filter applied. The method is claimed to require much less accurate specification of optical parameters than the through-focus method, and to achieve higher resolution. It assumes, however, that the transmission function of the sample is independent of tilt (unless the sample is also tilted with the beam), so that very thin samples must be used, and the noise level is poor at high spatial frequencies. Alignment of the images in the presence of noise is difficult. For a second application, see Kirkland *et al.* (1997).

7.3. Off-axis electron holography

Holography was originally proposed by Gabor (1949) in the in-line geometry described in Section 7.6 as a means for overcoming the aberrations of electron microscopes. Early work using optical reconstruction and either collimated or diverging illumination (to provide magnification) was hampered by limited coherence. Gabor's aim was finally achieved in 1995 (Ochowski *et al.* 1995), largely because of the availability of the field-emission source, and through use of the latter's off-axis geometry to deal with the twin-image problem. The off-axis mode uses a beam splitter to pass a coherent portion of the beam around the sample, so that it can interfere at the detector with the beam which has passed through the sample. The development of optical holography (Collier *et al.* 1971) in the commonly used off-axis geometry came several years after Gabor's original work, and was greatly facilitated by the invention of the laser. In this section, we consider only off-axis electron holography, mainly for the purposes of resolution improvement—super-resolution schemes based on in-line holography are discussed in Section 7.6.

For off-axis holography, the reference wave is normally obtained using an electrostatic biprism (Mollenstedt and Wahl 1968), which crosses the wavefield passing through the sample with a set of 'carrier fringes', described by the second term on the right of the first equation below. The reconstruction is then simple, and no twin-image problem is encountered. In one dimension, the intensity on a plane just below the sample is

$$\begin{aligned}
I(x) &= |q(x) \otimes t(x) + \exp(-2\pi i u_0 x)|^2 \\
&= 1 + |q(x) \otimes t(x)|^2 + \exp(-2\pi i u_0 x)\left(q(x) \otimes t(x)\right) \\
&\quad + \exp(+2\pi i u_0 x)(q^*(x) \otimes t(x)^*)
\end{aligned} \quad (7.10)$$

where $t(x)$ is the Fourier transform of $\exp(i\chi(u))$ (describing propagation of the image over the defocus distance), $\otimes$ denotes convolution, and $q(x)$ is the transmission function of the sample. This intensity is Fourier transformed for reconstruction, resulting in distributions centred at $u = 0$ and $u = \pm u_0$. By isolating one of these satellite spectra (say the last term), multiplying it by $\exp(-i\chi(u))$, and inverse transforming, the required transmission function (the unaberrated image) may be recovered. Thus the holographic reconstruction method gives access to the complex wavefunction at the specimen exit-face, and so allows correction of aberrations. The normal method of implementing this in TEM is through a biprism placed below the sample, so that the wavefield leaving the sample is magnified before it is crossed by the reference wave. This greatly relaxes the stability requirements needed to record the carrier fringes, which, referred to the object plane, are usually taken to be one-third of the spacing of the resolution required. This produces adequate separation of the satellite spectra. Lichte (1991) has analysed the optimum conditions for off-axis HREM holography—he finds that the number of pixels needed in a hologram to allow reconstruction at resolution d_H is a minimum of

$$N = [d_\mathrm{p}/(0.3 d_\mathrm{H})]^4$$

at the 'Lichte focus', which is three-quarters of the stationary phase focus (eqn 5.66), and so is given by

$$\Delta f_\mathrm{L} = -0.75\, C_\mathrm{s} (\lambda/d_\mathrm{H})^2$$

Thus, for example, to improve resolution from a point-resolution of $d_\mathrm{p} = 0.17$ nm to $d_\mathrm{H} = 0.1$ nm by off-axis holography requires 1031 pixels (linear dimension) and a biprism capable of generating 0.03 nm fringes referred to the object space. The final magnification needed may be so large (e.g. 4×10^6) as to require instrumental modification; it is fixed by the need for, say, four CCD pixels per carrier fringe (see Section 9.9 for discussion of CCD pixel cross-talk). Using this method, the Tubingen group have obtained reconstructed images of silicon crystals showing 0.104 nm resolution, using an electron microscope whose point-resolution was 0.198 nm (Ochowski et al. 1995). More details can be found in Lichte (1988). A method of strain mapping based on off-axis holography has been described by Hytch et al. (2011). An analysis of the extremely fine elastic energy-filtering effects of off-axis holography can be found in Van Dyck et al. (1999). In simple terms, for the most part, only the elastic scattering from the sample interferes with the off-axis reference wave, even for energy losses smaller than the spread of energies in the beam. This could have application for the removal of inelastic background from weak-beam images (Howie et al. 2006). This problem is similar to the decoherence problem of mesoscopic physics, which has also been studied using low-energy electron interferometry.

7.4. Imaging with aberration correction: STEM and TEM

If C_s could be reduced by electron-optical methods, a straightforward improvement in point-resolution d_p could be obtained according to eqn (6.17). This long-standing goal of electron-optical design, aberration correction, was finally achieved experimentally at the end of the 20th century. Because the aberration coefficients of rotationally symmetric electron lenses are always positive, there was thought to be little hope of cancelling out aberrations using round lenses. As discussed in Sections 2.10 and 3.3 in more detail, Scherzer (1947) was the first to suggest the use of multipole lenses for this purpose. Pioneering work since then by Seeliger, Mollensted, Deltrap, Crewe, Rose, Koops, and Hely described various schemes for reducing both spherical and chromatic aberration (see, e.g., Rose (1990)). But little could be achieved as long as the information limit to resolution due to electronic instabilities (eqn 4.11) was more severe than the Scherzer aberration limit (eqn 6.17). The situation changed dramatically in the 1990s, when useful improvements in resolution for both STEM (Krivanek et al. 1999) and TEM (Haider et al. 1998a,b) were published using new aberration-corrected instruments. (Earlier workers were frequently able to improve the performance of sufficiently poor instruments using correctors—the real challenge has been to exceed the performance of the very best uncorrected instrument, or to improve on other super-resolution schemes, such as the through-focus method.) One of the main difficulties has always been the determination of a systematic procedure for the alignment of the large number of electrostatic and magnetic optical elements. Human alignment takes a prohibitively long time, so that computer optimization techniques must be used. It has been said that aberration correction only become a reality when the speed of computers made the auto-alignment time less than the stability time of the electron microscope. Examples of the auto-alignment procedures used are given in Section 12.5. For STEM, the Ronchigrams of Section 10.10 are more likely to be used for auto-alignment of aberration correctors.

For HREM, a hexapole corrector (Rose 1990) which corrects for third-order spherical aberration $C_s = C_3$ was fitted to a 200 kV field-emission TEM (Haider et al. 1998a,b). Resolution remained limited by the information resolution limit d_i, and by chromatic aberration, which was not corrected ($C_c = 1.3$ mm). Improvements in d_i would expose a resolution limit of 0.05 nm, due to fifth-order aberrations, with $C_5 = 4$ mm for this instrument. The corrector uses two hexapoles to introduce negative spherical aberration, which just cancels that of the round objective lens. The result is an improvement in the point-resolution from 0.24 to about 0.13 nm. Figure 7.2 shows a corrected image of a thin

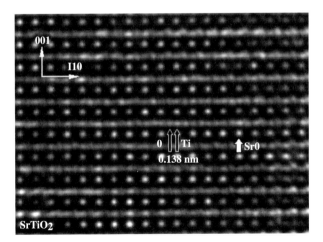

Figure 7.2 HREM image of $SrTiO_2$ down [110] obtained using the Haider/Rose spherical aberration corrector fitted to a Philips CM200 FEG ST electron microscope. Within error measurement, $C_s = -10$ μm, $\Delta f = 8$ nm. Every atomic column, including oxygens, are correctly resolved. The oxygen–titanium distance of 0.133 nm is indicated (Jia et al., personal communication, 2002).

SrTiO$_2$ crystal obtained from this instrument, projected in the [011] direction. The pairs of closely spaced atomic columns (Ti, oxygen) are clearly resolved. The researchers comment that the improvement in image sharpness and contrast (due to the concentration of energy into a taller, narrower lens impulse response) is more striking than the improvement in resolution. Images are recorded by a CCD camera, analysed to produce diffractograms for measurement of aberrations, and the results fed back to the correctors for adjustment of their currents and voltages. The inevitable errors in measurement mean that both C_s and Δf are not exactly zero—the best match with simulated images occurs for $C_s = -10$ μm, $\Delta f = 8$ nm. (Recall that for a phase object with $C_s = \Delta f = 0$ (eqn 3.30), no contrast at all is expected. Hence the contrast must arise either from additional multiple scattering (leading to a scattering phase different from $\pi/2$) or non-zero values of Δf and C_s.)

Aberration correction has brought new challenges and solutions to the field of HREM. In addition to improved resolution, the size of the region which can be studied by selected area diffraction is usefully reduced, since C_s is reduced. The much wider passbands and oscillation-free transfer function eliminate the damping effects of spatial coherence on the transfer function (eqn (4.8), with $C_s \approx 0$). Tilting the incident beam by up to 30 mrad was found to introduce negligible image shift, defocus, astigmatism, or coma, so that the use of hollow-cone illumination is greatly facilitated, and, for the thinnest samples, a more intense lattice image may be obtained by using a larger illumination angle. A larger pole-piece gap may be possible, allowing greater access for X-ray detectors (now up to about 0.8 sr, and capable of single-atom detection) and other exotic sample holders for *in situ* work. For STEM, aberration collection means the possibility of collecting emission from a field-emission source over a much larger angle, both for 'point' and larger sources, resulting in much higher probe currents (Fig. 8.7).

The interpretation of phase-contrast images from C_s-corrected HREM instruments requires renewed analysis, since the conventional method of balancing-off the spherical aberration phase shift against the defocus phase shift to give the 90° shift required for phase contrast imaging cannot be used. Since the information resolution limit d_i (eqn 4.11) is known, one may reintroduce an optimum amount of spherical aberration which makes the point-resolution limit d_p (eqn 6.17) equal to d_i. (O'Keefe 2000). There is also no need to use the usual Scherzer optimum defocus $\Delta f = 1.2\sqrt{(1.5 C_s \lambda)}$; we can instead choose the defocus to be one Scherzer unit of $-\sqrt{(C_s \lambda)}$ and so avoid the 'dip' in the contrast transfer function (CTF; see Fig. 4.3(a) and discussion following eqn (4.12)). C_s is optimized when the crossover of the CTF falls at the information limit of the microscope. The optimum CTF can be derived from the lens phase shift χ (eqn 3.24), since, at crossover, $\chi = 0$ and we get $u^2 = -2\Delta f/(C_s \lambda^2)$. Thus at the Scherzer defocus condition of $\Delta f = -\sqrt{(C_s \lambda)}$, CTF crossover occurs at $u_C = \sqrt{2} C_s^{-1/4} \lambda^{-3/4}$. On the other hand, the information limit is determined by the temporal-coherence damping envelope (third-term in eqn (4.8a)) and cuts off at $d_i^{-1} = u_0(\Delta) = [2/\pi \lambda \Delta]^{1/2}$. Equating u_C with u_0, the optimum value of C_s is found as

$$C_{s_{opt}} = \pi^2 \Delta^2 / \lambda \qquad (7.11)$$

where Δ is the root mean square (rms) value of the total spread of focus due to electronic instabilities (eqn 4.9). The objective aperture should be matched to u_C.

An alternative interpretation of images with $C_s = 0$ is possible using the projected charge density (PCD) approximation (eqn 5.26) (Lynch *et al.* 1975). It predicts an image from thin crystals whose contrast reveals directly the projected charge density $\rho_p(x, y)$ (rather than the electrostatic potential) in the sample. The images are symmetrical in focus, and depend on the total PCD, including the nuclear contribution (unlike X-ray diffraction). Note also that the derivation of eqn (5.26) does not require the first-order expansion of the exponential needed for the weak phase object approximation. All terms, describing higher orders of multiple scattering, are retained, so that the PCD approximation takes better account of multiple scattering than the weak phase object approximation, limited only by the approximation of a flat Ewald sphere. However, the very large angles involved with aberration-corrected lenses eventually result in failure of this approximation according to eqn (5.62), with the result that the depth of focus becomes small. (At 200 kV and a resolution of 0.14 nm, eqn (5.62) allows a thickness of about 8 nm for the validity of the PCD approximation with $C_s = 0$.) At sub-angstrom resolution, this creates interesting possibilities for 'optical sectioning' in both TEM and STEM, as discussed in Sections 5.16 and 8.6. It is also found that the impulse response of the aberration-corrected lens lacks the oscillating 'feet' of the Scherzer image, so that much cleaner images of atoms are seen from glasses and interfaces, with less confusing Fresnel noise.

In summary, aberration-corrected TEMs, operating at 200 keV, have now achieved the angstrom (and sub-angstrom) level of resolution previously obtainable only from high-voltage machines operating at 1 MeV (Ichnose *et al.* 1999), with obvious advantages regarding cost, space requirements, and radiation damage reduction. The information resolution limit can be improved further by decreasing the energy spread of electrons from the source to about 0.2 eV. This may require the fitting of an electron monochromator, similar to those already in use for energy-loss spectroscopy. With a spread of 200 meV full width at half maximum, an aberration-corrected 300 keV FEG TEM would have $\Delta(E)/E$ of 0.28 ppm, which combines with the currently attainable $\Delta(V)/V$ of 0.3 ppm to give $\Delta = 6.2$ Å, leading to an information limit of 0.44 Å. A CTF drawn out at optimum $C_{s_{opt}}$ and $\Delta = 6.2$ Å for such an instrument (O'Keefe 2000) shows clear information transfer beyond 0.5 Å, while a CTF for the equivalent uncorrected microscope shows that it does not reach the information limit at the Scherzer focus (due to the beam convergence of 0.1 mrad). The effect of beam convergence on the CTFs of C_s-optimized microscopes is negligible. Thus HREM in corrected instruments at resolution levels beyond 0.5 Å now appears feasible. The use of aberration correctors in STEM (Krivanek *et al.* 1999) is in many ways even more exciting than those for TEM, in view of the benefits of correctors for ELS, energy-dispersive X-ray (EDX), and ADF imaging. Correctors can provide much greater current into probes, due to the larger illumination angles possible (Fig. 8.7), given sufficient environmental and electronic stability, as shown in Chapter 8. Similar correctors have also been incorporated into the photoelectron emission microscopes built for synchrotrons.

7.5. Combining diffraction and image data for crystals

In structural biology, an important breakthrough occurred in 1975, when the structure of purple membrane was solved by a combination of electron diffraction and imaging data, using two-dimensional crystals (Unwin and Henderson 1975). (We discuss progress in solving structures by electron microdiffraction alone in Section 13.3.) Bragg intensities were used to obtain structure factor magnitudes, and the Fourier transform of the high-resolution image was used to obtain their phases. This work has served as the model for much work since, as described in Section 6.9. By comparison with inorganic structures, the membrane protein crystals offer the advantage of uniform (single molecule) thickness over large areas, but the disadvantages of film bending (Cowley 1961), orientation changes, sample motion due to damage, severe radiation sensitivity, and the need for a hydration medium.

Several groups have now applied these combined methods, originally developed for organic crystals, to inorganic crystals, mainly layer structures. If HREM images and spot diffraction patterns can be obtained from the same region of crystal, thin enough for kinematic conditions to be assumed, then, using the Fourier transform of eqn (4.2) (corrected for transfer function effects as described in Section 5.1), the amplitudes and phases of all the structure factors which lie within the information resolution limit of the microscope might be found. (A useful test for kinematic scattering when a crystal is known to be acentric lies in the fact that $I(\mathbf{g}) = I(-\mathbf{g})$ for such a crystal only if multiple scattering is negligible.) The amplitudes of higher-order reflections can then be obtained from the spot diffraction patterns, and direct methods, phase extension, and/or maximum entropy used to estimate the phases of these higher orders (which extend beyond the resolution of the image), based on the known low-order amplitudes and phases. If the symmetry of the crystal is also known as a result of diffraction pattern analysis, then these symmetry operations may be applied to the HREM images to improve statistics. If regions of constant thickness can be found, the images of many unit cells may be added together for the same purpose. Examples can be found in Li (1998), Dorset (1995), and Sinkler and Marks (1999).

Using combinations of diffraction and imaging data a number of structures have been solved. The explosion of literature on mesoporous silicates and their applications (e.g. for storing gas, or as catalysts), for example, has been accompanied by remarkable examples of the solution of these highly radiation-sensitive structures with large unit cells by a most impressive combination of electron diffraction, high-resolution imaging, and powder X-ray diffraction (Liu et al. 2002). The work by O. Terasaki and colleagues in the more recent literature can be taken as a model for this powerful approach (see, e.g., Deng et al. (2012) for application to the important class of metal-oxide framework structures, or MOFS). For other examples Dong et al. (1992) have extended the resolution of images of a radiation-sensitive organic film from 0.32 to 0.10 nm by applying the maximum entropy algorithm to diffraction pattern intensities. An impressive application of the use of spot diffraction patterns combined with HREM images has been the solution of the structure of a crystalline magnesium–silicon particle in an aluminium matrix (Zandbergen et al. 1997). This group has also used direct least-squares refinement, combined with the multislice algorithm, to analyse spot patterns, so that multiple-scattering effects are included in the analysis (Zandenbergen and Jansen 1998). *Ab initio* structure analysis of a carbonitrile compound using maximum entropy methods is described in Voigt-Martin et al. (1995), resulting in good agreement with X-ray results and with HREM images. The application of direct methods to thickness-integrated electron diffraction data is analysed in Sinkler and Marks (1999) (see Woolfson and Fan (1995) for a textbook treatment of direct methods). The channelling interpretation of imaging (Section 8.4) suggests that the exit-face wavefield will preserve peaks at atom positions despite some multiple scattering, thus extending the thickness range of direct methods. For partially ordered crystals such as oxides with variable stoichiometry the diffuse scattering in selected area patterns can greatly assist the interpretation of HREM images (see Yagi and Roth (1978) for a fine example of this approach, from which students can learn much).

In the field of zeolite catalyst structure determination, a combination of HREM imaging, powder X-ray diffraction, iterative phasing (by the 'charge-flipping' method; Ozlányi and Sütő 2008) and the imposition of a density map histogram constraint has had a major impact on the field, solving large complex structures which could not be solved by any other method (Gramm et al. 2006, Sun et al. 2009). The charge-flipping method, an iterative phasing method for crystals, has been extended for use in powder diffraction by Wu et al. (2006).

In Section 7.6 we describe how the phase problem can usually be solved by iterative methods if the shape (or 'support' function) of an object is known, together with measured diffraction intensities. The most powerful method of combining diffraction and imaging information is therefore based on this iterative algorithm, using the HREM image to provide the support (e.g. the shape of a nanocrystal) to phase an electron microdiffraction pattern. (Ptychography, described in Section 7.6, is an extension of this approach using a coherent nanoprobe.) An early attempt to implement this program was described by Weierstall et al. (2002), but spectacular atomic-resolution results were first obtained by Zuo et al. (2003) for a single carbon nanotube. The general method, and a description of the required software for combining HREM images and nanodiffraction patterns from the same nanocrystal, in order to improve image quality and resolution, is outlined in Zuo et al. (2011). A typical result is shown in Fig. 7.3. The software executes the following steps:

(1) background subtraction in the diffraction pattern;
(2) centring of the diffraction pattern;
(3) processing of the HREM image;
(4) matching the power spectrum of the image with the diffraction pattern;
(5) scaling and orienting the image to the diffraction pattern;
(6) obtaining support from image;
(7) implementing the iterative phasing algorithm (Millane and Stroud 1997), as discussed in Section 7.6, to force consistency between the HREM image, the measured diffraction intensities, and a set of structure factor phase estimates.

The paper by Ran et al. (2012) also treats the case of microdiffraction from a single carbon nanotube, which is merged with a high-resolution image to produce a greatly improved image. It is also shown here how to obtain both images and nanodiffraction patterns from single C60 molecules trapped inside a single-walled carbon nanotube. The damage resistance of C60, due to the nature of the carbon bonds, compared to that of the five-times averaged image of a single DNA molecule obtained by Fujiyoshi and Uyeda (1981) is striking. The use of both high-resolution images and nanodiffraction patterns from the same gold nanocrystal is described by Huang et al. (1996) in their study of surface atom strain relaxation in catalyst particles.

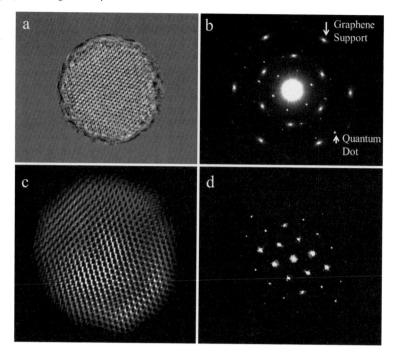

Figure 7.3 A CdS quantum dot (diameter 9 nm) imaged down [0001], and reconstructed with sub-angstrom resolution by combining the diffraction pattern and image. (a) Experimental HREM image after removing the background, contrast inversion, scaling, and rotation. (b) Diffraction pattern recorded on the image plate. (c) Reconstructed image. (d) Power spectrum of the reconstructed image. (From Huang *et al.* (2009), by permission from Macmillan Publishers Ltd., © 2009.)

While there are many difficulties to be resolved with these approaches, this combination of microdiffraction and HREM is probably the most promising approach to super-resolution TEM for crystalline samples, and the ability to scale and merge these data is a severe test of our ability to quantify high-resolution electron microscope data. The use of image plate detectors is important for quantification of the diffraction data. Clearly the trend is toward the incorporation of all the a priori knowledge we have about a structure, such as its symmetry, Bravais lattice, and composition (from EDX and ELS analysis). Information on the statistical distribution of interatomic bond lengths which occur in nature has not been used in the past, and is available. Perhaps the most important point is this: until the experimental electron diffraction patterns produce *reproducible* intensities, the use of a data analysis method (such as X-ray crystallography software) which does not depend on sample thickness is not justified.

Section 13.3 contains more discussion of the electron microdiffraction and precession techniques, and gives examples of solved structures. With the use of aberration-corrected images we can expect the merging of diffraction and image data to become even more powerful, and to be extended to three-dimensional reconstruction.

7.6. Ptychography, Ronchigrams, shadow images, in-line holography, and diffractive imaging

The super-resolution method of electron ptychography uses a STEM to record coherent nanodiffraction patterns for many probe positions. For periodic samples, the diffracted orders may overlap and interfere (thus solving the phase problem), while for non-periodic samples the probe positions may be arranged to overlap, with similar effect. We consider first the simpler case where the sample is periodic over a distance somewhat larger than the probe. The phases of the structure factors can then be extracted from the interference between overlapping orders. Ptychography is unique amongst resolution improvement schemes in being the only method capable of improving resolution beyond the information limit d_i—this apparent miracle is possible because it combines elements of diffraction with imaging. The idea originated in optics, and the first experiments in electron microscopy were due to Hoppe (1969). An elegant formulation of the complete theory has been given in a series of papers by Rodenberg, Bates, and others (see Rodenberg (2008) for a review). To simplify matters, we use results from Chapter 8 on STEM: eqn (8.1) (Fig. 8.3a) for the case of a probe-forming lens of unit magnification, with Bragg beams of equal amplitude at thickness $t = \xi_g/2$ diffracted by a centrosymmetric crystal. With α the angular deviation from the mid-point of the disk overlap at D, s the excitation error, x/a the fractional probe coordinate, $\Delta\theta$ the phase difference between reflections and $F(x)$ the source intensity distribution function, the intensity near the mid-point of overlapping orders becomes

$$I(u, c, \Delta\theta) = 1 + \int_{\text{source}} F(x - c) \cos\{2\pi x/a + \Delta\chi(\alpha) + \Delta\theta(s)\} dx \qquad (7.12)$$

At the achromatic mid-point D, $\alpha = s = \Delta\chi = 0$. (In two dimensions, the intensity is achromatic on a circle.) This important result therefore shows that there are points in the coherent CBED pattern which are aberration free (Spence and Cowley 1978). By positioning the probe on a centre of symmetry such that $I_D = I_F$, an origin may be defined with $c = 0$. For the idealized case of a point source and point detector, the phase difference $\Delta\theta_{g0}$ may thus be determined with respect to this origin. The d-spacing which has been phased is approximately equal to the source size in this case. In the general dynamical case, one can never determine the phase difference between beams which differ in scattering angle by an angle greater than the angular coherence of the source (i.e. the optimum illumination aperture).

Within the approximation of a flat Ewald sphere and a multiplicative object transmission function, however, all diffracted beams refer to the same axial incident beam direction and the phase differences may therefore be linked. Then, for a probe on a centre of symmetry the phases of all beams may be deduced if a phase of zero is assigned to the central beam. The important point is that this procedure allows d-spacings to be phased (for very high-order beams) which are much smaller than the effective source size. The geometrical image of the effective source need be no smaller than the first-order lattice spacing. The highest-order reflection which can be phased is limited only by the Debye–Waller factor and the flat Ewald sphere approximation, and phase determination becomes possible for beams which lie beyond the information limit of the instrument as set by the partial coherence envelope. Note that the function $F(x)$ in eqn (7.12) must also incorporate the effects of tip vibration. These experiments are reciprocally related to the formation of dark-field TEM lattice images with the optic axis in the centre of the objective aperture, but the STEM arrangement provides more information.

The remarkable thing about this geometry is that, although interference is needed only between *adjacent* orders, this suffices (by an obvious 'stepping out' process) to determine the phases of very high-order reflections. Thus the resolution of the method is not limited by the size of the electron source (or its image, the probe), which need only be small enough to provide coherence over an angular range as large as twice the Bragg angle. Unlike HREM imaging, the high-order reflections at the resolution limit never need interfere with the zero-order beam. Resolution in ptychography is ultimately limited by noise and by the Debye–Waller factor, and excitation error effects, which attenuate high-order Bragg beams, rather than by any electron-optical parameters or diffraction limits. Once the phases (and amplitudes) of all the orders have been determined (if single-scattering conditions can be assumed), a map of the crystal potential may be found by Fourier synthesis. By collecting the intensity at many such mid-points simultaneously, Nellist and co-workers have been successful in exceeding the information limit d_i^{-1} using this method (Nellist et al. 1995). Intensities were collected from a thin silicon crystal using a Vacuum Generators HB501 STEM at 100 kV with a point-resolution of 0.42 nm. The synthesized image clearly shows the silicon 'dumbells' projected along [011] with a resolution of 0.136 nm. This is finer that the information limit of $d_i = 0.33$ nm. The errors in this method due to multiple scattering have been calculated as a function of sample thickness and accelerating voltage (Spence 1978). The extension of this ptychography method to non-periodic samples is described at the end of this section.

There is another important interpretation of this experimental geometry, which relates it to Gabor's in-line holography and shadow imaging. We show by a simple ray diagram that the intensity distribution in the region of overlapping orders also consists of a magnified shadow lattice-image of the grating or crystal. This geometrical-optics interpretation of a coherent CBED pattern with overlapping orders is illustrated in Fig. 7.4 for the case where the probe is over-focused. The asymptotic extension of diffracted CBED cones toward the source defines virtual sources S', necessarily coherent with S (since they are images of the same source). By geometric construction it will be found that these virtual sources lie on one plane in the reciprocal lattice, and so form a virtual point diffraction pattern. The under-focus case (not shown) corresponds to the arrangement used to form 'Tanaka' large-angle CBED patterns with the beam-selecting aperture removed. Taken alone, two of these sources S and S' will produce Young's fringes on a distant screen with angular period $\Delta\alpha = \lambda/d_s$, where d_s is the separation of the sources. Now $d_s \approx 2\theta_B \Delta f$, so the period of the fringes is

$$\Delta\alpha = \lambda/(2\theta_B \Delta f) \tag{7.13}$$

where $\Delta\alpha$ is the angle subtended at the sample by one fringe on the plane of the CBED detector. The fringes thus become finer with increasing defocus, as observed. This expression is a good approximation if C_s is small for low-order reflections and large defocus. The fringes may thus be interpreted as a point-projection shadow lattice image, with magnification $M = L/\Delta f$. We can see this simply by noting that the period of the fringes on the screen is approximately $X = L\Delta\alpha = M(\lambda/2\theta_B) = Ma$, with a the lattice spacing. This is in

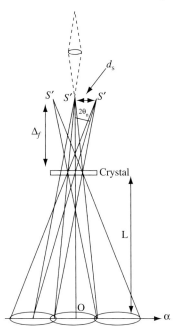

Figure 7.4 Ray diagram showing the formation of virtual sources S' when a STEM probe is overfocused. These sources produce interference fringes on the microdiffraction screen, which are a magnified shadow image of the crystal. The magnification is $L/\Delta f$.

accordance with the theory of Fourier or Talbot self-imaging, which predicts images periodic in both L and Δf (Cowley 1990), and such shadow lattice images have now been observed experimentally by several groups using field-emission STEM instruments (Cowley 1986, Tanaka 1988, Vincent et al. 1992). In the absence of spherical aberration, the magnification is infinite at zero defocus. The formation of shadow lattice images also follows from eqn (7.12), if we take $C_s = c = \Delta\theta = 0$ and solve for the period $\Delta\alpha$ of $I(\alpha)$ in α, the angular coordinate on the CBED screen.

If the illumination angle is made much larger than the Bragg angle, so that gross overlap of orders occurs, we have just the geometry proposed by Gabor for in-line holography. The patterns can be interpreted as holograms if the direct, central disc is much more intense than the others, providing a reference beam. The theory of this effect makes close connection with the theory of Fourier imaging, which uses the same geometry (Cowley 1981). Early experimental applications of Gabor's in-line proposal to electron microscopy were unsuccessful (Haine and Mulvey 1952), partly because of the twin image problem—it was found impossible to separate the real and virtual images in the reconstruction—and partly for lack of coherent, bright, field-emission sources. (See Collier et al. (1971) for an excellent review and history of both optical and electron holography.) Several solutions to the twin image problem have since been proposed—a summary can be found in the chapter by Spence in Voelkl (1998)—these are generally variants of the coherent detection method outlined earlier, in which some experimentally controllable parameter is varied as linear images are accumulated. The most successful solutions include: (1) the Fraunhofer (small object) method (see Matsumoto et al. (1994) for an application to HREM), in which one image is so far out of focus as to be undistracting; (2) summing images for different lateral source positions, after repeatedly bringing one of them back onto the axis in the computer (they have different dependences on source location); (3) integration of images over defocus or wavelength (this method is used extensively in photoemission holography; Barton 1988); (4) the transport of intensity method (Paganin and Nugent 1998) circumvents the twin image problem, requiring only one image recording for phase objects. Early work on in-line electron holography using STEM can be found in Munch (1975), Cowley and Walker (1981), and Cowley (1992a,b).

For a compact non-periodic object bathed in a coherent, diverging beam, it is a remarkable fact that the scattering pattern in the far-field becomes an image of the object. (If the source distance Δf in Fig. 7.4 is taken to infinity, by contrast, so that the illumination becomes planar, we know that the scattering pattern is a Fraunhofer diffraction pattern, bearing no obvious resemblance to the object.) The image formed at infinity due to illumination by a point source at distance Δf from the object turns out to be a conventional out-of-focus image, for which the focus defect is Δf and the magnification $L/\Delta f$. This equivalence between HREM out-of-focus images and point-projection images is explained in the language of Fourier optics in Spence (1992). Both of these are in-line holograms.

If the source-to-object distance in Fig. 7.4 is made very large, so that the illumination becomes collimated, and the source to detector distance L is made small, we have just the arrangement used for conventional HREM, if lenses are then added below the sample to magnify the image. Thus the through-focus methods of Section 7.1 may also be considered to be methods for reconstructing in-line electron holograms. A bright-field, out-of-focus HREM image is an electron hologram. The advantage of using a diverging, coherent beam is that it provides magnification without the need for lenses, which may introduce aberrations. The final resolution in the image reconstructed from a coherent diverging beam is then given approximately by the source size.

If the illumination angle is not limited, the effects of spherical aberration become dominant, and the pattern is known as a Ronchigram. Rochigrams were recorded from field-emission STEM instruments many years ago which clearly showed directly the crystal lattice planes (Cowley 1986). They may be used to measure the aberration constants of STEM instruments (Cowley 1979). Ronchigrams are coherent CBED patterns from thin samples (crystalline or non-periodic) with the objective aperture removed. They provide the best and most sensitive way to align a STEM for high resolution, and to measure the defocus and aberrations in STEM, as described in detail in Section 10.10. Figure 7.5 shows an experimental Ronchigram (Hembree et al. 2003). The fringes in the centre can be interpreted as a magnified lattice image in the manner described above (Fig. 7.4). The two eyes of the ellipses occur at the stationary phase positions in the transfer function (eqn 5.66). We will see that the area over which the Ronchigram is smooth is an indication of faithful information transfer. Ronchigrams bears a close similarity to shearing interferometer patterns.

We now show that such a coherent convergent-beam pattern is indeed a shadow image (or in-line hologram) by wave optics for a non-periodic object. Take the transmission function for the sample as $T(x)$ (with Fourier transform $t(K)$, where $K = \theta/\lambda = X/L\lambda$ as in Fig. 7.4), the electron wavefunction across the downstream side of the slab as $\Psi(x)$, and represent a spherical wave incident on the sample by $P(\Delta f) = \exp(-i\pi x^2/\Delta f \lambda)$ (with transform

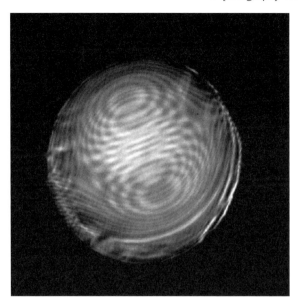

Figure 7.5 Ronchigram recorded on a 100 kV Vacuum Generators STEM using the 0.8 nm lattice planes of beryl in near two-beam conditions. The eyes of the two ellipses occur at stationary phase positions in the transfer function (eqn 5.66). Interference occurs due to gross overlap of coherent orders. These patterns are essential for accurate alignment and measurement of optical parameters in STEM as discussed in Section 10.10, where a Ronchigram from a non-periodic sample is shown (from Spence and Zuo 1992, Hembree *et al.* 2003).

$A(K) = \exp(i\pi \Delta f \lambda K^2)$, eqn (3.16)). Then, with magnification $M = L/\Delta f = X/x$, the exit-face wavefield is

$$\Psi(x) = T(x) P_{\Delta f}(x) \tag{7.14}$$

As in eqn (8.2) for STEM (with probe position $r_p = 0$), the wavefield in the far-field is

$$\Psi_D(X) = t(K) * A(K) \tag{7.15}$$

Evaluating this convolution, ignoring unimportant phase factors, we obtain

$$\Psi_D(X = L\lambda K) = T(X/M) * P_{\Delta f}(X/M) \tag{7.16}$$

But this is the same expression as for a defocused bright-field TEM image using collimated illumination (eqn 3.12). So this important result establishes that, for wide-angle coherent CBED in the absence of aberrations (or at large defocus), the shadow image within the central disc consists of a magnified 'ideal' image $T(X/M)$, which is 'out of focus' by the probe-to-sample 'defocus' distance Δf. This explains why in-line holograms of the edge of an opaque screen show the same Fresnel edge fringes we see in HREM. (This can be seen, for example, in the low-voltage field-emission point-projection images of Tobacco Mosaic Virus published by Weierstall et al. (1999).) The magnification has been achieved without the use of lenses. The effects of limiting the beam divergence are equivalent to limiting the semi-angle subtended by the objective aperture in HREM. The resolution of the shadow image is approximately equal to the effective source size responsible for the incident spherical wave. An in-focus image can only be obtained if $\Delta f = 0$, in which case M is infinite. In practice, an important disadvantage of this in-line holography geometry is that, at high magnification, with a limited beam divergence, the number of resolution-limited pixels (whose size is about equal to the source size) within the illuminated region is very limited, going to unity as the magnification becomes infinite. We should also be aware that the simple product in eqn (7.14) will fail if the phase-object approximation fails (Spence and Cowley 1978), which may occur because of the range of illumination angles involved. This geometry, and the effect of partial coherence on Ronchigrams, is discussed in more detail in Section 10.10.

The term 'diffractive imaging' (or 'lensless imaging') refers to methods which reconstruct an image from diffracted intensities and little other information. It has been found that the most important additional information needed is the shape (or 'support') of the sample. The support is the region in which the object transmission function is non-zero. Iterative methods have been developed which are capable of finding the phases in Fraunhofer diffraction patterns from non-periodic objects if this support of the object is known, together with the sign of the scattering potential. The algorithm is derived from work in X-ray diffraction (Sayre 1952) and electron microscopy (Gerchberg and Saxton 1972), to which a feedback feature has been added (Fienup 1982), and which together produce the successful hybrid input–output (HiO) algorithm. The method is often referred to as 'oversampling', since, as pointed out by Sayre (1952), for a periodic sample one requires half-order Bragg reflections in order that the number of Fourier equations relating complex structure factors to a real (not complex) density map is equal to the number of unknown phases. For an object whose transmission function is real (or pure imaginary), the support estimate can be approximate or 'loose'; if complex, it must be rather accurately known (Fienup 1987). The algorithms operate by iterating between real and reciprocal space, while imposing known information (constraints) in each domain. Note that this simple two-dimensional Fourier iteration is only possible in regions where the Ewald sphere is flat. Performance improves in three dimensions, where the three-dimensional diffraction volume can be filled from data collected on curved Ewald spheres. The method can be understood in terms of convex optimization, or projections onto convex and non-convex sets. A knowledge of diffracted intensities, support, and the sign of the scattering potential is found in practice to lead to convergence, given data of sufficiently high quality. The list of convex constraints, which greatly assist convergence, continues to grow. There are many developments of the original HiO algorithm, including the 'shrinkwrap' algorithm, which uses thresholding during iterations to update an estimate of the support (Marchesini et al. 2003). The method bears a conceptual similarity to the methods of solvent flattening used in direct approaches. This large and active field of research with applications in many fields is reviewed in Spence (2008); a comparison of the performance of algorithms can be found in Marchesini (2007), and a detailed experimental example using three-dimensional iterations is given in Chapman et al. (2006). While the HiO algorithm is intended for non-periodic samples, the charge-flipping algorithm applies similar ideas to the Bragg diffraction from crystals, using the space between atoms as the support, so that atomic resolution data are normally needed. Figure 7.3 shows an experimental example of diffractive imaging using HREM images and diffraction patterns to obtain a sub-angstrom resolution image of a CdS quantum dot. A spectacular example can also be found in the atomic-resolution image reconstruction of a single-walled carbon nanotube from its coherent electron nanodiffraction pattern (Zuo et al. 2003), and the imaging of oxygen atoms in a TiO_2 nanocrystal by numerical reconstruction from its electron diffraction pattern in De Caro et al. (2010).

Finally we consider the method of ptychography applied to non-periodic samples. An algorithm has been developed, the ptychographical iterative engine ('PIE', and its more recent variants), which seeks consistency by iteration between nanodiffraction pattern intensities, the sample structure estimate, and the probe wavefield (Faulkner and Rodenburg 2004). This uses coherent nanodiffraction patterns recorded from overlapping regions of the sample in order to solve the phase problem. For two-dimensional images, this immediately solves the problem of the limited field of view of in-line holography (and the twin-image problem) since the beam may be scanned over a large area. Oversampling of the beam area alone is required, so the number of detector pixels needed may not be large if the beam is small. In a remarkable application of the method, Humphry et al. (2012) have reconstructed atomic-resolution images of a small crystal from a field-emission SEM instrument operated in the transmission geometry. The resolution of the SEM is normally an order of magnitude worse, even if operated in the transmission geometry. The extension of this method to include the effects of Ewald sphere curvature and three-dimensional reconstruction (tomographic ptychography) are important challenges for the future (Marchesini et al. 2012). Simple Fourier iterations in two dimensions are only possible within the low-resolution region where the sphere is flat. At higher resolution, the entire diffraction volume must be filled before three-dimensional iterations between real and reciprocal space can be implemented, and information on probe position and sample orientation must be recorded. Resolution may be limited to the accuracy with which the probe position can be measured; however, corrections for this effect can be built into the algorithms (Maiden et al. 2012). Since lenses are not needed to form the probe, this super-resolution method, which allows truly quantitative phase-contrast imaging, has exciting applications to neutron, light, and X-ray microscopy, many of which have now been demonstrated. (For three-dimensional lensless ptychography of samples of bone using hard X-rays see Dierolf et al. (2010).) Optical applications of ptychography are under active development (Claus et al. 2012).

A variety of other super-resolution schemes have been proposed in electron microscopy. Many other geometries for electron holography in STEM are possible—one survey listed 20 modes (Cowley 1992a,b). The effects of aberrations enter in different ways in these geometries, allowing new possibilities for their removal. It has been shown by a related analysis (Spence and Koch 2001) that two-dimensional ALCHEMI patterns (see Section 13.1) consist of holograms of atomic strings in a thin crystal. The transport of intensity equations can be used to find the phase of a near-field wavefield if its intensity distribution is known on two adjacent planes. The approach is therefore similar to the through-focus method. If one of these planes is taken as the exit-face wavefunction for a phase object, then this intensity distribution (unity everywhere) is known a priori, so that a single out-of-focus plane is needed. Details and applications of the method are given in Paganin and Nugent (1998). It has also been proposed (Cowley et al. 1997) and demonstrated (Cowley 2001) that the strings of atoms in a thin crystalline sample may be used as a lens, to focus a STEM probe to even smaller dimensions beyond a thin crystal.

7.7. Direct inversion from dynamical diffraction patterns

Several schemes for direct inversion of dynamical diffraction pattern intensities to complex structure factors have appeared (Spargo *et al.* 1994, Moodie *et al.* 1996, Allen *et al.* 1998, 1999, Spence 1998, 2009, Spence *et al.* 1998). If successful, these methods could provide much higher resolution than imaging, while at the same time, by analogy with X-ray crystallography, they may be capable of identifying the species present in the unit cell from the height of the peaks in a potential map. Since they are based on analysis of CBED patterns they can be applied to nanoscale regions of material, which must, however, be crystalline and thick enough to produce significant multiple scattering. They could not easily be used to analyse the atomic structure of defects. *The essential idea of these schemes is that multiple scattering solves the phase problem, by allowing interference between different Bragg beams, or 'structure factors'.* Thus it is readily shown that the intensity of multiply scattered Bragg beams is highly sensitive to the phase of structure factors (and hence to atomic coordinates), unlike singly scattered beams. A powerful general approach has been to use reduction by symmetry of many-beam interactions to soluble two and three-beam forms (Moodie *et al.* 1996). By collecting a series of coherent CBED patterns with overlapping orders near to a zone axis orientation it is possible to find the complex amplitudes for all the entries in the scattering matrix **S** in eqn (5.50) (Spence 1998). It is also possible to find the structure matrix **A** directly from a knowledge of all the complex entries in **S**, and of the excitation errors (Allen *et al.* 1998, 1999). A different non-iterative approach to the direct determination of complex structure factors from CBED patterns from crystals of unknown structure has been based on Moodie's polynomial expansion (eqn 5.33) (see Koch and Spence (2003)). A scheme for removal of multiple scattering from electron diffraction patterns based on iterated projection between constrained sets is described in Spence *et al.* (1998). If two dynamical diffraction patterns can be phased at two different beam energies, the single-scattering can be recovered by division of these phased patterns (Spence 2009). For known crystal structures, optimization methods may be used to refine low-order bonding reflections in amplitude and phase. For a detailed analysis and measurement of structure-factor phases with tenth-degree accuracy from three-beam CBED patterns in acentric crystals see Zuo *et al.* (1989, 1993). A dynamical inversion scheme based on neural network theory is given in Van den Broek and Koch (2012).

References

Allen, L. J., Josefsson, T. W., and Leeb, H. (1998). Obtaining the crystal potential by inversion from electron scattering intensities. *Acta Crystallogr*. **A54**, 388.

Allen, L. J., Leeb, H., and Spargo, A. E. C. (1999). Retrieval of the projected potential by inversion from the scattering matrix in electron diffraction. *Acta Crystallogr*. **A55**, 105.

Anstis, G. R., Lynch, D. F., Moodie, A. F., and O'Keefe, M. A. (1973). n-Beam lattice images. III. Upper limits of ionicity in $W_4Nb_{26}O_{77}$. *Acta Crystallogr*. **A29**, 138.

Barton, J. J. (1988). Photoelectron holography. *Phys. Rev. Lett*. **61**, 1356.

Bevington, R. (2002). *Data reduction and error analysis for the physical sciences*. Third edition. McGraw-Hill, New York.

Blackman, M. (1939). Electron diffraction. *Proc. R. Soc. Lond., Ser. A* **173**, 68.

Chapman, H. N., Barty, A., Marchesini, S., Noy, A., Hau-Reige, S., Cui, C., Howells, M., Rosen, R., He, H., Spence, J. C. H., Weierstall, U., Beetz, T., Jacobsen, C., and Shapiro, D. (2006). High resolution *ab-initio* three-dimensional X-ray diffraction microscopy. *J. Opt. Soc. Am.* **23**, 1179.
Claus, D., Maiden, A., Zhang, F., Sweeney, F., Francis, G., Humphry, M., Schluesener, H., and Rodenburg, J. M. (2012) Quantitative phase contrast optimized cancerous cell differentiation via ptychography. *Opt. Express* **20**, 9911.
Coene, W., Janssen, G., Op de Beeck, M., and Van Dyck, D. (1992). Phase retrieval through focus variation for ultra-high resolution in field emission transmission electron microscopy. *Phys. Rev. Lett.* **69**, 3743.
Coene, W., Thust, A., Op de Beeck, M., and Van Dyck, D. (1996). Maximum likelihood method for focus-variation image reconstruction. *Ultramicroscopy* **64**, 109.
Collier, R. J., Burkhadt, C. B., and Lin, L. H. (1971). *Optical holography*. Academic Press, New York.
Cowley, J. (1961). Diffraction intensities from bent crystals. *Acta Crystallogr.* **14**, 920.
Cowley, J. M. (1967). Crystal structure determination by electron diffraction. In *Progress in Materials Science*, Vol. 13, No. 6, p. 269. Pergamon Press, Oxford.
Cowley, J. M. (1979). Adjustment of STEM instrument by use of shadow images. *Ultramicroscopy* **4**, 413.
Cowley, J. M. (1981). Coherent interference effects in SIEM and CBED. *Ultramicroscopy* **7**, 19.
Cowley, J. M. (1986). Electron diffraction phenomena observed with a high resolution STEM instrument. *J. Electron Microsc. Tech.* **3**, 25.
Cowley, J. M. (1990). *Diffraction physics*. North-Holland, New York.
Cowley, J. M. (1992a). *Techniques of transmission electron diffraction*. Oxford University Press, New York.
Cowley, J. M. (1992b). Twenty forms of electron holography. *Ultramicroscopy* **41**, 335.
Cowley, J. (2001). Comments on ultra-high resolution STEM. *Ultramicroscopy* **87**, 1.
Cowley, J. and Hudis, J. B. (2000). Atomic focusser imaging by graphite crystals in carbon nanoshells. *Microsc. Microanal.* **6**, 429.
Cowley, J. M. and Walker, D. J. (1981). Reconstruction from in-line holograms by digital processing. *Ultramicroscopy* **6**, 71.
Cowley, J., Spence, J., and Smirnov, V. (1997). The enhancement of electron microscope resolution by use of atomic focusers. *Ultramicroscopy* **68**, 135.
De Caro, L., Carlino, E., Caputo E., Cozzoli P., and Giannini, C. (2010). Electron diffraction imaging of oxygen atoms in nanocrystals at sub-angstrom resolution. *Nature Nanotechnol.* **5**, 360.
Deng, H., Grunder, S., Cordova, K., Valente, C., Furukawa, H., Hmadeh, M., Gándara, F., Whalley, A., Liu, Z., Asahina, S., Kazumori, H., O'Keeffe, M., Terasaki, O., Fraser Stoddart, J., and Yaghi, O. (2012). Large-pore apertures in a series of metal-organic frameworks. *Science* **336**, 1018.
Dierolf, M., Menzel, A., Thibault, P., Schneider, P., Kewish, C., Wepf, R., Bunk, O., and Pfieffer, F. (2010). Ptychographic X-ray computed tomography at the nanoscale. *Nature* **467**, 436.
Dong, W., Baird, T., Fryer, J., Gilmore, C., MacNicol, D., Bricogne, G., Smith, D., O'Keefe, M., and Hovmuller, S. (1992). Electron microscopy at 1-angstrom resolution by entropy maximization and likelihood ranking. *Nature* **355**, 605.
Dorset, D. L. (1995). *Structural electron crystallography*. Plenum, New York.
Dowell, W. (1963). Lattice imaging with inclined illumination. *Optik* **20**, 535.
Eades, J. A. (1988). Glide planes and screw axes in CBED: the standard procedure. In *Microbeam analysis*, ed. D. E. Newbury, p. 75. San Francisco Press, San Francisco.
Faigel, G. and Tegze, M. (1999). X-ray holography. *Rep. Prog. Phys.* **62**, 355.

Faulkner, H. and Rodenburg, J. M. (2004). Moveable aperture lensless transmission microscopy: a novel phase retrieval algorithm. *Phys. Rev. Lett.* **93**, 023903.

Fienup, J. R. (1982). Phase retrieval algorithms, a comparison. *Appl. Opt.* **21**, 2758.

Fienup, J. R. (1987). Reconstruction of a complex-valued object from the modulus of its Fourier transform using a support constraint. *J. Opt. Soc. Am.* **A4**, 118.

Fujiyoshi, Y. and Uyeda, N. (1981). Direct imaging of double-stranded DNA. *Ultramicroscopy* **7**, 189.

Gabor, D. (1949). Microscopy by reconstructed wavefronts. *Proc. R. Soc. Lond.* **A197**, 454.

Gerchberg, R. and Saxton, W. (1972). Phase determination from image and diffraction pattern in electron microscopy. *Optik* **35**, 237.

Gramm, F., Baerlocher, C., McCusker, L. B., Warrender, S. J., Wright, P. A., Han, B., Hong, S. B., Liu, Z., Ohsuna, T., and Terasaki, O. (2006). Complex zeolite structure solved by combining powder diffraction and electron microscopy. *Nature* **444**, 79.

Haider, M., Uhlemann, S., Schwan, E., Rose, H., Kabius, B., and Urban, K. (1998a). Electron microscopy image enhanced. *Nature* **392**, 768.

Haider, M., Rose, H., Uhlemann, S., Kabius, B., and Urban, K. (1998b). Toward 0.1 nm resolution with the first spherically corrected TEM. *J. Electron Microsc.* **47**, 395.

Haine, M. E. and Mulvey, T. (1952). The formation of the diffraction image with electrons in the Gabor diffraction microscope. *J. Opt. Soc. Am.* **42**, 763.

Hawkes, P. (1974). Constrained optimization in image deconvolution. *Optik* **41**, 64.

Hembree, G., Koch, C., Weierstall, U., and Spence, J. (2003). A quantitative nanodiffraction system for UHV STEM. *Microsc. Microanal.* **9**, 468.

Hoppe, W. (1969). Ptychography. *Acta Crystallogr.* **A 25**, 495.

Howie, A., Jiang, N., Spence, J. C. H., and Wu, J. (2006). Background intensity problems in high-resolution defect imaging. *Microsc. Microanal.* **12** (Suppl. S02), 900.

Huang, D. X., Liu, W., Gu, Y., Xiong, J., Fan, H., and Li, F. H. (1996). A method of electron diffraction intensity correction in combination with HREM. *Acta Crystallogr.* **52**, 152.

Huang, W. J., Zuo, J. M., Jiang, B., Kwon, K. W., and Shim, M. (2009). Sub-angstrom-resolution diffractive imaging of single nanocrystals. *Nature Phys.* **5**, 129.

Humphry, M. J., Kraus, B., Hurst, A., Maiden, A., and Rodenburg, J. M. (2012). Ptychographic electron microscopy using high-angle dark-field scattering from sub-nanometer imaging. *Nature Commun.* **3**, 730.

Hytch, M. and Stobbs, W. (1994). Quantitative comparison of HREM images with simulations. *Ultramicroscopy* **53**, 191.

Hytch, M., Cherkashin, N., Reboth, S., Houellier, F., and Claverie, A. (2011). Strain mapping in layers and devices by electron holography. *Phys Status Solidi (a)* **208**, 580.

Ichinose, H., Sawada, H., Takuma, E., and Osaki, M. (1999). Atomic resolution HVEM and environmental noise. *J. Electron Microsc.* **48**, 887.

Jia, C. L. and Thust, A. (1999). Investigation of atomic displacements at a Sigma 3(111) twin boundary in $BaTiO_3$ by means of phase-retrieval electron microscopy. *Phys. Rev. Lett.* **82**, 5052.

King, W., Campbell, G., Foiles, S., Cohen, D., and Hanson, K. (1998). Quantitative HREM observation of grain boundary structure in aluminium with atomistic simulation. *J. Microsc.* **190**, 131.

Kirkland, A., Saxton, W., Chau, K., Tsuno, K., and Kawasaki, M. (1995). Super resolution by aperture synthesis. *Ultramicroscopy* **57**, 355.

Kirkland, A., Saxton, O., and Chand, G. (1997). Multiple beam tilt microscopy for super resolved imaging. *J. Electron Microsc.* **1**, 11.

Kirkland, E. (1984). Improved high-resolution image processing of bright-field electron micrographs. *Ultramicroscopy* **15**, 151.

Kisielowski, C., Hetherington, C. J., Wang, Y. C., Kilaas, R., O'Keefe, M. A., and Thust, A. (2001). Imaging columns of light elements with sub-angstrom resolution. *Ultramicroscopy* **89**, 243.

Koch, C. and Spence, J. (2003). A useful disentanglement of the exponential of the sum of two non-commuting notices. *J. Phys. A*. **36**, 803.

Konnert, J., D'Antonio, P., Cowley, J., Higgs, A., and Ou, H.-J. (1989). Determination of atomic positions by CBED. *Ultramicroscopy* **30**, 371.

Krivanek, O. and Stadelmann, P. A. (1995). Effect of three-fold astigmatism on HREM. *Ultramicroscopy* **60**, 103.

Krivanek, O., Dellby, N., and Lupini, A. R. (1999). Toward sub-angstrom electron beams. *Ultramicroscopy* **78**, 1.

Li, F. (1998). Crystallographic image processing approach to crystal structure determination. *J. Microsc.* **190**, 249.

Lichte, H. (1988). A short glimpse at electron holography. In *Microphysical reality and quantum formalism*, ed. A. van der Merwe, p. 137. Kluwer, New York.

Lichte, H. (1991). Electron image plane off-axis holography of atomic structures. *Adv. Opt. Electron Microsc.* **12**, 25.

Lin, J. A. and Cowley, J. M. (1986). Calibration of the operating parameters for an HB5 STEM instrument. *Ultramicroscopy* **19**, 31.

Liu, Z., Fujita, N., Terasaki, O., Ohsuna, T., Hiraga, K., Camblor, M., Diaz-Cabanas, M., and Cheetham, A. K. (2002). Incommensurate modulation in the microporous silica SSZ-24. *Chem. Eur. J.* **8**, 4549.

Lynch, D. F., Moodie, A. F., and O'Keefe, M. (1975). The use of the charge-density approximation in the interpretation of lattice images. *Acta Crystallogr*. **A31**, 300.

McCallum, B. C. and Rodenberg, J. M. (1992). Two-dimensional demonstration of Wigner phase retrieval microscopy in the STEM configuration. *Ultramicroscopy* **45**, 371.

McKeown, J. and Spence, J. C. H. (2009). The kinematic convergent beam method for solving nanocrystal structures. *J. Appl. Phys.* **106**, 074309.

Maiden, A. M., Humphry, M. J., Sarahan, M. C., Kraus, B., and Rodenburg, J. M. (2012). An annealing algorithm to correct positioning errors in ptychography. *Ultramicroscopy* **120**, 64.

Marchesini, S. (2007). A unified evaluation of iterative projection algorithms for phase retrieval. *Rev. Sci. Instrum.* **78**, 011301.

Marchesini, S., He, H., Chapman, H., Hau-Reige, S., Noy, A., Howells, M., Weierstall, U., and Spence, J. C. H. (2003). X-ray image reconstruction from a diffraction pattern alone. *Phys. Rev.* **B68**, 140101(R).

Marchesini, S., Schirotzek, A., Maia, F., Yang, C. (2012). Augmented projections for ptychographic imaging. arXiv:1209.4924[physics.optics] <http://arxiv.org/abs/1209.4924>]

Matsumoto, T., Tanji, T., and Tonomura, A. (1994). Phase contrast visualisation of undecagold cluster by in-line electron holography. *Ultramicroscopy* **54**, 317.

Millane, R. and Stroud, W. J. (1997). Reconstructing symmetric images from their undersampled Fourier intensities. *J. Opt. Soc Am*. **A14**, 568.

Mobus, G., Schweinfest, R., Gemming, T., Wagner, T., and Ruhle, M. (1998). Interactive structural techniques in HREM. *J. Microsc.* **190**, 109.

Mollenstedt, G. and Wahl, H. (1968). A biprism for electron holography. *Naturwissenschaften* **55**, 340.

Moodie, A. F., Etheridge, J., and Humphreys, C. J. (1996). The symmetry of three-beam scattering equations: inversion of three-beam diffraction patterns from centrosymmetric crystals. *Acta Crystallogr*. **A52**, 596.

Munch, J. (1975). Experimental electron holography. *Optik* **43**, 79.

Nellist, P. D., McCallum, B., and Rodenburg, J. (1995). Resolution beyond the information limit in transmission electron microscopy. *Nature* **374**, 630.

Ochowski, A., Rau, W. D., and Lichte, H. (1995). Electron holography surmounts resolution limit of electron microscopy. *Phys. Rev. Lett.* **74**, 399.

O'Keefe, M. A. (2000). The optimum C_s condition for HREM. *Proc. Microsc. Soc. Am. 2000, Philadelphia*. Cambridge University Press, New York.

O'Keefe, M., Spence, J., Hutchinson, J., and Waddington, W. (1985). HREM profile image interpretation in MgO cubes. *Proc. 43rd EMSA Meeting*. San Francisco Press, San Francisco.

O'Keeffe, M. and Spence, J. C. H. (1993). On the average Coulomb potential and constraints on the electron density in crystals. *Acta Crystallogr.* **A50**, 33f.

O'Keefe, M. A., Hetherington, C. J., Wang, Y. C., Nelson, E., Turner, J., Kisielowski, C., Malm, O., Mueller, R., Ringnalda, J., Pan, M., and Thust, A. (2001). Sub-angstrom HREM at 300 kV. *Ultramicroscopy* **89**, 215.

Op de Beeck, M., Van Dyck, D., and Coene, W. (1996). Wave function reconstruction in HRTEM: the parabola method. *Ultramicroscopy* **64**, 167.

Oszlányi, G. and Sütő, A. (2008). The charge-flipping algorithm. *Acta Crystallogr.* **A64**, 123.

Paganin, D. and Nugent, K. (1998). Noninterferometric phase imaging with partially coherent light. *Phys. Rev. Lett.* **80**, 2586.

Pinsker, Z. G. (1949). *Electron diffraction*, transl. J. A. Spink and E. Feigl. Butterworths, London.

Plamann, T. and Rodenburg, J. M. (1994). Double resolution imaging with infinite depth of focus in single-lens scanning microscopy. *Optik* **96**, 31.

Press, W. H., Flannery, B., Teukolsky, S., and Vetterling, W. (1986). *Numerical recipes*. Cambridge University Press, New York.

Prince, E. (1982). *Mathematical techniques in crystallography*. Springer, New York.

Ran, K., Zuo, J.-M., Chen, Q., and Shi, Z. (2012). Electrons for single-molecule diffraction and imaging. *Ultramicroscopy* **119**, 72.

Rodenburg, J. M. (2008). Ptychography and related diffractive imaging methods. *Adv. Imag. Electron. Phys.* **150**, 87.

Rose, H. (1990). A spherically corrected semiaplanatic medium voltage TEM. *Optik* **85**, 19.

Saxton, O. (1994). What is the focus variation method? *Ultramicroscopy* **55**, 171.

Saxton, W. O. (1998). Quantitative comparison of images and transforms. *J. Microsc.* **190**, 52.

Sayre, D. (1952). Some implications of a theorem due to Shannon. *Acta Crystallogr.* **5**, 843.

Scherzer, O. (1947). Spherical and chromatic correction in electron lenses. *Optik* **2**, 114.

Schiske, P. (1968). Zur Frage der Bildrekonstruktion durch Fokusreihen. *Proc. 4th European Conf. on Electron Microscopy, Rome*, p. 145.

Schiske, P. (1973). Image processing using additional statistical information about the object. *Image processing and computer aided design in electron optics*, ed. P. W. Hawkes, p. 82. Academic Press, London.

Sinkler, W. and Marks, L. D. (1999). Dynamic direct methods for everyone. *Ultramicroscopy* **75**, 251.

Spargo, A., Beeching, M., and Allen, L. J. (1994). Inversion of electron scattering intensity for crystal structure analysis. *Ultramicroscopy* **55**, 329.

Spence, J. C. H. (1978). Practical phase determination of inner dynamical reflections in STEM. *Scanning Electron Microsc.* **1978** (I), 61.

Spence, J. C. H. (1992). Convergent-beam nano-diffraction, in-line holography and coherent shadow imaging. *Optik* **92**, 57.

Spence, J. C. H. (1998). Direct inversion of dynamical electron diffraction patterns to structure factors. *Acta Crystallogr.* **A54**, 7.

Spence, J. C. H. (2008). Diffractive (lensless) imaging. *Science of microscopy*, ed. P. Hawkes and J. C. H. Spence, p. 1196. Springer, New York.

Spence, J. C. H. (2009). Two-wavelength inversion of multiply-scattered soft X-ray intensities to charge density. *Acta Crystallogr.* **A65**. 28.

Spence, J. C. H. and Cowley, J. M. (1978). Lattice imaging in STEM. *Optik* **50**, 129.

Spence, J. and Koch, C. (2001). Atomic string holography. *Phys. Rev. Lett.* **86**, 5510.

Spence, J. C. H. and Zuo, J. M. (1992). *Electron microdiffraction*. Plenum, New York.

Spence, J., Calef, B., and Zuo, J. (1998). Dynamical inversion by the method of generalised projections. *Acta Crystallogr.* **A55**, 112.

Stark, H. (1987). *Image recovery: theory and applications*. Academic Press, New York.

Stout, G. H. and Jensen, L. H. (1968). *X-ray structure determination*. Macmillan, London.

Sun, J., Bonneau, C., Cantin, A., Corma, A., Diaz-Cabañas, M. J., Moliner, M., Zhang, D., Li, M., and Zou, X. (2009). *Nature* **458**, 1154.

Tanaka, N. (1988). Nanometer-area electron-diffraction of interface of semiconducting superlattices. *Japan. J. Appl. Phys.* **2**, L468.

Thust, A., Coene, W., Op de Beeck, M., and Van Dyck, D. (1996a). Focal-series reconstruction in HRTEM. *Ultramicroscopy* **64**, 211.

Thust, A., Overwijk, W., Coene, W., and Lentzen, M. (1996b). Numerical correction of lens aberrations in phase-retrieval. *Ultramicroscopy* **64**, 249.

Unwin, N. and Henderson, R. (1975). Molecular structure determination by electron microscopy of unstained crystalline specimens. *J. Mol. Biol.* **94**, 425.

Vainshtein, B. (1964). *Structure analysis by electron diffraction*. Pergamon, London.

Van den Broek, W. and Koch, C. T. (2012). Method for retrieval of the three-dimensional object potential by inversion of dynamical electron scattering. *Phys. Rev. Lett.* **109**, 245502.

Van Dyck, D., Lichte, H., and Spence, J. C. H. (1999). Inelastic scattering and holography. *Ultramicroscopy* **81**, 187.

Vincent, R. and Midgley, P. (1994). Double conical beam-rocking system for measurement of integrated electron diffraction intensities. *Ultramicroscopy* **53**, 271.

Vincent, R., Vine, W. J., Midgley, P. A., Spellward, P., and Steeds, J. (1992). Coherent overlapping LACBED patterns in SiC. *Ultramicroscopy* **50**, 365.

Voelkl, E. (ed.) (1998). *Introduction to electron holography*. Plenum, New York.

Voigt-Martin, I. G., Yan, D. H., Yakimansky, A., Schollmeyer, D., Gilmore, C. J., and Bricogne, G. (1995). Structure determination by electron crystallography using both maximum entropy and simulation approaches. *Acta Crystallogr.* **A51**, 849.

Weierstall, U., Spence, J. C. H., Stevens, M., and Downing, K. H. (1999). Point-projection electron imaging of TMV at 40 eV electron energy. *Micron* **30**, 335.

Weierstall, U., Chen, Q., Spence, J., Howells, M., Isaacson, M., and Panepucci, R. (2002). Image reconstruction from electron and X-ray diffraction patterns using iterative algorithms: experiment and simulation. *Ultramicroscopy* **90**, 171.

Woolfson, M. and Fan, H.-F. (1995). *Physical and non-physical methods of solving crystal structures*. Cambridge University Press, New York.

Wu, J., Spence, J. C. H., O'Keeffe, M., and Leinenweber, K. (2006). Ab initio phasing of X-ray powder diffraction patterns by charge flipping. *Nature Mater.* **5**, 647.

Yagi, K. and Roth, R. S. (1978). Electron microscope study of mixed oxides. *Acta Crystallogr.* **A34**, 773.

Zandbergen, H. and Jansen, J. (1998). Accurate structure determinations of very small areas. *J. Microsc.* **190**, 223.

Zandbergen, H., Andersen, S., and Jansen, J. (1997). Structure determination of Mg_5Si_6 particles in Al by dynamic electron diffraction studies. *Science* **277**, 1221.

Zuo, J. M. (1993). A new method of Bravais lattice determination. *Ultramicroscopy* **52**, 459.

Zuo, J. M., Spence, J. C. H., and Hoier, R. (1989). Accurate structure-factor phase determination by electron diffraction in noncentrosymmetric crystals. *Phys. Rev. Lett.* **62**, 547.

Zuo, J. M., Spence, J. C. H., Downs, J., and Mayer, J. (1993). Measurement of individual structure factor phases with tenth-degree accuracy: the (002) in BeO studied by electron and X-ray diffraction. *Acta Crystallogr.* **A49**, 422.

Zuo, J. M., O'Keeffe, M., Rez, P., and Spence, J. (1997). Charge density of MgO. *Phys. Rev. Lett.* **78**, 4777.

Zuo, J. M., Vartanyants, I. A., Gao, M., Zhang, M., and Nagahara, L. A. (2003). Atomic resolution imaging of a carbon nanotube from diffraction intensities. *Science* **300**, 1419.

Zuo, J. M., Zhang, J., Huang, W., Ran, K., and Jiang, B. (2011). Combining real and reciprocal space information for aberration-free coherent electron diffractive imaging. *Ultramicroscopy* **111**, 817.

Zvyagin, B. B. (1967). *Electron diffraction analysis of clay minerals*. Plenum, New York.

8
Scanning transmission electron microscopy and Z-contrast

8.1. Imaging modes, reciprocity, and Bragg scattering

The STEM is capable of sub-angstrom spatial resolution, and, in the annular dark-field mode (ADF), allows convenient integration with electron energy-loss spectroscopy (EELS), since the portion of the beam which passes through the central hole in the detector may be passed to an energy-loss spectrometer, giving information similar to soft X-ray absorption spectroscopy from sub-nanometre regions. As discussed in Section 13.2, this allows atomic-resolution images to be obtained in registry with EELS spectra from individual columns of atoms. The instrument, using a field-emission gun, was first developed by A. Crewe and co-workers at the University of Chicago (as reviewed in Crewe (1980)); in 1970 they used it to obtain the first electron microscope images of individual atoms (Crewe *et al.* 1970). (See Crewe (1963) for the first visionary presentation of the STEM concept.) Their images of isolated heavy atoms on a thin carbon film were obtained at 30 kV.

The STEM operates, as for a SEM, by scanning a fine probe over a sample; however, the sample is sufficiently thin that electrons transmitted through the sample and scattered may be detected and shown on a raster display, which is synchronized with the probe scan position. While the resolution of STEM is entirely competitive with, or better than, that obtainable by HREM, distortions in STEM images due to movements of the sample relative to the probe during scanning (or distortion in the scan raster) may make the measurement of local strain-fields more difficult, and a stationary beam may quickly drill a hole in the sample if it is uncontaminated. For movies, one cannot know what atoms within the scanned area, but not currently under the probe, are doing, until they are illuminated by the probe.

We may distinguish three main imaging modes in STEM. (1) Bright-field STEM, in which a detector is placed on the optic axis. (2) Dark-field STEM (ADF), in which an annular detector is used with a small inner cut-off (usually matched to the illumination aperture). Elastic (e.g. Bragg) scattering is collected to form the image. These two modes are simply related by reciprocity to the bright-field HREM and hollow-cone modes of TEM, respectively. (3) High-angle annular dark-field (HAADF) Z-contrast mode, in which larger inner and outer detector cut-offs are used to reduce the contribution of elastic Bragg scattering (and hence 'diffraction contrast' artefacts) and to increase the contribution of localized thermal diffuse scattering (TDS), sometimes called 'quasi-elastic scattering', and Rutherford scattering from the nucleii. (The term 'Z-contrast' was first used to denote STEM images formed from the ratio of high-angle elastic scattering to low-angle scattering, since these images showed a strong dependence on atomic number.) The last two modes are often not distinguished clearly in the literature, and the generic term 'ADF-STEM' has now come to be used for both; however, we will use it strictly for mode (2).

Imaging with localized TDS favours a simple interpretation in terms of an incoherent imaging mode. The question of how much purely elastic scattering should be allowed to contribute to Z-contrast images (by the choice of inner cut-off) has generated a large literature, complicated by the fact that the elastic scattering modulates the production of TDS, and that the elastic scattering of TDS cannot be suppressed. For all three modes, resolution is mainly determined by the angular range over which the illumination (objective) aperture is coherently filled, together with the aberration coefficients. The detector geometry also plays a role, and a variety of other detector shapes, such as a thin annulus, will be discussed. For all three modes, in crystals, the lattice is resolved if the coherent diffracted orders overlap.

It has taken two decades at least to clarify the dark-field STEM image contrast mechanisms, since both TDS and Bragg scattering make important contributions to the image. Lower-angle elastic scattering is the strongest, but, since the total elastic scattering is independent of probe position, it produces no contrast at all unless some is excluded by a central hole in the detector. The dependence of Bragg scattering on atomic number is also not simple. By the late 1970s, however, the theory for this elastic STEM imaging in bright- and dark-field (modes 1 and 2) was fairly fully worked out (Cowley 1976, Ade 1977, Spence and Cowley 1978, Fertig and Rose 1981). Soon afterwards, the first EELS spectra were obtained by STEM from regions within a single unit cell (Spence and Lynch 1982) and the first experimental ADF lattice images had appeared (Cowley 1984). However, it proved impossible to separate sample scattering effects from instrumental effects in any simple ADF theory because of the entangling effects of the detector hole. (A simple product representation of ADF-STEM imaging is only possible for amplitude objects.) The optimum choice of experimental conditions (detector and objective aperture sizes, choice of focus, etc.) for bright- and dark-field imaging of small molecules in STEM were discussed at length in Cowley (1976).

For mode (3), during work in the late 1970s at the Cavendish Laboratory, Cambridge, on imaging metal catalyst particles on a crystalline substrate (Treacy et al. 1978), Howie proposed that a sufficiently high-angle detector would minimize the unwanted Bragg scattering, giving a HAADF-STEM image with strong atomic number contrast (proportional to Z^2) according to the Rutherford formula for scattering from an unscreened nucleus (Treacy and Gibson 1993). An inner hole diameter of about 40 mrad was suggested (Howie 1979), corresponding to the angle at which the corresponding d-spacing becomes comparable with the thermal vibration amplitude. An important development was the first publication of atomic-scale images using high-angle scattering (Pennycook and Boatner 1988). A full treatment of this problem requires the inclusion of both Bragg scattering and TDS, and, with the development of sufficiently fast computers, these treatments began to appear in the 1990s, notably from the Cornell group (Hillyard and Silcox 1995) and elsewhere. Thus it was necessary to develop a full theory including both elastic scattering of the probe and the TDS which it generates. Sorting out all this—the relative importance of both types of scattering, the optimum detector shape, and its role in controlling coherence, both transverse and longitudinal—has generated a sizeable literature, from which the main points are summarized below. An authoritative review can be found in Allen et al. (2011). We begin with bright-field STEM, some general principles based on the theorem of reciprocity, and suggestions for aberration-free imaging in STEM.

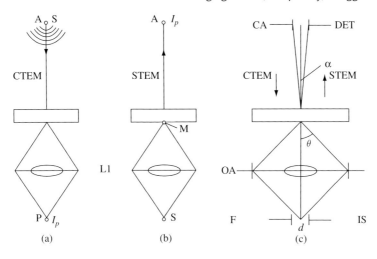

Figure 8.1 (a) The ray diagram for a simplified TEM instrument. One point within the illumination aperture at S emits a spherical wave, which becomes a plane wave at the distant sample. The lens L1 images the lower face of the sample onto the detector, one point of which lies at P. (b) Interchange of S and P produces the ray diagram for STEM. (c) Finite sources and detector pixels. On the left the condenser aperture (CA), objective aperture (OA), and detector (F) (with grain or pixel size d) are indicated for TEM. On the right, the reciprocal STEM detector (DET) and electron emitter (IS) of finite size d are indicated.

The principles of STEM imaging in all modes are most easily understood through the reciprocity theorem (Cowley 1969). Figure 8.1(a) shows a simplified diagram of a single-lens conventional transmission electron microscope (CTEM or TEM) instrument in which S represents a single point source within the final condenser aperture, while P is an image point on the detector. The reciprocity theorem of Helmholtz states that if a particular intensity I_p is recorded at P due to a point source S at A, then this same intensity I_p would be recorded at A in Fig. 8.1(a) if the source S were transferred to the point P and the system were left otherwise unaltered. (In fact the reciprocity theorem applies to complex amplitudes for elastic scattering, and may be extended to small inelastic losses if the intensities are considered (Pogany and Turner 1968).) The instrument has been redrawn in Fig. 8.1(b) with the source and image points of Fig. 8.1(a) interchanged, and now corresponds to the ray diagram for a STEM instrument. The spherical wave launched at A in Fig. 8.1(a) becomes a plane wave at the sample since A is at a large distance from the specimen, and lens L1 faithfully images all points near the optic axis simultaneously onto the detector plane in the neighbourhood of P. This 'parallel processing' feature of CTEM imaging makes the equivalent STEM arrangement seem very inefficient, since, according to the reciprocity theorem, the intensity I_p of Fig. 8.1(b) gives the desired CTEM image intensity for a single point only of that image. In order to obtain a two-dimensional image from the STEM arrangement it is necessary to scan the focused probe at M across the specimen—the reciprocity argument can then be applied to each scan point in turn. Conceptually, it may be simpler to imagine an equivalent arrangement in which the specimen is moved over the probe in order to obtain the image signal in serial form (like a television image signal) from the STEM detector at A. The inefficiency in STEM arises from the fact that, in order to obtain an image equivalent to the bright-field CTEM image, most of the electrons scattered by the specimen in Fig. 8.1(b) must be rejected by the small detector at A. These scattered electrons carry information on the specimen and can be used by forming a dark-field ADF image. In addition, it should be noticed that, since there are no lenses 'down-stream' of the specimen in Fig. 8.1(b), the resulting STEM image does not suffer from chromatic aberration due to specimen-induced energy losses.

This reciprocity argument can be extended to cover the case of instruments which use extended sources and detectors. Thus, as shown in Fig 8.1(c), consider the image formed by a CTEM instrument in which the illuminating aperture CA subtends a semi-angle α, the objective aperture subtends a semi-angle θ, and the image is recorded on a detector F with pixel size d. The illuminating aperture is taken to be perfectly incoherently filled. The reciprocity theorem can be used to show that an identical STEM image can be obtained from a STEM instrument fitted with a finite incoherent source IS of size d in which the 'illuminating aperture' (called the STEM objective aperture, OA) subtends a semi-angle θ, and in which the STEM detector, DET, subtends semi-angle α. (This is done by adding the effects of independent source points.) The equivalence means that images simulated for HREM (Section 5.11) can be used to assist in the interpretation of dark- and bright-field STEM images (modes 1 and 2 above). A computer program which calculates HREM images will accurately represent bright-field elastic STEM images of the same specimen so long as the incoherent sum over CTEM illumination angles (see Section 4.1) is reinterpreted as an incoherent sum over the STEM detector aperture. For HAADF-STEM images formed from a mixture of elastic and thermal scattering, a different method is used, as described later.

The STEM analogue for many-beam lattice imaging in the symmetrical zone axis orientation can be understood using the three-beam case as a simple example. Then the three rays emerging below the sample in Fig. 8.1(c) become Bragg beams. For STEM (now reading up the figure), we see that reciprocity requires that θ, the STEM objective (illumination) angle, must be sufficiently large to 'accept' three imaginary Bragg beams which would be diffracted from an imaginary point source placed at the STEM detector, DET. For a STEM instrument in which the electron gun is above the specimen, the objective is selected by imagining the specimen to be illuminated from a point below the specimen. In summary, we choose the STEM objective (illumination) aperture using the same criteria as in HREM with regard to resolution (but see Section 8.3).

Figure 8.1(c) is drawn out more fully in Fig. 8.2 for the case where the objective aperture semi-angle exceeds twice the Bragg angle, as needed for axial bright-field three-beam lattice imaging. The specimen is illuminated from below by a cone of radiation (semi-angle $2\theta_B$), which for a point source at P forms an aberrated spherical wave converging to the focused electron probe M on the specimen. Thus, around each scattered Bragg direction (for the illumination direction AM) we must draw a cone of semi-angle $2\theta_B$. The result is a set of overlapping coherent 'convergent beam' diffraction discs as shown in Fig. 8.2(b). Three of these overlap at the detector D in the three-beam case shown. As discussed below, it is the interference between these three discs at D which produces the lattice image as the probe is scanned across the specimen. Figure 8.3(a) shows the arrangement that would be used to form the analogue of tilted-illumination, two-beam fringes in CTEM (see Fig. 5.1b). Figure 8.3(b) shows an experimental coherent CBED pattern taken under similar conditions, showing the interference which occurs between overlapping orders (Tanaka *et al.* 1988).

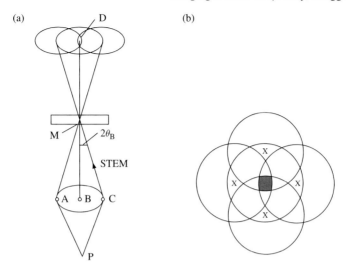

Figure 8.2 STEM ray diagram drawn in (a) with a cone of semi-angle $2\theta_B$ around each Bragg direction, assuming axial incidence along BD. In (b) the pattern is drawn out as it would appear in two dimensions for five-beam imaging; here the crosses indicate Bragg directions, while the hatched area indicates a possible bright-field STEM detector where the intensity is sensitive to probe position. For thick crystals these discs will show the intensity variations of the rocking curve.

A detailed analysis of these arrangements (Spence and Cowley 1978) has established the following general results, which take account of all multiple scattering effects.

1. For perfectly crystalline specimens, convergent-beam diffraction discs formed using an objective aperture of semi-angle $\theta < \theta_B$ (non-overlapping discs) show an intensity distribution which is independent of the probe position, the aberrations of the probe-forming lens, and its focus setting, provided the sample thickness does not vary. In this case the 'size' of the electron probe is necessarily greater than that of the crystal unit cell.
2. If $\theta > \theta_B$, the intensity within regions in which the discs do overlap depends on the probe position, the lens aberrations, and the focus setting of the probe-forming lens. The intensity at D in Fig. 8.3(a) varies sinusoidally as the probe moves across the specimen. We can see this by considering the interference along optical paths BMD and AMD. But we must first also consider the resolution limit due to limited spatial coherence. All spatial coherence effects are accounted for by summing the intensities detected around D over each point on the electron emitter. Since the probe is an image of this source, this corresponds to an integration over probe coordinate c, which must therefore be included in the analysis. For the systematic arrangement shown in Fig. 8.3(a), the intensity distribution near D for Bragg beams Ψ_0 and Ψ_g (originating at A and B) within the region where the orders overlap is given by (Spence 1978b)

$$I(u, c, \Delta\theta) = |\Psi_0(s)|^2 + |\Psi_g(s)|^2 + 2|\Psi_0(s)||\Psi_g(s)| \int_{\text{source}} F(x-c)$$
$$\times \cos\{2\pi x/a - \chi(u_B + \alpha/\lambda) + \chi(u_B - \alpha/\lambda) + \Delta\theta(s)\} \mathrm{d}x \quad (8.1a)$$

238 STEM and Z-contrast

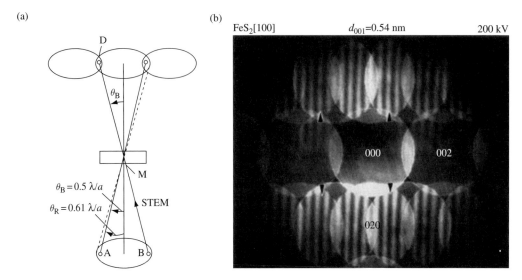

Figure 8.3 (a) The STEM analogue of the tilted-illumination, two-beam TEM lattice imaging shown in Fig. 5.1(b). The TEM conditions of Fig. 5.1(b) are obtained by placing the electron source at D, so that MA and MB become the two Bragg beams used for imaging. The illumination angle θ_R needed to make the electron probe width approximately equal to the first-order lattice spacing a is shown. As drawn, the diffraction discs subtend a semi-angle slightly larger than the Bragg angle, and hence overlap. (b) Experimental coherent CBED pattern from FeS_2 down [100], showing interference fringes between overlapping orders (Tanaka et al. 1988). Note the absence of fringes where discs do not overlap. Only the overlap intensity changes when the probe moves, so this is where we put a STEM detector. Equation (8.1) shows that these fringes are a magnified image of the lattice. The magnification is the camera length divided by the probe defocus. At overlap midpoints on the achromatic circle, all even aberrations (including spherical aberration) cancel.

Here α is a coordinate on the detector giving the angular deviation of a point in the CBED pattern from the mid-point Bragg condition D, $s = \alpha/2a$ is the excitation error, the first-order lattice spacing is a, c is the probe coordinate (so that c/a is the fractional probe coordinate), $\Delta\theta(s)$ is the phase difference between the Bragg beams (not their structure factors) and $u_B = \theta_B/\lambda = 1/(2a)$. The same origin must be used to determine c and $\Delta\theta(s)$. The incoherent source intensity distribution function is $F(x)$, and we assume an ideal point detector. Equation (8.1a) describes the two-beam interference at D due to coherent sources at A and B along optical paths BMD (for Ψ_0), and AMD (for Ψ_g). It gives the correct two-beam STEM lattice image using a point detector at D as the probe is scanned, varying sinusoidally with probe coordinate c (e.g. for $F(x) = \delta(x)$) and agrees with the expression for the reciprocally related TEM arrangement for two-beam lattice imaging (eqn 5.7) using inclined plane-wave illumination from D. (In that case $F(x)$ is interpreted as the area detector point spread function.) Note also that for a finite defocus, one sees a magnified lattice image within the region of overlap—this explains the appearance of the Ronchigrams discussed in Section 7.6. (For $C_s = c = \Delta\theta = 0$ and

$F(x) = \delta(x)$ in eqn (8.1a), the period $\Delta\alpha$ of $I(u, c, \Delta\theta)$ in α is $\lambda/(4\Delta f \theta_B) = a/(2\Delta f)$, producing fringes with infinite magnification at the in-focus condition $\Delta f = 0$.) These fringes are shown experimentally in Fig. 8.3(b) (Tanaka et al. 1988). In summary, within the region of overlap, the defocused coherent CBED pattern various sinusoidally across the screen (as shown experimentally in Fig. 8.3(b)), while any one point of it varies sinusoidally with the probe coordinate. Outside the regions of overlap, the diffracted intensity does not depend on the probe position c or lens aberrations.

The effect of enlarging the STEM detector is exactly analogous to the effect of enlarging the final condenser aperture in CTEM, resulting in a further integration over the detector coordinate α. So the final intensity is

$$I(u, c, \Delta\theta) = \int_{-\alpha_0}^{+\alpha_0} |\Psi_0(s)|^2 + |\Psi_g(s)|^2 + 2|\Psi_0(s)||\Psi_g(s)| \int_{\text{source}} F(x - c) \qquad (8.1b)$$
$$\times \cos\{2\pi x/a - \chi(u_B + \alpha/\lambda) + \chi(u_B - \alpha/\lambda) + \Delta\theta(s)\} \mathrm{d}x \mathrm{d}\alpha$$

for a thin detector extending from $-\alpha_0$ to $+\alpha_0$.

3. The intensity at the midpoint between overlapping discs where $\alpha = 0$ is a special case. Here, on the 'achromatic circle', the intensity depends on the probe position, but not those lens aberrations which are an even function of angle, such as focus setting and spherical aberration, as we see in eqn (8.1) for $\alpha = 0$. This suggests the ptychography schemes for aberration-free imaging, as discussed in Section 7.6.

For coherent bright- and dark-field imaging, the intensity distribution of the STEM probe is a complicated and not particularly useful function, since it is not detected. (For HAADF 'incoherent' imaging, the probe shape becomes a useful guide to resolution, as discussed in later sections.) Near the Scherzer focus setting for the probe-forming lens, the probe does, however, form a well-defined peak with rather extensive 'tails'. The width of this peak (the FWHM) is given approximately by $d = 0.61\lambda/\theta_R$ if we use the Rayleigh criterion and imagine the STEM lens to be imaging an ideal point field-emission source with $C_s = 0$ (see Fig. 8.2). If we set this probe 'size' equal to the unit-cell dimension (strictly the first-order lattice spacing) a, then the illumination angle (STEM objective aperture) needed to match the probe size to the unit-cell dimension is $\theta_R = 0.61\lambda/a$. But the Bragg angle is $\theta_B = 0.5\lambda/a$, so this matching condition becomes the condition that adjacent diffraction discs just begin to overlap, since we then have $\theta_R > \theta_B$. In this rather loose sense, lattice imaging become possible as the electron probe becomes 'smaller' than the crystal unit-cell dimension, and this is only possible if the diffraction discs are allowed to overlap. In fact, *a necessary but not sufficient condition for lattice imaging in STEM is that the exit pupil of the probe-forming lens be coherently filled over an angular range equal to that covered by the Bragg beams (or spatial frequencies) one wishes to image*.

As the probe becomes very much smaller than the unit cell, information on the translational symmetry of the crystal is progressively lost. It becomes difficult to distinguish the reciprocal lattice using the overlapping convergent-beam pattern, leaving only information on the point-group symmetry of the unit-cell contents, as reckoned about the current probe position. The symmetry information contained in convergent-beam patterns is discussed in Section 13.3, which should be read in conjunction with this section.

240 STEM and Z-contrast

We can conclude that for bright-field STEM, using a small central detector, under reciprocal aperture conditions, the STEM and TEM images will be identical if the Scherzer focus is used in both cases.

A modified bright-field STEM mode has proven useful for the imaging of light atoms (Findlay et al. 2010). Here an annular detector is used, but this falls entirely within the central direct beam, having an outer radius equal to that of the central disc of the diffraction pattern, producing an arrangement equivalent to hollow-cone illumination in TEM. The on-axis radiation is not detected. This detector is found to give images with better contrast and signal to noise ratio, especially for light atoms such as oxygen in the presence of heavy atoms, than HAADF images recorded on the same instrument, and which are similarly insensitive to sample thickness. Dramatic images of columns of hydrogen atoms in thin crystals of YH_2 have been imaged in projection by this method (Ishikawa et al. 2011). In practice these bright-field images, the HAADF image, and energy-loss spectra may all be collected simultaneously and in registry, together with EDX images.

A thin annular detector which spans the edge of the bright-field disc has also been shown to have advantages for samples containing both amorphous and crystalline material, by providing super-resolution and allowing the control of coherence conditions (Cowley 2001).

8.2. Coherence functions in STEM

The application of transfer theory to STEM is useful for understanding contrast, coherence, and resolution in bright- and dark-field. Since these are partly determined in dark-field STEM by the detector shape, this leads naturally to the question of the optimum detector geometry for resolution enhancement in STEM. We now show explicitly (rather than by the preceding reciprocity arguments) that the Fourier transform of the detector shape function $F(\mathbf{K})$ in STEM plays the same role as the coherence function $\gamma(\mathbf{r})$ (eqn 4.5) in HREM. For an ideal point field emitter, the STEM probe wavefunction is just the impulse response $\bar{A}(\mathbf{r})$ of the probe-forming lens (eqns 3.14, 3.15, and 4.3; Fig. 3.8 for bright-field), and the probe formation process (as opposed to overall image formation) is thus ideally coherent. The detector shape, by controlling the degree to which scattering from different atoms can interfere, determines the degree of coherence for scattering within the sample. For a probe centred at $\mathbf{r}_p$ and object transmission function $T(\mathbf{r})$ (eqn 3.2), the exit-face wavefunction is

$$\Psi_e(\mathbf{r}, \mathbf{r}_p) = \bar{A}(\mathbf{r} - \mathbf{r}_p) T(\mathbf{r}) \qquad (8.2a)$$

and the wavefunction at a distant detector is the Fourier transform of this

$$\psi_{\text{STEM}}(\mathbf{r}_p, \mathbf{K}) = \int \bar{A}(\mathbf{r} - \mathbf{r}_p) T(\mathbf{r}) \exp(2\pi i \mathbf{K} \cdot \mathbf{r}) d\mathbf{r} \qquad (8.2b)$$

$$= A(\mathbf{K}) \exp(2\pi i \mathbf{K} \cdot \mathbf{r}_p) * t(\mathbf{K})$$

$$= \int A(\mathbf{K}') \exp(2\pi i \mathbf{K}' \cdot \mathbf{r}_p) * t(\mathbf{K} - \mathbf{K}') d\mathbf{K}' \qquad (8.3)$$

$$= \bar{A}(\mathbf{r}) * T(\mathbf{r}) \exp(2\pi i \mathbf{K}' \cdot \mathbf{r}_p) \qquad (8.4)$$

where the asterisk denotes convolution, and $t(\mathbf{K})$ is the transform of $T(\mathbf{r})$. (All vectors in this section are two-dimensional.) The intensity collected by a detector described by a two-dimensional sensitivity function $F(\mathbf{K})$ is then

$$I(\mathbf{r}_p) = \int_{\text{Detector}} F(\mathbf{K}) \left| \int A(\mathbf{K}') \exp(2\pi i \mathbf{K}' \cdot \mathbf{r}_p) t(\mathbf{K} - \mathbf{K}') d\mathbf{K}' \right|^2 d\mathbf{K} \qquad (8.5)$$

We can compare this with the corresponding expression for HREM using an instrument with the same impulse response if we allow for inclined illumination, so that the intensities of images may be added for each illumination direction. Equation (4.2) describes the effects of plane-wave illumination inclined from direction $\mathbf{K}$ on a phase grating. Then, since $T(\mathbf{r}) = \psi_0(\mathbf{r}, 0)$, the HREM image is given by a convolution of the impulse response $\bar{A}(\mathbf{r})$ with the object transmission function in eqn (4.2) for inclined illumination

$$\psi_{\text{HREM}}(\mathbf{r}, \mathbf{K}) = \bar{A}(\mathbf{r}) * T(\mathbf{r}) \exp(2\pi i \mathbf{K} \cdot \mathbf{r}) \qquad (8.6)$$

This is identical to eqn (8.4) for STEM, which is written out in full as eqn (8.3). (This equivalence between HREM and STEM could, in hindsight, have been understood directly from eqn (8.2b), which describes both, but with a differing physical interpretation of the plane-wave factor, in one case describing illumination and in the other propagation to the STEM detector.) The integration of intensity over an incoherently filled illumination aperture $F(\mathbf{K})$ for HREM is therefore also described by eqn (8.5). Equation (8.5) is also equal to eqn (4.4), derived in a different way, so that the complex degree of coherence $\gamma(\mathbf{r})$, the Fourier transform of $F(\mathbf{K})$, is given by either the transform of the incoherently filled illumination aperture for HREM or the detector shape in STEM, both described by $F(\mathbf{K})$. The key approximation made (eqn 4.2) holds only for strong- and weak-phase objects. We therefore expect that, for thin samples, the use of an annular detector in STEM will produce images with the same coherence properties as those obtained under conical or hollow-cone illumination in TEM.

For the case of a small STEM detector, $F(\mathbf{K}) = \delta(\mathbf{K})$, $\gamma(\mathbf{r}) = 1$, and we obtain, from eqn (4.4)

$$I_{\text{STEM}}(\mathbf{r}) = \left| \bar{A}(\mathbf{r}) * T(\mathbf{r}) \right|^2 \qquad (8.7)$$

which is identical to the expression for bright-field HREM imaging. Resolution in bright-field STEM many be improved by a factor of about 1.6 through the use of a thin annular detector with an average radius equal to the objective aperture image (Cowley 1993). This mode is equivalent to the hollow-cone mode discussed in Section 4.4. For the study of electric (and magnetic) fields within polar crystals, the differential phase contrast mode has proven powerful, and this has now been extended to atomic resolution in ferroelectric BaTiO_3 using a segmented annular detector fitted to an aberration-corrected STEM (Shibata et al. 2012). If boundary conditions are correctly described, these fields are fully accounted for by the ionic state of the atoms (and appropriate scattering factors) in the finite TEM sample.

Although the reciprocity arguments of Section 8.1 are quite general and include all multiple-scattering effects, multiple scattering does destroys any simple interpretation in terms of coherence theory, since there is then no closed-form parameterization of the dynamical wavefunction in terms of $\mathbf{r}_p$ or $\mathbf{K}$ (which defines the centre of the Laue circle for a strongly diffracting thin crystal (eqn 5.40)). Physically, this means that if there are complicated dynamical 'rocking curve' effects within the region in which the coherent CBED orders overlap, one cannot use the above coherence theory.

The question of coherence along columns of atoms in ADF-STEM has been analysed by several authors, using a kinematic analysis which must now include the z coordinate of the atoms. In three dimensions, inside a crystal, the coherence volume (the three-dimensional analogue of $\gamma(\mathbf{r})$) becomes cigar-shaped, with its axis along the optic axis, and dimensions which depend on the detector shape, becoming narrower as the detector hole is enlarged. The longitudinal coherence is important, since it determines whether the total scattering from a column of atoms is proportional to the number of atoms in the column (with no interference along the column) or to the square of the number of atoms (full interference along the column). In the approximation that the TDS can be treated as Rutherford scattering, with total intensity proportional to Z^2, these two possibilities predict image intensities proportional to $Z^2 t$ and $Z^2 t^2$, respectively. One indication of coherence along the z-axis is the occurrence of higher-order Laue zone (HOLZ) rings in coherent nanodiffraction patterns—calculated patterns for ADF-STEM images, including TDS, show these clearly (Spence and Koch 2001). A detailed study (Treacy and Gibson 1993) shows that,

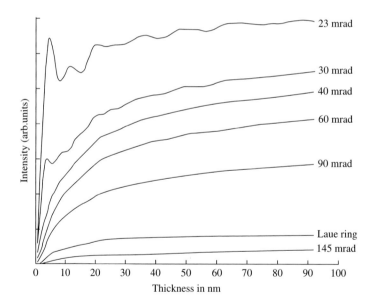

Figure 8.4 Thickness dependence of the ADF-STEM image at the centre of a [110] silicon atom column for several inner detector cut-off angles at 100 kV, with a large outer cut-off ($C_s = 0.05$ mm, $\Delta f = -12.2$ nm, $C_c = 1.5$ mm, $\Delta E = 0.26$ eV FWHM, $\theta_{ap} = 22$ mrad, source size $= 0.05$ nm, temperature $= 20$ K, mean displacement $= 0.0044$ nm, Debye–Waller factor $= 0.153$ Å^2). As the inner cut-off is made larger, excluding more Bragg beams, diffraction ripple decreases (100 kV, $C_s = 0.05$ mm, $\Delta f = -12.2$ nm, $\theta_{ap} = 22$ mrad) (Koch 2002).

for typical ADF detectors, the interference between different atomic columns can be made negligible (when $\gamma(\mathbf{r})$ becomes narrower than the lateral distance between atomic columns), but that to eliminate coherence along the atom strings requires extremely large inner cut-off detector angles, perhaps 100 mrad at 100 kV, resulting in considerable loss of signal. Figure 8.4 (Koch 2002) shows the intensity calculated at the centre of an atomic column of silicon in an ADF image at 100 kV as a function of thickness, including all TDS. The 'frozen phonon' method was used with several slices per atom thickness. The thickness dependence is seen to follow the linearity of the incoherent model only at small thickness, and to depend on the inner cut-off, as expected. With a small inner cut-off, more Bragg scattering is included, leading to oscillations in thickness (but a strong signal); a larger cut-off improves linearity at the expense of image intensity. In summary, transverse interference effects can readily be eliminated in ADF-STEM by the choice of detector shape (making the images much simpler to interpret than HREM images in terms of atom positions); however, as Fig. 8.4 shows, interference along the atom strings cannot easily be eliminated, making interpretation of the intensity of atomic images difficult in terms of atomic number and thickness. Thus, for the identification of substitutional atoms within an atomic column (see Fig. 8.9), detailed simulations would be needed. Increasing the source energy spread reduces coherence along the beam, but, through the chromatic aberration constant, also affects focusing.

8.3. Dark-field STEM: incoherent imaging, and resolution limits

This section concerns resolution limits and the approach to incoherent imaging in dark-field STEM (mode 2), in which we mostly ignore the contribution of thermal diffuse scattering. We consider only the contribution of elastic scattering, such as Bragg scattering. Consider first a STEM detector which collects all the scattered intensity, so that $F(\mathbf{K}) = 1$ and $\gamma(\mathbf{r}) = \delta(\mathbf{r})$. We then have, from eqn (4.7)

$$I_{STEM}^{DF}(\mathbf{r}_p) = \left|\bar{A}(\mathbf{r})\right|^2 * |T(\mathbf{r})|^2 \\ = \int \left|\bar{A}(\mathbf{r})\right|^2 |T(\mathbf{r}_p - \mathbf{r})|^2 d\mathbf{r} \quad (8.8)$$

This is the ideally incoherent imaging mode, such as that used, for example, by a camera, a 35 mm slide projector (which uses a transmission object, similar to our STEM, illuminated by a large incandescent source), or an optical microscope when used to image 'self-luminous' objects. The first term on the right-hand side is just the intensity distribution in the probe, and the object function is convoluted with this. For a phase grating, with $T(\mathbf{r}) = \exp(-i\sigma\phi_p(\mathbf{r}))$, this mode would give no contrast whatsoever since $T(\mathbf{r})T(\mathbf{r})^* = 1$, and the last term is then independent of probe coordinate. Equation (8.8) was used by the inventors of the STEM to interpret the first 30-kV STEM images of heavy atoms on a thin carbon substrate (Crewe 1980), with $T(\mathbf{r})$ taken to be the single-atom cross-section. We will see that this approximation works well for well-separated atoms, but fails if interference occurs between pairs of atoms or atom columns, or if the effects of the hole in the detector are considered. Equation (8.8) is entirely correct for imaging amplitude objects (eqn 6.2) and so can provide a map of the absorption function $\phi_i(\mathbf{r})$. Amplitude variation can also be obtained, for example, in energy-filtered images using characteristic inner-shell excitations (Section 13.2). This equation also describes the generation of X-ray production in STEM or TEM (e.g. ALCHEMI, Section 13.1) if $T(\mathbf{r})$ is reinterpreted as a generation function, related to the cross-section for ionization, and again to the imaginary part $\phi_i(\mathbf{r})$ of the relevant optical absorption potential (Heidenreich 1962, Cherns et al. 1973). This potential is assumed to be localized on the atoms. For inelastic electron scattering (such as thermal diffuse scattering or inner-shell ionization) localization can be imposed by collecting only the high-order scattering (corresponding to small impact parameters) using, for example, a high-angle annular detector (HAAD).

The introduction of a hole in the detector produces contrast for strong, weak-phase, and multiple-scattering objects. This can be seen at once by noting that the total elastic scattering is a constant, independent of probe position, so that the elastic contribution to an annular detector must be the complement of the bright-field STEM image discussed in Section 8.1. Thus

$$I_{\text{STEM}}^{\text{DF}}(\mathbf{r}_p) = 1 - I_{\text{STEM}}^{\text{BF}}(\mathbf{r}_p) \tag{8.9}$$

and $I_{\text{STEM}}^{\text{BF}}(\mathbf{r}) = I_{\text{HREM}}^{\text{BF}}(\mathbf{r})$, by reciprocity, if reciprocal aperture conditions are used. More directly, the use of an annular detector in STEM is equivalent to the use of dark-field hollow-cone illumination in TEM, as discussed in Section 4.4. The reader should sketch out ray diagrams for the two cases, with complementary apertures. The objective aperture for dark-field hollow-cone (conical) TEM illumination has an angular cut-off θ_H slightly less than the bright ring formed by the direct beam, so that a dark-field image is formed, as shown in Fig. 4.6. The TEM illumination, with distribution $F(\mathbf{K})$, covers all angles except those smaller than θ_H.)

In order to optimize the choice of instrumental parameters for highest resolution, we wish to separate instrumental and sample parameters. Just as this can only be done for HREM in the weak-phase object approximation, for dark-field STEM a full separation can only be achieved for amplitude objects. For phase objects and thicker multiply scattering objects, the central hole in the detector entangles these parameters, and the best that can be done is to lump together the detector function $F(\mathbf{K})$ with the object transmission function $T(\mathbf{r})$. We then isolate all other instrumental parameters in $\left|\bar{A}(r)\right|^2$ and its Fourier transform, $H(\mathbf{K})$, known as the optical function (OTF). Before doing this, we note a crude approximation which has been made, in which it is assumed that the scattered intensity outside the hole is proportional to the total intensity (and to that inside the hole), so that eqn (8.8) may be used for weakly scattering objects. This is a reasonable approximation for isolated atoms, but not if there are interference effects within the central disc, such as the Young's fringes which cross the diffraction pattern from a dimer. In this approximation, isolated atoms in ADF-STEM appear similar to dark-field HREM images of atoms (Section 6.5), so that, according to eqn (6.27), either a heavy atom on a light substrate or a small hole in a substrate would both give bright contrast. (The more exact analogy is with dark-field hollow-cone illumination (Section 4.4), which produces similar images.) The incoherent 'generation function' is $\sigma^2 \phi_p^2(r)/2$ (Cowley 1976).

It is instructive to consider how the hole in the detector creates contrast from a phase grating, by deriving an expression which shows how the coherence function operates in real space. This will indicate the conditions under which interference can occur in dark-field

STEM between different columns of atoms. Equation (8.5) may be Fourier transformed to give the spectrum of the image intensity as

$$\bar{I}(K_p) = \int A(\mathbf{K'})A(\mathbf{K'}+\mathbf{K}_p) \int F(\mathbf{K})t(\mathbf{K}-\mathbf{K'})t^*(\mathbf{K}-\mathbf{K'}-\mathbf{K}_p)d\mathbf{K}d\mathbf{K'}$$

$$= \int A(\mathbf{q}-\mathbf{K}_p/2)A^*(\mathbf{q}-\mathbf{K}_p/2) \int F(\mathbf{p}+\mathbf{q})t(\mathbf{p}+\mathbf{K}_p/2)t^*(\mathbf{p}+\mathbf{K}_p/2)d\mathbf{q}d\mathbf{p}$$

$$\approx \int A(\mathbf{q}-\mathbf{K}_p)A^*(\mathbf{q}+\mathbf{K}_p)d\mathbf{q} \int F(\mathbf{p})t(\mathbf{p}+\mathbf{K}_p/2)t^*(\mathbf{p}-\mathbf{K}_p/2)d\mathbf{p}$$

$$= [A(-\mathbf{K}_p)*A^*(\mathbf{K}_p)]O(\mathbf{K}_p) = H(\mathbf{K}_p)O(\mathbf{K}_p) \qquad (8.10)$$

where $\mathbf{K'} = \mathbf{q} - \mathbf{K}_p/2$ and $\mathbf{K} = \mathbf{p} + \mathbf{q}$. $O(\mathbf{K}_p)$ is the object function and $H(\mathbf{K}_p)$ the OTF (Loane et al. 1991), with $F(\mathbf{K})$ defining the detector shape. These approximations are applicable to thin samples, and ignore any dependence of the scattering by the object on the direction of incident illumination more complicated than that described by eqn (8.6). They preserve the 'product representation' and fail at thicknesses where a crystalline sample develops an appreciable 'rocking curve' variation of intensity in CBED discs. They do, however, achieve the desired separation of instrument function $A(\mathbf{K})$ and object function $O(\mathbf{K})$ in a similar way to the transform of eqn (8.8), but with the detector hole now included in the object function. The Fourier transform of $O(\mathbf{K}_p)$ gives the real-space representation

$$O_{ADF}(\mathbf{r}) = \int \left(\partial(\mathbf{s}) - \frac{J_1(2\pi u_1|\mathbf{s}|)}{2\pi|\mathbf{s}|}\right) t(\mathbf{r}+\mathbf{s}/2)t^*(\mathbf{r}-\mathbf{s}/2)d\mathbf{s} \qquad (8.11)$$

Here the term in the large brackets is the transform of a detector of infinite outer diameter $F(\mathbf{K}) = 1$, minus that of a hole of diameter $u_1 = \theta_H/\lambda$, as for the treatment of hollow-cone illumination in Section 4.4. Enlarging the hole decreases the width $x_{1,2}$ of the coherence function (Fig. 4.7), which indicates the distance $x_{1,2}$ between points in the sample which can interfere. If the phase grating expression eqn (3.30) is now used for $t(\mathbf{r})$ in (8.11), we obtain (Nellist and Pennycook 2000)

$$O_{ADF}(\mathbf{r}) = 1 - \int_{half-plane} \frac{J_1(2\pi u_1|\mathbf{s}|)}{2\pi|\mathbf{s}|} \cos(\sigma\phi\{\mathbf{r}+\mathbf{s}/2\} - \sigma\phi\{\mathbf{r}-\mathbf{s}/2\})d\mathbf{s} \qquad (8.12)$$

$$\approx \int \frac{J_1(2\pi u_1|\mathbf{s}|)}{2\pi|\mathbf{s}|} [\sigma\phi\{\mathbf{r}+\mathbf{s}/2\} - \sigma\phi\{\mathbf{r}-\mathbf{s}/2\}]^2 d\mathbf{s} \qquad (8.13)$$

if the phase shift is small (Ade 1977, Jesson and Pennycook 1993).

For the peaked functions describing the crystal potential projected down atomic columns, we see that the square of the difference between this function and a displaced copy adds to the image intensity for displacements $\mathbf{s}$ less than an 'incoherence width' $X_i = \lambda/\theta_H$ (eqn 4.17). The Bessel function with this width is the coherence function produced by an incoherent source the same size as the hole in the detector (eqn 4.16). Equation (8.12) correctly predicts the zero contrast in the absence of a hole—the desirable condition of independent incoherent imaging of atomic columns can be approached by enlarging the hole (with resulting loss of ADF image intensity) until the 'incoherence width' X_i is about equal to the distance between atomic columns. Then the integral in eqn (8.12) will contribute for values of s less than the width of the atomic peaks. Hence we require θ_H greater than twice the Bragg angle for the spacing d of interest, so that $X_c < d$. Alternatively, we might require that X_i be less than the probe width d_p, whose dimensions will be comparable with the atomic spacing. For an unaberrated probe, $d_p = 0.61\lambda/\theta_R$ (with θ_R the STEM objective aperture semi-angle) and we therefore require $\theta_H > 1.64\theta_R$. In either case, a hole significantly larger than the image of the STEM objective aperture in the detector plane is indicated.

We note that the dark-field imaging mode gives an image proportional to the square of the projected potential, and is therefore non-linear. (This leads, for example, to the appearance of the forbidden (002) Fourier coefficient in the transform of images of silicon in the 110 projection (Hillyard and Silcox 1995).) This non-linearity is the basis of the sensitivity of the method to atomic-number (Z) contrast. It arises because whereas bright-field HREM produces linear imaging due to interference between the direct beam and Bragg beams ADF-STEM excludes the direct beam, generating contrast by interference between pairs of scattered beams akin to dark-field HREM.

Within these approximations, eqn (8.10) indicates that the Fourier transform of the ADF image intensity is given by the product of an object function $O(\mathbf{K_p})$ (which nevertheless involves the detector hole) and the transform of the probe intensity $|A(\mathbf{r})|^2$

$$H(\mathbf{K_p}) = [A(-\mathbf{K_p}) * A^*(\mathbf{K_p})] \tag{8.14a}$$

This OTF is compared in Fig. 8.5 with the phase contrast transfer function for bright-field HREM given by $\text{Im}(A(\mathbf{K}))$. Both are shown for $C_s = 0.65$ mm (illumination angle $\theta_c = 0.25$ mrad, Scherzer focus, for TEM) at 300 kV. The OTF is shown in Fig. 8.5(a) at the 'most compact probe' condition for ADF-STEM, given by eqns (8.14b, c). (Note the similarity to HREM dark-field focus (eqn 6.26).) We see that the OTF has the desirable characteristics of a smooth fall-off at high spatial frequencies without contrast reversals, and extends to higher resolution than the Scherzer cut-off of eqn (6.17). Because $H(\mathbf{K})$ is the autocorrelation function of $A(\mathbf{K})$, it extends to twice the radius of the objective aperture, and in this sense, incoherent imaging has higher resolution than coherent imaging. However, the HREM mode analogous to Fig. 8.3 (STEM, with overlap of adjacent reflections) would be tilted illumination (Fig. 5.1b), which is also capable of 'doubling' resolution, and both these modes may introduce image artefacts. The comparisons in Figs 8.5 and 8.6 are also somewhat unfair because the effects of electronic instabilities, the energy spread in the beam, and tip vibration are not included in the STEM figures. In Chapter 7 we discussed the relative difficulty of extracting fine detail from images in each of these modes—the incoherent mode clearly has the advantage of allowing simple deconvolution of intensities, once the probe function is known, so that resolution is ultimately then limited by noise and its treatment under deconvolution. The ADF mode also has the advantage of less violent variations of image intensity with focus, so that experimentally it is not difficult to find a unique focus setting. However, the signal is weaker if a large hole is used in the detector to exclude diffraction-contrast effects. For the mapping of strain-fields, distortions due to probe drift during scanning or stray fields are disadvantages for STEM.

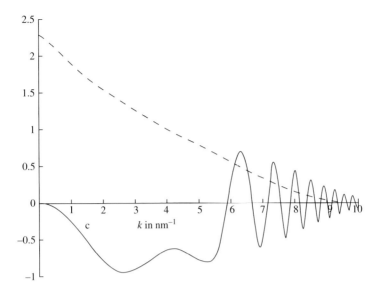

Figure 8.5 OTF for STEM (dashed) at the 'most compact probe' ADF focus ($\Delta f = -0.75(C_s\lambda)^{1/2}$, objective aperture $\theta_{\text{ap}} = 1.27(\lambda/C_s)^{1/4}$ given by eqns (8.14b, c) (Mory et al. 1987). These are compared with the CTF (continuous line) for bright-field HREM at the Scherzer focus for a 300 kV TEM-STEM instrument. $\lambda = 0.00197$ nm, $C_s = 0.65$ mm ($\theta_c = 0.25$ mrad for HREM). An information limit corresponding to $\Delta = 20$ Å has been assumed (O'Keefe et al. 2001). The HREM point-resolution is 0.17 nm (Koch 2002).

In Fig. 8.6(a) the real-space STEM probe intensity function $|A(\mathbf{r})|^2$ is compared with the HREM bright-field amplitude impulse response for a weak-phase object under the Scherzer conditions (cf. also Fig. 3.8). The STEM probe (at the conditions of eqns (8.14b, c)) is seen to be narrower for the same objective aperture and aberrations. In general, bright-field coherent image resolution depends on the phase shifts introduced on scattering, and scattering phases at point scatterers in the sample may be chosen which reverse this conclusion favouring incoherent STEM imaging (Goodman 2004). The focus and aperture settings which minimize the radius of the circular area containing 70% of the probe intensity are

$$\Delta f = -0.75(C_s\lambda)^{1/2} \tag{8.14b}$$

with aperture semi-angle

$$\theta_{\text{ap}} = 1.27(\lambda/C_s)^{1/4} \tag{8.14c}$$

(Mory et al. 1987), and are useful conditions for forming the most compact probe (most similar to an Airy disc) for microanalysis and ADF imaging. This focus setting differs from the optimum dark-field HREM focus (eqn 6.26) by a small numerical factor. If strong

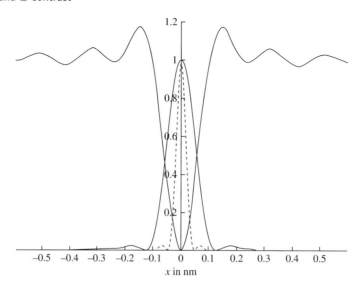

Figure 8.6 The impulse response for bright-field HREM (upper curve, FWHM = 0.123 nm) compared with the probe intensity distribution for a 300 kV TEM/STEM (FWHM = 0.11 nm) without aberration correction. The conditions are the same as in Fig. 8.5 ($C_s = 0.65$ mm, $\Delta f = -26.8$ nm, $\theta_{ap} = 9.4$ mrad for STEM). The STEM probe is given for the ADF focus. The dashed curve shows the stem probe for an aberration-corrected instrument (FWHM = 0.038 nm), scaled to unit height ($C_s = 0.01$ mm, $\Delta f = -3.3$ nm, $\theta_{ap} = 26.8$ mrad, eqn (8.14)). The intensity in the corrected probe is actually 63 times greater than that in the uncorrected probe. The comparisons in Figs 8.5 and 8.6 do not consider the effects of electronic instabilities, the energy spread in the beam, and tip vibration in the STEM figures.

low-order Bragg beams contribute to the ADF image, as in HREM, there will, however, be a tendency for the operator to choose the stationary phase focus for the low-order beams (eqn 5.66). Since, in the incoherent approximation, the effect of the probe can be deconvoluted from the image (at the expense of noise amplification), the possibility arises of using a large under-focus to preserve the transfer of high spatial frequencies out to 0.078 nm (Nellist and Pennycook 1998). Resolution can be greatly improved in STEM instruments fitted with aberration correctors (Krivanek et al. 1999, Batson and Krivanek 2002), as shown in Fig. 8.10. The arguments given in Section 7.4 can be used to choose the optimum value of C_s. Of equal importance is the additional current which a spherical aberration corrector can provide for EELS and EDX in STEM, in view of the much larger illumination angle possible, instabilities permitting. Figure 8.7 compares current with probe size for corrected and uncorrected optimized probes at 200 keV (D. Muller, personal communication, 2002). Again, the essential requirement for lattice imaging in ADF-STEM in crystals is that the objective aperture be coherently filled over an angular range greater than the Bragg angle, and that there is overlap at least of the first-order diffraction discs, that is, that $\theta_c > \theta_B$.

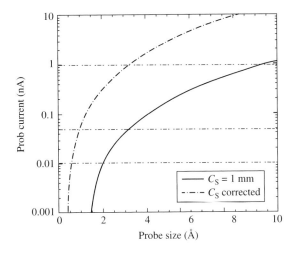

Figure 8.7 Current available in an optimized STEM probe plotted against its size, comparing the aberration-corrected and -uncorrected ($C_s = 1$ mm) cases at 200 keV (D. Muller, personal communication, 2002).

8.4. Multiple elastic scattering in STEM: channelling

Can an object function, which is independent of thickness, be usefully defined for ADF-STEM in the case where only strong multiple elastic scattering occurs? Strictly no, but some understanding of this problem can be obtained by writing the dynamical wavefunction $\Psi(\mathbf{r})$ in a crystalline sample as a product of a plane-wave term for the z-dependence (with wavevector $k_z^{(i)}$), multiplied by a transverse eigenstate $B^{(i)}(\mathbf{r})$, following practice in the field of particle channelling (Howie 1966). Channelling is the tendency of charged particles to run along lines of lowest potential energy in a crystal; these lines are the strings of atomic nuclei for electrons. We may distinguish two types of channelling—that which occurs with plane-wave illumination, as for ALCHEMI (Section 13.1), and that which occurs with a fine probe, as in STEM. In the first case the laterally periodic wavefield tends to pile up on the atomic nuclei; in the second the probe tends to focus along the atomic strings to a greater depth and intensity when it is situated over an atomic column, or to split into two probes, as shown in Fig. 8.8, if it is not. The relativistic increase in mass of the electron (and the increasing density of the bound states described below) leads to increasingly classical behaviour with increasing electron energy—a classical approach is justified when the energy-broadened bound states start to overlap (Spence 1988). The second form of channelling is relevant to STEM, since it is found that the probe 'channels' to different depths along atomic strings of different atomic number.

Associated with each eigenstate $B^{(i)}(\mathbf{r})$ is a transverse kinetic energy $\varepsilon^{(i)}$. The wavefunction is written

$$\Psi(\mathbf{r}) = \sum_j \alpha^{(j)} \exp(2\pi i k_z^{(j)} z) B^{(j)}(\mathbf{r}) \tag{8.15}$$

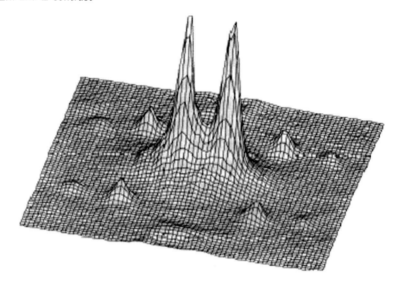

Figure 8.8 Shape of a 100 keV STEM probe of width 0.18 nm at a depth of 10 nm into a silicon crystal when situated midway between two atom columns in the [110] projection. The channelling effect causes it to split into two probes, one on each atomic column (Hillyard and Silcox 1995).

The one-electron Schrödinger equation becomes

$$\frac{-h^2}{8\pi^2 m}\left[\frac{\partial^2 B^{(j)}}{\partial x^2} + \frac{\partial^2 B^{(j)}}{\partial y^2}\right] - |e|\phi_p(x,y)B^{(j)} = \frac{h^2}{2m}[\mathbf{k}^2 - \mathbf{k}_z^{(j)2}]B^{(j)} \tag{8.16}$$

The transverse energies are related to the $\gamma^{(i)}$ of the TEM Bloch wave theory (eqn 5.38a) by

$$\varepsilon^{(i)} \approx -\frac{h^2 k_0^2 \gamma^{(i)}}{m} - |e|V_0$$

These transverse energies may be considered to be bound states (when ε is negative), lying within the negative-going crystal-projected potential-energy peaks $\phi_p(\mathbf{r})$ around each atom. The corresponding eigenstates are then labelled by analogy with the states of the hydrogen atom (Spence 1988). The lowest 1s bound state corresponds to the highest dispersion curve of the TEM Bloch wave theory. At 100 kV only two states are bound for silicon, and the corresponding transverse eigenstates are highly localized on the atom sites. (Coherent brehmsstrahlung (Section 13.1) results from transitions between these states.) More extended 'free' states also exist, with positive values of ε in the continuum.

In ADF-STEM by using a detector with a large hole we enhance the contribution from the more localized Bloch wave states, which scatter into larger angles. In addition, these states are the least dispersive, and hence are not washed out by the large beam divergence used in STEM. In such a Bloch wave model it may then be argued that the image is formed predominantly from the most localized 1s transverse eigenstates (Pennycook and Jesson

1991). These have the highest transverse kinetic energy, occupying regions of the lowest potential energy. The Virial theorem can be used to relate this energy to the potential energy, suggesting a linear dependence on atomic number, rather than the Z^2 dependence of high-angle Rutherford scattering from the nucleus. An expression for the ADF image may be derived along similar lines to eqn (8.13) (Nellist and Pennycook 2000), and an approximation developed which includes the probe coordinate and the predominant 1s state. The result may be interpreted similarly—now the coherence envelope (the transform of the hole) determines the degree to which different Bloch waves can interfere. Again, no image contrast is predicted without a hole in the detector, and a hole larger than the objective aperture image is indicated. The images do depend on thickness, but the analysis is more complicated than the thickness fringes of eqn (5.47), and may best be understood by following the propagation of the probe down atomic columns containing different species (Humphreys and Spence 1979, Fertig and Rose 1981, Hillyard and Silcox 1995). For example, in InP, Hillyard and Silcox found that when the probe was located over an indium atomic column, a channelling peak quickly appeared, disappearing after a thickness of about 10 nm. Locating the probe over a P column, however, generated a peak up to a thickness of 59 nm. In these simulations, which include TDS, we might expect the thickness dependence to be much less rapid than for elastic scattering images, since the TDS contribution is generated continuously throughout the sample, and so must be summed from all depths.

8.5. Z-contrast in STEM: thermal diffuse scattering

If the inner cut-off of the detector is made larger than the scattering angle $\theta_i = \lambda/d_{\text{RMS}}$, which corresponds to the thermal vibration amplitude $d_{\text{RMS}} = \sqrt{\langle U^2 \rangle}$, one may form an image from the thermal diffuse and nuclear scattering alone, since at larger angles Bragg scattering cannot occur. While treatments of thermal diffuse scattering (TDS) in electron microscopy date back to the 1950s, the STEM problem requires the incorporation of the probe wavefunction at the outset, since we are primarily interested in the variation of TDS as the probe moves from one atomic column to the next. Lattice resolution still requires a probe of comparable dimensions to the lattice spacing d of interest and illumination which is coherent over the corresponding angular range $\theta_c = \lambda/d$. Nevertheless it is worth recalling the form of the TDS scattering distribution under plane-wave illumination conditions, since a probe focused to atomic dimensions may have plane-wave-like form at its waist. For a monoatomic crystal the TDS scattering in an Einstein multiphonon vibrational model is given by (Hall and Hirsch 1965, Amali and Rez 1997)

$$I_{\text{TDS}}(\mathbf{K}) = f(\mathbf{K})^2 [1 - \exp(-M\mathbf{K}^2/4)] \tag{8.17a}$$

where $f(\mathbf{K})$ is the atomic scattering factor, $M = 8\pi^2 \langle U^2 \rangle$ is the Debye–Waller factor and $\langle U^2 \rangle$ is the mean square displacement of the atoms. This function I_{TDS} is zero at the origin, rises to a maximum at about $d = |\mathbf{K}|^{-1} = \lambda/\theta = 0.5$ Å for typical values of M, where θ is the scattering angle, and then falls off. The atomic electron scattering factor part can be written as

$$f(|\mathbf{K}|)^2 = \left(\frac{2[Z - f^{\text{X}}(K/2)]}{a_\circ (K/2)^2} \right)^2 \tag{8.17b}$$

where f^X is the X-ray scattering describing scattering from the electron cloud, while the first term in the square brackets describes the Rutherford scattering from the nucleus. Here a_o is the Bohr radius. Since the electronic contribution f^X falls off at high angles, while the term in square brackets in equation (8.17a) tends to unity, the remaining contribution I_{TDS} to the HAADF-STEM image is the Rutherford scattering from the point-like nucleus, which we see is proportional to Z^2/K^4, giving the name 'Z-contrast' imaging to HAADF-STEM. Written in terms of a scattering cross-section, this nuclear scattering is

$$\frac{dI_{TDS}}{d\Omega} = \frac{\gamma^2 Z^2}{4\pi^4 a_o^2} \tag{8.17c}$$

where γ is the relativistic factor. When integrated over all angles, the resulting cross-section has a value of about 0.0012 Å² for a strontium atom. If we imagine that the 'waist' of the focused beam has about the same diameter as the atom and channels down a column of 250 atoms in a crystal, then about 30% of the electron beam would therefore be scattered away into the detector by this process.

Another simple approach to understanding the contribution of TDS to HAADF imaging which immediately takes into account the probe position may be based on the work by Heidenreich (Heidenreich 1962, Cherns et al. 1973, Pennycook and Jesson 1991), who showed that the total inelastic scattering from process i generated in volume element $d\tau$ is just equal to $\Psi(r)\Psi(r)^*\phi_i^i(r)d\tau$, where $\phi_i^i(r)d\tau$ is the contribution to the imaginary part of the optical potential from process i for a centric crystal, and $\Psi(r)$ will here be the STEM probe wavefunction. Since TDS scattering is zero in the forward direction, the scattering not collected in the central hole in the detector might be neglected. Then, assuming that all the TDS is collected, the STEM image formed from this scattering alone, for a probe intensity $|\bar{A}(r)|^2$ displaced to scan coordinate r_p, becomes

$$I_{STEM}^{DF}(r_p) = \int |\bar{A}(r - r_p)|^2 \left|\phi_{i,p}^i(r)\right|^2 dr \tag{8.17d}$$

where $\phi_{i,p}^i(r)$ is the contribution to the TDS absorption potential projected in the beam direction. Values of the absorption potentials may be found for atoms in Radi (1970) and for crystals in Bird and King (1990) and Weickenmeier and Kohl (1991). For the localized TDS collected at high angles, $\phi_{i,p}^i(r)$ is a narrow peak centred on the atom sites. Thus the HAADF-STEM image is a map of the phonon absorption potential in this approximation. A thickness integration might also be included, by using the multislice superlattice method to calculate the probe intensity $|A(r)|^2$, and summing the scattered intensity at each slice. This simplified approach does not include the size of the detector hole as a parameter, nor does it include the Bragg scattering of TDS, which gives rise to Kikuchi lines.

Initially, attempts were made to exclude Bragg scattering and so make the ADF image 'more incoherent' and less sensitive to diffraction contrast effects, by increasing the size of the hole in the detector beyond the suggested figure of 40 mrad (Howie 1979). In this way, it was hoped to eliminate the thickness oscillations which are known to accompany multiple Bragg scattering but which are smoothed out by the thickness integrations involved in the theory of TDS (as in Fig. 8.4). Understanding this requires a detailed theory of TDS in STEM, which goes well beyond the scope of this book (see Wang (1995) for a textbook treatment). The following papers provide essential reading for experts: Allen et al. (2002, 2011) give a Bloch wave treatment of the multiple elastic scattering of the probe, and use their mixed dynamic form factor to relate TDS in ADF-STEM to EELS and EDX; Amali and Rez (1997) also provide a Bloch-wave treatment and discuss in detail the influence of different phonon models for the TDS (Einstein, Debye, single phonon, multiphonon) and the influence of HOLZ rings; Wang and Cowley (1989) provide a multislice treatment in which the TDS generation function is related to the derivative of the elastic scattering potential, and atomic columns are assumed to vibrate independently; Jesson and Pennycook (1995) also analyse the effects of replacing Rutherford scattering with a multiphonon model; while Loane et al. (1991) describe the 'frozen-phonon' multislice supercell approach on which one popular software algorithm is based (Kirkland 2010). (In the high-angle limit, multiphonon scattering using an Einstein model is equivalent to Rutherford scattering from the nucleus, which is proportional to the square of the atomic number.) The general result from all this work is that the simple incoherent imaging model (eqn 8.8) is a useful approximation for locating atoms, but that detailed image simulations are needed for quantitative analysis of intensities, and image artefacts (e.g. contrast reversals between medium- and heavy-weight atoms) may occur (Rez 2000). The thickness dependence of the images is found to saturate (Hillyard and Silcox 1995), as shown in Fig. 8.4.

For practising microscopists, the important questions then are: (1) what is the most efficient algorithm for simulating HAADF images which includes both elastic and TDS scattering and (2) what general principles can guide the choice of instrumental parameters, such as the detector hole size and choice of defocus and illumination aperture? Because it is well documented in book form (with accompanying source code), the multislice 'frozen phonon' method appears most suitable at present for simulations (Kirkland 2010). Briefly, a superlattice of, say 6 × 6 unit cells is used in a multislice calculation, in which the first slice contains the probe wavefunction $A(\mathbf{r})$ (Spence 1978a,b). A slice thickness of one atomic layer is typical. For each probe position, a complete multislice calculation of the scattered intensity from the entire crystal must be completed for several sets of atomic displacements, generated from the appropriate thermal vibration factors, and the results added together (e.g. for silicon at room temperature the mean vibration amplitude $\langle U \rangle$ is 0.0076 nm). An Einstein model for vibrations is adequate, since the detector averages over any mode structure. The results of these calculations (e.g. Hillyard and Silcox 1995) show the following.

1. Increasing the inner detector cut-off beyond the value $\theta_i = \lambda/d_{RMS}$ of about 40 mrad at 100 kV at which d-spacings become comparable with the thermal vibration amplitude has the effect of reducing the image intensity, but produces negligible change in the form of the image, and this finding holds for all illumination angles (probe sizes). As a rough guide, the inner detector cut-off should be at least three times the objective aperture size for incoherent imaging, and needs to be perhaps five times as large to eliminate thickness oscillations in crystals (see Fig. 8.4). As expected, the effect of varying the sample temperature is significant only if a large inner detector cut-off is used. However, a lower inner cut-off angle, including, for example, the lowest-order CBED disc overlaps (see Fig. 5.34(b)), is found to introduce greater thickness oscillations, as expected (Fig. 8.4). So it seems to be useful to exclude most of the low-order Bragg discs. Those that are included are then more likely to be kinematic, and have longer extinction distances. Although

for a typical ADF detector TDS makes the largest contribution to the signal for all but the thinnest samples at room temperature, by comparing calculations with and without TDS, Hillyard and Silcox (1995) concluded that, 'for almost all current observations, the influence of TDS on the detailed structure of the scaled images is minor'. In real space, the absorption-potential model (eqn 8.17d) provides a useful physical picture of this process—a probe, which is wider than the spacing between the periodically arranged peaks in $\phi^i_{i,p}(\mathbf{r})$ cannot be expected to produce an image showing these peaks. The effects of Bragg scattering are reduced, however, in ADF-STEM by the fact that TDS is generated continuously throughout the crystal, and its main effect on TDS is to redistribute it over the large detector.

2. A probe of atomic dimensions penetrates to different depths along atomic columns consisting of different species. The characteristic scattering angle depends on the atomic number of the column, and so changes as the probe is moved. (In the Wentzel model—see also eqn (6.34)—this angle is $\theta_w = \lambda Z^{1/3}/2\pi a_H$, with a_H the Bohr radius. This scattering angle is convoluted with the illumination cone.) It follows that a smaller inner cut-off may be useful for imaging lighter atoms.

3. It is useful to think of the probe in real space, in connection with channelling effects. For a probe narrower than the intercolumn spacing, a strong channelling effect ('tendency to run along paths of low potential energy') is observed, and the interference effects between different atomic columns becomes negligible. Figure 8.8 shows what happens when a probe is large enough to span two atomic columns—it splits into two focused 'probes' at a depth of 10 nm in silicon.

4. The effects of strain may be detected in ADF images, particularly at low temperatures, where the TDS background can be minimized. These strain effects may dominate those of atomic number. Forbidden lattice periodicities (e.g. the (200) in the [110] projection of silicon) have also been observed in the power spectrum of ADF images (Liu and Cowley 1990). Under particular conditions, other artefacts have also been simulated. For example, Wang and Cowley (1989, 1990) have calculated the separate contributions of TDS and Bragg scattering to the ADF image for various conditions, and show how the large vibrational amplitude of light atoms may enhance their contribution to the ADF image beyond a simple Z^2 law, and how diffraction effects in the HOLZ rings may reverse the contrast of the images (Spence et al. 1989). (The HOLZ rings indicate coherent scattering along atomic columns, but may be a small contribution to total scattering unless the unit cell is large and the thickness small.)

In summary, these dynamical tests of the incoherent imaging model have generally shown that it works reasonably well. However, in view of the dependence of ADF images on strain, on the inner detector cut-off, and on temperature, and of the observation of forbidden periodicities and stacking faults, we can conclude that the interpretation of these images, although more straightforward than HREM images, may still contain artefacts. While transverse interference effects can be eliminated by choice of detector shape (simplifying atom location), interference along the atom strings cannot, making interpretation of the intensity of atom images difficult in terms of atomic number and thickness. Thus, for the identification of unknown substitutional atoms within an atomic column, very detailed simulations would be needed. That substitutional atoms can readily be distinguished from atomic-scale roughness at surfaces was convincingly demonstrated for the

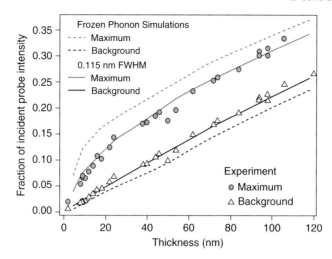

Figure 8.9 Comparison of experimental (symbols) and frozen-phonon simulated (line) image intensities for HAADF STEM imaging for various thicknesses of PbWO$_4$ (tetragonal, scheelite) down [100]. The maximum intensity in the image and background intensity are shown, and the improved fit is shown when a 0.115 nm FWHM electron source size is incorporated to limit the spatial coherence of the beam. The agreement is within 5%. (From LeBeau et al. 2009.) Compare with this with Fig. 8.4 for the effect of detector inner cut-off angle.

first time by work in Muller's group (Voyles et al. 2002). Here a silicon multilayer was formed in which alternate layers contain substitutional Sb atoms. The cross-section sample shows the boundary of the implanted region, confirming that the bright dots in this ADF-STEM image are individual Sb atoms. (Further confirmation comes from their known concentration.)

The accurate quantification of HAADF-STEM images thus has a long history. Figure 8.9 shows that good agreement can now be obtained between experimental and computed images over a wide range of thicknesses in favourable cases, such as the tetragonal PbWO$_4$ structure, using the frozen-phonon simulation method (Hillyard and Silcox 1995, Kirkland 2010, Allen 2011) to model the contributions of thermal diffuse scattering to STEM images (Wang and Cowley 1989), together with full multiple elastic Bragg scattering. Along [100], the structure shows separate columns of tungsten and lead atoms, which are resolved in the images. Agreement within about 5% can be obtained between theory and experiment, for the maximum and minimum image intensities, provided account is taken of the finite size of the STEM electron source, provided thickness can be measured independently, and provided that Debye–Waller factors are accurately known (a serious difficulty when analysing defects). In this case thickness was measured using EELS spectra and spatially averaged CBED patterns. Considerable attention was paid to ensuring linearity of the angle-integrating detector, which, unlike in the HREM case, is unaffected by pixel cross-talk. Images were recorded at 300 kV, with $C_s \sim 1.2$ mm (resolution 0.136 nm), using an inner detector cut-off at 65 mrad and an outer cut-off at 350 mrad. Image intensities were normalized against the incident beam intensity. Oxygen atoms were not resolved in the

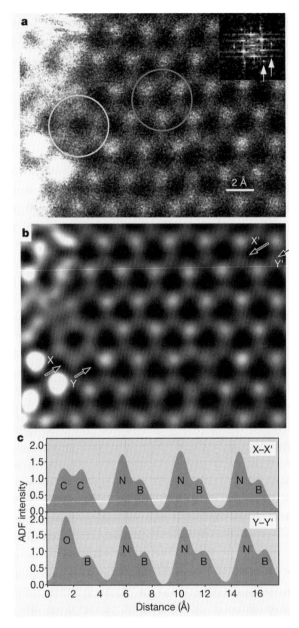

Figure 8.10 HAADF aberration-corrected STEM image of monolayer boron nitride (BN), as recorded (a), and smoothed and probe-deconvoluted (b). The FWHM of the probe is about 0.12 nm. Part (c) shows line traces along X–X' and Y–Y'. Histogram analysis of peak heights suggests the presence of substitutional atoms oxygen and carbon as shown (see also Plate 4). Inset: Fourier transform of this incoherent image shows sharp spots extending to 1.09 Å spacing. (From Krivanek et al. (2010), by permission from Macmillan Publishers Ltd., © 2010.)

images. Temperature factors were obtained from single-crystal X-ray work, and the focus setting used was that which provided maximum contrast. The result is substantially better than the agreement usually reached for HREM images (for a discussion of this comparison see Section 5.13). Despite this fit, contrast reversals were seen at some thicknesses, with tungsten appearing brighter than lead, so that an intuitive interpretation based on atomic number is not possible. The use of aberration correction cannot be expected to address this problem.

Images obtained from thin samples, however, can be interpreted using the simple incoherent imaging model (eqn 8.8) to provide a simple first estimate on which to base multiple-scattering calculations. The resolution expected in ADF-STEM can be understood by plotting out figures similar to Fig. 8.5. The optimum focus setting will be close to eqn (8.14b), and a coherently filled aperture whose size is given by eqn (8.14c) should be used for non-periodic samples.

Figure 8.10 shows an atomic-resolution image of a monolayer of boron nitride recorded by HAADF-STEM in an aberration-corrected STEM, using a beam energy of 60 kV to minimize knock-on radiation damage (Krivanek et al. 2010; see also Plate 4). Histogram analysis of the atom-image intensities shows the presence of substitutional atoms, as indicated. Applications of ADF-STEM to superconductor interfaces are reviewed in Browning and Pennycook (1999), and a comprehensive summary of the STEM method can be found in the text edited by Pennycook and Nellist (2011).

8.6. Three-dimensional STEM tomography

At nanometre or lower resolution, the methods of back-projection and related algorithms may be used to reconstruct a three-dimensional image of an inorganic particle from images taken in many projections, following a similar approach to that used in cryo-EM (De Rosier and Klug 1968). For a review of this method applied to problems in materials science, and an instructive history of the field of tomography, see Midgely and Weyland (2011). For reconstruction at higher resolution, the alignment and registration of images of the same particle taken in different orientations becomes extremely difficult because it requires atomic-scale fiducial marks, independent of optical conditions, because the effects of radiation damage may alter the particle structure during the experiment, and because the limited range of available tilts results in loss of information. As a solution to the problem of registration of images recorded in different orientations, Scott et al. (2012) have proposed the use of the centre of mass, together with an iterative algorithm, achieving 0.24 nm resolution in a three-dimensional reconstruction. The same method has produced an image of a dislocation core at near-atomic resolution in a Pt nanoparticle about 7 nm thick, using 104 HAADF-STEM projections over a 73° range (Chen et al. 2013). Detailed simulations are needed to understand the disruption of channelling which occurs at defects such as a dislocation core in projections other than those which present plane strain, the effects of half-period fringes discussed above when HAADF scattering is modulated by Bragg or elastic multiple scattering (which may produce forbidden reflections (Hillyard and Silcox 1995), the accuracy of the registration method, and the effect of Fourier filtering. However, this encouraging preliminary result does indicates the importance of automating this difficult data collection process.

In Section 5.16 we outlined the principle of discrete tomography, which can then be used to determine the external shape of nanocrystals at atomic resolution, if reasonable assumptions can be made about their crystal structure. Here the number of atoms in each atomic column seen in projection is estimated based on the image intensity, and the process repeated for several projections (LeBeau *et al.* 2010). For this discrete tomography approach, the ADF-STEM mode has the advantage of providing an imaging signal which is more linearly related to sample thickness over a larger range of thickness (as in Figs 8.4 and 8.9) than the bright-field HREM mode, which, as shown in Fig. 12.4(a), may oscillate rapidly with thickness. Its disadvantage lies in the possibilities of scan distortions and possibly greater radiation damage. Figure 8.11 shows a three-dimensional image of a Ag cluster lying within an aluminium matrix, which is assumed to have constant thickness (Van Aert *et al.* 2011). This three-dimensional reconstruction was assembled from just two HAADF-STEM image projections, recorded in an aberration-corrected instrument using discrete tomography and statistical parameter estimation theory. Additional projections were found to be consistent with this model, and although the uniqueness of the reconstruction from so few projections has been questioned, we can expect rapid development of these early pioneering achievements. Histogram analysis of the image intensities was used together with image simulation. The crystal lattice is assumed known, and vacancies are not considered. An earlier reconstruction using five corrected STEM projections can be found in Jiang *et al.* (2008).

Using multiple views of the same or similar particles, and noting the tendency of nanocrystals to facet, with information on local crystallite thickness provided by aberration-corrected ADF-STEM imaging, it is certainly possible to build up three-dimensional

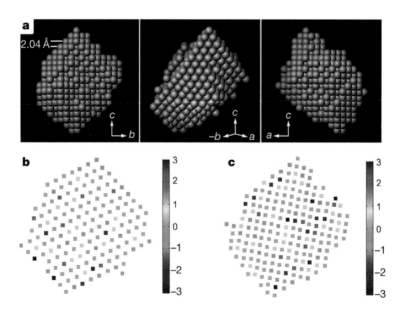

Figure 8.11 Three-dimensional image of a silver nanocrystal embedded in an aluminium matrix reconstructed from several ADF STEM images in different projections. (From Van Aert *et al.* (2011), by permission from Macmillan Publishers Ltd., © 2011.)

images at atomic resolution of a typical nanoparticle. This has been done in a detailed study of gold nanoparticles lying on rutile, in order to determine their equilibrium shape (Sivaramakrishnan et al. 2010). Here it was found that the equilibrium shape depended on particle size, an effect attributed to the interfacial triple line energy, which was thus measured to be 0.28 eV Å^{-1}.

In summary, discrete tomography offers the best path to three-dimensional reconstruction of particle shapes at atomic resolution; however, since the bulk is modelled as a perfect crystal, it cannot yet give us three-dimensional images of defect structures at atomic resolution.

Alternative approaches to three-dimensional imaging in STEM have been based on 'optical sectioning' or 'depth sectioning', which relies on the much shorter depth of focus of the electron probe in a corrected instrument (Xin and Muller 2010), and on the scanning confocal mode (SCEM) applied using an instrument in which both the probe-forming and imaging lenses are corrected (Wang et al. 2011). Figure 8.12 shows the depth of focus for an electron probe with (a) and without (b) aberration correction. The FWHM of the probe intensity measured along the optic axis is $L_D = 1.77\lambda/\theta^2$ (eqn 2.11c, with a more accurate numerical factor based on wave-optical calculations). This can be reduced to a few nanometres on current machines. A three-dimensional OTF for incoherent imaging can thus

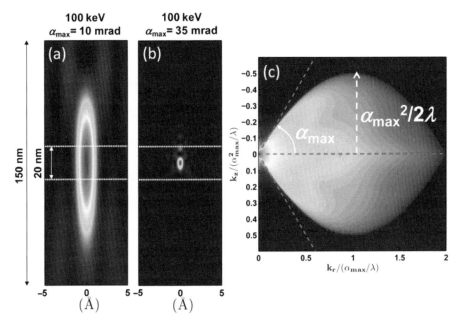

Figure 8.12 Simulations of probe intensity on a plane containing the optic axis for (a) uncorrected and (b) aberration-corrected instruments at 100 kV. Note the greatly increased illumination angle $\theta_{ap} = \alpha_{max}$ in (b), resulting in reduced depth of focus $L_D \sim 1.77\lambda/\alpha_{max}^2$. (c) The three-dimensional OTF for incoherent imaging obtained from (b). The optic axis runs along the vertical axis. (From Xin and Muller (2010).)

be derived by Fourier transform of this probe intensity distribution (Behan *et al.* 2009), as shown in Fig. 8.12(c) for the aberration-corrected conditions of Fig. 8.12(b). The doughnut shaped OTF has a very large missing 'cone', with a large semi-angle $(90 - \alpha_{max})$, since α_{max} is at most about 2°, even in corrected instruments. (See Xin and Muller (2010) for a discussion of elongation factors in this imaging mode and a comparison of the relevant three-dimensional transfer functions with the confocal mode, which we now discuss.)

If both probe-forming and image-forming lenses are aberration corrected, one can perform scanning confocal microscopy (SCEM) in a TEM/STEM electron microscope (see Cosgriff *et al.* (2010) for a review). The advantage of this mode is that the missing cone of scattering can then be filled (Wang *et al.* 2010), while scattering from points away from that at which the probe is focused is neither well illuminated nor focused onto the small detector. While there is no first-order phase contrast, useful contrast is obtained with inelastic scattering (e.g. from surface plasmons), showing nanoscale depth resolution (Wang *et al.* 2011).

References

Ade, G. (1977). On incoherent imaging in STEM. *Optik* **49**, 113.

Allen, L. J., Findlay, S. D., Oxley, M. P., and Rossouw, C. J. (2003). Lattice resolution from a focused coherent electron probe. *Ultramicroscopy* **96**, 47.

Allen, L. J., Findlay, S. D., and Oxley, M. P. (2011). Simulation and interpretation of images. *Scanning transmission electron microscopy: imaging and analysis*, ed. S. Pennycook and P. Nellist, p. 247. Springer, New York.

Amali, A. and Rez, P. (1997). Theory of lattice resolution in HAADF images. *Microsc. Microanal.* **3**, 28.

Batson, P. and Krivanek, O. (2002). Sub-angstrom resolution using aberration corrected electron optics. *Nature* **418**, 617.

Behan, C., Cosgriff, E. C., Kirkland, A. I., and Nellist, P. D. (2009). Three dimensional imaging by optical sectioning in the aberration-corrected scanning transmission electron microscope. *Phil. Trans. R. Soc., Lond.* **A367**, 3825.

Bird, D. and King, Q. A. (1990). Absorptive form factors for high energy electron diffraction. *Acta Crystallogr.* **A46**, 202.

Browning, N. and Pennycook, S. J. (1999). *Characterisation of high-T_c materials and devices by electron microscopy*. Cambridge University Press, New York.

Chen, C., Zhu, C., White, E., Chiu, C., Scott, M., Regan, B., Marks, L. D., Huang, Y., and Miao, J. (2013). Three-dimensional imaging of dislocations in a nanoparticle at atomic resolution. *Nature* **496**, 74.

Cherns, D., Howie, A., and Jacobs, M. H. (1973). Characteristic X-ray production in thin crystals. *Z. Naturforsch.* **28a**, 565.

Cosgriff, E., Nellist, P., D'Alfonso, A., Findlay, S., Behan, G., Wang, P., Allen, L., and Kirkland, A. (2010). Image contrast in aberration-corrected scanning confocal electron microscopy. *Advances in imaging and electron physics*, Vol. 162, ed. P. W. Hawkes, p. 45. Academic Press, Burlington, MA.

Cowley, J. M. (1969). Reciprocity in electron microscopy. *Appl. Phys. Lett.* **15**, 58.

Cowley, J. (1976). STEM of thin specimens. *Ultramicroscopy* **2**, 3.

Cowley, J. (1984). Dark-field lattice imaging in STEM. *Bull. Mater. Sci.* **6**, 477.

Cowley, J. M. (1993). Configured detectors for STEM imaging of thin specimens. *Ultramicroscopy* **49**, 4.

Cowley, J. M. (2001). STEM imaging with a thin annular detector. *J. Electron Microsc.* **50**, 147.
Crewe, A. V. (1963). Scanning transmission electron microscopy. *J. Microscopie* **2**, 369.
Crewe, A. (1980). The physics of the high resolution STEM. *Rep. Prog. Phys.* **43**, 621.
Crewe, A., Wall, J., and Langmore, J. (1970). Visibility of single atoms. *Science* **168**, 1338.
De Rosier, D. and Klug, A. (1968). Reconstruction of three-dimensional structure from electron micrographs. *Nature* **217**, 130.
Fertig, J. and Rose, H. (1981). Resolution and contrast of crystalline objects in high resolution STEM. *Optik* **59**, 407.
Findlay, S. D., Shibata, N., Sawada, H., Okunishi, E., Kondo, Y., and Ikuhara, Y. (2010). Dynamics of annular bright field imaging in STEM. *Ultramicroscopy* **110**, 903.
Frank, J. (1992). *Electron tomography*. Plenum, New York.
Goodman, J. W. (2004). *Introduction to Fourier optics*. Second edition. McGraw Hill, New York.
Hall, C. R. and Hirsch, P. B. (1965). Effect of thermal diffuse scattering on propagation of high energy electrons through crystals. *Proc. R. Soc. Lond.* **A286**, 158.
Heidenreich, R. D. (1962). Attenuation of fast electrons in crystals and anomalous transmission. *J. Appl. Phys.* **33**, 2321.
Hillyard, S. and Silcox, J. (1995). Detector geometry, TDS and strain in ADF STEM. *Ultramicroscopy* **58**, 6.
Howie, A. (1966). Diffraction channelling of fast electrons and positrons in crystals. *Phil. Mag.* **4**, 223.
Howie, A. (1979). Image contrast and localized signal selection techniques. *J. Microsc.* **117**, 11.
Humphreys, C. J. and Spence, J. C. H. (1979). Wavons. *Proc. Micros Soc. Am.* Claitors, Baton Rouge, LA.
Ishikawa, R., Okunishi, E., Sawada, H., Kondo, Y., Hosokawa, F., and Abe, E. (2011). *Nature Mater.* **10**, 298.
Jesson, D. E. and Pennycook, S. (1993). Incoherent imaging of thin crystals using coherently scattered electrons. *Proc. R. Soc. Lond.* **A441**, 261.
Jesson, D. E. and Pennycook, S. (1995). Incoherent imaging of crystals using thermally scattered electrons. *Proc. R. Soc. Lond.* **A449**, 273.
Jiang, L., Wang, P., Bleloch, A., and Goodhew, P. (2008). Three-dimensional reconstruction of Au nanoparticles using five projections from an aberration-corrected STEM. *Microsc. Microanal.* **14** (Suppl. 2), 1048.
Kirkland, E. J. (2010). *Advanced computing in electron microscopy*. Second edition. Springer, New York.
Koch, C. (2002). Determination of core structure periodicity and point defect density along dislocations. PhD Thesis, Arizona State University.
Krivanek, O., Dellby, N., and Lupini, A. R. (1999). Toward sub-angstrom electron beams. *Ultramicroscopy* **78**, 1.
Krivanek, O. L., Chisholm, M. F., Nicolosi, V., Pennycook, T. J., Corbin, G. J., Dellby, N., Murfitt, M. F., Own, C. S., Szilagyi, Z. S., Oxley, M. P., Pantelides, S. T., and Pennycook, S. J. (2010). Atom-by-atom structural and chemical analysis by annular dark-field electron microscopy. *Nature* **464**, 571.
LeBeau, J. M., Findlay, S. D., Wang, X., Jacobson, A. J., Allen, L. J., and Stemmer, S. (2009). High angle scattering of fast electrons from crystals containing heavy elements. Simulation and experiment. *Phys Rev.* **B79**, 214110.
LeBeau, J. M., Findlay, S. D., Allen, L. J., and Stemmer, S. (2010). Standardless atom counting in STEM. *Nano Lett.* **10**, 4405.
Liu, J. and Cowley, J. M. (1990). High angle ADF and high resolution SE imaging of supported catalyst clusters. *Ultramicroscopy* **34**, 119.
Loane, R., Xu, P., and Silcox, J. (1991). Thermal vibrations in CBED. *Acta Crystallogr.* **A47**, 267.

Midgely, P. A., Weyland, M., Thomas, J. M., and Johnson, F. G. (2001). Z-contrast tomography. *Chem. Commun.* **2001**, 907.

Midgely, P. and Weyland, M. (2011) STEM tomography. *Scanning transmission electron microscopy: imaging and analysis*, ed. S. Pennycook, and P. Nellist, p. 353. Springer, New York.

Mory, C., Colliex, C., and Cowley, J. M. (1987). Optimum defocus for STEM imaging and microanalysis. *Ultramicroscopy* **21**, 171.

Nellist, P. and Pennycook, S. (1998). Subangstrom resolution by incoherent STEM. *Phys. Rev. Lett.* **81**, 4156.

Nellist, P. and Pennycook, S. (2000). The principles and interpretation of annular dark-field Z-contrast imaging. *Adv. Imaging Electron. Phys.* **113**, 148.

O'Keefe, M. A., Hetherington, C. J., Wang, Y. C., Nelson, E., Turner, J., Kisielowski, C., Malm, O., Mueller, R., Ringnalda, J., Pan, M., and Thust, A. (2001). Sub-angstrom HREM at 300 kV. *Ultramicroscopy* **89**, 215.

Pennycook, S. and Nellist, P. (ed.) (2011). *Scanning transmission electron microscopy: imaging and analysis*. Springer, New York.

Pennycook, S. and Jesson, D. E. (1991). High resolution Z-contrast imaging of crystals. *Ultramicroscopy* **37**, 14.

Pennycook, S. J. and Boatner, L. A. (1988). Chemically sensitive structure-imaging with a scanning transmission electron microscope. *Nature* **336**, 565.

Pogany, A. P. and Turner, P. S. (1968). Reciprocity in electron diffraction and microscopy. *Acta Crystallogr.* **A24**, 103.

Radi, G. (1970). Complex lattice potentials in electron diffraction calculated for a number of crystals. *Acta Crystallogr.* **A26**, 41

Rez, P. (2000). Can HAADF scattering be represented by a local operator? *Ultramicroscopy* **81**, 195.

Scott, M. C., Chen, C., Mecklenburg, M., Zhu, C., Xu, R., Ercius, P., Dahmen, U., Regan, B., and Miao, J. (2012). Electron tomography at 2.4-angstrom resolution. *Nature* **483**, 7390.

Shibata, N., Findlay, S., Kohno, Y., Sawada, Y., Kondo, Y., and Ikuhara, Y. (2012). Differential phase contrast microscopy at atomic resolution. *Nature Phys.* **8**, 611.

Sivaramakrishnan, S., Wen, J., Scarpelli, M., Pierce, B. J., and Zuo, J. M. (2010). Equilibrium shapes and triple line energy of epitaxial gold nanocrystals supported on TiO_2 (110). *Phys. Rev.* **B82**, 195421.

Spence, J. C. H. (1978a). Approximations for dynamical calculations of micro-diffraction patterns and images of defects. *Acta Crystallogr.* **A34**, 112.

Spence, J. C. H. (1978b). Practical phase determination of inner dynamical reflections in STEM. *Scanning Electron Microsc.* **1978** (I), 61.

Spence, J. C. H. (1979). Uniqueness and the inversion problem of incoherent multiple-scattering. *Ultramicroscopy* **4**, 9.

Spence, J. C. H. (1988). Inelastic electron scattering. *High resolution transmission electron microscopy and associated techniques*, ed. P. Buseck, J. Cowley, and L. Eyring, p. 349. Oxford University Press, New York.

Spence, J. C. H. and Cowley, J. M. (1978). Lattice imaging in STEM. *Optik* **50**, 129.

Spence, J. C. H. and Koch, C. (2001). Measurement of dislocation core periods by nanodiffraction. *Phil. Mag.* **B81**, 1701.

Spence, J. C. H. and Lynch, J. (1982). STEM microanalysis by ELS in crystals. *Ultramicroscopy* **9**, 267.

Spence, J. C. H. and Zuo, J. M. (1991). *Electron microdiffraction*. Plenum, New York.

Spence, J., Zuo, J., and Lynch, J. (1989). On the HOLZ contribution to STEM lattice images formed using high angle dark-field detectors. *Ultramicroscopy* **31**, 233.

Tanaka, M., Terauchi, M., and Kaneyama, T. (1988). *Convergent beam electron diffraction II*. JEOL Company, Tokyo.

Treacy, M. and Gibson, J. (1993). Coherence and multiple scattering in 'Z-contrast' images. *Ultramicroscopy* **52**, 31.

Treacy, M. M. J., Howie, A., and Wilson, C. J. (1978). Z-contrast imaging of platinum and palladium catalysts. *Phil. Mag.* **A38**, 569.

Van Aert, S., Batenburg, K., Rossell, M. D., Erni, R., and Van Tendeloo, G. (2011). Three dimensional atomic imaging of crystalline nanoparticles. *Nature* **470**, 374.

Voyles, P. M., Muller, D. A., Grazul, J. L., Citrin, P. H., and Gossmann, H. L. (2002). Atomic-scale imaging of individual dopant atoms and clusters in highly n-type bulk Si. *Nature* **416**, 826.

Wang, Z. L. (1995). Dynamical theories of dark-field imaging using diffusely scattered electrons in STEM and TEM. *Acta Crystallogr.* **A51**, 569.

Wang, Z. L. and Cowley, J. M. (1989). Simulating HAAD STEM images including TDS. *Ultramicroscopy* **31**, 437.

Wang, Z. L. and Cowley, J. M. (1990). Dynamic theory of HAAD STEM lattice images for a Ge/Si interface. *Ultramicroscopy* **32**, 275.

Wang, P., Behan, G., Takeguchi, M., Hashimoto, A., Mitsuishi, K., Shimojo, M., Kirkland, A. I., and Nellist, P. D. (2010). Nanoscale energy-filtered scanning confocal electron microscopy using a double-aberration-corrected transmission electron microscope. *Phys. Rev. Lett.* **104**, 200801.

Wang, P., Behan, G., Kirkland, A., Nellist, P., Cosgriff, E., D'Alfonso, A., Morgan, A., Allen, L., Hashimoto, A., Takeguchi, T., Mitsuishi, K., and Shimojo, M. (2011). Bright field scanning confocal electron microscopy using a double aberration-corrected transmission electron microscope. *Ultramicroscopy* **111**, 877.

Weickenmeier, A. and Kohl, H. (1991). Computation of absorptive form factors for high energy electron diffraction. *Acta Crystallogr.* **A47**, 590.

Xin, X. and Muller, D. (2010). Three-dimensional imaging in aberration-corrected electron microscopy. *Microsc. Microanal.* **16**, 445.

9
Electron sources and detectors

In selecting an electron source for an experiment on a particular sample we are attempting to match the phase-space properties of the source (its size and emission solid angle (emittance), brightness, and beam energy) to those required by our experiment in a way which maximizes the signal we wish to detect. The degeneracy of a beam, δ, is equal to the mean number of particles per coherence time traversing a coherence area, and is also proportional to the brightness of the source. It is equal to unity when two electrons of opposite spin fill every cell in phase space, where the cell dimensions are defined by the uncertainty principle (Spence et al. 1994). Even for field-emission sources $\delta \sim 10^{-5}$, compared with values of around 10^{12} for the photons of a laser beam, where any number of these bosons may fill a cell. So for high-resolution imaging, where a given degree of coherence is required, this quantity determines the rate at which useful image-forming electrons traverse our sample.

The electron scattering we detect is therefore limited, for a given emittance, by the source brightness, coherence requirements at the sample, scattering strength, and detector efficiency. But most static electron microscope imaging experiments are sample, rather than source, limited, in the sense that radiation damage limits the maximum possible dose which can be delivered within the mechanical stability time of the instrument, for a given resolution. For dynamic processes, where snapshots of fast processes are required, the source may become limiting due to the short exposure times, as Coulomb interactions between electrons at crossovers in the column start to introduce unwanted beam divergence and energy spread (Boersch effect), thereby limiting coherence.

The three common electron sources are tungsten hair-pin filaments, lanthanum hexaboride sources, and field-emission sources. Laser-driven photocathodes are used for time-resolved imaging. Detectors include direct and indirect detection types—film, complementary metal-oxide semiconductors (CMOS), image plates, and Medipix in the first category, and charge-coupled devices (CCDs) coupled to a scintillation screen in the second. Scintillation detectors are commonly used as detectors for STEM, where the very large dynamic range of a photomultiplier coupled to a scintillator has advantages when spatial resolution on the scintillator screen is not required. This chapter gives brief notes on the suitability of all of these for high-resolution TEM and summarizes their characteristics. For electron sources, the main requirements are as follows:

1. High brightness (current density per unit solid angle). This is important in high-resolution phase-contrast experiments where a small illumination aperture is required (to provide coherence), together with sufficient image current density to allow accurate focusing at high magnification.

2. High current efficiency (ratio of brightness to total beam current). This is achieved through a small source size. Reducing the area of the specimen that is illuminated reduces unnecessary specimen heating and thermal movement during the exposure.
3. Long life under the available vacuum conditions.
4. Stable emission. Exposures of up to a minute are not uncommon for high-resolution dark-field imaging and ADF-STEM.

An ideal illumination system for high-resolution CTEM would give the operator independent control of the area of specimen illuminated, the intensity of the illumination, and the coherence conditions. Only the field-emission source approaches this ideal.

The area detectors used for imaging in TEM may be compared as follows. In cryo-EM one wishes to record images of a very large number of similar particles in the same field of view, each of which produces a very low-contrast image. In order to achieve a sufficient number of pixels per particle, the requirement is therefore for the largest possible number of pixels in a detector which offers linear recording over a small dynamic range. Film is well suited to this purpose and also provides an excellent archival medium, avoiding the prospect of warehouses filled with RAID arrays of digital data which must never be powered down. It also has better detective quantum efficiency (DQE) than CCDs at medium and high spatial frequencies. However, chemical development introduces delay, and film transport mechanisms may be unreliable. In materials science, the image analysis is rarely quantitative, and results must be seen immediately, with real-time image analysis crucial for aberration correction, focusing, and movie recording. For this purpose the modern scintillator screen (single crystal or powder phosphor) coupled to a CCD using either lenses or fibre-optics is ideal, and quantitative analysis is also possible if considerable additional effort is devoted to detector characterization. For two-dimensional crystalline samples in biology, this CCD system has also been found preferable to film for recording the diffraction patterns. For tomographic imaging in biology, CCDs are used because of the need for rapid alignment and focusing at each tilt, followed by fast orientation determination and data merging in three dimensions, a barely practical procedure using film recording. The new direct recording devices discussed below, despite their high cost, are even better. For recording diffraction patterns, where the largest possible dynamic range is needed, together with a large number of pixels, and a minimum of 'blooming' (cross-talk between pixels), image plates (a kind of electronic film) have proven ideal. Finally, the new generation of direct recording CMOS and related solid state detectors can detect the arrival of each beam electron and are able to accumulate many frames rapidly with fast read-out for high dynamic range, and so offer the highest performance possible for low-dose imaging of radiation-sensitive materials.

9.1. The illumination system

The two condenser lenses of a simple electron microscope illumination system are shown in Fig. 9.1. The microscopist usually has independent control of the focal lengths of both these lenses (C1 and C2). The excitation of the first condenser lens is sometimes called 'spot size'. The instrument is normally used with these excitations arranged so that planes S, S' and the specimen plane are conjugate, that is, with a focused image of the source formed at the specimen (focused illumination).

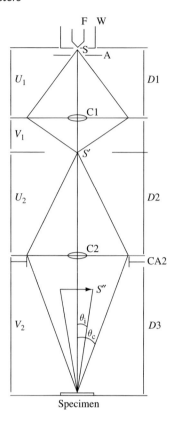

Figure 9.1 The illumination system of a simple electron microscope. There are two condenser lenses C1 (a strong lens) and C2 (a weak lens). F, filament electron source; W, Wehnelt cylinder; S, virtual electron source, with images S′ and S″; CA2, second condenser aperture. The distances U_1, U_2, V_1, and V_2 are electron optical parameters, while distances $D1$, $D2$, and $D3$ are easily measured from the microscope column. Values of these distances are given for a particular microscope in Table 2.1.

For a hair-pin heated tungsten filament the source size is about 30 µm. To prevent unnecessary specimen heating and radiation damage, a demagnified image of this is required at the specimen. The working distance $D3$ must also be large enough to permit specimen-change manoeuvres. These conflicting requirements (magnification M_1 small, $D3$ large) would necessitate an impractically large distance $D1$ if a single condenser lens were used. The usual solution is to use a strong first condenser lens C1 to demagnify the source by between 5 and 100, followed by a weak second lens C2 providing magnification of about 3 and a long working distance. Demagnification values for the two lenses and the corresponding focal lengths are given in Table 2.1. Since $M_1 = V_1/U_1$ is small the image of S formed by C1 is situated close to the focal plane of C1.

Table 9.1 The properties of four modern electron sources. These figures are approximate. See Tuggle and Swanson (1985) for details.

Source	Virtual source diameter	Measured brightness at 100 kV (A cm^{-2} sr^{-1})	Current density at specimen with 0.8 mrad illuminating aperture (A cm^{-2})	Energy FWHM (eV)	Melting point (°C)	Vacuum required (T)	Emission current (μA)
Heated field-emission, ZrO/W	100 nm	10^7–10^8	20	0.8	3370	10^{-8}–10^{-9}	50–100
RT field emission	2 nm	2×10^9	4000	0.28	3370	10^{-10}	10
Hair-pin filament	30 μm	5×10^5	1	0.8	3370	10^{-5}	100
LaB$_6$	5–10 μm	7×10^6*	14	1.0	2200	10^{-6}	50

*Measured at 75 kV (A. Broers 1976, personal communication). RT, room temperature.

As discussed in Section 2.9, that portion of the objective lens field on the illuminating side of the specimen (the pre-field) also acts as a weak condenser lens. The total demagnification of the source at the specimen is

$$M = M_1 M_2 M_3$$
$$= \left(\frac{V_1}{U_1}\right)\left(\frac{V_2}{U_2}\right) M_3$$
$$\approx M_3 \left(\frac{f(C1)D3}{D1 D2}\right)$$

where M_1, M_2, and M_3 are the magnifications of C1, C2, and the objective pre-field, respectively. A typical value of M_3 is $1/3$. The minimum value of $f(C1)$ is the minimum projector focal length discussed in Chapter 2, and may be about 2 mm. If the condition of focused illumination is maintained, the size of the illumination spot on the specimen can be reduced from a large value by moving the image S′ towards S, making V_1 smaller and U_2 larger.

A simple method of measuring the source size is to image S using lens C2 alone with lens C1 switched off. The magnification $M_2 = D3/(D1 + D2)$ is then about unity and can be obtained from measurements taken from the microscope column. If the pre-field demagnification is not known, a shortened specimen holder should be used on top-entry stages to bring the specimen out of the lens field.

If the aperture CA2 is incoherently filled (as it is when using a hair-pin filament), the field of view or illuminated specimen region is limited by the size of source S and the lens demagnification, while the coherence at the specimen is solely determined by the semi-angle θ_c. This method of focused illumination is known in light optics as critical illumination. Another method, where the image S′ is placed in the focal plane of C2 ($U_2 = f(C2)$) is also possible and is known as Köhler illumination in light optics. Here the coherence is limited by the source size, while the field of view depends on the size of the aperture which limits lens C1. On many electron microscopes the pole-piece of C1 forms this aperture.

Finally, the use of unfocused illumination must be considered. This is treated in most of the electron optics texts mentioned in Chapter 2 (e.g. Grivet 1965), where it is shown that the beam divergence θ_i (see Fig. 9.1) for unfocused illumination is given by the semi-angle subtended by the source image at the specimen if this angle is less than the angle subtended by the illumination aperture. Otherwise, the beam divergence θ_c is given by the angle subtended by the illumination aperture (see Fig. 9.1).

9.2. Brightness measurement

The electron current I passing through area A is proportional both to A and to the solid angle subtended by any filled illuminating aperture at the source. The constant of proportionality β is known as the beam brightness and is measured in amperes per square centimetre per steradian. The brightness is strictly defined for vanishingly small areas and angles as

$$\beta = I/\pi A \theta^2 \tag{9.1}$$

for an aperture of semi-angle θ. The effect of a lens is to reduce the current density (proportional to M^2) and increase the angular aperture (proportional to $1/M^2$) for a magnification M, leaving the brightness constant at conjugate planes in the microscope if aberrations are neglected. Values of β for four modern sources are given in Table 9.1. For a hair-pin filament with $\beta = 5 \times 10^5$ A cm^{-2} sr^{-1}, a 400 µm final condenser aperture in the plane of CA2 (Fig. 9.1) gives a current density in the illumination spot focused on the specimen of

$$j_0 = \pi \beta (r/D3)^2 \sim 1 \text{A cm}^{-2}$$

where r is the aperture radius. This is a typical value for high-resolution electron microscopy of radiation-insensitive materials when using a tungsten heated filament. The corresponding image current density on the viewing phosphor at a magnification of 500 K is

$$j_i = j_0/M^2 = 5 \times 10^{-12} \text{A cm}^{-2}$$

which is sufficient to expose the Ilford EM4 emulsion to an optical density of unity after 6 s, assuming that no specimen is present, or pre-field effect. Corresponding figures for a LaB$_6$ filament are $j_0 = 14$ A cm^{-2} and $j_i = 5.6 \times 10^{-11}$ A cm^{-2} with an exposure time of 0.5 s at the same magnification.

The theoretical upper limit to brightness is (Spence et al. 1994)

$$\beta_m = \rho e V_r / \pi k T$$
$$= 3694 \; \rho e V_r / T \text{A cm}^{-2}\text{sr}^{-1} \qquad (9.2)$$

where ρ is the emission current density at the cathode (filament) in A cm^{-2}, T (in K) is the filament temperature, and k is the Boltzmann constant. For thermionic sources the brightness increases with temperature, since ρ increases exponentially with temperature. Note that the brightness is proportional to the relativistically corrected accelerating voltage. Brightness measurements at high voltage have been given by Shimoyama et al. (1972), who found $\beta = 1.6 \times 10^6$ A cm^{-2} sr^{-1} at 500 kV.

Since the early work of Haine and Einstein (1952), many workers have shown that the maximum theoretical brightness could be achieved in practice for suitable combinations of filament height (h), bias voltage (V_b), and filament temperature (T). For $T = 2800$ K, a temperature giving reasonable filament life, eqn (9.2) gives $\beta = 4.4 \times 10^5$ A cm^{-2} sr^{-1} at 100 kV, in rough agreement with measured values ($\rho = 3$ A cm^{-2} at this temperature).

In designing a gun one aims for the highest current efficiency, that is the highest ratio of brightness to total beam current, as well as for high absolute brightness. Current efficiency is increased by making the Wehnelt hole small and h small. A typical beam current for a tungsten hair-pin filament is 150 µA, while many excellent high-resolution lattice images have been taken using pointed filaments with a beam current as low as 1 µA (Hibi and Takahashi 1971).

In order to compare electron sources or determine the optimum conditions for operating a source, it is necessary to be able to measure the source brightness. This may be done by the following method, which is useful for determining the best height for a thermal filament. A direct-reading image current electrometer is needed, connected to a small electron collector in the plane of the viewing phosphor. The solid angle, $\Omega \approx \pi\theta^2$, subtended by the second condenser aperture at the specimen must be measured from the microscope. The electron-optical magnification M must be known and the current I passing through a small circular area of radius r at the centre of the focused illumination spot should be noted. The brightness is then approximately

$$\beta = M^2 I / \pi^2 \theta^2 r^2$$

The following points are important:

1. The procedure should be repeated for successively smaller condenser apertures until the value of β obtained does not change with aperture size. This ensures that the aperture is filled with radiation for the measurement.
2. The current measuring area must be small compared with the focused spot size. If the full spot size is used the result will be a small fraction of the true brightness.
3. A correction for electron backscattering from the collection screen may be needed. Alternatively, a specially shaped collector with low backscattering coefficient can be used.
4. A specimen such as a holey carbon film must be used to define the object plane. Bear in mind that the screen intensity (not β) will depend on the strength of the objective pre-field, that is, on the specimen height and objective focus (Section 2.9). To ensure that the condenser aperture actually limits the incident beam divergence, it will be necessary to work with the specimen outside the objective lens field and a correspondingly weakly excited objective. Alternatively, with the object plane immersed within the lens field, the angle θ_c can be found from the size of diffraction spots as described in Section 10.7.

An example of a theoretical calculation of the brightness of a triode electron gun can be found in Kamminga (1972).

9.3. Biasing and high-voltage stability for thermal sources

A common problem in recording images at high resolution when using a tungsten thermal emitter is high-voltage fluctuation during the exposure. Tests for high-voltage stability are described in Section 11.2. On modern instruments the high voltage is stabilized against fluctuations by the negative feedback circuit shown in Fig. 9.2. The Wehnelt is biased a few hundred volts negative with respect to the filament and tends to reduce the electron beam current. The circuit is most easily understood by analogy with the thermionic triode, and can be redrawn as shown in Fig. 9.3. While the anode of an electron microscope is kept at ground potential for convenience, point A of the triode circuit is usually grounded. Figure 9.4 shows the relationship between grid (Wehnelt) voltage and plate (beam) current obtained by experiment. The presence of a condensed oil vapour of evaporated tungsten on the gun chamber walls will encourage microdischarge in the gun. If the discharge is small, the resulting increase in beam current will increase the bias voltage developed across R_b, which in turn, from Fig. 9.4, will tend to oppose the initial increase in beam current, the result being a stabilization in high voltage and electron wavelength.

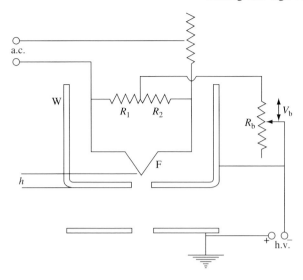

Figure 9.2 The self-biasing gun. The alternating current filament heating supply is connected to a.c. (a direct current supply is sometimes used). Balancing resistors R_1 and R_2, which keep the filament tip at a constant potential, are also shown. A bias resistor R_b is shown with bias voltage V_b. The anode is labelled A, the Wehnelt W, and the filament F. The feedback circuit shown provides improved current and temperature stability over a fixed-bias arrangement.

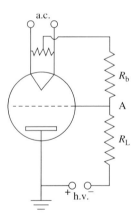

Figure 9.3 Equivalent triode circuit for an electron gun. In the conventional cathode biased triode amplifier, point A, rather than the anode, would be approximately at ground potential. R_L is an equivalent load resistor. The similarity with Fig. 9.2 should be noted.

The three triode constants can be defined and measured for an electron gun (Munakata and Watanabe 1962), and the stabilization analysed by applying Maxwell's loop equation to the small signal triode equivalent circuit for Fig. 9.3. This treatment shows that the stabilization depends both on the amplification factor of the equivalent triode and on the value of the bias resistor R_b.

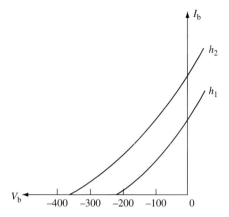

Figure 9.4 Approximate total beam current I_b as a function of gun bias for two different filament heights h_1 and h_2. Here h_2 is greater than h_1. The bias required for optimum conditions (high brightness) is less for the h_1 case than the h_2 case.

In summary, the effects of an increase in R_b are as follows:

1. Reduced total beam current as shown in Fig. 9.4. The bias required for cut-off depends on the filament height h, becoming larger as h is reduced.
2. Improved high-voltage stability.
3. Reduced chromatic aberration. Figure 9.5 shows the measured energy distributions for three field emitters. For a tungsten filament the FWHM varies between 0.7 and 2.4 eV with gun bias setting.
4. By treating the gun as an electrostatic lens the size and position of the virtual electron source can be found (Lauer 1968). This is shown in Fig. 9.6—its size varies only slightly with gun bias, while its position moves toward the anode with decreasing bias.
5. For a particular filament height and temperature there is a weakly defined maximum in brightness as a function of bias for self-biased guns.

The bias resistor R_b (several tens of megaohms) can be adjusted by the operator. On many microscopes a clockwise rotation of the bias control (turning it 'up') actually decreases the bias, resulting in an increased beam current.

An important feature of the self-biasing system is that the bias is controlled by the filament temperature, since this limits the beam current. Electrons are emitted only within the circular zero equipotential on the hair-pin cap; the size of this area grows smaller with increasing bias, vanishing to a point at the filament tip at cut-off bias. This explains the changing appearance of the focused illumination spot as the filament current is increased, increasing both the filament temperature and gun bias. The spot contracts as electrons are drawn from a progressively smaller region of the filament tip. A full treatment of self-biased guns can be found in Haine *et al.* (1958).

The bias setting may have an important bearing on the quality of high-resolution images of the thinnest specimens, where energy losses in the specimen are less important than the

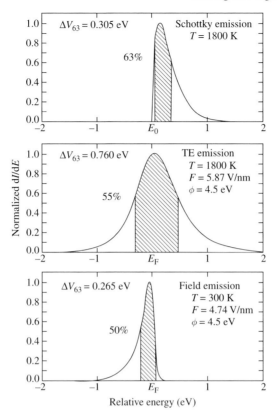

Figure 9.5 Electron emission energy distributions for a ZrO/W(100) Schottky emission tip, an uncoated thermal field emission tip (TE; W(100)), and a cold W(100) tip. Most commercial field-emission microscopes use the heated Schottky mode, with $\Delta V_{63} = 0.8$ eV (Tuggle and Swanson 1985).

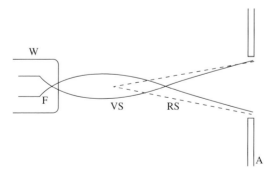

Figure 9.6 Formation of a virtual source (VS) in the region between the Wehnelt (W) and the anode (A). A real crossover is formed at RS.

energy spread of the beam leaving the filament. In Section 4.2 and Appendix 3 the effect of gun bias on image quality is also assessed (see, in particular, Fig. 4.3).

9.4. Hair-pin tungsten filaments

Most manufacturers supply pre-centred, annealed filaments at a pre-set height with replacement instructions. The thickness and length of the tungsten wire have been chosen with care since these determine the load applied to the filament supply. The filament life is limited by the effects of gas attack (mainly water vapour) and thermal evaporation. With a vacuum poorer than 10^{-5} Torr the first failure mechanism is more likely. On many modern instruments filament lives well in excess of 50 h are common. An hour meter actuated by the high-voltage relay can be used to measure the filament life, and provides an inexpensive monitor for the microscope vacuum. As small leaks develop in the microscope column, the filament life rapidly deteriorates. The two failure mechanisms can be distinguished as follows. Most filaments fail when a break develops near the top of either arm where the temperature is greatest. If the wire has retained its original thickness with a blunt, rounded appearance in the region of the break, the likely cause of failure is evaporation. Gas attack is characterized by a narrower tapering toward the break.

The brightness and coherence properties of hair-pin filaments are indicated in Table 9.1. The intensity of the final image depends on a large number of factors apart from the source brightness. Three of these, often overlooked, are the age of the viewing phosphor (which should be replaced at least once a year for high-resolution work), the correction of condenser lens astigmatism, and the extent of the objective lens pre-field used. A more intense final image is obtained as the specimen is immersed deeper into the objective lens field (Section 2.9). These filaments are marginally adequate for high-resolution lattice imaging of radiation-insensitive materials; their chief disadvantage is the large specimen area which must be illuminated, leading to unnecessary heating and specimen movement and lack of brightness.

9.5. Lanthanum hexaboride sources

These sources offer an increase in brightness of about 10 times over hair-pin filaments (see Table 9.1). They have the great advantage of operating at moderate vacuum conditions (better than 10^{-6} Torr), which can be obtained in most microscopes without modifications. A marginal gun vacuum can often be improved by attending to the gun vacuum seal. Other leaks are easily traced with a partial pressure gauge or mass spectrometer (see Section 12.8). Details of one experimental system are given in Ahmed and Broers (1972), in which a rod of LaB$_6$ with a ground tip of radius about 10 µm is heated indirectly with a tungsten coil. Another system, available commercially uses a thin strip of carbon containing a slot to support and heat the highly reactive LaB$_6$ emitter. This carbon strip is supported on the pins of a conventional hair-pin base, so that the filaments can be used as a plug-in replacement for hair-pin filaments. Their heating requirements are similar to those of a hair-pin filament (about 5 W). These sources are more expensive than hair-pin filaments but have a longer life so that the cost per hour may be comparable. To prevent oxidation of the tip, a minimum cooling time of 10 s must be allowed before admitting air to the gun after turning down the heating current. Since these filaments operate at a lower temperature than tungsten filaments, electrons are emitted within a narrower range of energy (Batson et al. 1976) so that high-resolution images recorded using these sources will show reduced chromatic aberration (see Sections 2.8.2 and 4.2).

9.6. Field-emission sources

The article by Crewe (1971), the inventor of the field-emission electron gun, provides a useful introduction to the subject of field-emission sources. The emitter usually consists either of a heated or an unheated single crystal of tungsten. A high electrostatic field (about 4×10^7 V cm^{-1}) in the region of the tip lowers the potential barrier at the tungsten surface, allowing electrons to tunnel into vacuum states. (In the scanning tunnelling microscope, they tunnel similarly from the tip into states of the sample above the Fermi level.) The (111) or (310) tip orientations are often used. The theory of field emission is well reviewed by Dyke and Dolan (1956). The high brightness of these sources (see Table 9.1), the small illumination spot possible at the specimen, the small energy spread, and the high spatial coherence of the radiation makes this source the best currently available for high-resolution electron microscopy in both the TEM and STEM. Fresnel fringes obtained with such a source and demonstrating this high degree of coherence, are shown in Fig. 10.6 (see also Section 10.7). Since the coherence width in the plane of the final condenser aperture may be comparable with or greater than the size of this aperture, the coherence width at the specimen cannot be found using the simple formula (4.11), since the illumination is then fully coherent. The large coherence width (up to 100 nm) obtainable allows the largest and smallest spatial periods to contribute to phase-contrast images and so can be expected to improve their quality (see Chapter 4).

Resolution is improved when using a field-emission source, for two reasons. First, the final illumination aperture becomes coherently filled, so that the complex amplitudes of the final images must be added for each illumination direction rather than their intensities The transfer theory of Section 4.2 must then be modified accordingly (Humphrey and Spence 1981, O'Keefe et al. 1983), resulting in improved resolution for the same beam divergence, as shown in eqn (5.53a). Second, from eqn (4.11) we see that the information limit goes inversely as the square root of the source energy spread Δ, which is reduced in field-emission guns. For heated (Schottky) field emitters, measurements give a FWHM of about 0.8–1 eV, and 0.3 eV for cold field emission. More usefully, the value of Δ needed to match experimental diffractograms at 300 kV, including both the source energy spread from a heated field emitter and high-voltage fluctuations, has been found to be 0.37 eV (O'Keefe et al. 2001). The virtual source size for a heated emitter is estimated to be 20 nm, or 5 nm for a cold field emitter. The source size and brightness for nanotip cold field emitters is measured and discussed in Qian et al. (1993), and the relationship between coherence, brightness, and beam degeneracy is further discussed in Spence et al. (1994). A comparison of the brightness, coherence, and degeneracy of synchrotrons, lasers, and field emitters is given in Spence and Howells (2002). Unlike X-ray sources, the number of electrons which can occupy a single cell in phase space (the degeneracy) is limited to two by the Pauli exclusion principle; this sets an upper quantum limit on brightness for electron sources. Current field emitters provide several orders of magnitude less than this limit, due to the limitations imposed by Coulomb interactions and the Boersch effect, but there is much current research aimed at producing brighter electron sources, some based on ionization of laser-trapped atoms. Although the brightness of field emitters is currently greater than that of storage ring synchrotron X-ray sources, their degeneracy is less (about 10^{-5} for a field emitter, about 10 for an undulator and about 10^{12} for an optical laser).

The possibility of illuminating only a very small area (about 20 nm) with a field-emission source enables a modified focusing technique to be used with important advantages for radiation-sensitive materials. A structurally unimportant area of the specimen can be illuminated and used for focusing at high magnification, while an adjacent area, unaffected by radiation damage during focusing and at the same height and focus condition, may be subsequently illuminated and used for the final image recording. The dose for the second area can be accurately controlled (independently of the coherence condition) and no refocusing is necessary. These are important advantages for high-resolution cryo-EM, where the introduction of field-emission sources has been found generally to result in much higher contrast images of proteins in ice.

The environmental requirements for a field-emission source are more severe than for other sources. In particular, magnetic shielding in the region of the gun and specimen is important to reduce stray a.c. magnetic fields. This is discussed in Chapter 10. The tips must be 'flashed' before use, that is, heated briefly to remove contaminants and smooth the tip profile. For a field-emission gun the 'Wehnelt' is at positive potential with respect to the tip, but nevertheless can be used to stabilize the beam current. The tips can be fabricated by electrolytic etching using sodium hydroxide in a similar way to that used for making thermionic pointed filaments.

The most important practical problem in producing commercial field-emission sources has been the control of emission stability. Before purchasing a field-emission microscope it is most important to examine the focused illumination spot at maximum imaging magnification to check for a stable intensity distribution. Methods for making field-emitter tips are given in Dyke et al. (1953) and Melmed (1991).

For STEM imaging at high resolution, a field emission gun is essential (Crewe 1971). Three main types have been developed—the 'cold' (room temperature) emitter (FE), the heated thermal-field-emission type (TF), and the Schottky emitter (SE). The FE and TF modes operate by tunnelling through a potential barrier near the Fermi energy, established by an extraction potential, while the SE type relies on a coating to reduce the work function of thermal emission over the barrier established by the vacuum level. Most commercially available systems now use a coated TF system. The energy spread and properties of these emitters are given in Table 9.1 and Fig. 9.5. More details can be found in Tuggle and Swanson (1985) and references therein. Electron monochromators have been in development since the 1930s; those currently available for commercial TEM instruments can provide about 1 nA of current into a bandwidth of 90 meV, and a monochromator is now available for the Nion cold field-emission STEM instrument.

9.7. The charged-coupled device detector

Detectors are characterized by their modulation transfer function (MTF), DQE, gain, dynamic range, and noise performance. Resolution is defined by the width of the point spread function (PSF), which is the image of an object much smaller than one detector pixel (referred to the object space). The MTF is then the modulus of the Fourier transform of the PSF, and so would be constant for all spatial frequencies for an ideal delta-function PSF. The Nyquist spatial frequency, for a pixel spacing x_p, is half the pixel spatial frequency (equal to $0.5/x_p$, in cycles mm^{-1}), so that, according to Shannon's theorem, two pixels are needed to sample each periodicity in intensity variation at the detector in order to avoid aliasing. DQE is defined in Section 6.7, and measures the noise introduced by the recording medium (a perfect detector has DQE = 1). Perhaps the earliest development of a CCD camera for direct electron imaging in TEM was described by Spence and Zuo (1988)—this used a liquid nitrogen-cooled CCD camera with fibre-optic coupling to a yttrium aluminium garnet (YAG) single crystal. These 'slow-scan' cameras, with much larger dynamic range than video systems, made rapid progress possible in electron tomography. Detailed analysis of these systems can be found in Daberkow et al. (1991), Krivanek et al. (1991), de Ruijter and Weiss (1992), Ishizuka (1993), and Zuo (1996). Reviews by Fan and Ellisman (2000), Faruqui and Henderson (2007), and Mooney (2007) also provide excellent overviews of digital imaging in TEM using CCDs.

An ideal detector would record the arrival of every beam electron (DQE = 1) into independent pixels (without blooming or 'cross-talk' between pixels). All real detectors add detector noise to this intrinsic shot or quantum noise of the electron beam, thereby lowering the DQE. DQE can then be improved by increasing the gain (the number of output pulses per beam electron), at the expense of dynamic range. DQE is proportional to the area under the MTF.

Because of its capacity for on-line image processing, real-time display of the transfer function of the microscope, and automated electron-optical alignment, the CCD system, which we discuss first, has become the most popular device for quantitative imaging of inorganic samples. The image plate, however, has advantages for holography (more pixels), radiation-sensitive materials (better DQE, more pixels), CBED (more pixels, better linearity), and for the recording of diffuse scattering in diffraction patterns (less blooming near Bragg peaks). The pixel-to-pixel registry between successive CCD images is a unique advantage of that system, allowing simple image algebra, as is the ability to view images immediately at the microscope, working until a desired result is obtained. CCD cameras consist of either a phosphor or a thin single-crystal YAG scintillator (Autrata et al. 1978, Chapman et al. 1989) bonded to a fibre-optics vacuum window face plate, with a Peltier- and water-cooled optical CCD detector operating at about -40 °C pressed against the window on the ambient pressure side. Alternatively, the scintillator may be coupled to the CCD by a lens, at inferior numerical aperture. The YAG scintillator may be coated on the electron-beam entrance side, both to provide electrical conductivity and to reduce optical reflections inside the YAG crystal. Detectors are available in sizes up to $10^4 \times 10^4$ pixels, with read-out speeds of 10 frames per second or more, using 18 µm pixels. Dynamic range increases with well depth and pixel size.

The CCD signal contains many artefacts, including read-out noise, dark current, and non-uniform gain variations across the image field. Most image-processing software will therefore treat CCD images initially as follows. A dark current image D is recorded with the shutter closed for the same time as the exposure of interest I_0. A flat-field or gain image F is also recorded with the shutter open using uniform illumination (no sample), and an exposure time which gives about the same average count as that expected in the final image. This image records variations in gain across the image field, due, for example, to 'chicken wire' patterns in the fibre-optics, or spatial variations in the doping levels in the YAG. A final image

$$I = (I_0 - D)/(F - D)$$

is then formed to be used for subsequent analysis, such as deconvolution and comparison with image simulations.

Any quantitative comparison of simulated and experimental images first requires a careful characterization of the detector system. This involves measurement of the gain, g, resolution (MTF), and DQE of the CCD system. (Linearity and uniformity may also be important.) This analysis is most important; the goodness-of-fit index obtained, for example, will depend sensitively on the way in which the PSF of the CCD detector is 'deconvoluted'. A typical CCD camera might contain 1024×1024 square pixels, each 25 μm on a side, with 14- or 16-bit capacity. The CCD characteristics are most simply measured using the noise method (Dainty and Shaw 1974, Zuo 1996), in which the shot noise from the electron beam is used to provide an equally weighted spectrum of spatial frequencies.

The gain, g, is defined as the ratio of the average number of CCD output counts I to the electron dose per pixel, N_e

$$g = \frac{\bar{I}}{N_e}$$

Its measurement therefore requires an absolute measurement of beam current, using, for example, a Faraday cup or the known response of image plates, which must be converted to number of beam electrons per pixel to obtain N_e. $\bar{I}$ is given by the output of the software used to control the CCD camera. A typical measured gain value for a YAG/CCD combination is 1.55—much higher values may be obtained if a fine-grained phosphor such as P43 is used; however, a high gain is not necessarily advantageous, since it results in loss of dynamic range. The gain of a CCD camera has been measured as a function of beam energy between 60 and 400 keV, where it is found to vary between 0.4 and 0.27, with a maximum of 0.6 at about 100 keV. Gain can be optimized by choice of YAG thickness at the expense of spatial resolution.

The resolution of the slow-scan CCD camera is described by its PSF, defined as the image of an object which is much smaller than one detector pixel. Unlike film and image plate detectors, the YAG/CCD combination spreads the light generated by a single beam-electron over several CCD pixels, so that the PSF has extended tails, and this effect must be removed by image processing. (This can be done with much noise amplification because this pixel 'cross-talk' affects mainly low, rather than high, spatial frequencies.) The MTF is defined as the modulus of the Fourier transform of the PSF, and would be constant (equal to c below) for all spatial frequencies for an ideal delta-function PSF. The MTF has been modelled for early Gatan CCD cameras by the function (Zuo 1996)

$$M(\omega) = \frac{a}{1 + \alpha\omega^2} + \frac{b}{1 + \beta\omega^2} + c$$

Here ω is spatial frequency in units of pixel^{-1} (with a maximum value of 0.5 pixel^{-1}), and the remaining quantities are fitting constants (e.g. $a = 0.40$, $b = 0.46$, $c = 0.15$, $\alpha = 3328$, $\beta = 13.51$). The first two terms model the head and tails (due to reflected photons in the YAG) of the PSF, while c represents an ideal response. At medium and high dose the MTF is found to be independent of dose; at low dose, read-out noise and dark current may dominate. The MTF can be measured by taking the Fourier transform of an image of a uniformly illuminated field, averaged over many exposures. The papers by Zuo (1996, 2000) provide practical details. Figure 9.7 compares the MTF of a CCD with a YAG single-crystal scintillator with that of the Fuji image plate system.

The DQE is defined as the square of the ratio of the output signal-to-noise ratio to the input signal-to-noise ratio. Three unwanted sources of noise are important—Poisson-distributed dark current noise and conversion noise, and Gaussian read-out noise, in addition to the irreducible quantum noise of the electron beam itself (shot noise). A DQE of unity is only possible if every beam electron is detected, with no other noise added to the shot noise, and if one has in addition an ideal PSF. Read-out and dark current noise are independent of the beam intensity. DQE can be shown (Ishizuka 1993) to be given by

$$\mathrm{DQE}(I) = \frac{mg\bar{I}}{\mathrm{var}(I)}$$

where m is a mixing factor measuring cross-talk between pixels (obtainable from the area under the MTF), g is the gain, $\bar{I}$ is the average number of CCD counts per pixel, and $\mathrm{var}(I)$ is the variance in that number. These quantities are obtained by recording a uniformly illuminated field at different average intensities. By applying the rules for the manipulation of variance to the processes of scintillation and subsequent electron–hole pair production in the CCD (Fano noise), an expression can be obtained for the DQE as a function of dose, including all sources of noise (Zuo 2000). Figure 9.8(a) shows the measured DQE for a YAG/CCD system at several voltages, compared with a simple theory which does not take account of linear noise. Figure 9.8(b) shows the measured DQE for the Fuji image plate system.

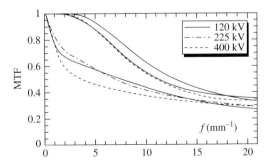

Figure 9.7 MTF of Gatan model 679 SSC CCD camera with YAG scintillator (lower three curves) at different beam energies, compared with Fuji the image plate system (upper curves). (From Zuo (2000).)

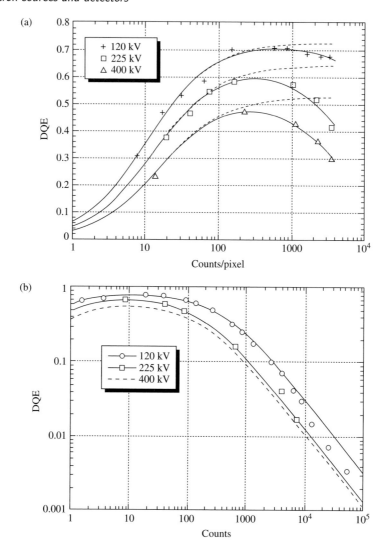

Figure 9.8 (a) DQE of Gatan 679 CCD camera with YAG at three beam energies. Dotted lines show theoretical results without linear noise. (b) DQE of a Fuji image plate system for comparison. (From Zuo (2000).)

The dynamic range of a CCD camera is usually defined as $R = 20\log(I_{max}/\sigma)$ in decibels, where σ is the detector read-out noise. In practice, lower figures will be obtained due to the effect of the PSF. The linearity (gain versus dose) of a Gatan model 679 CCD YAG camera with antireflection (AR) coating has been measured and found to be highly linear over its full 12-bit range (Zuo 2000).

The design and performance of an optimized CCD-based slow-scan cooled-area detector for 100 kV electrons with 16-bit dynamic range and 2000 × 2000 pixels, using a P43 phosphor and bakeable fibre-optic coupling is described in Hembree *et al.* (2003). This is suitable for uncompromising recording of nanodiffraction patterns from STEM instruments like the Nion machine, where high baking temperatures are used to minimize contamination.

Having obtained an accurate measurement of the MTF, this must then be used to remove the spreading effect of the PSF from HREM images recorded using a CCD camera. A number of methods have been evaluated, including the simple Fourier deconvolution routine provided with commercial CCD camera software. For the particular type of noise present in CCD images, the best results have been obtained using the Richardson–Lucy maximum-likelihood algorithm, as used for the analysis of images from the Hubble telescope. Details are given in Zuo (2000). An alternative approach to deconvolution, with its attendant noise amplification problems, is to convolute simulated images with the measured PSF before comparing them with experimental images.

A comparison of the strengths and weaknesses of film, CCD, and image plate detectors can be found in Zuo (2000), where the gain, linearity, resolution, dynamic range, and DQE of these systems are compared. In summary, the image plate system offers more pixels, similar or better dynamic range, excellent linearity, and better DQE at low dose than the CCD camera, but lacks the capacity for on-line image analysis. Image plates also avoid the undesirable 'blooming' feature associated with the CCD, which must be removed by image deconvolution. CCDs and image plates have similar gain.

9.8. Image plates

Image plates are flexible, reusable electronic film sheets which fit the standard TEM film cassette. They consist of a photostimulable phosphor which stores a latent image in trapped F-centres (for a couple of days at room temperature) with a large dynamic range and at least 3000×3700 pixels per sheet. At 120 keV, each beam electron creates about 2000 charged F-centres, which during later read-out, generates about 650 detected photons. Unlike film they are highly linear, reusable, and may be exposed to room light before electron exposure. The pixel size (about 19 μm) is determined by the laser-scanning read-out device, such as the Fuji FDL5000. The sheets are relatively inexpensive; thus a single reader servicing several TEMs provides a system for quantitative analysis at much lower cost than several CCD cameras. The dynamic range (also limited by the partially destructive read-out device) can be extended beyond 14 bits by reading out the film more than once. The plates are highly linear up to a saturation of about 10^5 beam electrons pixel^{-1}, with a gain of about 0.85 at 120 keV (Zuo *et al.* 1996, Zuo 2000). The PSF is determined both by the plates and the spot size of the laser reader—the MTF is shown in Fig. 9.7, and changes little at beam energies up to 400 keV. Unlike CCD cameras, noise is dominated by quantum noise in the beam, and the PSF lacks the undesirable tail due to light scattering in the YAG crystal. It is therefore ideal for the measurement of weak diffuse scattering beside a strong Bragg peak in diffraction patterns. The reduced 'blooming' of the image plate compared with the CCD means that HREM images recorded on image plates have a granular appearance, due to fluctuations in ionization along different electron paths and the large pixel size compared with film. The DQE of the image plate system is shown in Fig. 9.8(b)—this is seen to be much higher than that of a CCD camera at low dose. CBED patterns and diffuse scattering recorded on image plates lack the blooming effect of CCDs, showing greatly improved fine detail, resulting in a better match with calculations (Zuo 2000). These properties mean that IP are also well suited for image recording with radiation-sensitive materials or where large numbers of pixels are needed (e.g. electron holography).

9.9. Film

For cryo-EM an image detector is required with a very large number of pixels (to ensure a sufficient number of pixels per particle in each many-particle image) with modest dynamic range (since the images have low contrast). Film is well suited to these requirements, and is also an ideal archival storage medium, offering protection against the loss of digital information, the cost of powered digital storage, and changes in digital media and formats.

The response of film to light and high-energy electrons is quite different (Valentine 1966). While the cooperative action of many photons is needed to render the micron-sized grains of film developable, an interaction with only a single kilovolt electron is sufficient (Fig. 6.16). After development and fixing, read-out in transmission transmits a fraction T of the read-out light intensity. The optical density is defined as $D = \log_{10}(1/T)$, and this is related to the original electron dose E by

$$D = D_\text{s}(1 - \exp(-\alpha E)) + D_0 \approx \beta E + D_0$$

where D_s is the saturation density ($= 7.8$), D_0 ($= 0.1$) the background (fog), and $\beta = \alpha D_\text{s}$ is the speed of the film (0.41). (Figures in parentheses are for Kodak SO-163. Ilford Ortho Plus film is also used.) Speed is partly determined by the development conditions—faster film has a larger grain size. Thus film is linear up to an optical density of about $D_\text{s}/4$ and is used in cryo-EM around an optical density of about unity, at a magnification about 50 000 and incident electron dose of perhaps 20 electrons Å^{-2}. The dynamic range may be estimated as 30 dB for 6 µm pixels, or 45 dB for the 24 µm pixels used by CCDs (Zuo 2000). The MTF of Kodak SO-163 film is shown in Fig. 9.9. We see that film has much smaller pixels than either CCDs or image plates, resulting in a total number of pixels several times greater than even the image plate. The DQE of film is limited by granular noise (Zeitler

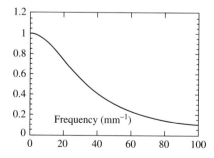

Figure 9.9 MTF of Kodak SO-163 film, showing fine pixel size possible with film. (From Zuo (2000).)

1992), and rises from 0.4 to about 0.75 at a density of about unity (Kodak SO-163) with a gradual fall-off at higher density. Because of the inconvenience of chemical processing and the general decline of the film-based photographic industry (leading to shortage of supplies) it is to be expected that direct-detection cameras, despite their very high cost, will soon replace film recording in most laboratories for high-resolution imaging in structural biology.

9.10. Direct detection cameras

We can think of solid-state detectors as operating a rolling shutter, which continuously adds a few hundred low-dose images per second into computer memory. They operate by recording the charge due to electrons or holes generated as the electron beam traverses a thin, biased, silicon membrane. In this way the resolution loss and pixel cross-talk due to light scattering in the scintillator and fibre-optic interfaces of a CCD/YAG system is eliminated and near noiseless image recording is obtained. However, radiation damage and consequent limitations on the lifetime of the device become a concern, depending on the cost of the active element. Radiation-hard devices have therefore been developed, and tested in TEMs. Two main types have been developed, the Medipix device and the monolithic active pixel device (MAPS) based on CMOS technology (Denes et al. 2007, Faruqi and McMullan 2011). For the Medipix chip, each pixel has its own read-out electronics (amplifier, discriminator, counter) capable of accepting count rates of up to 1 MHz. Thus the pixels in the Medipix count independently, whereas the MAPS data must be processed frame by frame. This detector shows noiseless read-out, high dynamic range, and a DQE at zero spatial frequency and low dose rates approaching unity for beam energies below about 120 kV, but falling off beyond that energy (McMullan et al. 2007). In typical use for a radiation-sensitive sample, the total dose from a sum of 100 frames was 160 electrons pixel^{-1} (4 electrons Å^{-2}), but the rotovirus imaged in ice could be seen clearly after the first frame, at 1.6 electrons pixel^{-1} (0.04 electrons Å^{-2}) due to the lack of a noise contribution.

Figure 9.10 shows detail of a MAPS CMOS device for electron microscopy. Unlike the Medipix, the spatial frequency performance of this detector improves with beam energy above 120 kV. It consists of an array of capacitors formed from reverse bias diodes (labelled N). These are discharged by electrons and holes generated by the passage of a beam-electron through the detecting P epilayer shown. The capacitor voltages and read-out are controlled by n-MOS transistors (shown with wires attached) on p-wells, beside the n-wells diodes (both are formed by heavy doping, which confines the motion of electrons). One example of this type of detector (DE-12 with 3000 × 4000 6-μm pixels) is reported to achieve good contrast at 4/5 of the Nyquist frequency, a considerable improvement over all other detectors (Bammes et al. 2012).

The limited dynamic range of a single recording with these detectors is compensated for by rapid 'continuous' read-out and accumulation of frames in a computer. This rapid read-out also allows for drift correction by sub-frame averaging (movement of the sample during the exposure) within the limits of additional noise due to multiple read-outs. Dynamic range increases as separately read out images are summed. If we consider a sum of M exposures, with a small fixed background count N_b per exposure per pixel, and N_s signal counts per

284 Electron sources and detectors

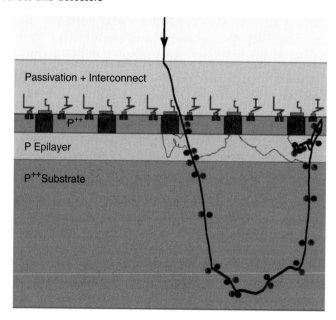

Figure 9.10 Schematic diagram of a single pixel in a MAPS detector. Beam electrons in the P epilayer generate secondary electrons which discharge capacitors in the N layer, next to transistors and interconnects shown. A Monte Carlo simulation for the track of one 300 kV beam electron is shown—the backscattering event which reverses its direction causes it to be detected twice. The occurrence of these double-counting events can be reduced by using a thinner detector layer. (From Faruqi and McMullan (2011).)

exposure per pixel, the signal to noise ratio S, assuming Poisson fluctuations (shot noise) on the beam electron arrivals, is

$$S \sim MN_s/\sqrt{MN_s + MN_b} \sim \sqrt{MN_s}(1 - N_b/2N_s)$$

Hence S increases as the square root of the number of exposures, a law of diminishing returns. Taking $S = 5$, and given N_b and N_s, this then gives the number of frames M needed for statistical significance of the accumulated image. We see that if read-out and electronic noise become large, a limit is set to the maximum dynamic range possible—hence the importance of getting low noise per exposure.

Figure 9.11 shows a comparison of the MTF and DQE for CMOS, CCD, and film detectors at 300 kV. We see that the CMOS detector offers the best performance overall (the Medipix has even better performance, limited to 120 kV), and that film has the best DQE in the mid-spatial-frequency range. For most materials science applications, however, the use of CMOS or Medipix detectors may be unnecessary, given the very high doses used—even at the highest read-out rate, these detectors would saturate in each frame. Monte Carlo simulations have been used to track the energy loss of beam electrons through these detectors in order to improve the design by optimum choice of detector-layer thickness—a

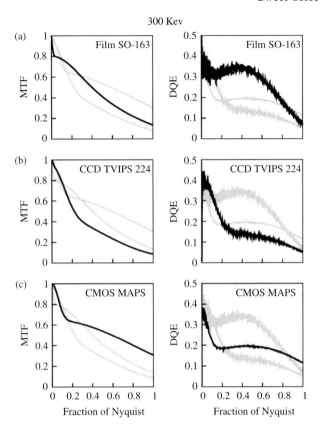

Figure 9.11 Experimental MTF plots (left column) and DQE plots (right column) for the labelled film, CCD, and MAPS detectors at 300 keV. The heavy line in each figure is the detector specified by the label. (From Faruqi and McMullan (2011).)

typical particle track from one beam electron results in between 260 and 1500 analogue-to-digital converter (ADC) values, depending on how much energy is lost in the epilayer.

Microscopists using these detectors must expect very large volumes of data. A typical CMOS detector (Gatan K2 Summit) has a 4000 × 4000 pixel read-out at 400 frames s^{-1} or 83 Gbit s^{-1}. A review of this detector and its use for single-electron counting and sample motion correction can be found in Li et al. (2013). In this work, using sub-frame averaging, the use of this direct-injection detector is shown to extend near atomic-resolution (3.3 Å) single-particle three-dimensional cryo-EM imaging from larger particles with high symmetry to a smaller protein (700 kDa) with low symmetry, showing side-chain density. The DQE is shown to improve on scintillation-based detectors, and beam-induced motion compensated for by summing many brief exposures brought into registry. The total dose was about 31 electrons Å^{-2}, divided into 20 shorter 0.2 s exposures, with 1.2 Å pixels, and a Nyquist cut-off of 2.4 Å. Three modes of operation are possible—automatic integration of frames, single-electron counting, and a mode in which the location of each beam electron is determined with sub-pixel accuracy. In the single-electron counting mode, significant improvements in DQE and MTF are possible. The damage-limited lifetime of the detection element for one such system is expected to be a few years with average use.

For radiation-sensitive materials, the ultimate goal of single-electron detection, rapid read-out, large dynamic range, low noise, and a large number of pixels has now been achieved using these direct-detection devices, with the MAPS device most suitable for higher beam energies. For the reduction of sample drift, however, we note that it has been found that a more rigid and conductive support film is more effective than rapid read-out and accumulation of post-aligned images (Glaeser et al. 2011). The detectors have shown the advantages and flexibility provided by rapid read-out with few electrons per frame (thus extending the life of the detector) for radiation-sensitive materials. The accumulation of many of these images to produce an image with large dynamic range has been made possible by the low-noise performance per frame. Larger and faster detectors of this type can be expected in future.

Solid-state detectors have also found uses for time-resolved imaging in TEM. Beacham et al. (2011) have used a Philips CM20 FEG TEM/STEM fitted with a variant of the Medipix2 chip to obtain useful image exposures down to 500 µs in single exposures, while noisy images can be recorded at 500 ns using the integral mode exposure. At a beam current of 1 nA, one has about 6250 electrons µs^{-1}. For their 256 × 256 pixel detector, an exposure of 11 µs is therefore required to obtain one electron per pixel. The dominant source of noise is the Poisson shot noise of the beam itself, and there is no read-out noise. Improvements in time resolution, to perhaps 20 s, should be possible using the time-of-arrival mode.

References

Ahmed, H. and Broers, A. (1972). Lanthanum hexaboride electron emitter. *J. Appl. Phys.* **43**, 2185.

Autrata, R., Schauer, P., Kvapil, J., and Kvapil, J. (1978). A single crystal of YAG. New fast scintillator in SEM. *J. Phys. E* **11**, 707.

Bammes, B. E., Rochat, R. H., Jakana, J., Chen, D. H., and Wah Chiu, J. (2012). Direct electron detection yields cryo-EM reconstructions at resolutions beyond 3/4 Nyquist frequency. *J. Struct. Biol.* **177**, 589.

Batson, P. E., Chen, C. H., and Silcox, J. (1976). Use of LaB$_6$ for a high quality, small energy spread beam in an electron velocity spectrometer. *Proc. 34th Ann. EMSA Meeting*, p. 534. Claitor's, Baton Rouge, LA.

Beacham, R., Mac Raighne, A., Maneuski, D., O, Shea, V., McVitie, S. and McGrouther, D. (2011). Medipix2/Timepix detector for time-resolved TEM. *13th International Workshop on Radiation Imaging Detectors* [online only] <http://m.iopscience.iop.org/1748-0221/6/12/C12052?rel=sem&relno=5>

Chapman, H., Craven, A. J., and Scott, C. P. (1989). YAG scintillators. *Ultramicroscopy* **28**, 108.

Crewe, A. V. (1971). High intensity electron sources and scanning electron microscopy. *Electron microscopy in materials science*, ed. U. Valdre and A. Zichichi, p. 163. Academic Press, New York.

Daberkow, I., Herrmann, K.-H., Liu, L., and Rau, W. D. (1991). Performance of electron image converters with YAG single crystal screen and CCD sensor. *Ultramicroscopy* **38**, 215.

Dainty, J. C. and Shaw, R. (1974). *Image science*. Academic Press, New York.

Denes, P., Bussata, J., Leeb, Z., and Radmillovic, V. (2007). Active pixel sensors for electron microscopy. *Nucl. Instrum. Methods* **A579**, 891.

Dyke, W. P. and Dolan, W. W. (1956). Field emission. *Adv. Electron. Electron Phys.* **8**, 89.

Dyke, W. P., Trolan, J. K., Dolan, W. W., and Barnes, G. (1953). The field emitter: fabrication, electron microscopy and electric field calculations. *J. Appl. Phys.* **24**, 570.

Fan, G. Y. and Ellisman, M. H. (2000). Digital imaging in transmission electron microscopy. *J. Microsc.* **200**, 1.

Faruqi, A. R. and Henderson, R. (2007) Electronic detection systems for electron microscopy. *Curr. Opinion Struct. Biol.* **17**, 549.

Faruqi, A. R. and McMullan, G. (2011). Electron detectors for electron microscopy. *Q. Rev. Biophys.* **44**, 357.

Glaeser, R. M., McMullan, G., Faruqi, A. R., and Henderson, R. (2011). Images of parafin monolayer crystals with perfect contrast: minimization of beam-induced specimen motion. *Ultramicroscopy* **111**, 90.

Grivet, P. (1965). *Electron optics*. Pergamon, London.

Haine, M. E. and Einstein, P. A. (1952). Characteristics of the hot cathode electron microscope gun. *Br. J. Appl. Phys.* **3**, 40.

Haine, M. E., Einstein, P. A., and Bochards, P. H. (1958). Resistance bias characteristics of the electron microscope gun. *Br. J. Appl. Phys.* **9**, 482.

Hembree, G., Koch, C., Weierstall, U., and Spence, J. (2003). A quantitative nanodiffraction system for UHV STEM. *Microsc. Microanal.* **9**, 468.

Hibi, T. and Takahashi, S. (1971). Relation between coherence of electron beam and contrast of electron image of biological substance. *J. Electron Microsc.* **20**, 17.

Humphrey, C. J. and Spence, J. C. H. (1981). Resolution and coherence in electron microscopy. *Optik* **58**, 125.

Ishizuka, K. (1993). CCD cameras. *Ultramicroscopy* **52**, 7.

Kamminga, W. (1972). Numerical calculation of the effective current density of the source for triode electron guns with spherical cathodes. *Image processing and computer aided design in electron optics*, ed. P. Hawkes, p. 120. Academic Press, New York.

Krivanek, O. L., Mooney, P. E., Fan, G. Y., and Leber, M. L. (1991). Slow scan CCD camera for TEM. *J. Electron Microsc.* **40**, 290.

Lauer, R. (1968). Ein einfaches Modell fur Electronenkanonen mit gekrummter Kathodenoberflache. *Z. Naturforsch.* **23a**, 100.

Li, X., Mooney, P., Zheng, S., Booth, C., Braunfeld, M., Gubbens, S., Agard, D., and Cheng, Y. (2013). Electron counting and beam-induced motion correction enable near-atomic resolution single-particle cryo-em. *Nature Methods*, **10**, 584.

McMullan, G., Cattermole, D., Chen, S., Henderson, R., Llopart, X., Somerfield, C., Tlustos, L., and Faruqi, A. R. (2007). Electron imaging with Medipix2 hybrid pixel detector. *Ultramicroscopy* **107**, 401.

Melmed, A. (1991). Making nanotips. *J. Vac. Sci. Technol.* **B9**, 601.

Mooney, P. (2007). Optimization of image collection for cellular electron microscopy. *Methods Cell Biol.* **79**, 661.

Munakata, C. and Watanabe, H. (1962). A new bias method of an electron gun. *J. Electron Microsc.* **11**, 47.

O'Keefe, M. and Saxton, O. (1983). *Proc. 41st. EMSA Meeting, Phoenix, AZ*, p. 288. Claitor's, Baton Rouge, LA.

O'Keefe, M. A., Nelson, E. C., Wang, Y. C., and Thust, A. (2001). Sub-angstrom resolution of atomistic structures below 0.08 nm. *Phil. Mag.* **B81**, 1861.

Qian, W., Scheinfein, M., and Spence, J. C. H. (1993). Brightness measurements of nanometer sized field-emission sources. *J. Appl. Phys.* **73**, 7041.

de Ruijter, W. J. and Weiss, J. K. (1992). Methods for measuring properties of CCD cameras for TEM. *Rev. Sci. Instrum.* **63**, 4314.

Shimoyama, H., Ohshita, A., Maruse, S., and Minamikawa, Y. (1972). Brightness measurement of a 500 kV electron microscope gun. *J. Electron Microsc.* **21**, 119.

Spence, J. and Howells, M. (2002). Synchrotron and electron field emission sources: a comparison. *Ultramicroscopy* **93**, 213.

Spence, J. C. H. and Zuo, J. M. (1988). Large dynamic range parallel detection charge-coupled area detector for electron microscopy. *Rev. Sci. Instrum.* **59**, 2102.

Spence, J. C. H., Qian, W., and Silverman, M. P. (1994). Electron source brightness and degeneracy from Fresnel fringes in field emission point projection microscopy. *J. Vac. Sci. Technol.* **A12**, 542.

Tuggle, D. W. and Swanson, L. W. (1985). Emission characteristics of the ZrO/W thermal field electron source. *J. Vac. Sci. Technol.* **B3**, 220.

Valentine, R. C. (1966). Response of photographic emulsions to electrons. *Advances in optical and electron microscopy*, Vol. 1, ed. R. Barer and V. E. Cosslet, p. 180. New York, Academic Press.

Zeitler, E. (1992). The photographic emulsion as analog recorder for electrons. *Ultramicroscopy* **46**, 405.

Zuo, J. M. (1996). Electron detection characteristics of slow scan CCD camera. *Ultramicroscopy* **66**, 21.

Zuo, J. M. (2000). Electron detection characteristics of slow-scan CCD camera, imaging plates and film. *Microsc. Res. Techn.* **49**, 245.

Zuo, J. M., McCartney, M., and Spence, J. (1996). Performance of imaging plates for electron recording. *Ultramicroscopy* **66**, 35.

10
Measurement of electron-optical parameters

This chapter describes several simple techniques for measuring the important electron-optical constants which affect the quality of high-resolution images. The methods include the use of digital diffractograms for the measurement of defocus, spherical aberration astigmatism, and 'resolution'. For aberration-corrected instruments, electron interferograms (Ronchigrams or Thon ring diffractograms) are used to provide on-line estimates of aberration coefficients, which can then be corrected to desired values.

10.1. Objective-lens focus increments

On a well-designed instrument the finest objective-lens focus step Δf should be of the same order as the variation in focal plane position due to electronic instabilities. From eqn (2.38) an upper limit is

$$\Delta f = C_c \left(\frac{\Delta V_0}{V_0} + 2 \frac{\Delta I}{I} \right) \tag{10.1}$$

This gives 4.2 nm for a lens with $C_c = 1.4$ mm and $\Delta I/I = \Delta V/V$ 10^{-6}. There are several methods for measuring these steps—three are given in this book, two in this section and one in Section 10.6. The first method uses the displacement with focus of diffraction images (Hall 1949).

1. Prepare MgO crystals by passing either a continuous or holey carbon film through the faint smoke of some burning magnesium ribbon. This can be ignited, with some difficulty, using a Bunsen burner. Examine the crystals in the microscope.
2. MgO has the sodium chloride structure of two displaced face-centred cubic (FCC) lattices. The nanocrystals will appear as small cubes of sizes down to a few tens of nanometres in width. The presence of Pendellösung fringes indicates that the beam runs in the [110] direction of a diagonal of the cube face. The diffraction pattern for the [100] case is shown in Fig. 10.1. Crystals in the [111] orientation appear hexagonal, the projection of a cube along a body diagonal. Identify the reflections in the diffraction pattern of a suitably thin crystal for a crystal near the zone-axis orientation, observed without an objective aperture in place, a set of images of the crystal will be seen (one 'diffraction image' for each strong Bragg reflection) which merge near the Gaussian focus

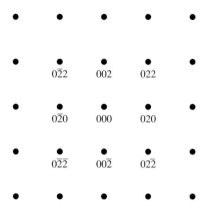

Figure 10.1 The (100) reciprocal lattice plane for MgO. The lattice spacing is 0.21 nm for the 002 reflection. Amongst MgO smoke crystals this pattern is easily recognized from the perfect squares formed in the diffraction pattern.

and separate laterally with increasing defocus. Using the objective aperture to isolate the central beam and one other strong beam, identify the Miller indices of a particular diffraction image. By measuring this lateral image displacement for a known number of focal steps a value of Δf can be obtained.
3. Align the microscope as described in Chapter 12. Align the illuminating system, obtain the current centre, check the cold-fingers for continuity, and correct astigmatism on the carbon background without an objective aperture.
4. Without touching the objective aperture holder, record a double exposure of the selected crystal at two focus values close to the Gaussian focus. The two focus values chosen should fall within the span of a single focus control (say n steps apart) and the magnification should be chosen such that some movement of the diffraction image is observed with this focus change. The full range of the finest focus control may have to be used.
5. The separation of diffraction images is next measured from the images. The electron-optical magnification must be measured using the methods discussed in Section 10.3. For the finest focal steps one is measuring distances of 1–2 nm.
6. If the lateral separation of the images recorded at magnification M is X, corresponding to a defocus change in the object space $n\Delta f$, for n focal steps, then an analysis similar to that shown in Fig. 3.5 (for defocus rather than spherical aberration) gives

$$\Delta f = X/(n\theta M)$$

Here θ is the angle between the central beam and the reflection used, which can be obtained using Bragg's law. At 100 kV, $\theta = 17.55$ mrad for the MgO (002) reflection.

The second method is simpler than the first, although some machine shop work may be required. Make up a specimen holder in which the specimen grid is held at a fixed angle of, say, 45° to the incident beam. Place a thin carbon film in the holder and examine it in the microscope. Select a region near the centre of the grid and record an image at the minimum contrast focus. All points on a line across the specimen at the same height will be imaged at this focus. Advance the finest focus control by several clicks, and, without touching the translation controls, record a second image. Measure the distance between the two lines of minimum contrast at right angles to the lines and divide by the magnification. For a specimen at 45° to the incident beam, the focus change is then equal to the resulting distance.

A third method for measuring focus increments is described in Section 10.6.

10.2. Spherical aberration constant

Figure 2.13 shows the effect of spherical aberration on a ray leaving the specimen at a large angle. For an axial point on a crystalline specimen, a high-order Bragg reflection is such a ray. The resulting displacement between bright- and dark-field images at Gaussian focus can be used to measure C_s (Hall 1949). Alternatively, the in-focus displacement of a bright-field image formed with no objective aperture as the illumination is tilted can be used to give a value for C_s. Equation (2.30), referred to the image plane, gives the image translation as

$$d = MC_s\theta^3 \qquad (10.2)$$

Two bright-field images are recorded successively at exact focus with the central beam aligned with the optic axis and at a small known angle to it. The resulting image translation gives d and hence C_s. The method in detail is as follows.

1. Evaporate a thin layer, less than 20 nm thick, of any low-melting-point FCC metal onto a holey carbon film to form a fine-grained polycrystalline film. With the thinnest films grains will be roughly spherical. The film should be thin enough to leave exposed areas of carbon for astigmatism correction and the grains should be small enough to leave holes between some grains.
2. Examine the film and align the microscope with particular attention to current centring. Use an uncovered area of carbon to correct the astigmatism as discussed in Chapter 12. Next, take a double exposure of a hole in the film with the central beam successively on the optic axis and then tilted, so that an identified diffraction ring falls on the optic axis. The position of the axis should be marked after current centring using the screen pointer. Two tilt conditions can usually be pre-set with the microscope bright- and dark-field channels. Take both images at exact focus, without an objective aperture. A small error in focusing is not important if a high-order diffraction ring is chosen, ensuring that the phase shift due to spherical aberration dominates that due to defocus.
3. To ensure that the central beam has been accurately aligned, repeat step 2 four times with the beam tilted into four orthogonal directions. If the microscope has been accurately aligned, and there is no astigmatism, the four image displacements will be equal. Recordings of the tilted and untilted diffraction patterns (double exposed) are also required to measure θ. This makes a total of eight double exposures. A magnification of 25 000 is adequate for the images.

292 Electron-optical parameters

4. For FCC metals the tilt angle is obtained from

$$\theta = \frac{r}{r_0}\frac{\lambda}{a_0}(h^2 + k^2 + l^2)^{1/2}$$

where a_0 is the cubic unit-cell constant ($a_0 = 0.408$ nm for gold), r_0 is the measured distance between the central beam and the diffraction ring with indices (h, k, l), and r is the distance between the two central beams appearing on the doubly exposed diffraction pattern. A small condenser aperture should be used. Equation (10.2) can then be used to obtain a value for C_s with d the image separation under tilt θ. An indexed ring pattern for FCC metals is given in Fig. 10.2.

Note that the spherical aberration constant C_s depends on the lens focal length according to Fig. 2.12. A shorter focal length reduces spherical aberration. Thus the value of C_s obtained applies only for the particular objective-lens current and specimen height used for the calibration—these should be close to the values found to give images of the highest resolution. A second method of measuring C_s is given in Section 10.6.

Where values of C_s are needed for computer-simulated images, it is important to estimate the sensitivity of the images to C_s. For a simple three-beam image, a black–white contrast reversal occurs for a change ΔC_s given by $\pi \Delta C_s \lambda^3 U_h^4 /2 = \pi$ (see eqn 3.24), that is, values of C_s are required to within an accuracy of $\Delta C_s = 2/\left(\lambda^3 U_h^4\right)$ for a spatial frequency U_h. At high resolution ($U_h > 3.3$ nm^{-1}, $\Delta C_s < 0.3$ mm at 100 kV) this accuracy is difficult to achieve; however, the method described above will be found to be more accurate than the optical diffractogram technique described in Section 10.6, but less accurate than automated

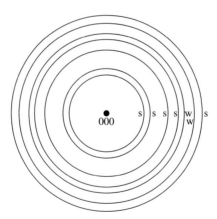

Figure 10.2 The ring diffraction pattern from a polycrystalline face-centred cubic material used for determining the spherical aberration constant of a microscope. To assist in identifying the rings they are labelled w (weak) and s (strong). Their radii have the ratio of the square roots of 3, 4, 8, 11, 12, 16, 19, 20, 24, ... The indices of the reflections shown, from the centre, are (000), (111), (200), (220), (311), (222), (400), (313). A beam stop is useful when recording the diffraction pattern to prevent the central beam from overexposing the detector.

refinement of Ronchigrams (Section 10.10). Note that the error ΔC_s is independent of C_s so that the allowable fractional error is increased by reducing C_s. The major source of error in the above method (accurate to within perhaps 10%) arises from the neglect of focusing effects. An improved method which is independent of Δf has been described by Budinger and Glaeser (1976) (accurate to within perhaps 3%), and that work contains a fuller discussion of errors.

10.3. Magnification calibration

At high resolution, crystal lattice fringes form the most convenient length standard, but an accuracy in magnification calibration of better than 5% is very difficult to achieve. The important sources of error are the accuracy with which the lens may be reset to a particular value and changes in specimen height between the specimen and the calibration specimen. For many periodic specimens, a cell dimension will be known from X-ray work (Wyckoff 1963). A tilted specimen may introduce large height changes and consequent error. From eqn (2.9) the effect of a small change Δf in specimen height at large magnification is

$$\frac{\Delta f}{f_0} = -\frac{\Delta M}{M} \tag{10.3}$$

where ΔM is the magnification change and f_0 is the focal length of the objective lens. Thus, if a change in height equal to the thickness of a grid is allowed on specimen change, the result is a 6% error in magnification for $f_0 = 1.6$ mm.

Partially graphitized carbon is particularly convenient in that it can be dusted on to fragile biological specimens (Heidenreich et al. 1968). Lattice fringes obtained from this specimen in all directions are a sensitive assurance of accurate astigmatism correction, as shown in Fig. 10.3. The spacing is 0.34 nm. The asbestos fibres crocidolite and chrysotile (Yada 1967) are also suitable standards. Using fine tweezers, these can be teased apart and placed onto a grid already containing a specimen. Crocidolite has spacings of 0.903 nm along the fibre direction and 0.448 nm for fringes at 60° to the edge, while the spacings of chrysotile asbestos are 0.45, 0.46, and 0.73 nm for a specimen observed normal to the fibre axis. Chrysotile appears to be more radiation sensitive than crocidolite, which is available commercially.

All these specimens have the advantages that their orientation is simple to determine and they may be added to a prepared specimen. The important point for accurate high-resolution length measurement is that the calibration specimen must be on the same grid and as near to the unknown specimen as possible, and that lens currents are not altered between recordings of these specimens, apart from necessary refocusing. The change in height between the calibration specimen and the unknown specimen is easily measured from the focus change needed to bring each into focus.

Lattice images obtained from any of the test specimens available commercially will provide magnification calibration for a particular specimen holder, stage, pole-piece, magnification setting, and high voltage within about 5%. Where a variable-height stage is used, a correction chart must be drawn up giving the true magnification in terms of indicated magnification and measured objective-lens current needed to focus the specimen at a particular height. From Section 2.3, the magnification change due to variation in focal length

294 *Electron-optical parameters*

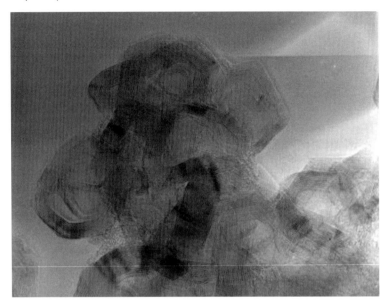

Figure 10.3 Lattice fringes suitable for magnification calibration obtained from partially graphitized carbon. The presence of clear fringes in all directions is an indication of accurate astigmatism correction. The fringe spacing is 0.34 nm. A specimen of partially graphitized carbon can be kept permanently on hand as a preliminary check on the microscope stability and resolution.

of the objective lens is seen to be inversely proportional to the focal length of the lens or proportional to the square of the objective-lens current. Owing to magnetic hysteresis, the field strength of the lens (and hence the magnification) will depend on whether a particular set of lens currents are approached from a high or a lower lens current. Some instruments include an automatic lens-current cycling device which improves the resetting accuracy of magnification to about 3%.

There is an increase in magnification between the viewing phosphor (if fitted) and the electronic detector on most microscopes. The ratio of magnifications on the two planes is

$$\frac{M_1}{M_2} = \frac{v_1}{v_2}$$

where M_1 and M_2 are the magnifications on the phosphor and detector planes, respectively, and v_1 and v_2 are distances between the projector lens pole-piece and the phosphor and detector planes. These distances can be measured approximately from the microscope. At high magnification Section 2.2 indicates that the depth of focus is sufficiently great for all these images to be treated as 'in focus'.

Measurements of fringe spacing should be taken over as large a number of fringes as possible to reduce errors. A list of lattice spacings useful for magnification calibration is given in Table 10.1. The optimum magnification setting for high-resolution microscopy can be obtained by observing the digital diffractogram (Thon rings) from a thin amorphous

Table 10.1 List of useful larger lattice spacings for magnification calibration.

Specimen	Symbol	Lattice (hkl)	Spacing (nm)	Method*
Cu-phthalocyanine	$C_{32}H_{16}N_8Cu$	001	1.26	(a)
Cu-phthalocyanine	$C_{32}H_{16}N_8Cu$	201	0.98	(a)
Crocidolite	—	Along fibre	0.903	(c)
Potassium chloroplatinite	K_2PtCl_4	100	0.694	(a)
Cu-phthalocyanine	$C_{32}H_{16}N_8Cu$	001	0.63	(c)
Potassium chloroplatinate	K_2PtCl_6	111	0.563	(a)
Pyrophyllite	$Al_2O_3 \cdot 4SiO_2 \cdot H_2O$	020	0.457	(a)
Potassium chloroplatinite	K_2PtCl_4	001	0.412	(a)
Molybdenum trioxlate	MoO_3	110	0.381	(a)
Potassium chloroplatinite	K_2PtCl_4	100	0.347	(c)
Gold	Au	111	0.235	(b)

*The letters refer to the imaging conditions shown in Fig. 5.1

carbon sample, and adjusting the magnification until the point-resolution limit is about two-thirds of the Nyquist spatial frequency, for single binning.

10.4. Chromatic aberration constant

A simple technique for the determination of C_c follows from eqn (2.34)

$$\Delta f = C_c \left(\frac{\Delta V_0}{V_0} - 2\frac{\Delta I}{I} \right)$$

If the high voltage V_0 is kept constant and the method described in Section 10.1 is used to measure the defocus Δf for a measured change in objective-lens current, the chromatic aberration constant can be found using

$$C_c = \frac{I \Delta f}{2 \Delta I}$$

where I is the mean value of lens current used and ΔI is the change in lens current for defocus Δf. A measurement of C_c using the variation of focal length with high voltage is also possible, but less convenient. For most high-resolution work, however, it is the quantity Δ defined in eqn (4.9) which is of interest, and this is most easily measured using diffractogram analysis (see Section 10.6). The effects of changes in C_c on image quality are indicated in Fig. 4.3.

10.5. Astigmatic difference: three-fold astigmatism

A technique for correcting astigmatism is discussed in Chapter 12. Occasionally, however, it may be necessary to measure the astigmatic difference C_a which appears in eqn (3.23). When astigmatism is severe, C_a can be obtained from measurements of Fresnel fringe widths. The

296 *Electron-optical parameters*

width of the first Fresnel fringe formed at the edge of, say, a small circular hole in a thin carbon film is very approximately

$$y = (\lambda \Delta f)^{1/2} \tag{10.4}$$

for defocus Δf. From a measurement of y, the focus defect can thus be obtained. The effect of astigmatism is to separate longitudinally by an amount C_a the position of exact focus for edges at right angles. This can be seen from eqn (3.23), which contains in addition to the usual focus term an azimuthally dependent 'focus' describing astigmatism. If the maximum and minimum fringe widths measured in perpendicular directions are $y_{\max}$ and $y_{\min}$, referred to the object plane, then eqn (10.4) gives

$$C_a = \frac{y_{\max}^2 - y_{\min}^2}{\lambda}$$

The following practical considerations are important:

1. The clearest fringes will be obtained using well-defocused illumination ($\theta_c < 1$ mrad), and a correspondingly extended exposure time. From Chapter 4 the coherence condition for observation of fringes of width d is

$$\theta_c < \frac{\lambda}{2\pi d}$$

2. The micrograph should be taken close to Gaussian focus, where the fringe asymmetry is most pronounced. Over-focused fringes (first fringe dark, lens too strong) show more contrast than under-focused fringes (Fukushima *et al.* 1974).

A more sensitive measurement of C_a can be obtained from a diffractogram of a thin carbon film (Section 10.6). Three-fold astigmatism can also be measured by fitting eqn (3.23) to diffractograms (O'Keefe *et al.* 2001).

10.6. Diffractogram measurements

Diffractograms are the most useful and important tool at the disposal of the microscopist for fault diagnosis and for the measurement of instrumental parameters. They were developed by Thon and co-workers (Thon 1971). A diffractogram is the modulus of the Fourier transform of the HREM image of a very thin film of amorphous carbon, as described in Section 3.5. Then the quantity observed is given by eqn (3.41). In high-resolution electron microscopy diffractograms have six main uses:

1. To detect and measure astigmatism.
2. To detect specimen movement during an exposure.
3. To measure the focus defect at which a micrograph was recorded.
4. To measure the microscope's spherical aberration constant.
5. To measure the damping envelope constants Δ and θ_c (see Sections 4.2 and 10.9).
6. For automated alignment and focusing of the microscope (Section 12.5).

Diffractograms have also been used to provide information on the structure of specimens, since they expose the phase of the scattered electron wave. However, the main use of on-line diffractograms is for training microscopists to correct image astigmatism and to find the Scherzer focus. Astigmatism may be difficult to detect in an image, but shows up readily in a diffractogram as an elliptical distortion of the characteristic ring pattern. Equally important is the sensitivity of diffractograms to image movement during an exposure (drift), which may also be difficult to detect in the image. Figure 10.4 shows the effect of astigmatism on a diffractogram, while the effect of drift is indicated in Fig. 10.5. These two effects are clearly distinguished. The diffractogram provides an excellent differential diagnosis of these common image imperfections.

Diffractograms can also be used quantitatively. A simple FORTRAN program is given in Appendix 1 to perform the necessary data analysis. Given the radii of the rings appearing in a diffractogram, this program calculates both the focus setting Δf and the spherical aberration constant C_s using the method of Krivanek (1976). By measuring the ring radii in two orthogonal directions it is also possible to find the coefficient of astigmatism C_a. In order to use the program given in Appendix 1 it is necessary to know the radii of the rings in the optical diffraction pattern at their maximum (or minimum) intensity. These must be supplied in units of $S_n = \theta_n/\lambda$, where θ_n is the electron scattering angle which

Figure 10.4 The diffractogram of an electron micrograph of a thin amorphous carbon film. (On-line digital image analysis software presents these patterns directly at the microscope.) The electron image was recorded with an incorrectly adjusted stigmator and this produced the elliptical distortion of the optical diffraction pattern. Figure A1.1 gives an example of correct stigmator adjustment, resulting in a circular ring pattern. The procedure for adjusting the stigmator is given in Section 12.1. The bar across the pattern is a beam stop used to exclude the central bright spot from the image. Notice the difference between this diffraction pattern (which contains astigmatism but no drift) and that shown in Fig. 10.5, which contains drift with little astigmatism.

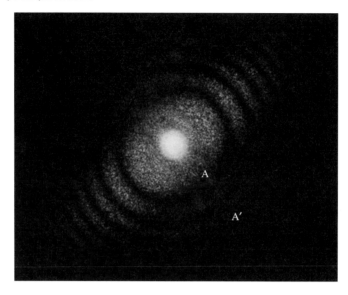

Figure 10.5 The diffractogram of a micrograph of an amorphous carbon film in which the specimen has moved during the exposure. The specimen movement is in the direction AA' at right angles to the direction of maximum ring contrast and results in low contrast along this line.

gives rise to a particular ring n. The quantity S_n is simply related to the radii r_n of the diffractogram rings. Digital image analysis software performs this analysis on-line at the microscope.

A simple method of finding S_n is to include some lattice fringes in the original micrograph. These fringes produce sharp spots in the diffractogram which can be used to calibrate the ring patterns. If the Bragg angle at the original specimen for a spot observed on the optical diffractogram is θ_B, then the S_n for the rings is given by

$$S_n = \frac{2r_n \theta_B}{r_B \lambda_{\text{electron}}} \tag{10.5}$$

where r_B is the distance measured on the optical diffractogram from the central spot to the sharp spot arising from the Bragg reflection θ_B. If the diffractogram rings appear elliptical, the coefficient of astigmatism is easily found by measuring two sets of ring radii along the major and minor axes of the ellipse. The difference between the two focus values obtained for these two sets of data is then equal to C_a (see Fig. 2.20).

The accuracy of the diffractogram technique for measuring C_s is, unfortunately, rather poor. It increases with both the order and the total number of rings used. An error analysis of the method shows that best results are obtained at a large under-focus value where many high-order rings are seen. In small-unit-cell crystal imaging (few beams) the determination of C_s is critical to image interpretation—for example a change of 0.4 mm in C_s reverses the black–white contrast of silicon images ($\lambda = 0.037$, $\Delta f = -300$ nm, $C_s = 2.2$ mm). At the point-resolution limit (eqn 6.17) the change in C_s needed to reverse contrast is 38%, independent of C_s and λ. Since the diffractogram analysis may give values of C_s varying by more than a millimetre within the same through-focus series, the determination of C_s is then best done by the more accurate methods of Section 10.2 for a known specimen position (focusing current). In large-unit-cell crystals of known structure, C_s can be refined by quantitative image matching, since, with many beams contributing to the image, a unique image match is readily obtained. (In small-unit-cell crystals there are many pairs of values of C_s and Δf which give the same image, as shown in Fig. 5.1(d).) Aberration-corrected instruments allow the value of C_s to be set by the operator.

An ingenious method has also been developed by Krakow and co-workers for obtaining the entire through-focus transfer function from a single diffractogram, using an inclined amorphous carbon specimen. An application of this technique can be found in Frank et al. (1978/9). One such diffractogram will frequently allow the resolution-limiting effects of chromatic aberration and beam divergence to be distinguished as a function of focus. A tableaux of diffractograms, covering a range of conditions, is shown in Fig. 12.8.

In general, then, to ensure that a high-resolution image can be calibrated with approximate values of C_s, Δf, and C_a it is only necessary to include in the image a small region of thin amorphous carbon. A region of contaminant will also serve.

An interesting and very accurate method for measuring the aberration coefficients using the phase correlation function between pairs of HREM images has been described by Meyer et al. (2002).

10.7. Lateral coherence width

Spatial coherence in electron microscopy is characterized by $\gamma(r, t)$, the complex degree of coherence. As discussed in Chapter 4, the width X_c of this function is directly related to the incident beam divergence θ_c if the illuminating aperture (C2 of Fig. 2.3) is incoherently filled. For hair-pin filaments this is normally the case (see Chapter 4). For a field-emission gun it may be coherently filled. The semi-angle subtended by the illuminating aperture can either be measured directly from the microscope or obtained from the diameter of diffraction spots. In the first case, $\theta_c = r/L$ with r the radius of the aperture C2 and L the distance between this and the specimen. The transverse (spatial) coherence width for a hair-pin filament is then given by

$$X_c = \frac{\lambda}{2\pi\theta_c} \tag{10.6}$$

If the second method is used, care must be taken to avoid exceeding the linear range of the detector. If the measured diameter of a diffraction spot is d and the distance between the central spot and a spot with scattering angle β is D, then

$$\theta_c = \beta\left(\frac{d}{2D}\right)$$

Here β is twice the Bragg angle. The diameter of weak high-order spots will be found to be equal to that of the central beam, and these are less likely to saturate the detector. An advantage of this method over direct measurement is that full account is taken of the

300 *Electron-optical parameters*

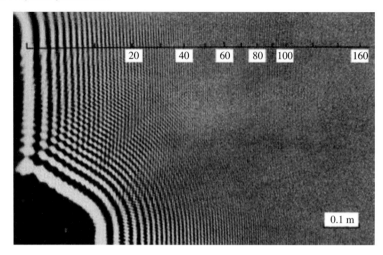

Figure 10.6 The very large number of Fresnel fringes which can be observed using a field-emission source at 100 kV. More than 160 fringes can be counted at the edge of this molybdenum oxide crystal. The high source brightness allows a very small illuminating aperture to be used (see also Section 4.5 for a discussion of the effect of source size on coherence).

objective lens pre-field (see Section 2.9). This is important for an immersion lens operated at short focal length, as used at high resolution.

A more direct impression of coherence effects can be obtained from recordings of Fresnel fringes. Figure 10.6 shows Fresnel fringes obtained from an instrument fitted with a field-emission source. The number of fringes observed is related to the coherence width X_c, as indicated in a simplified way below. This simplified model of Fresnel diffraction allows the fringes to be thought of as arising physically from interference between an extended line source along the specimen edge (the wave scattered from the specimen edge) and the unobstructed wave. The correct detailed form of Fresnel fringes must, however, be obtained from the analysis of Chapter 3 using the Fresnel propagator. A plane wave incident on an edge can be represented by the radiation emitted by a point source in the plane of an illuminating aperture. As shown in Fig. 10.7, a translation of this point source from A to B has the effect of shifting the fringe system. Independent point sources separated by the diameter of the aperture will establish fringe systems which are slightly out of register. The resultant sum of intensities produces a set of fringes whose contrast is thereby degraded. Sources separated by d introduce a fringe shift $\Delta z \theta$ where Δz is the focus defect. An approximate expression for the positions of the fringe maxima with a single source point is

$$x_n = \sqrt{2\Delta z \lambda}(n-k)^{1/2} \tag{10.7}$$

where k is a constant ($k < 1$) such that $2\pi k$ specifies the phase shift introduced on scattering at the specimen edge. The 'period' of the nth fringe is $(x_n - x_{n-1})$. For two sources with separation d, fringes will be more or less washed out beyond the nth fringe where

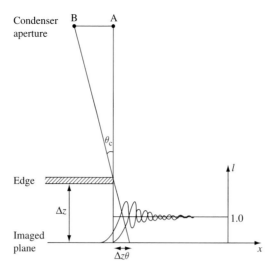

Figure 10.7 Fresnel fringes formed by two distant separated point sources. The intensity is not zero at the specimen edge. The fringes formed by source B alone (shown by the broken line) would be displaced with respect to those formed by source A alone (continuous line). The fringes become finer with increasing distance from the edge, which is taken to be opaque. The more realistic case for electron microscopy of an edge with a complex transmission function is treated in Chapter 3. A line of independent point sources along AB sets up many fringe systems slightly out of register and therefore blurred beyond a certain distance from the edge.

$$(x_n - x_{n-1})/2 = \Delta z \theta \tag{10.8}$$

Equations (10.7) and (10.8) can be solved for n. Equation (10.7) then gives the distance x_0 over which fringes will be clearly seen. This is

$$x_0 = A_0 \lambda / \theta_c$$

where A_0 is a constant of the order of unity involving k. An accurate value of this constant can only be obtained from the full treatment for a continuous extended source given in Chapter 3, which unfortunately does not give a simple result. Note that through the constant k the fringe visibility depends on the scattering properties of the specimen. A comparison of this result with eqn (10.6) suggests that within the limits of this simple model, the transverse coherence width X_c is proportional to the distance over which Fresnel fringes are observed, measured normal to the specimen edge. A comprehensive analysis of Fresnel diffraction has been given by Komrska (1973).

As shown in many texts, the angle θ_c falls sharply as the illumination is defocused. The secret of recording many clear Fresnel fringes is to use a well-defocused second condenser, the smallest spot (maximum demagnification) on the first condenser, low magnification (about 15 000), and long exposure. So long as the final illuminating aperture remains incoherently filled, the use of a smaller, brighter source does not increase the coherence in the specimen plane but simply produces a more intense image. The coherence depends only on the illumination angle θ_c.

Three other techniques have been used to give an indication of coherence length. The electron biprism gives a direct measurement of $\gamma(x_{1,2})$ from the contrast of the central fringe. This instrument developed by Mollenstadt and co-workers, is discussed in Gabor (1956), Hibi and Takahashi (1969), and Merli et al. (1974). Another technique uses the translation of lattice fringes with defocus to estimate X_c (Harada et al. 1974). Interference fringes have also been observed within the region of overlap of two diffraction spots (discs) using crossed crystals to obtain sufficiently close orders (Dowell and Goodman 1973). These fringes arise from interference between waves emitted from separated points within the illuminating aperture, and thus indicate a degree of coherence across the illuminating aperture (see Sections 4.5 and 7.6).

10.8. Electron wavelength and camera length

The electron wavelength and microscope camera length can be obtained from two measurements of Kikuchi line spacings on the diffraction pattern of a specimen of known structure. The camera length L is related to the measured distance D between two identified reflections or Kikuchi lines by

$$Dd = L\lambda \qquad (10.9)$$

where d is the crystal lattice spacing giving rise to the observed Bragg reflection. Physically, the camera length corresponds to that distance between the specimen and a detector plane (without lenses) which would be needed to produce a diffraction pattern of the same size as that observed in the microscope, where lenses are used to magnify the diffraction pattern formed in the objective-lens image focal plane.

Normally eqn (10.9) is used to find d from a measurement of D with λ and L known. However, if a diffraction pattern could be recorded in the correct orientation, containing two identified zone axes A and B measured at distance R apart, the camera length L could be found from $L = R/\theta$, where θ is the angle between A and B (assumed small). This angle is known if the crystal structure is known. Zone axes are revealed by a characteristic pattern of Kikuchi lines. Thus, in principle, L can be found without knowing the electron wavelength. This value of L then enables the wavelength λ to be found from eqn (10.9).

In practice it is not necessary to record two zone axes on one micrograph, since the problem can be recast in the form of known angles and two measured distances between Kikuchi lines by selecting a favourable orientation. Simple expressions for λ and L in terms of these measurements are given together with a worked example in the paper by Uyeda et al. (1965). This is probably the simplest wavelength determination method available for electron microscopy. The error $\Delta\lambda/\lambda$ is typically only about 5% of the error in the ratio of Kikuchi line separations. The overall accuracy may be considerably less that 0.5%. Somewhat simpler variations of this method are described by Høier (1969) and Fitzgerald and Johnson (1984), together with a worked example. The method allows the ratio d/λ to be determined to about 0.1%. Higher accuracy in wavelength measurement can be obtained by fitting the HOLZ lines which cross the central disc of a CBED pattern to dynamical simulations (Section 13.3; Spence and Zuo 1992).

10.9. Resolution

There is no simple way to define, let alone measure, the resolution of an electron microscope under the most general conditions. The essential difficulty arises from the fact that, under coherent conditions, the ability of a microscope to distinguish or resolve neighbouring object points depends on the scattering properties of those object points. Thus, resolution depends on the choice of specimen (in particular, on the magnitude of any phase shift introduced on scattering) and it is strictly necessary, even for weakly scattering objects, to specify the nature of the specimen when specifying resolution. The ability to resolve the individual atoms of a particular molecule is thus no guarantee that those of a different atomic species will be resolved, even given similar interatomic spacings and instrumental conditions. Since the resolution of a microscope should be a property of the instrument alone (not the sample), a simple definition of resolution cannot be given. Under the normal conditions of strong multiple scattering in electron microscopy, even greater difficulties arise.

The Rayleigh criterion cannot be applied to HREM since it was devised for incoherent imaging conditions, where it is possible to define point-resolution in a way which is independent of the object. It may, however, apply to STEM in the HAADF mode if this is ideally incoherent. The situation in HREM is more akin to coherent optical imaging, in which it is customary to give the transfer function of a lens rather than to specify a single number for its resolution. Only by assuming a phase shift on scattering of exactly $-\pi/2$ from an idealized weakly scattering point object is it possible to plot out an idealized impulse response for the microscope, using the Fourier transform of the transfer function as in Fig. 3.8. This 'impulse response' is the image of an object chosen to be sufficiently small to make the image independent of small changes in the size of the object. Since the 'impulse response' of a modern medium-voltage (300–400 kV) electron microscope is about 0.17 nm, a point test object much smaller than this, i.e. of subatomic dimensions, would be required in practice to observe the impulse response directly. Thus direct observation of the impulse response is clearly impractical, quite apart from the problems of finding a suitable invisible substrate (for the STEM mode discussed in Chapter 8, the impulse response is the probe intensity in the incoherent imaging approximation, and this may be imaged directly on modern TEM/STEM instruments). Several proposals for characterizing the resolution performance of electron microscopes have, however, been put forward, both for strongly and weakly scattering specimens, and these are now briefly discussed.

For the strongly scattering majority of electron microscope specimens it is generally useful to think in terms of the specimen lower 'exit'-face wavefunction as the 'object', and to consider the resolution-limiting effects of the electron lenses which image this wavefunction. Both theory and experiment support the idea (essentially the column approximation of diffraction contrast theory) that this exit-face wavefunction is locally determined by the specimen potential within a narrow cylinder less than 0.5 nm in diameter for thin specimens. However, there is no guarantee that the dynamical exit-face wavefunction will contain simple peaked functions in one-to-one correspondence with object point scatterers. Unfortunately, this exit-face wavefunction is not observable, so that a comparison of this function with the final image is not possible. All that is known a priori about the dynamical exit-face wavefunction is that it is locally determined and preserves the point group and any translational symmetry of the specimen potential projected in the beam direction, under most common operating conditions (the image may not preserve symmetry if astigmatism is present and resolution is limited). However, reliable methods do exist for computing the exit-face wavefunction (see Section 5.7) for specimens of known structure. The most general methods, therefore, for specifying the resolution of an electron microscope would be based on a comparison of computed and experimental images of a specimen of known structure, thickness, and orientation. This comparison is based on a 'goodness-of-fit' index or R factor similar to that used in X-ray diffraction and allows the extraction of structural detail out to (or beyond) the information-resolution limit of the microscope, as described in Chapter 7. The limits to the accuracy with which atomic positions in crystals of known structure can then be determined by this method are set ultimately by noise (see Section 5.13). The variation of this R factor between the computed and experimental images of a known structure with small changes in the atom positions of the computed images would then give an impression of instrumental resolution. If the specimen were crystalline, the choice of a large unit cell ensures that the transfer function is sampled at a large number of points.

The situation for weakly scattering objects in HREM is generally simpler, and the definition and measurement of resolution under these conditions has received a great deal of attention in the literature. Before discussing these methods, it is worth considering how the microscopist can be assured that his or her images were recorded from a specimen region satisfying the weak-phase object approximation. At least four tests are possible: (1) In an instrument fitted with convergent-beam or STEM facilities, the specimen thickness can be measured from the microdiffraction pattern (see Section 13.4). In very thin regions, the central disc of the microdiffraction pattern must be much stronger than any other for single-scattering conditions. (2) If wedge-shaped specimens are used, the kinematic scattering region can be selected at will (see Fig. 12.4a). (3) For crystals, one can be reasonably sure that weak-scattering conditions obtain if the measured Bragg electron diffraction intensities accurately predict the corresponding diffractogram intensities (see eqn 3.42). The damping envelope constants must also be known (see Section 4.2). (4) For a crystal whose projection is known a priori to be non-centrosymmetric, the observation of a centrosymmetric microdiffraction pattern is an assurance of single scattering.

The damping envelope constants θ_c and Δ exercise the dominant influence on the form of the microscope impulse response under weak-scattering conditions. Thus, rather than citing a single number for the point- or information-resolution limit of an instrument, microscopists will be chiefly concerned with determining the constants C_s, Δf, θ_c, and Δ for each image recorded. All of these depend, through the specimen position ('height'), on the objective lens current (see Section 2.7 and Appendix 3). As briefly mentioned in Section 1.2, an early task of the microscopist is to determine the specimen position which gives the most favourable combination of these constants. The measurement of C_s, Δf, and C_a has already been discussed, and an ingenious method of measuring Δ and θ_c using two micrographs recorded under identical conditions has been described by Frank (1976) (see also Saxton 1977). This relies on the fact that the Young's fringes seen in the optical diffraction pattern from a similar pair of superimposed, slightly displaced micrographs extend only to the band-limit of information common to both micrographs. The fringe contrast thus falls to zero at the onset of electron noise, which is in principle the only differing contribution to the two micrographs. By plotting this band limit of common information against defocus, it is possible to determine both Δ and θ_c. Since $\Delta = C_c Q$ (eqn 4.9), where Q involves the quadrative addition of instabilities due to high voltage and lens-current fluctuations, together with the effect of the thermal spread of electron energies, any one of these quantities can also be determined if all the others are known. As discussed in Appendix 3, the band limit due to partial spatial coherence moves to higher spatial frequencies with increasing defocus. A detailed example of the measurement of these constants is given in O'Keefe et al. (2001).

The problem of measuring the point-resolution of an instrument without using a priori information about either the object or the instrument has also been discussed by Frank (1974). His suggestion is that the half-width of the cross-correlation function formed between two successive recordings of the same amorphous carbon film can be used to define the point-resolution of an electron microscope. The method is limited to weak-phase objects or incoherent imaging. Nevertheless, *this appears to be the best practical proposal for a standardized point-resolution measurement yet put forward*, since all instrumental aberrations and instabilities are included, while the effects of electron noise are minimized through cross-correlation. An improvement in this method has been described by Barthel and Thust (2008), who introduce illumination tilts to improve accuracy. An asymmetrical correlation function indicates astigmatism (Frank 1975). The formation of a cross-correlation function is much easier in the STEM mode.

The method of seeking the smallest point separation in the image of a thin carbon film is unsatisfactory, owing to the effects of electron and Fresnel 'noise'. At high resolution the correlation found between point separations measured on successive images is found to be poor. For point separations greater than 0.5 nm, atomic clusters formed by evaporation of PtIr alloy are useful. These specimens are available commercially. A light evaporation of gold on carbon is also easy to prepare (see Section 1.3) and forms similar small atomic groups. Both these specimens are useful for practising alignment and astigmatism correction and to enable the user to become familiar with through-focus contrast effects on small particles, as discussed in Chapter 6. The most convincing point-resolution test is a recording of a molecule containing heavy atoms in a known arrangement.

Lattice images are also widely used as a resolution test (Heidenreich et al. 1968). This subject is fully discussed in Chapters 5 and 12. It is also possible to estimate instrumental resolution by comparing computer-simulated images with experimental many-beam structure images recorded without using an objective aperture. If the specimen structure and thickness are known (as for the case of graphene or a MgO cube wedge), together with the microscope focus setting, illumination aperture, and spherical aberration constant, images can be computed for a range of values of Δ until a good match is obtained. This will then determine the resolution limit due to incoherent instabilities.

Note that for axial three-beam fringes recorded at the focus settings of eqn (5.10), half-period fringes may appear and give a spurious impression of very high-resolution detail, beyond the true band limit set by incoherent instabilities (see eqn 4.11). It is also important to stress that the analysis of Section 4.2 (eqn (4.8) in particular) assumes either kinematic scattering or, more generally, a strong zero-order diffracted beam (which may also occur as a result of Pendellösung oscillations). Unless a wedge-shaped crystal is used, or microdiffraction facilities are available, it is not usually possible to be certain that this condition has been satisfied for a particular image region and therefore that the 'damping envelope' concept can be applied. These very fine fringes are unlikely to be recorded at the Scherzer focus since the microscopist can maximize their contrast by selecting a focus setting which makes the slope of $\chi(u)$ zero in the neighbourhood of the important inner Bragg reflections (see Section 5.8 for more details). By equating this condition (eqn 5.66) to the focus condition for obtaining half-period fringes (eqn 5.10), we find

$$a_0^4 = C_s \lambda^3 / n = u_0^{-4}$$

as the crystal periodicity which will produce axial half-period fringes of highest contrast in an instrument operating at particular values of C_s and λ. Fringes obtained under these conditions allow the largest illumination aperture to be used (producing the most intense final image) for the smallest contrast and 'resolution' penalty, and it is likely that many of the sub-angstrom fringe spacings reported in the literature were recorded under these conditions. *Such fringes do not measure instrumental resolution in any useful sense*, and must be thought of as, at best, a technique for measuring the instability limit due to mechanical vibration. Note that for half-period, axial three-beam fringes of this type, the damping envelope concept described in Section 4.2 does not apply (despite the presence of a strong central beam), since it neglects terms such as $\phi_h \phi_{-h}$ of eqn (5.9).

The comments in Section 5.13 on the accuracy with which atomic positions can be determined for crystals of known structure should be read in conjunction with this section. Section 10.10 on auto-tuning is also relevant. Finally, if an image peak is known a priori to represent a single atomic column, D. Van Dyck has pointed out that we can consider that every beam electron measures its position. The error in our measurement of the mean position (the peak width) is therefore divided by the square root of the number of beam electrons contributing to the peak (standard error).

10.10. Ronchigram analysis for aberration correction

The Ronchigram was introduced in Section 7.6, which provides essential background. For our purposes here we consider it to be the far-field diffraction pattern formed by a small electron probe focused slightly upstream (or downstream, as shown in Fig. 10.8), using a large coherently filled cone of illumination—a wide-angle coherent convergent beam pattern. The sample, for aberration measurement, is usually a very thin carbon film or crystal. In practice, these conditions can be achieved by removing the STEM objective aperture entirely. (For a crystalline sample, as used in the Ronchigram shown in Fig. 7.5, this semi-angle is much larger than λ/d, where d is the largest periodicity in the sample, so that there would be gross overlap of coherent orders.) It will be seen that this is the geometry originally proposed for in-line holography by Gabor, which produces a coherent shadow image of the sample, projected from a point (Fig. 7.4).

The Ronchigram was first developed for the testing of astronomical mirrors (Ronchi 1964). The text by Malacara (1978) first suggested its use to this author for the measurement of aberrations in STEM, and it was first used for this purpose by Cowley and Disko (1980). Figure 10.8 shows the essential principle in its simplest form. We ignore diffraction by the sample, but assume that it produces well-defined features in the shadow image. Consider the case where spherically aberrated rays arriving at high angles are focused on the sample, but paraxial rays A focus beyond it (under-focus, Δf negative). The shadow image magnification $M \sim L/\Delta f$ is clearly a function of scattering angle, and so will become infinite as Δf tends to zero for paraxial rays. At large defocus, where spherical aberration is negligible,

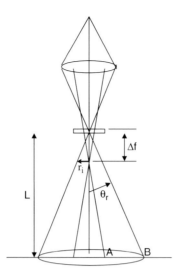

Figure 10.8 Ray paths for a Ronchigram with spherical aberration. The magnification is $M \sim L/\Delta f$. Bragg scattering from a sample would occur within this coherent cone. The shadow image magnification varies from A to B as given by eqn (10.11).

there is little distortion of the shadow image. The working distance L can be found from measurements of the position of diffraction spots, and a shadow image of the lattice planes in a thin crystal (formed as shown in Fig. 7.4) of known spacing can then be used to find M, and hence calibrate the medium and coarse focal steps of the objective lens Δf.

Rays for some higher angles θ_r are strongly affected by spherical aberration and may be focused on the sample, if the paraxial focus point lies downstream as in Fig. 10.8. Then we can apply an equation similar to eqn (2.30) to this figure, using $r_i = 3 C_s \theta_r^3$ and $\theta_r = r_i/\Delta f$, so that

$$\theta_r^2 = \Delta f / 3 C_s \tag{10.10}$$

Then, near B in the Ronchigram the magnification of the shadow image will become infinite in the radial direction, since, for those rays at θ_r, the point of projection for the shadow image lies at the sample. This produces an identifiable ring in the pattern at scattering angle θ_r. Measurements of the radius of this ring therefore provide a value of C_s from the known defocus. In more detail using a wave-optical analysis, Cowley (1979) has shown that the magnification in the Ronchigram varies for under-focus with angle as

$$M(\theta) = \frac{M_0}{[1 + (4C_s/\Delta f) \sin^2(\theta/2)]} \tag{10.11}$$

where M_0 is the paraxial focus. In this way we are led to the idea (as shown by eqn (7.16)) that a Ronchigram is a high-resolution shadow image whose magnification varies with angle. It is remarkable then that the Fourier transform of local patches of a Ronchigram give Thon diffractogram rings (Section 10.6), as shown by Lupini et al. (2010). We will see later that a measurement of this local magnification is proportional to the second derivative of the aberration function, which is needed in order to correct aberrations in the STEM.

This simple analysis shows the sensitivity of the Ronchigrams to spherical aberration. We now demonstrate their sensitivity to higher-order aberrations, using a crystalline sample to simplify the analysis. The diffraction pattern with overlapping orders has already been shown to be sensitive to aberrations in eqns (7.12) and (8.1a) and an analysis of coherent overlapping orders using two-beam dynamical theory can be found in Spence and Cowley (1978). A kinematic analysis was given for the three-beam axial case with application to aberration measurement by Lin and Cowley (1986). In that case one obtains a series of elliptical rings whose radii may be used to measure defocus, spherical aberration, and the off-axis aberrations coma, astigmatism, and distortion. We take an exit-face wavefunction of the form

$$\Psi(\mathbf{r}) = T(\mathbf{r})A(\mathbf{r}) \tag{10.12}$$

where $A(\mathbf{r})$ is the impulse response of the lens (eqns 8.2a, 3.14, and 3.15), now including the effect of spherical aberration. For a thin periodic sample we take the weak phase approximation

$$T(\mathbf{r}) = 1 - i\sigma\phi_0 \cos(2\pi\mathbf{g}\cdot\mathbf{r} + \theta) \tag{10.13}$$

where θ is a linear phase shift which depends on probe position. The Fourier transform $\Psi_D(\mathbf{K})$ of eqn (10.12) gives the scattered amplitude; the Ronchigram intensity is found to be

$$I(\mathbf{K}) = \Psi_D(\mathbf{K})\Psi_D^*(\mathbf{K}) = 1 + 2\sigma\phi_0 \sin[E(\mathbf{K},\mathbf{g})]\cos[O(\mathbf{K},\mathbf{g}) + \theta] \tag{10.14}$$

where

$$\begin{aligned} E(\mathbf{K},\mathbf{g}) &= [\chi(\mathbf{K}+\mathbf{g}) - \chi(\mathbf{K})] \text{ even in } g \\ &= \pi g^2 \lambda \left[C_s \lambda^2 (3K_x^2 + K_y^2 + g^2/2) + \Delta f \right] \end{aligned} \tag{10.15}$$

and

$$\begin{aligned} O(\mathbf{K},\mathbf{g}) &= [\chi(\mathbf{K}+\mathbf{g}) - \chi(\mathbf{K})] \text{ odd in } g \\ &= 2\pi\lambda K_x \mathbf{g} \left[C_s \lambda^2 (K_x^2 + K_y^2) + \Delta f + C_s \lambda^2 \mathbf{g}^2 \right] \end{aligned} \tag{10.16}$$

Here the cosine term in (10.14) gives similar patterns to those used for the testing of large telescope mirrors, and is sensitive to probe position and hence sample drift. For large defocus, this term produces fringes of spatial frequency $[g\lambda(\Delta f + C_s\lambda^2 g^2)]^{-1}$. The sin term produces patterns similar to a lateral shearing interferometer. Setting this term to zero defines the loci of ellipses of zero contrast when $E(\mathbf{K},\mathbf{g}) = n\pi$, as seen in the experimental pattern in Fig. 10.9. The ellipses n are defined by

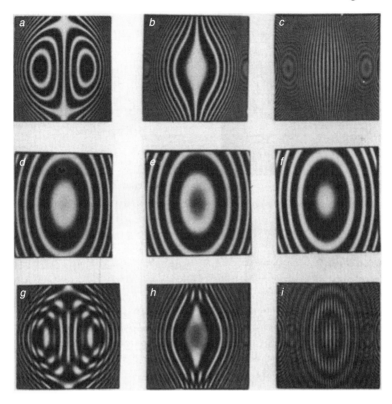

Figure 10.9 Simulated Ronchigrams with $\theta = 0$, $C_s = 0.8$ mm, from a thin crystal with lattice vector $|\mathbf{g}| = 1.25$ nm^{-1} in the symmetrical three-beam case. The top row is the cosine term in eqn (10.14) for defocus -300 nm, 0, $+450$ nm; the middle row is the sin term; the bottom row is their product. (From Lin and Cowley (1986).)

$$\frac{K_x}{a_n^2} + \frac{K_y}{b_n^2} = 1 \tag{10.17}$$

where

$$b_n = 3^{1/2} a_n = \left(\frac{n/(g^2 \lambda) - \Delta f}{C_f \lambda^2} + \frac{g^2}{2} \right)^{1/2} \tag{10.18}$$

If the magnification of the Ronchigram is obtained from the position of Bragg spots from a known crystal structure, both defocus and spherical aberration constants (and higher-order aberration coefficients for coma, astigmatism, and distortion) can then be obtained by fitting these ellipses to the data (Lin and Cowley 1986). This important paper thus established the usefulness of Ronchigrams for calibrating the operating parameters in STEM.

With the appearance of aberration correctors, a more general method was needed for fast real-time measurement of all higher-order aberrations in order to generate feedback signals which could be fed to aberration-correcting electron-optical elements. This method uses non-periodic samples, such as nanogold dots on amorphous carbon, in order to extract measured values of the aberration coefficients. (We might think of such a sample as crystal with a very large unit cell, so the Bragg angle becomes vanishingly small.) The theory of this method, which we now outline, is given in papers by Rodenburg and Macack (2002), Lupini and Pennycook (2008), Lupini et al. (2010), and Sawada (2008).

Equation (3.19) relates the position r_i, at which an aberrated ray crosses the paraxial focus at $z = V$, to the gradient of the wavefront aberration function of a lens, using

$$r_i = V\alpha = V\frac{d\gamma}{dr} = \left(\frac{\lambda}{2\pi}\right)\nabla\chi(\theta_x, \theta_y) \tag{10.19}$$

at unit magnification, where the gradient is taken with respect to the angles used to define the ray aberrations in eqn (3.27). We can imagine the rays in Fig. 3.5 originating from a field emitter at P and passing through a sample at P′, continuing on to the right toward a Ronchigram detector in the far field. In the iconal approximation energy flows along classical rays, and we may consider that a local shadow image of an identifiable sample feature is laid down around each ray as it strikes the Ronchigram detector in the direction $(\theta_x, \theta_y) = \boldsymbol{\theta}_1$. If we now allow for a general probe coordinate (or sample position) at $\mathbf{R}_1$, then the ray at C in Fig. 3.5 crosses the paraxial focus plane (the sample) at

$$\mathbf{X}_1 = (\lambda/2\pi)\nabla\chi(\boldsymbol{\theta}_1) + \mathbf{R}_1 \tag{10.20}$$

and arrives at the detector at the two-dimensional angle $\boldsymbol{\theta}_1$. A similar equation may be written for a second ray at $\mathbf{X}_2$. Writing an expression for $(\mathbf{X}_1 - \mathbf{X}_2) = d\mathbf{X}$ and using the Taylor expansion, we have

$$d\mathbf{X} = (\lambda/2\pi)\mathbf{H}(\boldsymbol{\theta})d\boldsymbol{\theta} + d\mathbf{R} = d\mathbf{C} + d\mathbf{R} \tag{10.21}$$

where the matrix of second derivatives $\mathbf{H}$ is defined by

$$\mathbf{H}(\boldsymbol{\theta}) = \begin{bmatrix} \frac{\partial^2 \chi}{\partial \theta_x^2} & \frac{\partial^2 \chi}{\partial \theta_x \partial \theta_y} \\ \frac{\partial^2 \chi}{\partial \theta_y \partial \theta_x} & \frac{\partial^2 \chi}{\partial \theta_y^2} \end{bmatrix} \tag{10.22}$$

The local magnification is then the inverse of $\mathbf{H}(\boldsymbol{\theta})$. Where the determinant of the matrix is zero, the magnification becomes infinite. If a sharp feature in the shadow image is moved (e.g. by moving the sample or probe by $d\mathbf{R}$) and the corresponding distance $d\mathbf{C}$ by which its shadow image moves is measured, then $d\mathbf{X} = 0$ and we have

$$d\mathbf{R} = -(\lambda/2\pi)\mathbf{H}(\boldsymbol{\theta})d\boldsymbol{\theta} \tag{10.23}$$

One method of aberration correction is therefore to determine the local magnification at $\boldsymbol{\theta}$ using the change $d\boldsymbol{\theta}$ in the position of the shadow image of a sample feature, due to a measured change in its position $d\mathbf{R}$ at the sample. The Ronchigram is therefore divided up into small patches for analysis. This then gives a matrix of second derivatives of the aberration function, to which an estimate of the aberration function may be fitted. Several image features at different angles may be used, and several Ronchigrams. During alignment, signals based on these measurements may be sent back to the optical elements, allowing iterative refinement of the fit between measured and estimated parameters, as in the methods of adaptive optics.

For the simple case of defocus Δf alone, we have $(\lambda/2\pi)\chi(\theta) = C_1\theta^2/2$, where $C_1 = \Delta f$, so that, from (10.23), $d\mathbf{R} = -\Delta f d\boldsymbol{\theta}$, and image shift is proportional to probe shift, with the proportionality constant giving defocus directly. If we consider first-order aberrations alone (eqn 3.27), we find

$$d\mathbf{R} = \begin{pmatrix} C_1 + C_{12a} & C_{12b} \\ C_{12b} & C_1 - C_{12b} \end{pmatrix} d\boldsymbol{\theta} \tag{10.24}$$

The effect on a small sample (or probe) motion is then to both scale and rotate this motion in the Ronchigram. Two probe shifts (four components) will be needed to find these four aberration constants. For higher-order coefficients, the shifts vary with position in the Ronchigram, since $\mathbf{H}$ then contains $\boldsymbol{\theta}$.

The effects of partial spatial coherence on Ronchigrams can be understood from eqn (7.12), where the integration of intensity over the electron source is seen to degrade fringe contrast. Since each point on the source is imaged onto the sample as the STEM probe, this integration may also be interpreted as an integration over fluctuations in scan coordinate, sample vibration, or field-emission tip vibration. An integration over the component energies in the beam is also possible, in order to include the effects of chromatic coherence. Measurements of the electron source size, based on damping effects on Ronchigrams from thin diamond crystals, can be found in Dwyer et al. (2010). Here the interference fringes between overlapping coherent orders (similar to those shown in Fig. 8.3) were fitted to dynamical simulations, and the size of the focused beam at the sample adjusted for best fit. This image of the physical field-emission area, demagnified by about 50, depends on the lens settings. In fact emission occurs from a virtual source inside the tip, whose size can also be determined from coherence measurements using a field-emission microscope (Spence et al. 1994). The aberrations of this virtual lens at the tip can also be modelled and measured, and aberration coefficients for this 'tip lens' obtained (Scheinfien et al. 1993).

The effects of partial coherence on Ronchigrams can also be expressed in terms of the derivative matrix $\mathbf{H}(\boldsymbol{\theta})$ (Lupini et al. 2010). Then it is found, in the weak phase-object approximation, that the entire Ronchigram is damped in a similar way to the contrast transfer function, with the important difference that the damping depends on the local magnification. The spatial coherence envelope is

$$D(\boldsymbol{\tau}) = \exp\left[-(w\pi\lambda^{-1}\mathbf{H}(\boldsymbol{\theta})^{-1}\cdot\boldsymbol{\tau})^2\right] \tag{10.25}$$

where w is the $1/e$ width of the Gaussian beam at the sample, and $\boldsymbol{\tau}$ (with dimensions of length) is a vector conjugate to a scattering vector originating within one local patch

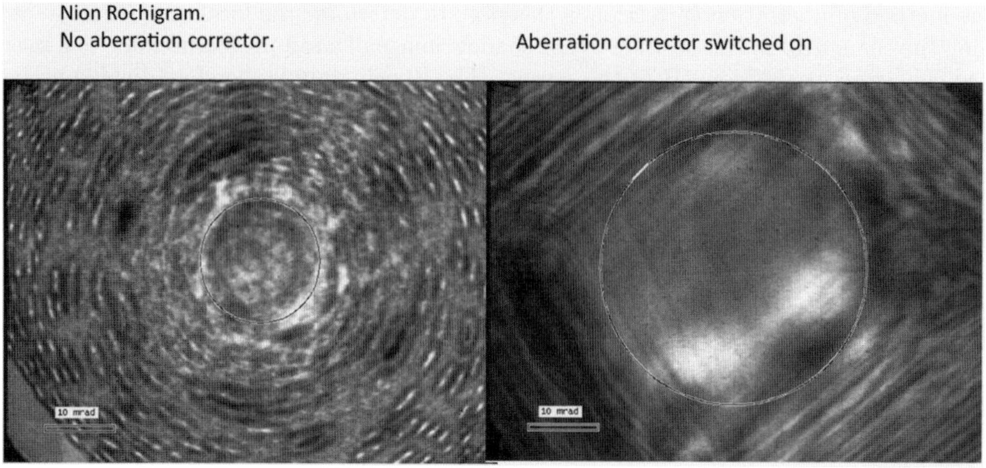

Figure 10.10 Ronchigram from an amorphous sample (left) with the aberration corrector switched off and (right) with it switched on. Circles indicate the resolution improvement obtained, which improves from 2.4 to 0.8 Å. (Courtesy Nion Corp.)

of the Ronchigram. The situation is complicated by the fact that Ronchigrams exist in a kind of mixed real and reciprocal space. At large defocus, for example, we have seen (eqn 7.16) that the Ronchigram is equivalent to a bright-field image, so that the roles of the scattering vector and τ become reversed, and in this case we can think of τ as a spatial frequency. An envelope for temporal coherence can also be derived in terms of $\mathbf{H}(\mathbf{\theta})$, which becomes dominant at high angles. The possibility then arises of estimating $\mathbf{H}(\mathbf{\theta})$ from measurements of the damping.

A practical account of the manual alignment of a high-resolution STEM instrument using Ronchigrams is given in James and Browning (1999). In general terms, one wants simply to make the central region of the Ronchigram as large as possible, since the smooth intensity variation of the interferogram here indicates faithful information transfer. The dramatic effect of an aberration corrector on a Ronchigram is shown in Fig. 10.10.

References

Barthel, J. and Thust, A. (2008). Quantification of the information limit of transmission electron microscopes. *Phys. Rev. Lett.* **101**, 200801.

Boothroyd, C. B. (1997). Quantification of energy filtered lattice images and coherent convergent beam patterns. *Scan. Microsc.* **11**, 31.

Budinger, T. F. and Glaeser, R. M. (1976). Measurement of focus and spherical aberration of an electron microscope objective lens. *Ultramicroscopy* **2**, 31.

Cowley, J. M. (1979). Coherent interference in convergent-beam electron diffraction and shadow imaging. *Ultramicroscopy* **4**, 435.

Cowley, J. M. and Disko, M. M. (1980). Fresnel diffraction in a coherent convergent electron beam. *Ultramicroscopy* **5**, 469.

Dowell, W. C. T. and Goodman, P. (1973). Image formation and contrast from the convergent electron beam. *Phil. Mag.* **28**, 471.

Dwyer, C., Erni, R., and Etheridge, J. (2010). Measurement of the effective source distribution and its importance for quantitative interpretation of STEM images. *Ultramicroscopy* **110**, 952.

Fitzgerald, J. D. and Johnson, A. W. S. (1984). A simplified method of electron microscope voltage measurement. *Ultramicroscopy* **12**, 231.

Frank, J. (1974). A practical resolution criterion in optics and electron microscopy. *Optik* **43**, 25.

Frank, J. (1975). Controlled focusing and stigmating in the conventional and scanning transmission electron microscope. *J. Phys. E* **8**, 582.

Frank, J. (1976). Determination of source size and energy spread from electron micrographs using the method of Young's fringes. *Optik* **44**, 379.

Frank, J., McFarlane, S. C., and Downing, K. H. (1978/9). A note on the effect of illumination aperture and defocus spread in bright field electron microscopy. *Optik* **52**, 49.

Fukushima, K., Kawakatsu, H., and Fukami, A. (1974). Fresnel fringes in electron microscope images. *J. Phys. D* **7**, 257.

Gabor, D. (1956). Theory of electron interference experiments. *Rev. Mod. Phys.* **28**, 260.

Hall, C. E. (1949). Method of measuring spherical aberration of an electron microscope objective. *J. Appl. Phys.* **20**, 631.

Harada, Y., Goto, T., and Someya, T. (1974). Coherence of field emission electron beam. In *Proc. 8th Int. Congr. Electron Microsc., Canberra*, Vol. 1, p. 110. Australian Academy of Science, Canberra.

Heidenreich, R. D., Hess, W., and Ban, L. L. (1968). A test object and criteria for high resolution electron microscopy. *J. Appl. Crystallogr.* **1**, 1.

Hibi, T. and Takahashi, S. (1969). Relation between coherence of electron beam and contrast of electron image. *Z. Angew. Phys.* **27**, 132.

Høier, R. (1969). A method to determine the ratio between lattice parameter and electron wavelength from Kikuchi line intersections. *Acta Crystallogr.* **A25**, 516.

James, E. and Browning, N. (1999). Practical aspects of atomic resolution imaging and spectroscopy in STEM. *Ultramicroscopy* **78**, 125.

Komrska, J. (1973). Intensity distributions in electron interference phenomena produced by an electrostatic bi-prism. *Opt. Acta* **20**, 207.

Krivanek, O. L. (1976). A method for determining the coefficient of spherical aberration from a single electron micrograph. *Optik* **45**, 97.

Lin, J. A. and Cowley, J. M. (1986). Calibration of the operating parameters for an HB5 STEM instrument *Ultramicroscopy* **19**, 31

Lupini, A. (2011). The electron Ronchigram. *Scanning transmission electron microscopy: imaging and analysis*, ed. S. Pennycook and P. Nellist, p. 117. Springer, New York.

Lupini, A. and Pennycook, S. (2008). Rapid autotuning for crystalline specimens from an inline hologram. *J. Electron Microsc.* **57**, 195.

Lupini, A., Wang, P., Nellist, P., Kirkland, A., and Pennycook, S. (2010). Aberration measurement using the Ronchigram contrast transfer function. *Ultramicroscopy* **110**, 891.

Malacara, S. (1978). *Optical shop testing*. Wiley, New York.

Merli, P. G., Missiroli, G. F., and Pozzi, G. (1974). Electron interferometry with the Elmiskop 101 electron microscope. *J. Phys. E* **7**, 729.

Meyer, R. R., Kirkland, A. I., and Saxton, W. O. (2002). A new method for the determination of the wave aberration function. *Ultramicroscopy* **92**, 89.

O'Keefe, M. A., Hetherington, C. J., Wang, Y. C., Nelson, E. C., Turner, J. H., Kisielowski, C., Malm, J., Mueller, R., Ringnalda, J., Pan, M., and Thust, A. (2001). Sub-angstrom HREM at 300 kV. *Ultramicroscopy* **89**, 215.

Rodenburg, J. and Macack, E. (2002). Optimizing the resolution of TEM/STEM with the electron Ronchigram. *Microsc. Anal.* **90**, 5.

Ronchi, V. (1964). Forty years of history of a grating interferometer. *Appl. Opt.* **3**, 437.

Saxton, W. O. (1977). Spatial coherence in axial high resolution conventional electron microscopy. *Optik* **49**, 51.

Sawada, H. (2008). Measurement method of aberration from Ronchigram by autocorrelation function. *Ultramicroscopy* **108**, 1467.

Scheinfein, M., Qian, W., and Spence, J. C. H. (1993). Aberrations of emission cathodes. Nanometer diameter field-emission electron sources. *J. Appl. Phys.* **73**, 2057.

Spence, J. and Cowley, J. M. (1978). Lattice imaging in STEM. *Optik* **50**, 129.

Spence, J. C. H. and Zuo, J. M. (1992). *Electron microdiffraction*. Plenum, New York.

Spence, J. C. H., Qian, W., and Silverman, M. (1994). Electron source brightness and degeneracy in field emission point projection microscopy. *J. Vac. Sci.* **A12**, 542.

Stroke, G. W. and Halioua, M. (1973). Image improvement in high resolution electron microscopy with coherent illumination (low contrast objects) using holographic deblurring deconvolution, III, Part A, Theory. *Optik* **37**, 192.

Thon, F. (1971). Phase contrast electron microscopy. In *Electron microscopy in materials science*, ed. U. Valdre and A. Zichichi, p. 570. Academic Press, New York.

Uyeda, R., Nonoyama, M., and Kogiso, M. (1965). Determination of the wavelength of electrons from a Kikuchi pattern. *J. Electron Microsc.* **14**, 296.

Wyckoff, R. (1963). *Crystal structures.* Second edition. Wiley, New York.

Yada, K. (1967). Study of chrysotile asbestos by a high resolution electron microscope. *Acta Crystallogr.* **23**, 704.

11
Instabilities and the microscope environment

The most successful HREM and STEM instruments of the future will be those which are most favourably sited. One famous recent installation has been made near Dresden in a remote rural location. Aberration correction is pointless if electronic and mechanical instabilities limit performance. These instabilities produce an incoherent broadening of the microscope point-spread function which limits the resolution and can be likened to the imposition of a limiting objective aperture. The measurement of instabilities is an essential first step in setting up a high-resolution laboratory, and much time will be saved if measurement facilities are installed at the outset so that when a subtle fault develops it becomes a simple matter to rule out the easily measured instabilities such as magnetic-field interference and vibration. Overall, the digital diffractogram and Ronchigram are probably the best general diagnostic aids for incoherent instabilities.

The book by Alderson (1975) contains useful information which complements this chapter. The papers by Muller and Grazul (2001) and James and Browning (1999) describe in detail the environmental requirements for sub-angstrom resolution in STEM, while Turner *et al.* (1997) discuss the HREM case.

11.1. Magnetic fields

The effect of a.c. magnetic fields on high-resolution images can be seen by applying a known field while observing lattice fringes. A small transformer is a suitable source of 50 Hz magnetic field which will produce image blurring when held near the specimen chamber. A field applied near the illuminating system results in a loss of intensity owing to the resulting oscillatory beam deflection. Some manufacturers quote a sensitivity for their machines to a.c. fields; a value of 0.2 nm μT^{-1} (horizontal component at 50 Hz, 1 μT = 0.01 G) is typical for the effect of a field on image resolution. A typical figure for the sensitivity of the illumination spot to deflection is 1 μm μT^{-1}. Note that the effect of the stray field is greatest where the electron velocity is lowest, for example near the filament. The microscope lenses themselves have a shielding effect, and leakage is most likely at the junction between lens casings. A small homogeneous constant magnetic field such as that due to the Earth is of no consequence in CTEM; however, an inhomogeneous constant field may introduce astigmatism. Permanent magnets should be kept well clear of the microscope.

For high-resolution TEM (0.2 nm resolution) an a.c. field of less than 0.03 μT rms (horizontal component at 50 Hz) is required in the vicinity of the specimen chamber; 0.01 μT

may be needed for 0.1 nm resolution. This may be difficult to achieve since the fields commonly found in laboratories are often greater than 0.5 µT owing to mains electrical wiring. Stray fields can be reduced by ensuring that only one ground return for the microscope exists, by moving ancillary equipment, and, if necessary, by re-routing the mains wiring. The essential requirement for avoiding stray fields is that the same current enters on the active line as returns via neutral, and that only one path to ground exists. A common mistake is to place, say, an electrically operated vacuum gauge on the microscope cabinet connected to a different socket and ground return from that used by the microscope, which may be a special high-quality ground. If the microscope and vacuum gauge chassis are in contact, a ground loop can result producing a large a.c. field. Similarly, all additional safety equipment, such as electrical water-valve solenoids (which will be grounded through the water plumbing) should be checked for ground loops by disconnecting the ground and observing the effect on the focused illumination spot and measured field strength near the specimen chamber. A complete analysis of ground loops and shielding techniques can be found in the text by Morrison (1967).

The simplest test for 50 Hz a.c. fields is as follows. Form a small focused illumination spot at a magnification of about 15 000 times. Run the spot quickly across the viewing phosphor using the beam-translate controls. A zig-zag after-image on the phosphor indicates interference of stray field with the illumination system.

While the microscope itself, used as an oscilloscope in this way, is probably the most sensitive indicator of stray fields, an a.c. field can easily be measured with a coil connected to an oscilloscope. This will also indicate the important frequency components of the interference. For maximum sensitivity, use the largest possible number of turns of very fine wire—the windings of a mains transformer have been used. It is important to use a long, twin-core shielded lead between the coil and oscilloscope to ensure that the field set up by the transformer in the oscilloscope is not included in the measurements. The oscilloscope should be used in the balanced input mode with the braid and the user grounded. For calibration, the Biot–Savart law gives the field directed concentrically around a long straight wire as

$$B = 20\frac{1}{D}\mu T$$

where I is the current in amperes passing through a wire at a distance D cm from the coil. This gives the sensitivity of the coil in, say, millivolts per microtesla. This equation also shows that a current of 0.5 A running 1 m from the microscope produces a field of 0.1 µT, sufficient to prevent atomic resolution.

Severe a.c. field interference causes image distortion (see Fig. 11.1) and a unidirectional blur on Fresnel fringes. If the interfering field is homogeneous and cannot be eliminated at its source, a magnetic field compensator can be used which applies a field of opposite phase to the interfering field. A sensor coil is used to measure the ambient field, which is amplified and applied, with reverse phase, near the specimen chamber. A description of this equipment can be found in Gemperle et al. (1974). These devices are most effective against interference by low-frequency homogeneous fields, but can only compensate fields at one point.

The microscope itself should not contribute more than about 0.03 µT to the ambient a.c. field. While ground loops are a common cause of a.c. field interference, other causes include

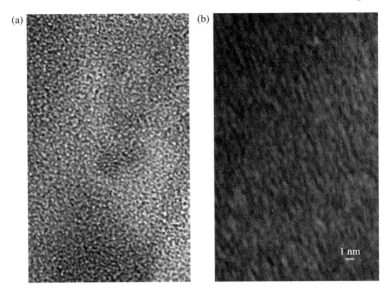

Figure 11.1 The effect of 50 Hz magnetic field interference on the high-resolution image of an amorphous carbon film. The field strength is 3 µT (30 mG) measured outside the specimen chamber in the direction of maximum strength and would be considerably reduced at the specimen itself. This field is due to the current in the goniometer motor drives, which must be switched off before recording images on some models of electron microscope. The field is off in (a) and on in (b).

electric motors (in air conditioners, elevators, rotary pumps, etc.) and transformers. These can often be rotated to a position of minimum interference. The undesirability of siting a microscope near an elevator motor, autotransformer, fluorescent light, electric welding unit, railway track, tram track, or NMR unit is discussed further in Alderson (1975), which should be read with care before choosing a site for a new instrument.

Since the amplitude of image blurring due to a.c. magnetic field interference depends on the accelerating voltage, it is easily distinguished from mechanical vibration using image recordings taken at, say, 50 and 100 kV. If the source of the interference is mechanical, it will not change with voltage.

For STEM the environmental field requirements are more stringent. Slowly varying fields produce image distortions, since much longer exposure times are involved in ADF-STEM or EELS than HREM. Stray fields of less than 0.01 µT may be needed at both the gun and the sample, and the viewing chamber glass in TEM/STEM instruments is a common source of leakage. For aberration-corrected instruments, the requirements will be more stringent still, since aberration correction to the 0.1 nm level is pointless with 0.3 nm probe broadening due to stray fields. It is always preferable to track down the sources of strong fields (e.g. a chilled water pump motor) and eliminate them, rather than to use active field compensators or attenuation by magnetic shielding. Strong fields at EELS spectrometers are particularly important—many researchers have reported the need to use wooden chairs, since the movement of steel furniture visibly affects the spectrum.

Von Harrach (1995), James and Browning (1999), and Muller and Grazul (2001) give more details of environmental requirements and problems for STEM.

11.2. High-voltage instability

The incoherent image broadening due to high-voltage fluctuations, resulting in variations in all the lens focal lengths, must be reduced to well below the spatial coherence width for it to be possible to obtain a high-resolution phase-contrast image. A high-voltage stability of a few parts in 10^6 is just sufficient, as discussed in Section 2.6.2. The important sources of instability are:

1. noise in the reference voltage;
2. the stability of the high-voltage divider resistor chain;
3. dirt in the oil tank;
4. the cleanliness of the gun chamber;
5. the vacuum in the gun chamber.

Microdischarge reaches a local maximum at about 10^{-4} Torr, so that a stable gun must be operated well above or below this pressure. A vacuum of 5×10^{-7} Torr (measured in the rear pumping column) is generally adequate for high-resolution imaging using a conventional filament, and this will only be obtained with a microscope in excellent condition.

High-voltage instability is revealed by flickering on the beam current meter and by fluctuations in the size of the caustic image. If either of these are encountered, dirt in the gun chamber is the most likely cause. Small amounts can be removed by conditioning, that is, by operating the gun for a short period above the voltage at which it is intended to be used. Major cleaning of the gun electrodes must be done using fine metal polish and alcohol, finishing with a burst of dry nitrogen gas under pressure. For thermionic filaments, the Wehnelt hole must be examined after cleaning every time it is replaced. The Wehnelt should be cleaned by ultrasonic agitation. Examine the main gun chamber O-ring for hair-line cracks.

For thermionic electron sources, much of the labour of high-resolution electron microscopy consists of time spent cleaning the microscope gun chamber. Instability which persists after cleaning the gun is likely to be due to poor vacuum. This may be caused by leaks elsewhere in the column, or, in earlier times, by outgassing from photographic plates. Outgassing may also be due to an old specimen fragment lodged in the column, or to the microscope interior itself. A method for distinguishing high-voltage instabilities from those due to objective-lens current fluctuations is described in Agar et al. (1974). This is based on the fact that while a change in objective-lens current tends to rotate the image, the image blurring due to high-voltage fluctuations acts in a radial direction.

A sensitive permanent monitor of high-voltage stability is an oscilloscope. On most instruments a connection is available (on the oil tank or energy-loss spectrum analyser) from which a signal may be taken off through shielded leads to an oscilloscope. The oil-tank monitor voltage is proportional to the microscope accelerating voltage. A ground loop is possible if the metal case of the oscilloscope is in contact with the metal of the microscope cabinet and if the wall socket used for the oscilloscope has a separate ground return from that used by the microscope.

11.3. Vibration

Mechanical vibration is rarely a problem on well-sited modern instruments, and specimen movement during an exposure is more likely to be due to thermal expansion of the stage or specimen (see Section 11.4). Alderson (1975) contains a full discussion of the effects of vibration and the choice of site for a microscope. Seismologists are the experts in this field and usually have access to vibration-measuring equipment.

The microscope and its supports can be thought of as a mechanical oscillator. The resonant frequency of a mechanical oscillator is inversely proportional to the square root of the product of mass and compliance. The aim in minimizing vibration of an electron microscope is to keep the resonant frequency of the microscope system lower than any likely exciting vibrations, say less than 20 Hz. This is done by making the mass and compliance large and the microscope structure very rigid to minimize the number of internal coupled oscillators. The mass can be made large by siting the microscope in a basement on a massive concrete block set into the ground and insulated from the building structure. Alternatively, the compliance may be increased by supporting the microscope on air bags, the system used on most modern machines.

Roughing pumps, water pumps for building heating and cooling, and turbulent cooling water flow in the microscope are common causes of vibration. A change in the water pressure can be used to diagnose this latter cause. The digital diffractogram of a carbon film will indicate the presence of drift or vibration with the characteristic rings washed out in the direction of any vibration or drift. This is a useful test if a double image exposure does not indicate the presence of systematic drift, yet vibration is suspected. It enables the effects of astigmatism and vibration to be distinguished, as shown in Fig. 10.5. A useful diagnostic technique is to attempt to exaggerate the fault observed by applying known fields or vibrations. If the microscope is fitted with an electron detector, a sensitive indication of vibration effects or magnetic fields is obtained by feeding the output from this detector into an oscilloscope or spectrum analyser. A small focused spot is formed across a fixed specimen edge, so that movement of the spot appears as a fluctuation in the output of the detector.

A laser light-beam reflected onto a wall from the surface of a dish of mercury placed on a vibrating surface can be used to give a rough measure of vibration amplitudes; however, only differences between the displacement of the dish and that of the laser can be detected in this way. The construction of an inexpensive vibration monitor using a crystal microphone is described by Nicholson (1973). These vibration measurements frequently reveal an appreciable difference between vibration levels during the day and those recorded at night. The human hand is also a sensitive vibration detector.

For STEM, field-emission tip vibration plays a similar role to sample vibration in TEM, suggesting that tip mounts should be designed to be as stable as TEM stages.

11.4. Specimen movement

Specimen movement during an exposure (drift) is the most common cause of spoiled image recordings at high resolution. The amount of drift is referred to the object plane and so depends only on the exposure time and not on the magnification. In high-resolution STEM the effect of drift is to cause a distortion of the image. The common causes of drift are discussed here.

1. The specimen grid must make good electrical contact with the specimen holder, and any retaining screw or clip must be tight. The specimen itself must either be a conductor or coated with conducting material such as carbon to prevent it from accumulating charge under the action of the beam. A charging specimen is easily identified by the erratic movement or jumping which occurs. A specimen picked up from paper with the finest tweezers will occasionally include a paper fibre collected by the tweezers. If dust particles are frequently found with the specimen a fan should be installed in the microscope room, drawing air in through a dust filter and thereby maintaining the microscope room at a slight positive pressure. Dust is carried out under doors and cracks by the continuous draught established by the fan. Following practice in the semiconductor industry, clean-room systems are available to maintain less than a given density of particulates of a given size. Smoking should not be allowed in the microscope room. Always examine a specimen first at low magnification to check for the presence of charging dust particles or fibres which may be out of the field of view at high magnification. These may introduce a variable amount of astigmatism.
2. It is important on top-entry tilting holders that there is good electrical contact between the portion of holder which is free to tilt and that which is not. Since jewelled bearings are often used, a metal spring must be used to provide electrical contact, and this must be effective.
3. There must be no contact between the specimen holder and the anti-contamination cold-finger. A cold-finger whose height is adjustable is an advantage, since one may want to use the shortest possible focal length, thereby crowding the specimen, cold-finger, and objective aperture into a very small gap a few millimetres wide. An electrical continuity test is useful. There should be an open circuit between the frame of the microscope (connected to the specimen) and the external extension of the cold-finger into the nitrogen reservoir. The continuity should be checked through the full range of specimen tilt positions.
4. A clogged lens cooling-water filter is a common cause of drift. The restricted water flow causes the lenses to overheat, which in addition to causing drift may also cause the illumination spot to wander as the impedance of the lens winding varies with temperature. The external lens casing should not feel hot—if it does, a blocked water filter is a likely cause. Microscopes using internal recirculated water supplies generally do not have this problem. The duty cycle of the thermoregulator should be checked every week as this will indicate any reduction in cooling-water pressure. A refrigeration unit is avoided in some microscope designs by running the entire system somewhat above room temperature.
5. The translation mechanism must be operating properly. The O-rings surrounding the push rods to the specimen stage must contain the correct amount of vacuum grease. Too little or too much can cause drift. Any pull-off springs must supply the correct tension over their full range. The mechanism can be tested by operating one translation control (say the X direction) several turns clockwise, then the same number of turns anticlockwise to bring the specimen to its original position. If the Y control has not been touched, there should be no movement in the Y direction.

For bright-field images, where exposure times are short (a few seconds), specimen drift is not generally a problem. For high-resolution dark-field images and STEM exposures of up to 30 s are common, and for these, if the conditions mentioned above are met, the main requirement is complete thermal equilibrium of the specimen and stage. This may take several hours to achieve if the lenses have been turned off. The longer the microscope can be left running, with all liquid-nitrogen reservoirs full, the better for long-exposure high-resolution work. Before taking such a TEM recording, examine the specimen at the highest magnification available for signs of drift using an area of specimen adjacent to that of interest. Use the translate controls to 'home in' on the area of interest, using small oscillatory X and Y excursions of the translation controls to relieve any strain in the mechanism. A modest jolt to the microscope bench will be seen (at high magnification) to further settle the specimen.

For even longer exposures, a procedure which has been found to be successful is to allow the microscope to stabilize, set up all the experimental conditions exactly, then turn down the filament and leave the microscope for several hours with an adequate supply of liquid nitrogen. On returning, it is only necessary to turn up the filament, focus, and record the image without touching the specimen-translate controls. The thermal expansion of the specimen holder rod used on side-entry stages generally makes these less suitable for long-exposure high-resolution imaging.

Specimen movement is easily measured by taking a double exposure with several seconds separation. It also appears in an optical diffractogram of a thin carbon film as a fading of the ring pattern in one direction, thereby distinguishing the effect of drift from that of astigmatism (see Fig. 10.5). A slow unidirectional drift in focus is usually due to thermal gradients in the microscope column. These can be reduced by eliminating draughts, by ensuring that the lens cooling-water temperature is close to room temperature and is supplied at the correct flow rate, and by leaving the magnification setting near maximum when the microscope is not in use.

11.5. Contamination and the vacuum system

Contamination rates well below $1 \text{ nm}^3 \text{ min}^{-1}$ are frequently quoted for modern instruments. While this figure depends on a variety of experimental conditions such as the specimen thickness and temperature, no serious difficulty should be experienced with contamination on TEM instruments if plasma cleaning of the sample and holder is used, and if the machine is fitted with a modern design of anti-contamination cold-finger held at liquid-nitrogen temperature. Contamination refers to the build-up of decomposed carbon on a specimen. There is evidence (Knox 1974) that this forms in a crown or annular mound with a diameter equal to that of the illumination spot if this is small. In a clean instrument, the main sources of contaminant are the specimen, the specimen holder (both frequently introduced from outside the machine), the diffusion pump vapour, photographic plates, if used, and O-ring grease. Mass spectrometer analysis shows a marked increase in vacuum components likely to cause contamination when a specimen holder which has been touched is introduced into the microscope, owing to finger grease. For this reason, the specimen holder should always be handled with disposable or nylon gloves. It can be cleaned using pure alcohol and ultrasonic agitation, or better, by using a plasma-cleaning device. Many O-rings will not require any grease, on those that do (movable and frequently changed parts) use a minimum and carefully check for dust particles trapped in the grease on the O-ring. With side-entry holders the vacuum-seal O-ring should be examined for dust under an optical microscope every time before the holder is inserted into the microscope. Once an O-ring has established a good seal it should not be interfered with.

The contamination rate can be reduced by heating the specimen with a large illumination spot, or by improving the overall microscope vacuum. The rate is often measured by noting the reduction in the size of small holes in a carbon film after several minutes. A great deal of research has been done on contamination, particularly in scanning microscopy, and the article by Hart et al. (1970) contains references to much earlier work. To obtain a sufficiently low contamination rate for STEM, it is usually necessary to pre-heat or 'bake-out' the specimen under vacuum.

Specimen etching is occasionally experienced if oxygen or water vapour are present near the specimen. If this is observed, the most likely cause is a vacuum leak in the specimen exchange mechanism. Some specimens themselves release oxygen under the action of the beam. Water vapour is well condensed by the anti-contamination device if this is at liquid-nitrogen temperature (where the water vapour pressure is 10^{-9} Torr); however, it quickly vaporizes as the temperature is allowed to rise (water vapour pressure is 1 Torr at $-20\ °C$).

A microscope vacuum of at least 10^{-7} Torr is required for high-resolution work, measured in the pumping column at the rear of the microscope. The vacuum affects both the contamination rate and the high-voltage stability. The vacuum will be inferior in the region of the photographic plates if used (unless they have been very well dried) and may be greatly improved within the specimen anti-contaminator owing to the cryopump action of these cold surfaces. A good vacuum in the gun chamber is important, though this is difficult to measure directly. Apart from leaks, the main limitation on the microscope vacuum is adsorbed molecules on the internal walls of the instrument. On an instrument with a low leak rate, an order of magnitude improvement in vacuum can generally be achieved by heating the rear pumping line with electrically heated tapes for several hours. The pumping lines must not be heated beyond a temperature which will damage any components. The lens casings cannot be baked-out in this way, but by adjusting the thermoregulator they can safely be run rather warm for some hours with a similar effect.

A useful simple test for the microscope leak rate can be made by timing the decline in measured vacuum to a particular point after the instrument has been closed down—that is after closing the main diffusion pump valve to the microscope and all roughing lines. A weekly check will indicate the development of any new leaks which are easily traced using a mass spectrometer or partial pressure gauge as a leak detector. If one is not fitted, a large leak can sometimes be found by squirting small quantities of acetone around the suspected area and watching for a change in the vacuum gauge. A log book kept of the number of hours between filament changes will also serve to indicate the quality of the vacuum when using a thermionic or LaB_6 source. Thus an hour meter, actuated by the high-voltage relay, makes an excellent vacuum monitor.

A major advance in contamination reduction has been the development of plasma-cleaning devices, which are now essential for both HREM and STEM. These dramatically effective devices must be applied to the entire portion of the holder which enters the microscope, and to the sample holder tip and sample. The process is so effective in removing carbonaceous material that, in STEM, the problem of contamination has now been replaced by that of hole-drilling by the beam, which is prevented by contamination. With the very high currents possible using aberration correctors this problem has become more severe. A probe left focused on the same place for a long time eventually either drills or contaminates. A plasma cleaner could be incorporated into the sample exchange mechanism.

11.6. Pressure, temperature, and draughts

Air flow from air-conditioning systems can be a serious problem, and an air velocity of less than 5 m min^{-1} is desirable. Strips of tissue paper have been used to detect airflow—a 0.3 m × 0.6 cm strip should deflect by less than 2.5 cm. These 'tell-tales' can be placed around the room to monitor draughts. Draughts can be reduced while maintaining constant temperature by siting the heat-producing electronics in another room, by using radiative room cooling, or by surrounding the microscope with thick curtains. Temperature variations in the microscope room should be less than 0.25 °C. Some machines are wrapped in Neoprene or 'bubble-wrap' to improve isolation. Recently, ambient pressure changes due to doors opening, or even weather events, have been found to produce large distortions in ADF-STEM images at 0.2 nm resolution (Muller and Grazul 2001). Pressure changes produce time-dependent forces across the vacuum–sample holder interface. A sensitivity of 0.1 nm Pa^{-1} has been measured. A passing storm can create pressure fluctuations of 30 Pa. For this reason, a hermetically sealed airlock cover is commonly fitted to the highest-performance machines. Similarly, ordinary speech will be seen to blur HREM images during an exposure.

References

Agar, A. W., Alderson, R. H., and Chescoe, D. (1974). Principles and practice of electron microscope operation. In *Practical methods in electron microscopy*, Vol. 2, ed. A. M. Glauert. North-Holland, Amsterdam.

Alderson, R. H. (1975). The design of the electron microscope laboratory. In *Practical methods in electron microscopy*, Vol. 4, ed. A. M. Glauert. North-Holland, Amsterdam.

Gemperle, A., Novak, J., and Kaczer, J. (1974). Resolving power of an electron microscope equipped with automatic compensation of transverse a.c. magnetic field. *J. Phys. E* **7**, 518.

von Harrach, H. S. (1995). Instrumental factors in FEG STEM. *Ultramicroscopy* **58**, 1.

Hart, R. K., Kassner, T. F., and Maurin, J. K. (1970). The contamination of surfaces during high energy electron irradiation. *Phil. Mag.* **21**, 453.

James, E. M. and Browning, N. D. (1999). Practical aspects of atomic resolution imaging and analysis in STEM. *Ultramicroscopy* **78**, 125.

Knox, W. A. (1974). Contamination formed around a very narrow electron beam. *Ultramicroscopy* **1**, 175.

Lichte, H. (2001). The Triebenberg Laboratory—designed for HREM and holography. *Microsc. Microanal.* **1**, 894.

Morrison, R. (1967). *Grounding and shielding techniques in instrumentation*. Wiley, New York.

Muller, D. A. and Grazul, J. (2001). Optimizing the environment for sub-0.2 nm STEM. *J. Electron Microsc.* **50**, 219.

Nicholson, P. W. (1973). A vibration monitor for ultramicrotomy laboratories. *J. Microsc.* **102**, 107.

Turner, J. H., O'Keefe, M. A., and Mueller, R. (1997). Design and implementation of a site for a one-angstrom TEM. *Microsc. Microanal.* **3**, 1177.

12
Experimental methods

This chapter describes the practical problems in recording high-resolution images, mainly in TEM mode. The best way to learn to record and interpret good high-resolution images is to record a lot of images of gold particles on thin carbon films and analyse these using the digital diffractograms (for TEM) or Ronchigrams (for STEM) described in Sections 10.6 and 10.10. Then, for TEM, proceed with the project described in Appendix 5 ('The challenge of HREM').

To obtain point detail consistently on a scale less than 0.5 nm requires that the microscope be kept in excellent condition. Changes to the objective lens pole-piece and specimen stage must be kept to a minimum. Apertures must be clean. Of critical importance is a microscope site which is satisfactory with respect to vibration, magnetic fields, humidity, temperature, and dust (see Chapter 11). The spherical aberration constant, measured under the proposed experimental conditions, should be less than 2 mm (see Section 10.2) for uncorrected machines. *However, it is the electronic stability and reliability of the microscope which are most important for high-resolution work*. High-voltage stability must be $\Delta V/V = 3 \times 10^{-7}$ min^{-1} or better (depending on resolution), with a similar figure for the objective-lens current stability. A full discussion of the influence of electronic instability can be found in Section 4.2, Appendix 3, and Section 2.8.2. A contamination rate better than 1 nm^3 min^{-1} is required. Thermal drift of the specimen must be minimized (see Section 11.4). A vacuum of at least 10×10^{-6} Pa, measured at the top of the rear pumping line, is required to minimize contamination and high-voltage instability. Oil-free dry pumping must be used throughout. The interior of the microscope, in particular the gun chamber, must be scrupulously clean. Plasma cleaning of the sample holder (and sample) provides a big improvement in reducing sample contamination, and is now routine for the STEM mode, which is especially sensitive to contamination problems. The lattice fringes seen in small particles of partially graphitized carbon form a useful test specimen (see Section 10.3) and are available commercially. These test specimens can be kept permanently on hand, and, before starting work each day, clear images of the 0.34 nm lattice fringes in this specimen should be obtained using a single-tilt specimen holder, to confirm that the microscope is operating correctly.

Most high-performance machines now separate the electron microscope room from an electronic control room, in order to remove the 100 W human thermal source of heat and movement from the machine. Then, if a technician is responsible for loading samples, while users are only permitted to operate software in the control room, mechanical damage to the microscope during training can be reduced.

12.1. Astigmatism correction

The following comments apply to uncorrected microscopes not fitted with auto-tuning software. The setting of the objective stigmator is critical at high resolution. Some machines provide direction (azimuth) and strength controls for the stigmator, while others allow the strength to be varied in orthogonal directions (see Section 2.8.3). The astigmatism should be approximately corrected by adjusting the controls in a systematic way to produce a symmetrical Fresnel fringe around a small hole (Agar *et al.* 1974). The more accurate correction needed at high resolution can then be obtained by observing the structure of a thin amorphous carbon film at high magnification, and by study of the tableau of diffractograms provided by the image analysis software. An electron-optical magnification of at least 400 000 is needed to correct astigmatism accurately in 0.3 nm image detail using $\times 10$ viewing binoculars. Both the focus controls and the stigmator should be adjusted for minimum contrast, which should occur at a sharply defined setting (eqn 6.26). Under-focusing the objective by a few tens of nanometres should then show the grainy carbon film structure with no preferential direction evident in the pattern. This is a difficult procedure requiring considerable practice, and can be avoided using on-line digital diffractograms displayed at about 10 frames per second (see Section 10.6). This is fast enough for users to become familiar with the transition of these patterns from a Maltese cross via an ellipse to a circle as astigmatism is corrected. The following points are important.

1. Since both are well within the lens field, the position of the objective aperture and lower cold-finger and the state of cleanliness of both critically affect the required stigmator setting. Never touch either, after astigmatism correction or before recording an image. A small amount of contamination on the objective aperture may not matter so long as the field set up by it can be corrected using the stigmator. Test for the amount of contamination on the objective aperture by moving the aperture while viewing the image. There should not be a violent degradation in astigmatism with small aperture movements.
2. The astigmatism correction needed varies with the type of holder used, the high voltage, and the objective lens strength. It appears to depend weakly on the magnification, possibly owing to flux linkage between the lenses.
3. A common cause of serious astigmatism is a piece of dislodged or evaporated specimen in the objective pole-piece. If the machine is vented, these sample fragments can sometimes be seen using a dentist's mirror. But a service engineer may be needed to remove and clean the pole-pieces.
4. Unless the cold-finger reservoir is continuously topped-up with liquid nitrogen by a pump, the image will rapidly become astigmatic when the reservoir boils empty, which causes the cold-finger to move as it warms up. So there is no point in correcting the astigmatism until the cold-finger reaches liquid-nitrogen temperature. This may take at least half an hour after first filling the reservoir.
5. Astigmatism can result if an intense Bragg reflection is caught on the objective aperture edge. This situation should be avoided.

The sensitivity of high-resolution images to astigmatism can be judged by observing the lattice fringes of graphitized carbon, in bright field with untilted illumination, while adjusting the stigmator. To obtain fringes running in all directions (Fig. 10.3), both stigmators must be set very accurately. One-dimensional fringes are more easily obtained, but these do not ensure accurate astigmatism correction.

Astigmatism must be corrected each time the objective aperture is moved, and before recording every image.

12.2. Taking the picture

It has been said that the skill of high-resolution electron microscopy lies in knowing when to take the picture. Specimen drift is to some extent an unpredictable effect, so that a certain amount of luck is required for the longer-exposure recordings. The following procedure should minimize the number of spoiled image recordings—these precautions become increasingly important as the exposure time is increased. A list of conditions which should be checked before recording high-resolution images is given in Section 12.9.

1. Where binocular focusing is used, focus each eyepiece separately (closing one eye), then adjust one eyepiece with both eyes open until a sharp image of the phosphor grain is seen. Set up the digital image acquisition system and display.
2. The anti-contamination liquid-nitrogen Dewar should have been full for at least half an hour. Astigmatism is introduced if the temperature of the lower anti-contamination cold-finger is allowed to alter (owing to thermal expansion), so that this must be allowed adequate time to reach thermal equilibrium. The specimen must have been out of contact with the lower cold-finger for several minutes—this is important if a tilting stage is used with an extended top-entry holder operating at short focal length. At the extremes of tilt, the specimen may touch the lower cold-finger, resulting in specimen cooling and drift. The thermal stability and cleanliness of the three components immersed in the objective-lens field (cold-finger, objective aperture, and specimen) are of the utmost importance, in order to prevent charging effects. By switching to a low-magnification range where the field of view is limited by the lower cold-finger, the cleanliness of this aperture can be examined.
3. Check for high-voltage stability both by observing the electron beam current meter, which should not flicker, and by examining the caustic image. On sitting down at any microscope for high-resolution work, the first thing to do is to watch the beam current meter pointer closely for at least 20 s. Intermittent flickering of the meter during this time indicates that the instrument is unsuitable for high-resolution work. Using this test, most instruments in routine use will be found to be unsuitable without some maintenance (usually gun cleaning and leak repair). An occasional flicker (say, once every 20 s) may not be important, since these voltage spikes are often very short compared with both the exposure time and the persistence time of the human eye. An oscilloscope monitor (Section 11.2) can be used to distinguish fast spikes from an increase in the continuous background fluctuation which determines the resolution limiting factor Δ in eqn (4.9). The caustic image is obtained by switching to diffraction mode without a specimen or imaging apertures in place. Adjust the diffraction focus for a disc with a bright outer ring. This ring contracts or expands with changing high voltage and can be examined through the binocular. Imaging filters have a streak mode which provides the best way to analyse the frequency components (e.g. 50 or 60 Hz) in high-voltage instabilities.

4. The microscope must be correctly aligned as described in the instruction manual. Check the illuminating system, in particular the gun tilt, condenser astigmatism, and condenser aperture centring. For thermionic electron sources, gun tilt will need frequent readjustment during the first hour of operation. This is important, since for the highest resolution one is generally using a small illuminating aperture and so requires the maximum possible source brightness. Check the gun bias and make a note of the illumination conditions. These will be needed if a comparison with computed images is intended. If the microscope is not fitted with a high-voltage 'wobbler', obtain the current centre for the objective lens by adjusting the beam tilt until no image translation is observed at the centre of the screen with change in focus (see Agar et al. 1974). A difference at the object plane of about 0.5 μm is common between the current centre (the point about which the image rotates with change in lens current) and voltage centre (the point about which the image expands with change in high voltage). If a high-voltage wobbler is fitted, alignment by voltage centre is preferred since, for an axial image point, this confers some immunity to high-voltage instability due to microdischarge during long exposures. This, and the energy spread of electrons leaving the filament, is the main source of chromatic aberration for thin high-resolution specimens where energy losses in the specimen itself are generally negligible. The use of voltage-centre alignment minimizes chromatic aberration for axial image points due to these effects at the expense of image distortion due to fluctuation in lens current (which would be minimized by using current-centre alignment). Once aligned, modern instruments should not require any other major adjustments other than those mentioned above. An excellent review of the practical procedures used to correct misalignment and geometric and chromatic field aberrations can be found in Riecke (1975). A useful simple method for coma-free alignment is described in Kimoto et al. (2003).

5. The diffraction conditions required must be set up and recorded. For many-beam lattice images of small crystals, a motor goniometer is required since top-entry stages, particularly if operated at short focal length, are not eucentric. It is necessary to operate both the translate and tilt controls simultaneously to keep the same region of specimen in view while tilting. This is difficult for the smallest crystals, which are usually also the thinnest, and therefore most suitable. The method for doing this is described in Section 12.4. A zone-axis orientation is usually required and it is important to ensure that the region of crystal which is 'in focus' is that from which the recorded diffraction pattern is obtained, since this may form the basis of subsequent image simulation. After obtaining the required tilt condition, make a final check for the clearance between the specimen and the lower cold-finger using the electrical continuity check. The current centre obtained in (4) above ensures that the illumination direction coincides with the optic axis, so that the central diffraction spot can now be used to locate the objective aperture (if one is to be used) with respect to the optic axis (see Fig. 5.1).

6. Check the specimen for drift by observing the image at the highest magnification available for several seconds. Use a region of the specimen near that to be subsequently recorded.

7. Once a suitable region of the specimen is found, 'home in' with the translation controls (see Section 11.4) and give a modest jolt to the microscope bench to settle the specimen. The final adjustments will be to the focus and stigmator controls (see Sections 12.5 and 12.1). Where image plates are used, one must operate the plate advance mechanism

then wait for a few seconds to allow mechanical vibrations to decay. The final focus adjustment, in which a setting is established a certain number of 'clicks' from the minimum contrast condition, must be made *after* operating the plate advance mechanism. Vibration from this mechanism may alter the specimen height by up to 10 nm. Open the shutter (or expose the electronic detector) carefully without otherwise touching the microscope. Do not talk or move any metal objects during the exposure. Magnetic field variations of up to 0.4 μT have been recorded as a result of the movement of a wristwatch. The effects of loud speech on lattice images can easily be observed through the binocular. Finally, close the shutter. The disturbance caused by a plate transport mechanism, and the time taken for this to decay, can be seen by watching the lattice image after closing the shutter. Fringes generally remain obscured for several seconds after the transport has stopped, giving an indication of the sensitivity of the lattice fringe image to mechanical vibration. CCD cameras eliminate this problem. A common problem with side-entry stages is pressure pulses in the room due to doors being opened and closed somewhere in the building during an exposure.

When using field-emission sources, beam damage is a primary consideration, as described in detail in Section 6.12. This can appear as amorphization, hole drilling, or blotchy strain contrast. Reduce this by either reducing the beam intensity or (if knock-on damage is suspected) the accelerating voltage, or by cooling the sample. Coating the sample with carbon has been found to reduce surface sputtering and, by conduction, to prevent charging.

12.3. Recording atomic-resolution images—an example

The main requirement for finding lattice fringe images is patience. The following is a step-by-step outline of the procedure used to find and record the image shown in Fig. 12.4, which is used as an example of the many-beam lattice imaging method. Alternatively, crystals of graphitic carbon, gold, indium phosphide, gallium nitride, or MgO may be used (as suggested in Appendix 5) in place of silicon. MgO smoke cubes have the advantages of requiring very simple specimen preparation, while providing an internal thickness calibration. However, small crystals are more difficult to align than a large single crystal. The silicon image shown was obtained from a chemically thinned wedge-shaped specimen of silicon with the incident beam along the [110] crystal direction. Approximately seven Bragg reflections contribute to the image; however, this number is not sufficient to allow the individual silicon atom columns to be resolved. We assume that the optimum specimen position in the objective lens pole-piece has been found (see Sections 1.2 and 2.7). This is specified by the objective lens current needed to focus a specimen at the correct height, and is shown on the lens display. When using a top-entry stage with specimens surrounded by a thick rim such as chemically thinned silicon and germanium, spacers or washers placed above the specimen may be needed to obtain the correct focusing current. (This is not an issue with symmetric field geometries.) Consistent results between specimens can be obtained by thinning these specimens most of the way from one side, as shown in Fig. 12.1, so that only one special washer need then be kept for these specimens. No such problems arise with side-entry stages in which the specimen height is continuously adjustable. However, side-entry specimen holders should be examined under a low-power optical microscope to

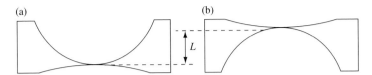

Figure 12.1 For chemically thinned specimens when using a top-entry holder, the specimen height in the microscope (and therefore the lens current needed to focus) may depend on which way up the specimen is when inserted into the instrument. Inverting the specimen may alter the height by an amount L which is comparable with the lens focal length. In order to work at small focal lengths, and so reduce aberrations, one usually wants the specimen very close to the lower anti-contamination cold-finger. This can be achieved by thinning most of the way from one side, and consistently inserting specimens in the holder in orientation (a).

ensure that the specimen retaining clip is accurately positioned and making firm contact with the specimen. For both types of holder the specimen must make firm mechanical and electrical contact with the specimen holder.

1. The preliminary checks given in Section 12.10 should be conducted. These procedures must be followed every day before commencing work if long fruitless searches for lattice fringes are to be avoided.
2. Load the standard specimen (partially graphitized carbon—see Section 10.3) into the microscope and confirm that the 0.34 nm fringes can be seen. A non-tilting specimen holder can be used.
3. Fit the silicon specimen to the tilting specimen holder, being careful not to damage it. *Wear gloves at all times when handling the specimen holder* both to reduce contamination and avoid collecting toxic BeO from beryllium components. Check the position of the retaining clip on side-entry type holders. On top-entry holders in which the specimen is retained by a threaded ring the tightening torque applied to this ring is critical, and brittle specimens are easily broken. Treat silicon and germanium as you would a small piece of glass.
4. Load the silicon specimen into the microscope and bring the specimen to a horizontal position. Find the hole in the centre of the specimen. The illumination system must now be adjusted. For a thermionic source, the filament image should be examined at a magnification of between ×20 000 and ×50 000 and made symmetrical by adjusting the gun-tilt controls. Figure 12.2(a) shows an image of a correctly adjusted pointed filament; however, a field-emission source is more likely to be used. In Fig. 12.2(b) we see the effect of introducing a stray a.c. magnetic field (see Section 11.1), in this case caused by switching on the power supply to the goniometer motor. The central tip of the filament image provides a sensitive test of the presence of stray fields. For all sources (thermionic, LaB_6, Schottky/field-emission gun) this source image should be symmetrical. The gun-bias setting corresponding to the lowest beam current should be used, to minimize chromatic aberration (see Fig. 4.3 and Section 2.8.2). For field-emission sources, coherence decreases with increasing extraction voltage (larger virtual source size) as energy spread and emission current increase. A clean condenser aperture must

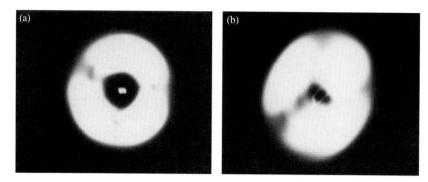

Figure 12.2 (a) Focused pointed-filament image used to form the images shown in Fig. 12.4. The form of this image depends on the gun-bias setting, the size of the Wehnelt hole, and the filament tip height. At the high magnification used to form lattice images, only a small portion of the bright ring falls over the plate camera. A great deal of time can be wasted looking for lattice fringes on an instrument in which the filament has not been set up correctly. (b) The effect of 50 Hz stray a.c. magnetic field on the filament image shown in (a). Note the alteration in the appearance of the filament tip. This field was due to a piece of ancillary equipment and washed out all fine lattice fringes. A simple test for these fields, in which the spot is run across the screen using the beam deflection coils, is described in Section 11.1.

be selected (examine it by switching the microscope to lowest magnification) and centred. The images shown in Fig. 12.4 were obtained with a condenser aperture which subtends a semi-angle of 1.16 mrad at the specimen, as can be measured from a recording of the diffraction pattern taken with focused illumination (see Fig. 12.7). A diameter of 200 μm is a typical condenser aperture size. Factors affecting the choice of condenser aperture size are discussed in Sections 4.2 and 4.6—a trade-off between the higher contrast due to improved coherence found with a small aperture but reduced intensity, is involved, which in turn affects the exposure time, sample drift, and choice of magnification. The condenser aperture is centred by swinging the condenser focus controls through the exact focus position and adjusting the aperture position to obtain an image which expands and contracts concentrically. The condenser astigmatism controls must also be set correctly, and the maximum strength for the first condenser lens must be selected (sometimes called 'spot size') which will subsequently provide sufficient intensity for focusing at high magnification (see Sections 4.5 and 9.1 for other factors affecting the choice of first condenser lens excitation). If a thermionic filament is being used for the first time it will be necessary to check the gun-tilt controls every 15 min or so for the first hour of operation.

5. The specimen must now be accurately tilted into the zone-axis or symmetrical orientation. Figure 12.4 was obtained from a silicon crystal in the [110] zone-axis orientation, that is, with the electron beam travelling in the [110] crystal direction. A method for doing this is described in Section 12.4 for top-entry stages. Check that the image is in focus at the correct (optimum) objective-lens current (see Section 1.2). When using tilted specimens there may be an appreciable difference between the lens current needed

to bring regions at the edge of the specimen into focus and that needed to focus the central region of the specimen. Intersecting bend contours may be used to find zone-axis orientations.

6. A preliminary check of astigmatism, drift, and focusing should now be made. Examine a thin edge of the specimen at a magnification of about 600 000 and find the minimum contrast setting of the stigmator and focus controls without an objective aperture, as described in Section 12.5. Once this has been found, weaken the objective lens by about 90 nm (see Section 10.1) to increase the image contrast and watch the image carefully for about 5 s. It should appear completely motionless, and the background graininess of the image should appear sharp. With experience, the appearance of this background becomes a useful rapid guide to image quality. If the image is not sharp or is unstable check the following:
   1. The anti-contaminator has reached thermal equilibrium.
   2. The high voltage is stable.
   3. The specimen is not touching either cold-finger.
   4. A dirty objective aperture has not accidentally been left inserted into the microscope.
   5. Intermittent erratic jumping of the image commonly indicates that it is charging up. Silicon and germanium are sufficiently good conductors to prevent this from happening so long as an earth return path exists for charge to leak away. If charge appears to be accumulating on the specimen, check that it is making good electrical contact with the specimen-holder ring, and that an electrical connection exists between this ring and the specimen-holder body across the jewelled bearings. Special spring-metal washers are provided for this purpose (see Section 11.4). It is a waste of time to look for lattice fringes unless a stable, sharp image of the background in a very thin specimen region can be obtained. Check that sufficient intensity is available to enable the stigmator and focus controls to be set correctly at a magnification of about 600 000.

7. A search must now be made for a suitable area of specimen. A uniform wedge is required if the change in appearance of the image with thickness is of interest—these are obtained from tripod polished samples; ion-milled samples have a less constant thickness gradient and tend to be parabolic if etching is just stopped at perforation. Only by comparing computer-simulated images for various thicknesses with experimental images of a wedge-shaped specimen can one be certain of the image interpretation and so select regions of crystal thickness in which a simple image interpretation is possible. A convenient range of thickness would include two bright-field Pendellösung thickness periods. The separation of these 'equal thickness' fringes in the image depends on the wedge angle, accelerating voltage, and electron-optical magnification. To obtain these fringes, insert the smallest objective aperture to include only the central diffracted beam (without altering the specimen orientation) and examine the image at a magnification of about 500 000. Search the wedge edge for a region in which approximately two dark Pendellösung bands fill the portion of the viewing screen lying above the photographic plates. The bands should be even and the wedge edge smooth. Figure 12.3 shows such a region from which the lattice image in Fig. 12.4 was obtained. Note that the image extends slightly beyond the edge of the wedge. It is a mistake initially to seek a wedge with the smallest possible angle (Pendellösung bands widely separated) since specimens are invariably bent in these regions. Remove the objective aperture after 'homing in' with the translation controls

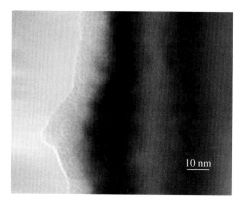

Figure 12.3 Pendellösung fringes (the dark vertical bands) formed in silicon in the symmetrical [110] zone-axis orientation (see Fig. 5.1c), formed using a small central objective aperture which excludes all but the central beam. Calculations give the increment in thickness between these bands as 23 nm; they therefore provide a useful method of calibrating the thickness scale for lattice images obtained from wedges. They also indicate the shape of the wedge—the thickness is constant along lines of constant intensity. Kinematic or single-scattering images are obtained on the thinner side of the first dark fringe, so that images of defects in this thickness range may be interpreted by the theory of Sections 6.1 and 6.2 (see also Section 5.3.1). The lattice image is obtained by removing the objective aperture, given the necessary instrumental stability, and is shown in Fig. 12.4(a) for this same specimen region.

(see Section 11.4) once a suitable region of specimen has been found. It may be necessary to examine several specimens in order to find a suitable specimen area.

8. If a stable image can now be seen showing sharp background contrast, make small adjustments to the stigmator orientation and fine-focus controls until a lattice image is seen. For Fig. 12.4, final adjustments to these controls were made at an electron-optical magnification of 640 000. (Higher magnification can take you into the region of 'empty magnification', where there are many detector pixels within the width of the electron microscope impulse response so that further magnification is pointless.) A simple way to set the magnification is to observe the digital diffractogram (Thon rings) and adjust the magnification until the point-resolution limit is about two-thirds of the Nyquist spatial frequency for single binning. Owing to the use of a non-standard objective lens current, a correction factor has been applied to the indicated magnification to obtain this true magnifications (see Section 10.3). The viewing binoculars usually provide an additional magnification of 10 times. Even without an objective aperture in place, a faint image of the Pendellösung fringes should be seen, whose contrast depends on the focus setting. Look for fringes initially along the thinner side of the first Pendellösung fringe. Check that both binocular eyepieces are focused (see Section 12.5) and that the image remains stable at least for the length of time needed for an exposure. If the image is too dim, switch to lower magnification and check the form of the filament image. Check also that the most intense part of the filament image is being used. Look 'into' the image rather than 'at' it. Knowing what you expect to see will soon lead to more rapid identification

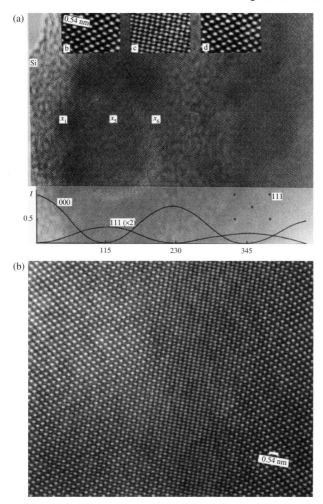

Figure 12.4 (a) Lattice image of silicon in the [110] symmetrical orientation taken at an accelerating voltage of 100 kV. The edge of the specimen can be seen top left, showing Fresnel edge fringes. Enlargements of regions x_2, x_5, and x_6 are shown at insets (b), (c), and (d). The image at x_5 gives a false impression of the structure (shown in inset b), while that at x_2 and x_6 faithfully represents this projection of the diamond structure to the resolution available. The intensity of the four (111) type Bragg beams contributing to the image, together with that of the central beam, is shown below in approximate registry with the image using a horizontal thickness scale in angstroms. The interpretation of these images is discussed further in Section 5.14 (see Spence et al. 1977, Glaisher and Spargo 1985). (b) Portion of the image shown in (a) around x_5 at higher magnification. The change in image appearance with thickness (which increases from left to right across the picture) is clearly shown. Near the centre of the image at a thickness of approximately 11.5 nm the central beam has been extinguished (see lower half of (a)) by Pendellösung fringes, resulting in a 'half-period' image (see Section 5.1) which does not reveal the crystal structure. This image results solely from interference between the four (111)-type reflections. The amplitudes *and phases* of all Bragg beams contributing to the image (see Fig. 12.8) must be studied in order to understand the thickness dependence of these images (see Section 5.14).

of a fringe image—this can only come with experience. Remember that the CCD will record far more detail than you can see through binoculars. The stigmator adjustment is critical—some experts find a point midway between orthogonal striations in the image. Making the smallest adjustments to the stigmator control, select the image showing sharp tunnels. Adjust the fine-focus control for maximum contrast.

9. Prepare the CCD shutter (or image plate or film advance mechanism for cryo-EM), make final focus adjustments, pause, and record the image (see Section 12.2). *Write down the objective lens current and focus defect.* A desk lamp with a dimmer is useful here.

For silicon images observed under the above conditions you will find several focus settings which give sharp, apparently identical images of high contrast. These are Fourier (or 'Talbot') self- images (see Section 5.14). Between these focus settings the lattice image seems to disappear. The focus increment between 'identical' images (images which appear identical through the viewing binocular) for silicon viewed in the [110] direction at 100 kV is $\Delta f_0 = 86$ nm. Apart from an unimportant half-unit-cell translation, these Fourier images are in fact identical in regions of perfect crystal. Only one image, however, will include a faithful high-contrast image of any crystal defects which may be present—this is the 'optimum focus' image (see Section 6.2).

As discussed in Section 5.15, for small-unit-cell crystals, the perfect crystal lattice image may show very low contrast (or be absent entirely) at this optimum focus if, as is common with older microscopes and metallurgical or semiconductor problems, the fundamental lattice spacing exceeds the point-resolution of the instrument. The highest-contrast lattice images may then occur at very large under-focus ($\sim$300 nm). The focus increment between Fourier images can be used to calibrate the focal steps of the microscope focus controls (see Section 10.1).

If a through-focus series is required, it is important to bear three points in mind.

1. Vibration from a plate advance mechanism following each image recording may alter the height, and thus the focus condition, of the specimen by an amount greater than the electronic focal increment selected. This problem is avoided with an electronic recording device.
2. Magnetic hysteresis may make it necessary to avoid reversing the direction of the lens current change, unless lens normalization has been implemented.
3. It may not be possible to return reproducibly to an earlier focus setting if this would involve changing ranges between the various focus knobs. This problem is eliminated on digital machines, which use a system of 'continuous' focus adjustment.

The exact crystal orientation used for the final image recording is difficult to determine, since the smallest selected area aperture will include a much larger region of specimen than that included in the lattice image. Problems associated with the design and construction of specimen-tilting devices, which will allow specimens to be tilted accurately and reproducibly to the same orientation, provide an important limit to the development of the methods of high-resolution electron microscopy. The need for accurate orientation setting increases with specimen thickness. Having set the crystal as close as possible to the zone-axis orientation using the method of Section 12.4, the best practical guide to orientation is probably the symmetry of the image. Since this depends on the stigmator setting, the possibility of inadvertently compensating for a small orientation error arises by using the stigmator. This 'cylindrical lens' (see Section 2.8.3) can be used to 'defocus' a strong pair of fringes and so reduce their contrast relative to a weaker pair of fringes running at right angles to these. The result may be a symmetrical two-dimensional image obtained from a specimen in which the five Bragg reflections contributing to the image do not have equal strength owing to an orientation error. A check on this unwanted effect is given by the diffractogram method described in Section 10.6.

For the finer fringes it may be necessary to search specimens for several days before a good image is obtained. It is difficult, however, to concentrate for more than about 2 h at a time. The anti-contaminator must not be allowed to warm up during the day (see Section 11.5), as this would result in a lengthy delay while the specimen area regains thermal equilibrium.

12.4. Adjusting the crystal orientation using non-eucentric specimen holders

Much of the time spent on high-resolution work is devoted to corrections of the specimen orientation. The tedious alignment process (which becomes increasingly critical in thicker crystals) is greatly facilitated on an instrument fitted with microdiffraction or 'nanodiffraction' mode. Here the symmetry properties of the convergent-beam pattern from bent specimens can be used to seek crystalline regions in the exact, required orientation.

A systematic method of bringing a small crystal into the zone-axis or symmetrical 'cross-grating' orientation is now outlined for instruments not fitted with microdiffraction facilities. It applies to top-entry stages fitted with a motor-driven goniometer, fitted with a pair of foot pedals to control the orientation and tilt of the specimen. The general principle of the method can also be applied to double-tilt side-entry stages if it is found that the specimen position which produces the highest-quality images does not coincide with the specimen height needed for eucentric operation.

1. Insert the specimen into the microscope and set the goniometer to the neutral position so that the specimen disc is approximately horizontal. We shall assume that the required zone-axis orientation is close to this position. Switch the microscope to diffraction mode and observe the focused diffraction pattern of a small crystal or thin region of specimen edge.
2. A pattern similar to that shown in Fig. 12.5 should be seen. The arc of intense spots is a portion of the Laue circle which traces the intersection of the Ewald sphere (see Fig. 12.5) with the reciprocal lattice plane of interest. We wish to alter the specimen orientation with respect to the incident beam so that this Laue circle contracts to a point at the centre of the pattern. The incident beam is taken to be aligned with the microscope's optic axis using the method of voltage or current centring described in Section 12.2. The difficulty is that any changes in specimen orientation will be accompanied by a translation of the specimen, thus bringing a new region of specimen into the area which contributes to the diffraction pattern. This difficulty is overcome by obtaining an *image* of the specimen (a shadow image) within each diffraction spot. Do this by adjusting

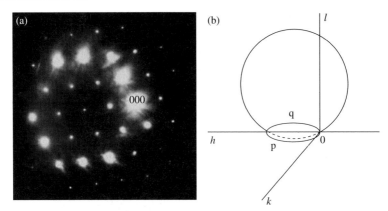

Figure 12.5 The Laue circle. The central beam is labelled (000). This circle, the bright ring of spots in (a), is formed by reflections which occur near the intersection of the Ewald sphere (shown in (b)) with the plane hk, whose normal is $0l$. To bring the crystal into the orientation needed for lattice imaging, the sphere (which always passes through the origin) must be rolled across to the right so that $0l$ forms a diameter and the Laue circle $0pq$ contracts to a point. The pattern in (a) was obtained from a silicon crystal near [110] orientation at 100 kV accelerating voltage.

the 'diffraction focus' control (taking the diffraction pattern out of focus) until a poor-quality image of the specimen region of interest is seen within the central diffraction disc, as shown in Fig. 12.6. The 'rule of thumb' mentioned in step (3) requires that the diffraction lens is always weakened, not strengthened, to take it out of focus.

3. Now adjust the specimen orientation using the foot controls while simultaneously adjusting the main specimen translation controls in such a way that the same region of specimen remains imaged within the central diffraction disc. This procedure requires practice and patience. A useful rule of thumb is that in order to contract the Laue circle,

Figure 12.6 Defocused central diffraction spot from a small triangular-shaped silicon crystal. The out-of-focus diffraction pattern makes each diffraction spot into a low-magnification shadow image of the specimen, allowing simultaneous observation of the crystal orientation and position. As the specimen is tilted, corrections can be made to the translation controls to keep the same region of specimen in view.

Figure 12.7 Diffraction pattern from the silicon crystal used to form the images in Fig. 12.4. The condenser lens is fully focused, showing the illumination conditions used when recording the images. The 'beam divergence' can therefore be obtained from a picture such as this (see Section 4.6). The crystal is close to the symmetrical [110] zone-axis orientation (see also Section 5.6).

the tilt and orientation controls should be altered in such a way that the specimen tends to move toward the centre of the Laue circle. Figure 12.5 shows the diffraction pattern as it appears for a specimen close to the zone-axis orientation, while Fig. 12.7 shows the focused zone-axis diffraction pattern. There is little point in setting the orientation more accurately than this for ductile materials, since local bending takes the specimen through a range of orientations.

12.5. Focusing techniques and auto-tuning

Focusing techniques differ according to the diffraction conditions used and the image interpretation required. The wobbler focusing aid cannot be used at high resolution, nor does the disappearance of the Fresnel fringe coincide either with Gaussian focus or a useful reference focus. The complexity of these fringes for a phase edge, for which fringes occur both inside and outside the specimen edge, makes them of little use at high resolution (see Section 3.1).

1. For dark-field images recorded using tilted illumination and a central aperture (Fig. 6.12), the optimum focus is best judged by eye, taking into account all the precautions mentioned in Section 12.2. Only by close optical examination of a lot of micrographs will the microscopist learn to recognize a 'sharp' image, free from drift, astigmatism, and inaccurate focusing. At a magnification of 300 000 using a ×10 binocular, the eye can resolve detail at the resolution limit of the microscope, but the contrast is generally too low to make this useful and one usually focuses on detail about the size of a

small molecule. This will ensure that the transfer function is 'flat' to a resolution of, say, 0.6 nm (Section 6.5). Obtaining point detail in dark-field images on a conventional instrument showing structure much finer than this is largely a matter of trial and error. Nevertheless, the change in image appearance with defocus in dark-field is far less dramatic than in bright-field, so that focusing may not be as critical. In practice it does appear possible to correct for the image distortion introduced by the asymmetric diffraction conditions by using the stigmator (see Fig. 6.8 and Section 6.5). A dark-field reference focus condition, based on the appearance of small holes, has been mentioned in Section 6.5. Bear in mind that detail smaller than the viewing phosphor grain size cannot be seen. A typical phosphor grain size is 15 μm; however, this varies from instrument to instrument and is easily measured using an optical microscope and a graticule. The resolution of a phosphor screen is limited by multiple scattering of the incident beam (the 'interaction volume'), and lies in the range 40–100 μm, depending on beam energy.

2. For bright-field phase-contrast images of small molecules on ultrathin carbon supports such as graphene (see Section 6.2) where the objective aperture has either been removed or chosen according to eqn (6.16b), the optimum focus can be obtained by counting focus steps in the under-focus direction (weaker lens) from a particular reference focus until the value of eqn (6.16a) is obtained. Several images are recorded in the neighbourhood of this focus. The focal steps are calibrated as described in Section 10.1. The reference focus is the focus condition at which a thin amorphous carbon film shows a pronounced contrast minimum when viewed at high magnification. This arises in the following way. If the transfer function (eqn 3.42) is drawn out for increasing under-focus (Δf negative), it will be seen that the band of well-transmitted spatial frequencies (Fourier components which suffer a 90° phase shift) becomes narrower and moves beyond the range of object detail which can easily be seen through the binocular. Thus the contrast decreases initially with increasing under-focus from the Gaussian focus ($\Delta f = 0$). At the minimum-contrast focus a second loop develops in the intermediate spatial frequency range ($d = 0.6$ nm for $C_s = 1.4$ mm), which increases the image contrast as it grows into the form of Fig. 6.1. The minimum contrast occurs at about 30 nm under-focus for $C_s = 1.4$ mm and can be found for other values of C_s by drawing out a series of transfer functions. Alternatively, it can be taken approximately equal to the optimum dark-field focus $\Delta f = -0.44\sqrt{C_s \lambda}$ (eqn 6.26), since this makes $\chi(u) \approx 0$ to the highest resolution possible. (Note that the PCD approximation would give minimum contrast for $\Delta f = 0$ from eqn (5.27).)

Subject to the limit set by radiation damage, several images are usually recorded close to the required focus, which can be chosen to suit either the PCD approximation or the projected potential image interpretation, depending on the specimen thickness and the resolution required (see Sections 3.4 and 5.3).

All these focusing techniques may result in severe radiation damage. The possibility of focusing on a region of specimen adjacent to that subsequently used for recording the image addresses this problem, but in cryo-EM specialized focusing methods must be used, as described in Glaeser *et al.* (2007). Here the development of the spot-scan (Downing and Glaeser 1986) and low-dose methods (see Section 6.9) have produced major advances. Another approach to minimizing damage for inorganic samples is to set up a continuously recording digital image loop of a few seconds' duration. After moving to a fresh region, the loop is stopped after good images are seen. The first frames of such a video recording may reveal undamaged material, which can then be summed.

3. The focusing technique used when recording many-beam lattice images of large-unit-cell crystals (structure images) has been outlined in Section 12.3. In practice the microscopist will adjust the focus control to produce the image of highest contrast. The selection of the Scherzer (optimum) focus can then be confirmed using a diffractogram (Section 10.6). Image calculations have, in most cases, confirmed that, in thin regions of large-unit-cell crystals (those whose unit-cell dimensions exceed the point-resolution of the microscope), these high-contrast images are those which may be interpreted straightforwardly in terms of the specimen's structure. For metals and semiconductors, however, where the unit-cell dimensions may be smaller than the microscope point-resolution (eqn 6.17), the lattice images of highest contrast will occur at the focus setting given by eqn (5.66) which is much larger than the Scherzer optimum setting (eqn 6.16). Optimum-focus images for the analysis of defects in these crystals can be obtained by counting calibrated focus steps from the minimum-contrast condition and confirming the choice of focus using a diffractogram. Regions of perfect crystal in these images, however, may show rather poor (or non-existent) contrast; nevertheless, any defects present will be imaged faithfully to the point-resolution limit of the instrument.

In a well-designed microscope, all images of interest should fall within the span of a single focus control knob. The focal increments should be comparable with the focusing stability limit set by high-voltage instabilities and lens-current instabilities (see Section 10.1). On aberration-corrected machines this increment will be less than 1 nm. The range of focus settings over which images of good contrast can be seen depends on the size of the illuminating aperture used (and so, indirectly, on the brightness of the electron source) as discussed in Section 5.2.

Automated computer alignment software is now more accurate than human alignment of HREM instruments, and, as described in detail in Section 10.10, the measurement and correction of aberrations would be impossible without it. At high resolution, the TEM alignment must be based on automated measurement of beam direction (with respect to optic axis), defocus, astigmatism, and astigmatism direction. The development of the CCD camera (Spence and Zuo 1988; see also Section 9.7), which provides on-line diffractograms, has made this automation (and the development of on-line tomography) possible. The algorithm used for coma-free TEM auto-alignment is described in detail in Krivanek and Fan (1992) and Krivanek and Mooney (1993). Note, however, that the auto-tuning software may assume operation at the factory-preset 'eucentric focus' lens currents, and will fail otherwise. Figure 12.8 shows the tableaux of diffractograms automatically acquired by a computer. In principal, by fitting eqns (3.23) and (3.42) to a diffractogram and solving for the parameters Δf, C_s, and C_a, suitable correction currents can be fed back to the stigmator and objective lens to bring these quantities to their desired values, for either an uncorrected (Section 6.2) or aberration-corrected (Section 7.4) instrument. Beam alignment (which takes no account of sample alignment) is more difficult, and requires the introduction of three tilts, as shown in Fig. 12.8(a) (before) and (b) (after). After removing the tilt-induced beam defocus, the tilt is estimated from the apparent astigmatism. The final astigmatism reduction is demonstrated in Fig. 12.9(a) and (b). Defocus is then set to the

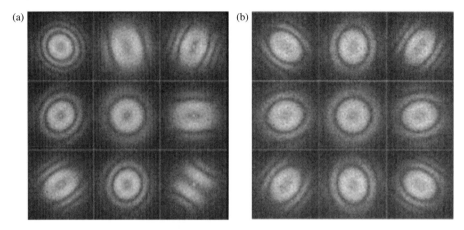

Figure 12.8 Beam tilt alignment automated at 200 kV. (a). Tableaux of diffractograms used during auto-alignment. The outer diffractograms were recorded after intentionally introducing beam tilts of 6 mrad into the directions shown. An analysis of the patterns then gives the magnitude of the misalignment. The mis-tilt of 4 mrad measured by this method from the tableaux was then reduced to less than 0.1 mrad in (b). (Picture by M. Pan, P. Mooney, H. de Ruijter, Gatan Corp., 2002, with permission.)

value requested by the operator, shown at Scherzer defocus in Fig. 12.9(c) for an uncorrected instrument. Auto-alignment software also provides feedback signals for aberration correction devices, for which human alignment would be impossibly time-consuming. As discussed in Sections 2.10 and 7.4, the development of aberration correctors was thus made possible when computer speeds brought the time for auto-alignment below the stability time of the microscope.

12.6. Substrates, sample supports, and graphene

The most important sample support films are amorphous carbon films (containing submicron holes), vitreous ice, Quantifoil, silicon nitride membranes, and graphene. The last of these solves many problems in high-resolution electron microscopy, as we shall see. The holes in the amorphous carbon may be spanned by ultrathin amorphous carbon, ice, or graphene flakes. Finally, many experiments have used carbon nanotubes to trap molecules, which can then be imaged and studied.

All substrates introduce unwanted contrast effects of some kind, unless samples are studied which, for example, span an open hole in a carbon film or otherwise extend into the vacuum from an edge. In dark-field experiments it is the elastic and inelastic scattering from the substrate which limits the image contrast (see Section 6.6), since it provides image background. Fluctuations of substrate thickness on an atomic scale, and Fresnel 'noise' are an even more important problem when using a thin amorphous substrate. (This is true also for samples thinned from the bulk, and, apart from the methods of surface science used in ultra-high vacuum machines, only one technique, chemo-mechanical tripod polishing,

Substrates, sample supports, graphene **341**

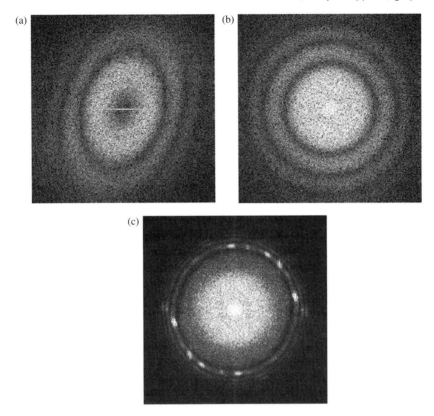

Figure 12.9 Automated astigmatism alignment. A 200 kV diffractogram (a) before astigmatism correction and (b) after automated correction. The astigmatism has been reduced to less than 1 nm, better than possible by human alignment. (c) The effect of then requesting the Scherzer focus setting. The first minimum of the diffractogram (point resolution) occurs at about 0.18 nm. (Courtesy M. Pan, Gatan.)

appears to be capable of reducing surface roughness to the 1 Å level). Carbon film thickness can be reduced in a partially controlled manner using the plasma cleaner. Fresnel diffraction effects can be minimized by reducing the coherence of the illumination, which unfortunately also reduces any pure phase contrast from the atomic structure of interest. These effects will be seen in the contamination which grows even on a thin sample extending into the vacuum.

Amorphous carbon films a few nanometres thick can be purchased commercially, or prepared by indirect evaporation. Techniques for manufacturing holey carbon films are described in Dowell (1970). The difficulty with all techniques for preparing holey carbon film is reproducibility. The Dowell technique, while more elaborate than most, appears to give more consistent results. Paraffin droplets are used to produce 20 nm holes in Formvar plastic, onto which the carbon is evaporated. Pure carbon is then indirectly evaporated onto freshly cleaved mica or clean glass, with a glass slide interposed between the carbon source and the mica. These films can then be floated off onto the surface of distilled water with care. The films are picked up onto the previously made holey carbon films by drawing the holey carbon film (held on a grid with tweezers) up from below the surface of the water, and allowing it to dry. On examination in the microscope, the holes in the carbon film should be covered with an ultrathin layer of amorphous carbon. A vibrating-crystal film thickness monitor can give an estimate of the thickness of evaporated films. (The density of amorphous carbon is about 1850 kg m^{-3}.) Thickness can also be determined using electron energy-loss spectra (Egerton 1976).

In the field of cryo-EM, samples are embedded in a thin film of vitreous ice about 50 nm thick, spanning the holes in an amorphous carbon film. Vitreous ice forms at atmospheric pressure only rapidly for small thicknesses, so that many home-build 'guillotine' devices have been constructed to dunk a water droplet on a holey carbon film rapidly into liquid ethane. (Crystalline ice damages biological samples.) Subsequent thinning of the ice can to some extent be continued by using the electron beam. Full details are given in Glaeser et al. (2007) (see also section 6.8), and commercial apparatus is available for automated production of these samples. For biologists an important development has been the production of holey films by lithography. These 'Quantifoil' supports provide a periodic array of identical holes, facilitating automated data collection (Ermantrut et al. 1998). For holography, silicon nitride membranes are useful. These may be as thin as 30 nm covering up to 500 μm square, and are available with clipped corners so that they fit TEM holders.

Thin flakes of graphite and magnesium oxide have been used to support small molecules from the beginning (Figs 6.6 and 6.13). Rhenium trichloride, which is soluble in the alcohol used for crushing the graphite flakes, is a useful heavy atom test specimen. But this approach has been revolutionized by the development of methods for making large areas of monolayer graphite sheets ('graphene'; Geim and Novoselov 2007), which provide truly ideal chemically inert substrates for electron microscopy. They are conductive, unlike ice, and so avoid charging effects, and have a known thickness which does not introduce significant multiple scattering. They form a vacuum seal, and so can provide the basis of a submicron liquid cell (Yuk et al. 2012), in which a 200 nm thick water droplet trapped between graphene sheets does not freeze in the microscope vacuum because there is negligible evaporative cooling through the graphene envelope. Most importantly, images of particles on graphene agree closely with simulations (with no evidence of a 'Stobbs factor'), so that the use of graphene may finally usher in an era of truly quantitative electron imaging. Although the earliest texts on TEM listed methods for making thin graphite films using adhesive tape to separate layers, only recently has this, and many other approaches, been refined to enable large areas of monolayer graphite to be made. Fabrication methods include chemical vapour deposition (Aleman et al. 2010) for polycrystalline graphene, and mechanical cleavage using adhesive tape (Meyer et al. 2008). Methods for making graphene hydrophilic are described in Pantelic et al. (2011). Note, however, that knock-on damage is severe if accelerating voltages above 60 kV are used, and that in poor vacuum oxygen etching will occur.

Attempts have also been made to image unsupported molecules across small holes in carbon films (see Plate 5 for DNA; Tanaka et al. 1975). So far the thinnest and most successful substrates used have been those prepared either by *in situ* electron beam thinning (see Section 6.2) or using graphene.

12.7. Film analysis and handling for cryo-EM

With the appearance of direct-injection detectors (Section 9.10), the cryo-EM community now has for the first time a detector offering better performance than film for the particular application of single-particle imaging, where large dynamic range is not needed (since the contrast must be weak to avoid multiple scattering) but the largest possible number of pixels is essential. (Each bright-field image field includes many molecules, and one requires many pixels per molecule in the reconstruction.) CCD detectors gained limited acceptance in cryo-EM because of their limited pixel number, poor DQE at low dose, and limited MTF due to cross-talk between pixels (see Section 9.7). However, direct-injection detectors are very expensive and have yet to achieve wide acceptance. We therefore give here some brief notes on image quantification and handling using film for cryo-EM users. Clear accounts of the response of photographic emulsions to electrons are given in Valentine (1966) and Agar et al. (1974). Briefly, the optical density D of an electron micrograph is proportional to electron exposure E up to an optical density of between 1 and 2. This contrasts with the situation in light photography where the response is more closely logarithmic. For electrons, then

$$D = kE$$

where the optical density D is defined as

$$D = \log_{10}(I_0/I)$$

with I_0/I the ratio of incident to transmitted light intensity for the developed micrograph. Thus a micrograph of density 2 transmits 1% of the light incident on it and so appears dense black in transmission. It is the reflected light which is important when viewing prints. The electron speed k is listed in Agar et al. (1974) for most modern emulsions. With E in electrons μm^{-2}, the speed k has units electrons^{-1} μm^2. The reciprocal of the speed is the emulsion sensitivity and gives the number of electrons μm^{-2} necessary to expose an emulsion to unit density. The sensitivity is typically about unity.

For high-resolution dark-field images recorded at high magnification using a LaB$_6$ source, a fast film with $k = 0.5$ at 100 kV is required with high DQE (discussed in Section 9.7). The resolution r in line pairs mm^{-1} is *approximately* $r = \frac{1}{2}d$, where d is the width of the emulsion point-spread function. This width is approximately equal to the electron penetration depth in the emulsion, and is typically between 5 and 10 μm for 100 kV electrons. It is thus commonly supposed to be independent of the film grain size (which increases with film speed) for all but the fastest speeds, and sets a limit to the maximum useful optical enlargement of about 30 times. In practice, for short electron exposures, statistical electron noise may set a lower limit.

At the magnifications commonly used for high-resolution images in CTEM (300 000–700 000), the recorded image will require further optical enlargement of between 15 and 30 times if 1 nm detail is to occupy approximately 1 cm on the final print.

It is sometimes possible for light from the electron filament to reach the electron emulsion. On very long exposures the result is a blurred over-exposed patch which cannot easily be avoided but is rarely troublesome.

344 *Experimental methods*

Some practical considerations when handling plates for high-resolution work are now listed.

1. Use a water rinse between developer and fixer and fix for twice the time taken for the emulsion to clear. Wash for a full 30 min in running water. Drying marks can be a serious problem and are avoided by giving the films a final rinse for about 10 min in a large bath of distilled water containing a magnetic stirrer and a few drops of photographic wetting agent. Dry the films in a dust-free cabinet at about 35 °C. Do not attempt to speed up the drying process in any way. The safest storage for film is in their original boxes—they should not be stored back to back, but in small plastic envelopes.
2. A fine corrugation or reticulation pattern on a scale not much larger than the film grain is sometimes seen. This is caused by too great a temperature difference between the photographic solutions and the water used for washing. Tap water is generally well below room temperature. Ideally the wash water should be at the same temperature as the chemicals used.
3. Over-exposed films or plates can frequently be salvaged by using Farmer's reducer, which slowly and uniformly reduces the optical density of a developed micrograph. This reduction in optical density results in an unavoidable loss of information since, for electron exposures, the contrast and signal-to-noise ratio both increase with exposure (Valentine 1966).

12.8. Ancillary instrumentation for HREM

In addition to the microscope itself, various pieces of ancillary equipment have been referred to throughout this book. Considerations regarding these are collected together in this section for reference.

1. A mass spectrometer or partial pressure gauge is an extremely useful adjunct to an electron microscope. The downtime spent tracing leaks which occur with increasing frequency as the machine ages can be greatly reduced by using the mass spectrometer as a leak detector, tuned to helium, argon, or a proprietary leak-detection gas. In addition to also functioning as an accurate vacuum gauge, a mass spectrometer gives a full analysis of contamination products in the microscope (Echlin 1975). An inexpensive unit will soon repay its cost through rapid diagnosis of vacuum faults. Some units contain magnets, others operate on the principle of the radiofrequency quadrupole. If magnets are present, these must be carefully positioned in order not to disturb the electron image.
2. A supply of 'technical grade' bottled dry nitrogen is needed. The microscope should be filled with dry nitrogen gas whenever internal maintenance is necessary to reduce the amount of water vapour allowed into the column. Water vapour is usually the largest gas component in a vacuum system and is an important source of specimen etching. The pump downtime for a microscope filled with dry nitrogen is also less because water vapour is pumped rather slowly.
3. For magnification calibration on an instrument used at various objective lens focal lengths, a digital objective lens current meter is useful.

4. A direct-reading image current density electrometer is useful both as an accurate exposure meter and for quantitative experiments. The dynamic range of these instruments greatly exceeds that of film, which makes them useful for obtaining the ratio of the intensity of a particular Bragg reflection to that of the incident beam. Similarly, only by taking two accurately timed exposures (one of which must be very short) can the ratio of the background intensity from a dark-field substrate to that of the incident beam be measured photographically. This is easily done using an electrometer. These instruments also simplify gun-tilt alignment (the focused spot is adjusted for maximum intensity) and are invaluable when setting up filaments or for making brightness measurements. On many instruments an electrometer is fitted by the manufacturers; a simple design is described by Hills and Garner (1973) for instruments not fitted with an electrometer, or a pico-ammeter may be connected to the coaxial connector on the spectrometer. For absolute measurements of electron flux a correction may be needed for electron backscattering, particularly at lower voltages. On most instruments the phosphor itself is used to collect electrons for current measurement, on others a separate screen must be fitted, which should be removable and can be fitted, for example, in place of the movable beam stop.
5. In view of the importance of thermal stability for long-exposure dark-field recordings, it is an advantage to arrange for a liquid nitrogen pump to keep the anti-contamination reservoir continuously filled.
6. A rubber bulb fitted with a nozzle is always useful for blowing away dust from a specimen, or the traces of cleaning agents after cleaning the microscope gun chamber. Alternatively the lab supply of dry nitrogen may be used. Never blow into the internal parts of a microscope—human breath contains a great deal of water vapour. The low temperature and high pressure of Freon from aerosol cans can easily fracture brittle samples.
7. The use of an oscilloscope for high-voltage stability testing has been mentioned in Section 11.2.
8. On some microscopes a simple electrical continuity check can be installed, connected between the interior of the cold-finger liquid nitrogen reservoir and the microscope frame. Since the specimen is also electrically connected to the microscope frame, an open circuit between frame and cold-finger reservoir indicates that the specimen is not in contact with the anti-contamination cold-fingers above and below the specimen.

12.9. A checklist for high-resolution work

The following checks are listed for rapid reference. They should be performed before commencing any high-resolution work. This list may be displayed in the EM room.

1. EXAMINE THE O-RING ON A SIDE-ENTRY HOLDER UNDER AN OPTICAL MICROSCOPE FOR DUST AND FIBRES CAUSING LEAKS AND CONTAMINATION. THIS IS IMPORTANT.
2. LENS COOLING-WATER THERMOREGULATOR. DUTY CYCLE CORRECT?
3. NEW OR RECENTLY INSTALLED SOURCES OF VIBRATION?
4. METER CHECKPOINTS ON MICROSCOPE ELECTRONICS.
5. READ VACUUM.

6. FOR CRYO-EM: PLATES/FILM. OUTGASSED? LOADED?
7. HAS SAMPLE HOLDER BEEN CLEANED IN PLASMA CLEANER?
8. LIQUID NITROGEN. COLD-FINGER IN THERMAL EQUILIBRIUM?
9. HIGH-VOLTAGE STABILITY CHECK. GUN CLEANING?
10. EXAMINE FILAMENT IMAGE—STRAY MAGNETIC FIELDS?
11. ALIGN ILLUMINATION SYSTEM.
12. ALIGN IMAGING LENSES.
13. SPECIMEN–COLD-FINGER CONTACT?
14. ALL COMPUTERS RUNNING AND COMMUNICATING WITH EACH OTHER CORRECTLY?

References

Agar, A. W., Alderson, R. H., and Chescoe, D. (1974). Principles and practice of electron microscope operation. *Practical methods in electron microscopy*, Vol. 2, ed. A. M. Glauert. North-Holland, Amsterdam.

Aleman, B., Regan, W., Aloni, S., Altoe, V., Alem, N., Girit, C., Geng, B., Maseraati, L., Crommie, M., Wang, F., and Zettl, A. (2010). Transfer-free batch fabrication of large-area suspended graphene membranes. *ACS Nano*. **4**, 4762.

Dowell, W. C. T. (1970). The rapid production of holey carbon formvar supporting films. In *Proc. 7th Int. Congr. Electron. Microsc., Grenoble*, p. 321.

Downing, K. and Glaeser, R. (1986). Improvement of HREM of radiation sensitive specimens with reduced spot size. *Ultramicroscopy* **20**, 269.

Echlin, P. (1975). Contamination in the scanning microscope. *Scanning electron microscopy 1975*, p. 679. IIT Research Institute, Chicago.

Egerton, R. F. (1976). Foil thickness determination by electron spectroscopy. *J. Phys. D* **9**, 659.

Ermantraut, E., Wohlfart, K., and Tichelaar, W. (1998). Perforated support foils with pre-determined hole size, shape and arrangement. *Ultramicroscopy* **74**, 75.

Geim, A. K. and Novoselov, K. S. (2007). The rise of graphene. *Nature Mater.* **6**, 183.

Glaeser, R., Downing, K., DeRosier, D., Chiu, W., and Frank, J. (2007). *Electron crystallography of biological macromolecules*. Oxford University Press, New York.

Glaisher, R. and Spargo, A. (1985). Aspects of HREM of tetrahedral semiconductors. *Ultramicroscopy* **18**, 323.

Hashimoto, H., Kumao, A., Hino, K., Endoh, H., Yotsumoto, H., and Ono, A. (1973). Visualization of single atoms in molecules and crystals by dark-field electron microscopy. *J. Electron Miscrosc.* **22**, 123.

Hills, G. J. and Garner, R. T. (1973). A design for a simple electronic exposure meter for use with an electron microscope. *J. Microsc.* **98**, 105.

Kimoto, K., Ishizuka, K., Tanaka, N., and Matsui, Y. (2003). Practical procedure for coma-free alignment using caustic figure. *Ultramicroscopy* **96**, 219.

Krivanek, O. L. and Fan, G. Y. (1992). Application of SSCCD to on-line microscopy control. *Scan. Microsc.* Suppl. 6, 105.

Krivanek, O. L. and Mooney, P. E. (1993). Applications of SSCD in TEM. *Ultramicroscopy* **49**, 95.

Meyer, J. C., Kisielowski, C., Erni, R., Rossell, M., Crommei, M., and Zettl, A. (2008). Direct imaging of lattice atoms and topological defects in graphene membrane. *Nano Lett.* **8**, 3582.

Pantelic, R. S., Suk, J. W., Hao, Y., Ruoff, R., and Stahlberg, H. (2011). Oxidative doping renders graphene hydrophilic. *Nano Lett.* **11**, 4319.

Riecke, W. D. (1975). Instrument operation for microscopy and microdiffraction. *Electron microscopy in materials science*, Part I, ed. U. Valdre and E. Rueld. Commission of the European Communities, Directorate General 'Scientific and Technical Information and Information Management', Luxembourg.

Spence, J. C. H. and Zuo, J. M. (1988). Large dynamic range parallel detection charge-coupled area detector for electron microscopy. *Rev. Sci. Instrum.* **59**, 2102.

Spence, J. C. H., O'Keefe, M., and Kolar, H. (1977). High resolution image interpretation in crystalline germanium. *Optik* **49**, 307.

Tanaka, M., Higashi-Fujime, S., and Uyeda, R. (1975). Electron microscope images of mercury atoms bound to DNA filament. *Ultramicroscopy* **1**, 7.

Valentine, R. C. (1966). Response of photographic emulsions to electrons. *Advances in optical and electron microscopy*, Vol. 1, ed. R. Barer and V. E. Cosslett. Academic Press, New York.

Yuk, J. M., Park, J., Ercius, P., Kim, K., Hellebusch, D. J., Crommie, M. F., Lee, J. Y., Zettl, A., and Alivisatos, P. (2012). High resolution EM of colloidal nanocrystal growth using graphene liquid cells. *Science* **336**, 61.

13
Associated techniques

For each new detector that is fitted to an electron microscope, a new subdiscipline of electron microscopy is created. This chapter provides a brief survey of some techniques which are compatible with HREM and STEM, and which provide complementary information. It also provides a review of the various techniques which can provided additional information on the atomic number, chemical form, local coordination, electronic structure, and bonding of the atoms present, namely EDS, EELS, microdiffraction or CBED, and cathodoluminescence (CL) in STEM. The equally important diffraction-contrast imaging methods, useful for imaging strain fields, are covered elsewhere (Williams and Carter 1996, DeGraef 2003).

These new subdisciplines are closely related to certain existing well-established fields. Thus, for example, the theory of CL closely parallels that of photoluminescence (PL), while EDX has close similarities with X-ray fluorescence spectroscopy. Electron energy-loss near-edge spectroscopy (ELNES) has much in common with soft X-ray absorption spectroscopy, both with X-ray absorption near-edge structure (XANES) and with the extended X-ray absorption fine structure (EXAFS), which, like EELS, probe the density of empty states. Photoelectron spectroscopy, using either incident X-rays (XPS, or X-ray photoelectron spectroscopy) or ultraviolet light (UPS, or ultraviolet photoelectron spectroscopy) probes filled states. The books by Joy *et al.* (1986), Garratt-Reed and Bell (2003), and Egerton (2011) contain much relevant background material for this chapter. In summary, while EELS can give the local density of empty electronic states (with unsurpassed spatial resolution and perhaps momentum selection) and emission spectroscopies provide the filled states, HREM and STEM can provide atom positions. ELNES, using a sub-nanometre probe is particularly powerful because of its sensitivity to the three-dimensional local arrangement of atoms about an identified species, provided that there are not many inequivalent atomic sites for that species irradiated by the probe. All of this information, collected from the same region, may be compared with theoretical models.

13.1. X-ray microanalysis and ALCHEMI

The conventional energy-dispersive X-ray spectrometer, when fitted to a transmission electron microscope, is capable of detecting the characteristic X-rays of all the elements present in the sample which have an atomic number greater than that of sodium. Lighter elements may be detected using 'windowless' detectors. The energy resolution of the X-ray spectrum produced is typically only about 150 eV; however, the overall efficiency is high, owing to the large collection solid angle and 'parallel' processing method used. This solid angle may extend to as much as 0.8 sr for the new silicon-drifted detectors used with an aberration-corrected TEM/STEM, in which the corrector allows much more space for 'wrap-around' X-ray detectors. The heights of pulses due to incoming X-ray photons are proportional to the X-ray energy. These pulses are sorted in a multichannel analyser to form a histogram or X-ray spectrum. Unlike a three-dimensional crystal spectrometer, which disregards photons whose energy falls outside the current slit setting, every photon arriving at such an EDX system is counted, regardless of its energy, so that this serial device has many of the advantages of a parallel detection system. The performance is limited, however, by the dead-time of the detector and by the maximum count rate which can be processed. This limit may be reached in a channel other than that of interest, thereby limiting the count rate in the channel of interest. Excitations in the silicon detector itself may also give rise to spurious peaks. An electron-probe-forming lens is required for high spatial resolution. Mechanical instabilities produced by the X-ray detector and its liquid nitrogen Dewar may conflict with the requirements for HREM work. Through the use of the symmetrical condenser–objective lens, together with a side-entry stage, however, some impressive design compromises have been achieved.

The reduction of the X-ray background is a primary concern for quantitative EDX. Background X-rays may originate either in the illumination system, and so appear in a 'hole count' spectrum (in which the electron beam is passed through a hole in the sample), or from post-specimen interactions. These may be due to high-energy backscattered electrons, to fluorescence of the general specimen surroundings, or to transmitted high-energy electrons scattered through large angles. All the evidence suggests that one must think of the pole-piece region as filled by a 'plasma' of kilovolt electrons travelling in all directions.

The spatial resolution of the EDX technique has been the subject of a considerable literature (see Joy et al. (1986) and Garratt-Reed and Bell (2003) for reviews). Because of its importance in the field of electron beam lithography, many authors have reported the results of detailed incoherent multiple-scattering calculations which follow the progress of inelastically scattered electrons through a solid, using 'Monte Carlo' computational techniques (see Joy (1995) for a textbook treatment and software). The tear-drop shaped envelope of scattered electrons whose energy exceeds the atomic ionization energies has a narrow neck at the entrance face of the sample—high spatial resolution for EDX in STEM is obtained using samples thinner than the length of this neck. An exact solution to the transport equation describing this process exists for thin foils (Fathers and Rez 1979). A resolution of a few nanometres is commonly achieved. These multiple-scattering calculations and experiments suggest that the spatial resolution b in EDS is given approximately, for thin samples, by

$$b = (1760/V)t^{3/2}$$

where b and t, the specimen thickness, are expressed in angstroms, and V is the accelerating voltage in electron volts. However, under favourable conditions in an aberration-corrected STEM instrument a single atom may be detected (Lovejoy et al. 2012, Suenaga et al. 2012), as shown in Fig. 13.1. This achievement was made possible by the large X-ray detection solid angle, small 0.1 nm probe, small sample thickness, and large probe current which the aberration corrector provides. Also shown is the corresponding EELS spectrum and HAADF image, all of which can be recorded simultaneously and with pixel-to-pixel registry. The narrow forward scattering angles contributing to the EELS peak produce

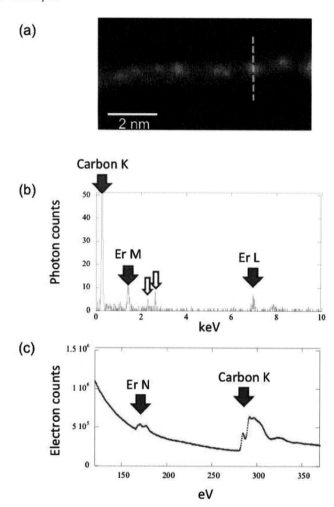

Figure 13.1 (a) HAADF-STEM image of single erbium single atoms inside a carbon nanotube about 1 nm apart. (b) Sum of 10 EDX spectra recorded at the linescan shown in (a). Er M and L lines are indicated, and the carbon K emission line is seen. Open arrows indicate the Mo grid line and S and Cl from chemicals used to prepare the sample. (c) EELS spectra from the same linescan position, showing much higher count and background at the Er N position. (From Suenaga et al. (2012), by permission from Macmillan Publishers Ltd., © 2012.)

much higher count rates than the X-ray count; however, an expression for the minimum number of detectable atoms in EELS (eqn (5.26) in Egerton (2011)) is proportional to the square root of the cross-section for processes contributing to the background, which is far larger than the EDX case.

In practice it is often the ability to distinguish a certain species in a given matrix which is important in EDX, rather than spatial resolution. A typical limit on the sensitivity of the method is about 0.05 at.%; however, this depends on many experimental parameters, including the shape of the sample (whether particles or bulk material are used) and the atomic number of other elements present.

Windowless detectors and ultrathin window detectors have been developed for the detection of elements as light as boron (Williams and Joy 1984, Garrett-Reed and Bell 2003). By comparison with EELS, we note that the EELS spectrometer need not influence HREM performance, unlike the EDX detector. As in Fig. 13.1, the background in an EELS spectrum is much greater than that in an EDX spectrum owing to the unfavourable shape of the EELS edges and to multiple scattering. Light-element lines in EDX may also be confused by the presence of low-energy, high-order emission lines from heavy elements present in the matrix. The strongly forward-scattering nature of EELS means that the angular collection efficiency in EELS is high (unlike X-ray emission, which is isotropic), and one has true parallel detection capability with EELS. This also gives EELS a significant advantage over X-ray absorption spectroscopy at synchrotrons, which is a serial technique.

The background in EDX spectra from crystalline samples arises from coherent bremsstrahlung, which therefore sets the limit to elemental sensitivity. This effect arises on modern machines which use a probe small enough to span regions of perfectly crystalline material; it introduces structure into the normally unstructured bremsstrahlung background produced by non-crystalline matter. In passing down along a string of atoms of separation L, an electron travelling with velocity v emits monochromatic, plane-polarized X-rays of frequency $\omega = 2\pi v/L$ (in the electron rest-frame) as they buzz through the crystal and energy $\hbar\omega$. In the laboratory frame, taking account of the relativistic Doppler effect, the emission energy is

$$\varepsilon = \frac{hc\beta}{L(1 + \beta \sin\theta)} = \frac{12.4\beta}{L(1 + \beta \sin\theta)}$$

where ε is in keV, L is in Å, $\beta = v/c$, and θ is the take-off angle of the X-ray detector measured from the horizontal plane. This effect has been studied in detail, and a full Bloch wave quantum-mechanical treatment has been given (Reese et al. 1984). The atom spacings L are further related to crystal reciprocal lattice vectors $\mathbf{g}$ by $L = (\mathbf{g} \cdot \mathbf{n})^{-1}$, where $\mathbf{n}$ is a unit vector in the beam direction. Thus the effect of crystallinity in a sample is to concentrate the otherwise continuous distribution of bremsstrahlung into a set of discrete peaks, one for each reciprocal lattice vector. The important reciprocal lattice vectors are those lying approximately antiparallel to the beam direction. Classically, the radiation results from the acceleration of the beam electron ('braking radiation'). The continuous background results from accelerations in all directions. The sharp coherent bremsstrahlung peaks result from accelerations in the beam direction alone, which correspond quantum mechanically to the umklapp process in which the crystal as a whole takes up momentum in units of $\hbar\mathbf{g}$ (Spence and Reese 1986). Coherent bremsstrahlung radiation may therefore be thought of loosely as dipole radiation which emerges (non-relativistically) at 90° to the beam direction, since the axis of the dipole lies parallel to the beam and $\mathbf{g}$. Radiation in the forward direction due to transverse acceleration (known as channelling radiation) has also been observed in electron microscopy (Fujimoto and Komaki 1986, personal communication). In the absence of inelastic phonon scattering, these discrete coherent bremsstrahlung and

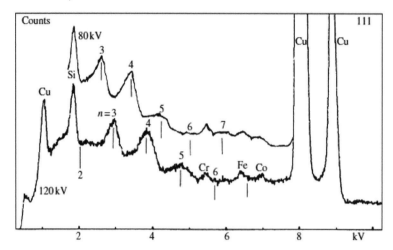

Figure 13.2 Coherent bremsstrahlung emission from a thin diamond crystal recorded by EDX on the Philips EM400 transmission electron microscope. Take-off angle $\theta = 20°$, accelerating voltages 80 and 120 kV as shown. Each numbered peak n results from a layer n of reciprocal lattice points $\mathbf{g}$ normal to the beam such that $\mathbf{g} \cdot \mathbf{H} = n$, where $H = [111]$ is the beam direction. Owing to the symmetry of the diamond lattice all structure factors in the $n = 2$ and $n = 6$ HOLZ layers are zero. The corresponding coherent bremsstrahlung peaks are therefore absent, confirming the crystallographic nature of this bremsstrahlung.

channelling radiation peaks (and combinations thereof) would be the only background seen from crystalline samples. It is likely therefore that in the past there have been mistaken identifications of lines due to this mechanism (Vechio and Williams 1986). Figure 13.2 shows a typical coherent bremsstrahlung spectrum obtained by EDX on a TEM instrument from a diamond sample. Since carbon X-rays are too soft to be detected by this system, no characteristic X-rays were expected from this sample. The effect is a useful one, since accelerating voltages may be chosen which position these peaks away from those sought in microanalysis, and therefore minimize the background contribution. CB peaks can easily be distinguished from characteristic X-ray emission by changing the energy of the electron beam—the CB peaks change in energy, but the characteristic peaks do not.

In addition to supplying microanalytical information, the EDX method can be used quantitatively to determine the fraction of substitutional impurity atoms which lie on particular sites in a crystal lattice. If characteristic X-rays from the trace elements of interest in crystalline regions of a TEM sample can be detected using the EDX system, then in most cases it is possible to determine the crystallographic site of the impurity. This method, known as 'atom location by channelling enhanced microanalysis' (or ALCHEMI), involves no adjustable parameters, the fractional occupancies of the substitutional impurities of interest being given in terms of measured X-ray counts alone. Since TEM or STEM is used, the method can be applied to areas as small as a few hundred angstroms in diameter, while the detection sensitivity is limited by that of the EDX system to about 0.1 at.%. Elements which are neighbours in the periodic table can normally be readily distinguished. The method uses the dependence on the orientation of the incident electron beam of characteristic X-ray emission, and does not require either the specimen thickness or the precise beam orientation to be known. In conventional microanalysis by EDX, a large illumination aperture is used. To obtain an orientation-dependent effect, a beam divergence smaller than the Bragg angle must be used. No dynamical electron diffraction calculations are required for the interpretation of this quantitative method. The classical problems of cation ordering in spinels, feldspars, olivine minerals, and jet turbine-blade alloys have all been studied by this method (for reviews see Jones (2002) and Wu and Veblen (2010)). The first applications of the method were described by Taftø and Liliental (1982) and Taftø and Spence (1982), while a tutorial review of the principles and applications of the technique can be found in Spence and Taftø (1983), which also contains a brief historical summary of the related effects (such as the Borrman effect in X-ray diffraction) on which it is based.

The technique depends on the fact that characteristic X-ray emission from crystals is in fact modulated by the transverse electron wavefield in the crystal, as shown in Fig. 13.3. This intensity distribution is given by

$$I(r, t_0) = \int_0^{t_0} I(r, t) dt \qquad (13.1)$$

where $I(r)$ is given by eqn (5.47) with $\chi(\mathbf{g}, \Delta f) = 0$ and t_0 is now the crystal thickness. Equation (5.47) has been integrated throughout the depth of the crystal to obtain a quantity proportional to the X-ray emission yield, in the approximation of perfect localization (see Cherns et al. (1973) for a detailed discussion). Figure 13.4 shows the experimental variation

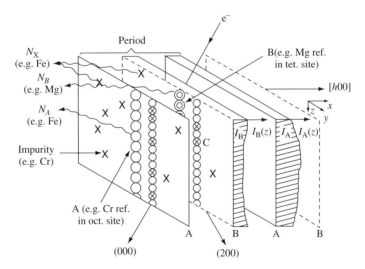

Figure 13.3 The principle of ALCHEMI. The period of the crystal stacking sequence along the systematics direction $[h00]$ is indicated. The wavefield is constant in the y direction, and varies with depth along z as shown. The characteristic X-ray production is proportional to the shaded areas. Note that there is no channelling effect for small thickness. The case of a spinel crystal containing iron impurities at X is shown.

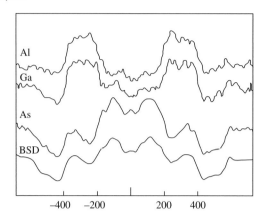

Figure 13.4 Variation in characteristic X-ray emission as a function of angle for a low-divergence electron beam rocked across the (200) planes of GaAlAs. For the three X-ray signals, equal changes in the height of the curve represent equal percentage changes in the number of counts. The backscattered electron signal (BSD) is included for reference. (From Christensen and Eades (1986).)

of X-ray emission with incident beam direction for GaAlAs (Christensen and Eades 1986). For ALCHEMI, the characteristic X-ray emission intensities from both the impurity atom and the atoms of the host crystal need be measured for only two or three crystallographic orientations of the collimated incident electron beam. A specimen thickness of about 1000 Å is sought, but the thickness is not critical and the illumination 'beam divergence' should be a fraction of the Bragg angle for a first-order reflection. In many problems it is possible to limit the substitutional sites of the impurity atom to a few likely possibilities. The crystalline specimen is then oriented in the 'systematics' or 'planar channelling' orientation, so that crystal planes with a reciprocal lattice vector normal to the beam contain alternating candidate sites for the impurity and alternating species of the host crystal. For example, in the zinc blende structure, the $[00h]$ systematics orientation would contain all the zinc atoms on the A planes of Fig. 13.3, while the B planes would contain only sulphur atoms. The idea is to compare the emission of the host and the impurity as the incident beam direction is changed and the wavefield inside the crystal moves sideways across the atomic planes. If the impurity emission follows that of the zinc atoms, they must both be on the A planes. In fact, by measuring the Zn, S, and impurity X-ray emission for just two incident beam directions (both approximately normal to $[00h]$) it is possible to determine the fractions of the impurity which lie on the A and B planes, respectively. By repeating this experiment for other sets of planes, the crystal site can be determined.

For each of these two incident beam orientations, the fast electron sets up a standing wave in the crystal whose intensity variation has the period of the lattice indicated in Fig. 13.2. For ideally localized X-ray emission, the characteristic X-ray emission intensity due to the ionization of the crystal atoms is proportional to the height of this standing wave at the atom concerned. The total X-ray emission from, say, the species on the A planes of Fig. 13.3 is then proportional to the thickness-integrated electron intensity I_A on the A planes, shown shaded in the figure. This area changes with changes in the incident beam direction. Because a systematics orientation has been chosen, the electron intensity is constant along the A and B planes. We let C_X be the concentration of impurities on the A planes and $(1 - C_X)$ be the concentration on the B planes. The X-ray emission from the host atoms in known sites on the A and B planes will be used to provide an independent monitor of the electron intensities I_A and I_B, which also excite X-rays from the impurities. Thus the principle of ALCHEMI is to use the host atoms as reference atoms or 'detectors', to 'measure' the thickness-integrated dynamical electron intensity distribution. We let $N_A^{(1,2)}, N_B^{(1,2)}$, and $N_X^{(1,2)}$ be the six X-ray counts from elements A, B, and X for a 'channelling' orientation (1) and a 'non-channelling' orientation (2) in which the electron intensities on the A and B planes are equal. Then we have the six relationships

$$N_B^{(1,2)} = K_B I_B^{(1,2)} \tag{13.2}$$

$$N_A^{(1,2)} = K_A I_A^{(1,2)} \tag{13.3}$$

$$N_X^{(1,2)} = K_X C_X I_B^{(1,2)} + K_X (1 - C_X) I_A^{(1,2)} \tag{13.4}$$

where the superscripts refer to X-ray counts obtained in the two successive orientations. In addition, for the non-channelling orientation, $I_A^{(2)} = I_B^{(2)}$. Here K_A, K_B, and K_X are constants which take account of differences in fluorescence yield, and other scaling factors. These equations can be solved for C_X in a way which eliminates I_A and I_B. If we define the ratio of counts for two orientations as

$$R(A/X) = \frac{N_A^{(1)}/N_X^{(1)}}{N_A^{(2)}/N_X^{(2)}} \tag{13.5}$$

then

$$C_X = \frac{R(A/X) - 1}{R(A/X)(1 - \beta)}$$

where

$$\beta = \frac{N_B^{(1)} N_A^{(2)}}{N_A^{(1)} N_B^{(2)}}. \tag{13.6}$$

Equation (13.6) can thus be used to find the concentration of species X on the A planes (see Fig. 13.3) in terms of the measured X-ray counts $N_A^{(1,2)}$, $N_B^{(1,2)}$, and $N_X^{(1,2)}$ alone. In practice, orientation (1) is usually chosen as slightly greater than the Bragg angle, and orientation (2) is one which avoids the excitation of Bragg beams.

This technique has been applied to a wide range of materials including minerals, ceramics, oxide superconductors, semiconductors, and intermetallic turbine-blade alloys (Jones 2002, Wu and Veblen 2010).

356 Associated techniques

A crucial assumption of the method is that the X-ray emission process is highly localized. (The meaning of localization is discussed in Section 13.2.) This assumption has been tested by Bourdillon *et al.* (1981), Self and Buseck (1983), Pennycook (1987), Rossouw and Maslen (1987), and Spence *et al.* (1988) (see also Bourdillon (1984)).

The localization problem can be minimized, and a great improvement in accuracy obtained, by collecting X-ray emission as a function of the two angular coordinates of the electron beam around a zone axis and using multivariate analysis to interpret the data. This two-dimensional statistical ALCHEMI approach (Rossouw *et al.* 1996) also allows a very simple interpretation, as shown in Fig. 13.5. If the pattern from host atom A closely resembles that from substitutional atom B, then B lies on the same (equivalent) site as A. In the figure we see immediately that for this γ-TiAl crystal, Ga lies on the Al site, Hf lies on the Ti site, while Mn lies on both. (The linear expansion coefficients which express the Mo map in terms of a sum of the Ti and Al maps give quantitative occupancies.)

Since crystal planes can be found which contain interstitial sites, the ALCHEMI method has been extended to include interstitial site-occupancy determination (Rossouw and Miller 1999). In addition, under the column approximation, any local change in diffraction

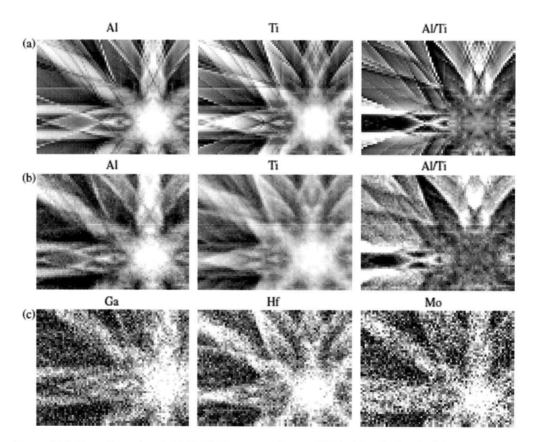

Figure 13.5 Two-dimensional ALCHEMI patterns from γ-TiAl: (a) calculated, (b) experimental, (c) experimental from dopants indicated. (From Rossouw *et al.* (1996).)

conditions due to crystal bending or thickness changes under the probe will not affect the ALCHEMI results, since these changes affect both the impurity emission and that from the reference host atoms in a similar way. A final important assumption of the method, however, is that the impurities are uniformly distributed in depth throughout the crystal. Thus, the method fails for a segregated sheet of impurities lying normal to the beam at a particular depth in the crystal.

Channelling effects have also been observed to influence characteristic energy-loss peaks (Spence et al. 1981, Taftø and Lehmfuhl 1982). Applications of this effect can be found in Taftø et al. (1982), where it is used to distinguish normal and inverse spinels, and in Taftø (1984). The additional complications introduced by multiple scattering of the beam electron following the energy-loss event make this a less generally useful technique.

It is important to emphasize that for modern analytical STEM instruments, both these coherent bremsstrahlung and channelling effects are unavoidable when studying crystalline material—the first sets the limit of detectability for trace elements by controlling the background, while the second may lead to errors in quantifying the stoichiometry of nanocrystals.

13.2. Electron energy loss spectroscopy in STEM

The extensive literature on this large subject has been reviewed by Disko et al. (1992), Reimer (1995), Brydson (2001), Spence (2006), Colliex (2011), and Egerton (2011). Here we briefly review those aspects of EELS most relevant to high-resolution electron microscopy. Electron energy-loss spectrometers record the number $F(E)$ of transmitted electrons which, in passing through the sample, give up an amount of energy in the range E to $E + \Delta E$ to the various elementary excitations of the sample. $F(E)$ is related to the dielectric function of the sample, and contains discrete peaks at the energies of all the important excitations, such as inner-shell atomic ionization edges and plasmon excitations. The spectra are similar to soft X-ray absorption spectra, and are related to the density of empty states.

Electron energy-loss spectrometers are well removed from the sensitive pole-piece region of an electron microscope and therefore need not degrade the ultimate performance of an atomic-resolution instrument, provided that stray fields and ground loops are avoided. Thus, it is possible to obtain both atomic-resolution images and electron energy-loss spectra from the same identifiable sub-nanometre region of the specimen. The size of the region from which spectra can be obtained will depend on the probe-forming capabilities of the instrument. In STEM mode, electron energy-loss spectra may be obtained from a sub-nanometre region at any identified image pixel at which the electron probe has been stopped.

Electron energy-loss spectra have been used chiefly to provide microanalytical information and information on the charge state and local coordination of ions, to extract 'optical constants' from materials, and to study plasmon excitations in small particles. (This can be done by forming images using plasmon-loss electrons.) We first briefly discuss microanalysis by EELS. Since the primary beam electron (following energy loss) is itself detected in EELS, rather than a secondary emission product, as in EDX or CL, this method has the great advantage of providing chemical information at the highest possible spatial resolution. Secondary emission products, however, may result from a de-excitation event some distance from the primary excitation.

The thickness-dependent contributions of multiple inelastic scattering must be removed from the spectrum if quantitative results are expected from any but the very thinnest samples. This may be done by the convenient 'logarithmic deconvolution' method (Johnson and Spence 1974; see also Wang et al. 2009), which solves the inversion problem of multiple inelastic scattering exactly. (Thus, for example, a low-loss spectrum containing multiple plasmon peaks will show a single peak following logarithmic deconvolution.) A list of computer programs that can be used for this method is given in Egerton (2011).

The theoretical basis of the EELS technique has its historical origins in the problem of stopping power for charged particles, which arises in many fields of physics. The most important early papers are those of Bethe (1930) and Kainuma (1955). The second of these addresses the very important problem of combined multiple elastic and multiple inelastic scattering (see, for example, Howie (1963)). Since electrons which have been inelastically scattered may also be elastically Bragg scattered, it is possible to form lattice images in energy-filtering instruments which select only those electrons which have lost a particular amount of energy in passing through the sample (Craven and Colliex 1977). The energy-filtered images may appear almost identical to the elastic lattice image. In images filtered for a particular inner-shell excitation, the 'wrong' atomic species may therefore appear, as shown by Spence and Lynch (1982). Since the probability of single inelastic scattering is proportional to t/λ for a process with mean free path λ, while that for Bragg scattering (at the Bragg angle) is proportional to t^2/ξ_g^2 (see eqn 5.20), either process may dominate, depending on the experimental conditions and material.

Image-forming electron energy-loss spectrometers have been developed which may be fitted to HREM instruments for the purposes of 'chemical mapping' at atomic resolution. The interpretation of these images will be complicated in all but the thinnest samples, due to both the unavoidable elastic scattering mentioned above and the effects of inelastic localization, which set a limit on the spatial resolution of energy-filtered images.

Spatial resolution in EELS is limited not only by the maximum size of the electron probe within the sample but also by inelastic 'localization'. Dynamical calculations for coherent probe spreading can be found in Humphreys and Spence (1979) and Fig. 8.8. If the inelastic electron scattering is taken to be confined to a cone of semi-angle $\theta_E = \Delta E/2E_0$ (ΔE is the energy loss), then this radiation must originate from a specimen volume whose transverse dimension is approximately $L = \lambda/\theta_E$, where L is known as the inelastic localization, and can be taken as a measure of the size of the region in which the inelastic scattering is coherently generated. It is therefore also the size of the image element which would be formed if all this scattering were passed through a perfect lens and focused. This simple result for L is consistent with the results of more elaborate quantum-mechanical calculations (Craven et al. 1978) and with an argument based on the time-and-energy uncertainty principle (Howie 1979).

This can be understood by considering the classical impact parameter b. A beam electron travelling at velocity v spends an amount of time $T = b/v$ in the neighbourhood of the target atom. If we think of the pulse of electromagnetic energy associated with the passage of the beam electron evaluated at this atom, then this will contain Fourier components over a range $\Delta \omega = 2\pi/\Delta T$, for which the range of associated quantum-mechanical excitation energies (the beam electron energy loss) is $\Delta E = h\Delta \omega = hv/b$. Using $v = hk/m$ and $E_i = mv^2/2$ we then find that $L = \lambda/\theta_E = 2\pi b$, so that the localization L is approximately equal to the classical impact parameter b. A more detailed calculation (Egerton 2009) gives $L = 0.52\lambda/<\theta>$, where $<\theta>$ is the median scattering angle which contains 50% of the

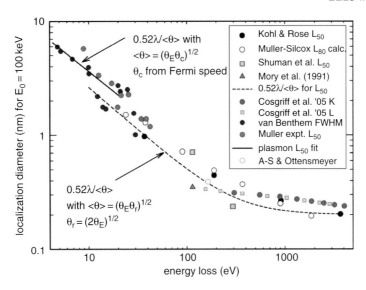

Figure 13.6(a) Delocalization measurements (dots) for inelastic scattering as a function of energy loss, for 100 kV incident electrons. The dashed curve is $L = 0.52\lambda/<\theta>$, as given in the text and based on the median scattering angle $<\theta>$ containing half the inelastic scattering (Egerton 2009).

inelastic scattering. Experimental measurements of spatial resolution in EELS as limited by this inelastic delocalization effect can be found in Shah et al. (2011). Figure 13.6(a) summarizes many such measurements, and compares them with theory.

Alternatively, the inelastically scattered amplitude can be written as the Fourier transform of an inelastic scattering potential in the Born approximation. The forces due to this potential affect the passing electron only when it is within a distance b of the atom, and for times of the order of ΔT. The theory of the photoelectric effect, on which the analysis of XANES and much ELNES are also based, assumes, however, almost 'instantaneous' excitation of an atom. Here the abrupt 'switching on' of the core hole (within less than 10^{-17} s) is known as the 'sudden approximation'. No evidence for the failure of this approximation has yet been produced in inelastic electron scattering.

Figure 13.6(b) shows the ELNES K-shell ionization edges for silicon, aluminium, and magnesium atoms in different crystal structures (Taftø and Zhu 1982). It is clear that the shape of these spectra from the same atom in different crystals depends strongly on the local coordination of the atom, and not on purely atomic properties. In this way we can obtain *three-dimensional* chemical bonding and local coordination information not readily obtainable by imaging. The type of atom present can be determined by comparing the ionization energy with tabulated values, while ELNES is also very sensitive to neighbouring interatomic distances (Spence 1985, Jiang et al. 2008). Plate 7 shows the dramatic elemental sensitivity and spatial resolution now possible using an aberration-corrected instrument, in which the three metal atom columns in this oxide crystal can be imaged separately in this [001] projection using electrons which have caused inner-shell excitations (Botton et al. 2010).

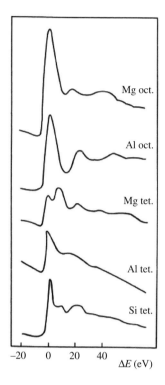

Figure 13.6(b) Electron loss near-edge structures of Si, Al, and Mg atoms in different local crystallographic environments. The K edges of magnesium in olivine (octahedral), aluminium in spinel (octahedral), magnesium in spinel (tetrahedral), aluminium in orthoclase (tetrahedral), and silicon in olivine (tetrahedral) are shown. (From Taftø and Zhu (1982).)

While some limited success can be expected in relating characteristic ELNES spectral shapes to local atomic structure by these 'fingerprinting' methods, a full understanding of this effect requires large numerical computations and a sophisticated theory. (For a review of methods see Rez *et al.* (1999) and Spence (2006).) A complete theoretical treatment must include the effects of multiple elastic scattering of the beam before and after the inelastic event ('double channelling'), the effect of using a coherent cone of illumination, and the detector geometry (Kohl and Rose 1985, Spence 1987). In addition, all the details of the excited state of the crystal electronic structure must be simulated. All of these processes are taken account of in modern software packages, such as that based on the mixed dynamic form factor, described in full in Witte *et al.* (2009). Multiple elastic scattering of the probe, and the generation of an inelastically scattered wavefield (which is itself then elastically scattered), may be described by the efficient multislice algorithm using an artificial supercell. The possibility of an inelastic event must be allowed for at every equivalent atom site and at every depth. We assume that the multislice algorithm is available for computation of the dynamical elastic wavefield $\Psi(\mathbf{R}, \mathbf{K}, z_n)$ at depth z_n, with two-dimensional scattering vector $|\mathbf{K}| = \theta/\lambda$. The result is that the dynamical inelastic scattering from a process, whose atomic scattering factor for inner-shell ionization is $f_{q,g}$, is (Cosgriff *et al.* 2005)

$$I_p(\mathbf{R}) = \frac{1}{AV} \sum_{n=1}^{N} \left[\sum_{g,q} \Psi^*(\mathbf{R}, \mathbf{K}, z_n) \Psi(\mathbf{R}, \mathbf{K}, z_n) \mu_{q,g} \right] \Delta z_n$$

This expression allows for multiple elastic scattering but a single inelastic event. (Two successive inelastic events would produce energy loss $2\Delta E$.) It differs from earlier inelastic multislice treatments (e.g. Spence and Lynch 1982) by the use of a non-local potential. The mixed dynamic form factor

$$\mu_{q-g} = \frac{\lambda}{2\pi\Omega} \sum_n \exp[-M(\mathbf{q}-\mathbf{g})] \exp[2\pi i(\mathbf{q}-\mathbf{g}) \cdot \mathbf{r}_i] f_{q,g}$$

involves such a non-local potential in a supercell of volume V and area A. The slice thickness is Δz_n, Ω is the volume of the host crystal subcell and $\mathbf{r}_i$ is the coordinate of the ionized atom. $M(\mathbf{q})$ is the Debye–Waller factor. The evaluation of $f_{q,g}$ requires numerical values of the appropriate atomic wavefunctions (e.g. Saldin and Rez 1987). It involves the scattering kinetics, constrained by momentum conservation, together with an integration of the resulting allowed scattered electron wavevector over the solid angle defined by the STEM detector shape. The FEF8 computer code (Ankudinov et al. 1998) has been extensively used for simulation of ELNES. Difficulties may arise due to excited state effects such as self-energy terms and core holes.

Several attempts have also been made to apply the methods of EXAFS to EELS spectra. While some success has been obtained (see Egerton (2011) for a review), the limited energy range of EELS spectrometers (0–2.5 kV) has frequently meant that in complex crystals one is dealing with many overlapping edges in a crowded spectrum. It is suggested that ELNES (which deals with the first 40 eV or so beyond an edge) be distinguished from extended energy-loss fine structure (EXELFS) (which deals with the remainder of the edge) at that energy at which the wavelength of the ejected secondary electron is approximately equal to the nearest-neighbour interatomic distance. Physically, these regimes are distinguished by the importance of multiple scattering for the secondary electron in the ELNES range, which makes ELNES sensitive to angular bond correlations.

We have seen in Fig. 13.1 that it is possible to detect single atoms by EELS in STEM. With the ability to synthesize artificial oxide crystal layer structures using layer-by-layer deposition, it has become possible to analyse the nature of the interlayer chemistry and bonding since here new possibilities arise, perhaps not found in nature, with new electronic and magnetic properties. Figure 13.7 (from Shah et al. (2012)) shows ELNES spectra taken from an $La_{2/3}Sr_{4/3}MnO_4$ 'digital superlattice'. The spectra, recorded using an aberration-corrected instrument, show directly, with spatial resolution of about 1 Å, the hybridized states formed between layers. The peaks marked 1 are Mn 3d orbitals, 2 are La 5d orbitals, 3 are Sr 4d orbitals, and 4 are metal 4sp states. This clearly demonstrates the power of the ELNES method to provides inner-shell spectra giving local chemistry with unrivalled spatial resolution, in registry with high-resolution images. Many similar studies have been made combining nanoscale ELNES spectra and ADF-STEM images at atomic resolution. For example, these can in principle distinguish between deep states at dislocation cores due to the intrinsic core structure, its reconstructions and defects, and impurities (for dislocations in GaN see, for example, Arslan and Browning (2002)).

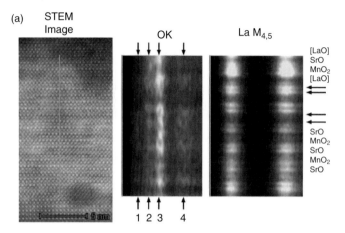

Figure 13.7 HAADF-STEM image of an artificial oxide superlattice, showing the line along which the ELNES spectra were taken, with layer stacking shown at the right. In the centre is shown the oxygen K edge, with energy loss running across the page and the position of the probe down the page. Features are noted. The La $M_{4,5}$ edge is shown at the right. (From Shah et al. (2012).)

More information on the oxidation state of atoms at defects and disordered matter can be obtained by comparing these near-edge spectra with calculations for different states of ionicity or with crystals of related stoichiometry. For example, Jiang et al. (2008) have compared nanoprobe ELNES in $CaSiO_3$ glasses with both theoretical and experimental ELNES from several crystals (related to wollastonite and an olivine structure), whose crystal structure contains similarly coordinated atoms to those in different regions of the glass. In this way local regions of chained, ringed and tetrahedrally coordinated SiO_4 structures could be found in the glass and mapped out on the nanoscale. The effects of beam damage were carefully evaluated using time-resolved ELNES data collection. Examples of the EELS approach used in correlation with HREM images recorded at 0.12 nm resolution can be found in Scheu et al. (2000). Here a $Cu/(1120)$ α-Al_2O_3 interface is studied, prepared by molecular beam epitaxy. Oxygen 2p states are found to hybridized with Cu 3d and/or Al 3p states at the interface. The interface structure was determined with atomic precision by HREM. Examples of EELS obtained with sub-nanometre spatial resolution can also be found in Batson (1993), for an analysis of the silicon $L_{2,3}$ edge as it passes into the oxide at a transistor gate oxide, and also in the work of Muller et al. (1999) on the oxygen K edge (more localized than the Si $L_{2,3}$ edge) across a similar interface. Nufer et al. (2001) combined HREM with EELS and electronic structure calculation in their analysis of a rhombohedral twin boundary in alumina. Bourret and Colliex (1982), detected oxygen at the cores of dislocations in germanium; while Spence and Lynch (1982) detected a variation in elemental concentration within a single unit cell of a barium aluminate crystal.

By correcting fifth-order aberrations in STEM and improving the spectrometer coupling optics, a 100-fold increase in signal intensity is possible. This has allowed the fine structure of the Ti and Mn L-edges to be mapped out in a transition-metal oxide multilayer, thus providing a chemical map at the angstrom scale of spatial resolution (Muller et al. 2008). Differences in local electronic and atomic structure between single substitutional silicon atoms in graphene have even been measured using STEM ELNES (Ramesse et al. 2013). The increased current also improves performance when using a monochromator, which now delivers better than 20 meV energy resolution (Krivanek *et al.* 2009). Developments in STEM EELS are reviewed more fully in Spence (2006), Varella *et al.* (2011), and Radtke and Botton (2011).

13.3. Microdiffraction, CBED, and precession methods

From the earliest days of electron diffraction there has been a continuous effort to solve crystal structures using electron diffraction data alone. At first it was hoped that a kinematic analysis might serve, and a certain number of layer structures which could be prepared as very thin films were solved using methods modified from X-ray crystallography practice. (The texts by Pinsker (1949), Vainshtein (1964), Cowley (1967), and Zvyagin (1967) review this early work—see also Avilov in Cowley (1992). A modern review for inorganic materials can be found in Zou *et al.* (2011).) But low R factors (e.g. less than 5%) could only be obtained if dynamical effects were included in calculations, since unrealistic sample thicknesses of less than about 4 nm were needed to obtain kinematic conditions for light elements at 100 kV, and contamination and thickness variation of the sample under the illuminated area limited progress. These early results are now in the process of being rediscovered, using better equipment and applications of the precession electron diffraction method. In particular, using automated goniometers and electronic detection under computer control, it is now possible to collected three-dimensional diffraction data sets from nanoparticles, despite the formidable problems of 'bringing back' the nanoparticle onto the eucentric axis after each tilt. (For a comprehensive review of this method of automated diffraction tomography (ADT) see Mugnaioli *et al.* (2012).) The validity domain of the kinematic approximation was investigated in detail for oxide crystals, for example, in 1975 (Lynch *et al.* 1975). The main difficulty is that X-ray structure analysis does not contain crystal thickness as a parameter, whereas electron diffraction results, because of multiple scattering, depend sensitively on thickness, which cannot easily be measured, leading to unreproducible results. Exceptional cases include organic monolayers of known thickness, such as those used in cryo-EM. Here the images are also insensitive to the mean potential of the parallel-sided ice film in which they are embedded, since, in the phase grating approximation, only changes in potential are seen.

Just as HREM imaging 'solves the phase problem' (by directly revealing atomic positions), by the 1960s it was understood that multiple scattering also solves the phase problem. The intensity of multiply scattered Bragg beams depends on structure factor phases. (This effect was used, for example, to measured structure factor phases for an acentric crystal, with $0.1°$ accuracy, by electron diffraction (Zuo *et al.* 1989), achieving far greater accuracy than is possible using many-beam X-ray methods.) This raised the hope that inorganic structures could be solved by taking advantage of dynamical effects and of the nanometre-sized probes available with field-emission instruments to obtain 'perfect crystal' patterns. However this has proven possible in very few cases—there are usually too many adjustable parameters. But the development of the CBED microdiffraction or nanoprobe method in the 1960s and 1970s, together with the automated refinement of HOLZ line positions, has meant that one can now usually determine both the space group (Eades 1988) and the Bravais lattice (Zuo 1993) of an unknown crystal. Spence and Zuo (1992) provide a textbook treatment with worked examples. Kim *et al.* (2013) describe a quantitative method

364 Associated techniques

of symmetry mapping in piezoelectric materials, based on nanodiffraction patterns, which allows the local monoclinic symmetry observed in some regions to be related to polarization switching. These are important first steps in structure analysis, which heavily constrain the determination of atom position. To benefit from the effects of multiple scattering, this analysis must be performed on a region of crystal thicker than that used for HREM. Where this is possible, these measured constraints can be included in the HREM image matching process described in Sections 7.1 and 7.5, and demonstrated in Fig. 7.3.

The differences between X-ray and electron diffraction must be understood before software intended for X-ray crystallography can be used with kinematic electron diffraction data. As noted above, for X-ray analysis, a small crystal is normally bathed within the X-ray beam (Bragg geometry), so that the analysis does not depend on thickness at all (except through extinction effects). The X-ray wavelength is much larger, resulting in a much smaller Ewald sphere. X-ray rocking curves cover a tiny fraction of the Bragg angle (unlike CBED patterns), so that X-ray data are usually 'angle-integrated' across the entire rocking curve. Transmission electron diffraction (TED) spot patterns, however, provide the intensity at one point within the rocking curve. This difference affects the thickness dependence of the intensities. In the two-beam theory at the Bragg condition, the electron diffraction intensity (not angle integrated) varies sinusoidally with thickness (as seen experimentally in Fig. 12.3)—for small thicknesses the thickness variation is therefore quadratic in the kinematic limit. The X-ray situation for angle-integrated data is quite different (Stout and Jensen 1968). The intensity of a reflected Bragg reflection may be equal to that of the incident beam (in the 'perfect crystal' two-beam theory, without absorption, if angle resolved) or, if angle integrated, proportional to some power x ($1 < x < 2$) of the relevant structure factor amplitude, depending on the degree of crystal perfection (mosaicity) and the extinction correction model used (eqn 13.7). There is no analogy for polarization effects in electron diffraction. The Direct Methods software used to solve crystal structures by X-ray crystallography assumes that each reflection has been recorded at the Bragg condition; however, electron diffraction patterns recorded in the zone axis condition contain many beams simultaneously excited with different excitation errors, which modify their phases. (Given a model for scattering theory, allowance may be made for this effect, since excitation errors are known.) For organic monolayers, the small thickness (producing rods in reciprocal space) and film bending may justify the use of an expression for diffracted intensity which is integrated across the rocking curve. With $w = s_g \xi_g$ and $U_g = 1/(\xi_g \lambda)$, this is given, using the two-beam eqn (5.20), as (Blackman 1939)

$$\bar{I}_g(t) = \int_{w=-\infty}^{w=\infty} |\Phi_g(t,w)|^2 dw$$

$$= \frac{\pi |U_g|^2 t}{2K^2 g} \quad \text{for } t < 2\xi_g$$

$$= \frac{\pi |U_g|}{2Kg} \quad \text{for } t < 2\xi_g \tag{13.7}$$

Note the different powers of the structure factor for different thickness ranges, and the absence of any thickness dependence for large thickness (absorption has been neglected).

The latter condition assists the interpretation of precession electron diffraction data (Vincent and Midgley 1994)which are partially angle integrated in this way.

Direct Methods requires about five strong reflections per atom, and atomic-resolution diffraction data extending to at least 0.1 nm (Dorset 1995). While Direct Methods is intended for application to Fourier coefficients of charge density F_x, kinematic electron diffraction measures coefficients of electrostatic potential $F_\mathbf{g}$ (eqn 5.18). These are related, through Poisson's equation, by the Mott–Bethe formula (Spence and Zuo 1992), which strictly cannot be inverted (to give F_x from $F_\mathbf{g}$) for a polyatomic crystal without knowledge of the atomic coordinates. The derivation of Direct Methods assumes a known sign for the scattering charge density and sharp atom peaks (atomicity)—these may also be reasonable assumptions which can be made for the potential, allowing the application of Direct Methods to Fourier coefficients of the potential rather than charge density. The sign of the total potential depends on knowledge of its mean value V_0 (O'Keeffe and Spence 1994); however, transmission diffraction is not sensitive to this. The potential for an isolated ion may change sign (Vainshtein 1964). (Electron scattering factors for ions may also change sign.) In general, the phases of electron structure factors are not formally equal to those of X-ray structure factors; however, we find from simulations that, both for centrosymmetric and non-centrosymmetric crystals, the phases are in fact equal for all but the weakest reflections. A comparison of various approximations (kinematic, two-beam (eqn 13.7), inclusion of excitation error effects etc.) applied to elastic energy-filtered electron diffraction spot patterns from thin organic films can be found in Wu and Spence (2002), where data are scaled using a Wilson plot. The use of the Kohler mode of selected area illumination in this work provides spot patterns from much smaller areas than conventional selected area diffraction, which minimizes bending effects. It is common in electron diffraction studies to find rather poor R-factors compared with X-ray work.

Traditionally, 'convergent-beam' (CBED) patterns, which complement HREM images, have been used for the following purposes:

1. For the determination of the crystal periodicity in the beam direction. This is given by

$$C_0 = 2/(\lambda U_0^2) \tag{13.8}$$

where λ is the electron wavelength and U_0 is the radius of the first-order Laue zone (FOLZ) ring seen in the microdiffraction pattern. The dimensions of a unit cell and its angle in the plane normal to the beam can also be obtainable from the ZOLZ pattern (Zuo 1993) and HOLZ lines in the central disc.
2. To align the crystal zone axis with the electron beam. The alignment of the region of interest for HREM is possible if no mechanical alteration to the specimen height is required in changing from the microdiffraction to the HREM mode.
3. In thicker specimens of small-unit-cell crystals, the two-beam CBED method of specimen thickness determination may be applied (Blake *et al.* 1978).
4. In favourable cases, it may be possible to determine the space group of the crystal (Goodman 1975, Eades *et al.* 1983, Steeds and Vincent 1983). However, a thicker region of crystal will be required for the CBED pattern than for the HREM image. A solution to this problem lies in the use of the Tanaka wide-angle CBED method (Eades 1984), which allows convergent beam patterns to be obtained over an angular range larger

than the Bragg angle, without overlap of orders. It may therefore be used to reveal contrast at higher angles in CBED patterns from very thin crystals, if sufficiently large areas of uniform thickness are available. The ability of this technique to isolate orders without overlap is particularly valuable for thin specimens of large-unit-cell crystals (with correspondingly small Bragg angles).

5. For the refinement of low-order structure factors in crystals, by comparison of experimental and computed CBED rocking curves (Zuo et al. 1999). From these, a charge density map may be synthesized to reveal the bonding, as shown in Plate 6 for cuprite.

6. HOLZ lines may be used to map out strains. Applications are described in Spence and Zuo (1992) with worked examples. Detailed mapping of the strain-field within a single transistor by CBED is described in Zhang et al. (2006).

7. Although there has been little success in solving new crystal structures by CBED in the presence of strong multiple scattering, there is no reason why CBED patterns cannot be obtained from very thin crystalline samples, or from the same area used for HREM imaging. This produces CBED patterns showing discs of uniform intensity, as shown in Fig. 13.8, which can then be analysed by standard crystallographic methods, as for X-ray diffraction. (The average intensity of the entire disc is used, rather than analysing the rocking curve within it. The discs are a consequence of using a nanoprobe.) Indeed, the appearance of uniform intensity across a CBED disc is a useful structure-independent indicator of the absence of multiple scattering. The 'charge-flipping' algorithm then provides a useful method of solving the phase problem *ab initio*, so that if several projections can be obtained, the structure might be solved entirely, or in conjunction with HREM images of neighbouring regions, to provide a model for refinement of the microdiffraction intensities. This method can be combined with elastic energy filtering and a detector with a large dynamic range, and is suited to structures with more isotropic symmetry than the layer structures to which the HREM method is usually limited.

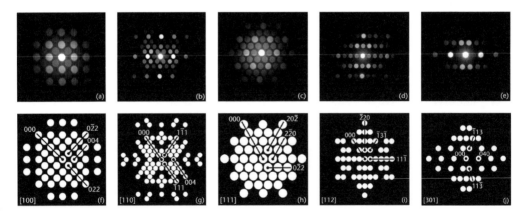

Figure 13.8 Kinematic convergent-beam patterns from several orientations from a crystal of spinel. The crystals are so thin (as for HREM) that the CBED discs show constant intensity, suggesting an absence of multiple scattering perturbations, so allowing a three-dimensional phasing and reconstruction of the electrostatic potential within the unit cell, while retaining the benefits of a small probe (McKeown and Spence 2009).

It has the great advantage that, in addition to allowing analysis of nanocrystals with a nanoprobe, since the direct beam is spread out into a disc, this disc can be recorded on the same intensity scale as the other beams, providing 'absolute' intensity calibration. For an example of this 'kinematic CBED' (KCBED) method, see McKeown and Spence (2009).

In general, the HREM method is most useful for crystals with a short period in the beam direction and large cell dimensions normal to the beam, since these give the maximum information in this projection. Such crystals are the least favourable for CBED (since orders are likely to overlap), for which the opposite relative dimensions are preferable, unless a wide-angle technique can be used. The HREM method is also preferable for the study of the defect structure of crystals and for the observation of superlattices, which give rise to very weak reflections in a CBED pattern but to intense points in a HREM image, particularly in thicker regions. The great power of the CBED method lies in its ability to provide symmetry information about the specimen structure which is independent of the HREM imaging parameters (e.g. astigmatism correction) and, in high-angle patterns, to provide information about the crystal structure in the electron beam direction from a single crystal setting. The limited tilting facilities available on modern HREM machines makes this information difficult to obtain otherwise. Thus, for example, the component of a dislocation Burgers vector in the beam direction can be determined in favourable cases from CBED patterns (Carpenter and Spence 1982). (See Morniroli (2002) for a review.)

The precession electron diffraction method (Vincent and Midgley 1994) has now been extensively developed for the purposes of directly solving inorganic structures by transmission electron diffraction, as demonstrated, for example, by Gjønnes et al. (1998).

The issue of how 'kinematic' the scattering is in precession electron diffraction has been analysed by several authors. Eggeman et al. (2010) used observation of kinematically forbidden reflections in silicon to show that the largest possible precession angle is needed to minimize multiple scattering effects. (This angle is limited by encroachment of the HOLZ and lens aberrations.) Ciston et al. (2008) suggested that forbidden reflections decay exponentially as the precession angle is increased. Sinkler et al. (2007) simulated the thickness dependence of Bragg beams after precession for α-(Ga, In)$_2$SnO$_5$. Figure 13.9 shows the result for the low-order reflections for various precession cone angles. The departures from ideal quadratic thickness dependence are clear at thicknesses greater than about 10 nm, in addition to the desirable smoothing effect on dynamical thickness oscillations which precession causes. This can be understood as an averaging over extinction distances, which depend on excitation error. Own et al. (2007) have reviewed the advantages of aberration correction for precession electron diffraction, and a detailed description of the experimental apparatus can be found in Own and Marks (2005).

We consider now the nanodiffraction patterns formed using field-emission STEM instruments, since the information these contain is closely related to that of a HREM image if an electron probe of near-atomic dimensions is used. This material should be read in conjunction with that of Chapter 8. The size of the microdiffraction probe depends on the demagnification of the probe-forming lenses, on the electron source size, and on the focus setting, aperture size, and spherical aberration constant of the probe-forming lens. For field-emission sources these last three factors are the most important, with the probe-broadening

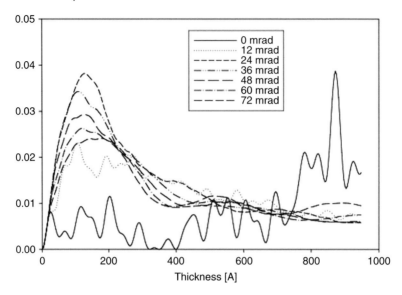

Figure 13.9 Thickness dependence of Bragg beams after precession for a doped tin oxide for various precession cone angles. (From Sinkler et al. (2007).)

effect of spherical aberration becoming most important for large aperture sizes. In fact, it is a useful approximation to consider that the electron source in field-emission instruments is an idealized point emitter, in which case the focused probe is 'diffraction limited' (i.e. it is the image of a point source, as formed by an imperfect lens). Thus, in the absence of spherical aberration, the probe has a size given approximately by $2r_s = d_p$ where

$$d_p = 1.22\lambda/\theta_c \tag{13.9}$$

where θ_c is the semi-angle subtended by the STEM objective aperture and shown in Fig. 13.10. In the presence of spherical aberration, there is no simple expression for probe size. However, computer calculations have shown (Spence 1978, Mory et al. 1987) that, except at the optimum focus setting, the probe intensity distribution in an uncorrected STEM contains rather extended 'tails' and oscillations. In practice, the size of the STEM probe can be roughly estimated from the resolution of a STEM lattice image, or by forming a probe image directly (Etheridge et al. 2011). Calculations have shown that the optimum focus needed to form the most compact probe (e.g. for microanalysis) is approximately $\Delta f = -0.75 C_s^{1/2} \lambda^{1/2}$. The principle of reciprocity nevertheless indicates that the Scherzer focus setting gives structure images under STEM lattice imaging conditions.

The assumption that the probe is diffraction-limited corresponds to the requirement that the objective aperture (C2 in Fig. 13.10) be coherently filled. This is so if the coherence width

$$X_a = \lambda/2\pi\theta_s \tag{13.10}$$

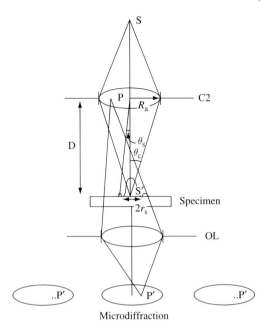

Figure 13.10 Ray diagram for microdiffraction. Note that P' is conjugate to P, and S is conjugate to the probe at S'. Partial coherence effects are incorporated by integrating over the source at S. (From Spence and Carpenter (1986).)

in the plane of the aperture is larger than the aperture. Here θ_s is the semi-angle subtended by the geometrical electron source image (formed on the specimen) at the aperture. Thus the requirement for a coherently filled aperture depends on the size of the electron source, which is proportional to θ_s. Since P in Fig. 13.10 is conjugate to the set of points P', we do not expect that CBED patterns from perfect crystals should differ in the two extreme cases where C2 is either coherently or incoherently filled. However, these points are conjugate only at exact focus, so that the through-focus behaviour of coherent and incoherent CBED patterns differs greatly. For defects, the coherent and incoherent CBED pattern may be entirely different, since in the absence of the Bragg condition each source point P gives rise to a displaced pattern of diffuse elastic scattering. For the coherent (incoherent) case the separate complex amplitudes (intensities) of all these patterns must be added together. Partial coherence effects can be accounted for by integrating over an extended source. Each different source point produces a probe at a slightly different position on the sample. However, we have seen in Chapter 8 that for perfectly crystalline samples the diffraction pattern is independent of probe position (if the orders do not overlap), and therefore of source size. For microdiffraction from defects (or the case of overlapping orders), the diffraction pattern will depend strongly on the size of the source and the coherence conditions.

For a discussion of the close relationship between coherent microdiffraction and lattice imaging in STEM, the reader is referred to Chapter 8. In the coherent case, a method for the computer simulation of microdiffraction patterns has been described (Spence 1978) in which the probe wavefunction is used as the boundary condition within an artificial superlattice at the entrance surface of the specimen, and the multislice multiple-Bragg-scattering algorithm is used. In a series of papers, experimental microdiffraction patterns from various materials have been compared with the results of these calculations (see, for example, Zhu and Cowley (1982)). A method of microdiffraction pattern simulation for defects has also been described for the incoherent case, based on the use of the popular diffraction-contrast software under the column approximation, by taking advantage of the reciprocity theorem (Carpenter and Spence 1982). A fuller discussion can be found in the text by Spence and Zuo (1992), which reviews electron microdiffraction and CBED with worked examples.

The similarities between the problems involved in the interpretation of HREM images and coherent CBED patterns become clear when we consider the case of CBED patterns formed using a probe which is 'smaller' than the crystal unit cell. Then, as shown in Chapter 8, the CBED orders will overlap, and interference effects will be seen within the region of overlap which depend on the focus setting and spherical aberration constant of the probe-forming lens (Cowley and Spence 1979). Thus, the problem of determining atom positions by a comparison of computed and experimental coherent microdiffraction patterns involves a similar set of adjustable parameters (atom coordinates and instrumental parameters) to those involved in HREM image matching. This applies both for defects and, in the microdiffraction case, for a particular identified group of atoms within a unit cell. Cowley (1981) showed such a pattern from $Ti_2Nb_{10}O_{29}$ in which, since the probe size is 'smaller' than the size of the unit cell, the symmetry of the pattern is not that of the crystal as a whole but rather possesses the local symmetry as reckoned about the probe centre (Cowley and Spence 1979). The patterns were seen to repeat as the probe was moved by a lattice translation vector (28 Å in length).

Since, as shown in Fig. 8.3(b), the intensity within the region of overlapping orders is sensitive to probe position, mechanical movement of the specimen or movement of the probe due to electronic instabilities is as important for coherent microdiffraction as for HREM. However, it can be shown that the loss of contrast in a coherent microdiffraction pattern due to vibration is a function of the scattering angle and focus setting. In general, for patterns with overlapping orders (or from defects) the finest detail which can be extracted is expected to be about the same as that which would appear in the corresponding STEM lattice image. High-angle scattering is most sensitive to instabilities.

For many problems in materials science, these interferometric effects and their dependence on instrumental parameters which result from the use of the smallest probes constitute an unnecessary complication. A more fruitful approach is then to observe characteristic disturbances in certain non-overlapping orders diffracted by interrupted or distorted planes of the crystal lattice. Then, in the spirit of '$\mathbf{g} \cdot \mathbf{b}$' analysis in TEM, by noting which of the microdiffraction spots are unaffected, it may be possible to determine fault vectors or to classify defect types using a probe only slightly 'larger' than the lattice period (Zhu and Cowley 1983). Several examples of this powerful approach have now appeared in the literature. Earlier dynamical calculations and experiments on coherent microdiffraction from metal catalyst particles (Cowley and Spence 1981) had shown that the normal CBED discs frequently appear as annular rings, possibly also broken up into segments of arc. These may appear simply as small blobs of intensity around the geometrical perimeter of the CBED disc. This effect is only observed when the probe lies near an edge or discontinuity in the

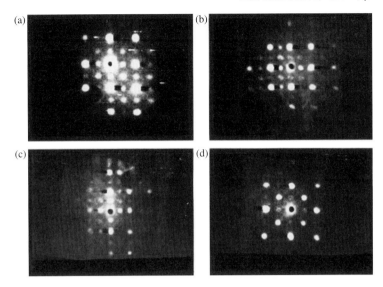

Figure 13.11 Microdiffraction patterns from antiphase domain boundaries in ordered Cu_3Au, obtained using a sub-nanometre probe: (a) from the crystal, probe 'not' on a boundary; (b) probe on a 'good' boundary; (c) probe on a 'too good' boundary; (d) probe on a 'bad' boundary. (From Zhu and Cowley (1982).)

lattice. This is a coherent interference effect, and is not due to dynamical scattering in the crystal. We now consider an example.

The patterns shown in Fig. 13.11 were obtained from samples of Cu_3Au containing antiphase domain boundaries of three possible types, depending on the near-neighbour coordination of atoms in the boundary. The electron probe size is about 5 Å, smaller than the average domain size. Since the domains represent a discontinuity in the Cu_3Au superlattice, the expected annular splitting is seen in the first-order superlattice reflections only. These are the reflections for which $\mathbf{g} \cdot \mathbf{R} \neq 0$, where $\mathbf{R}$ is the translation vector between neighbouring domains. By analysing all the possible boundary arrangements, Zhu and Cowley (1983) were able to draw up a classification scheme relating spot splitting to boundary type.

This work provides an example of a powerful general approach to a problem which would be insoluble using the HREM technique, since lattice images of Cu_3Au antiphase boundaries have proven extremely difficult to interpret. A similar approach has been applied to the case of stacking faults and twin boundaries in FCC metals, where the determination of fault vectors from spot splittings observed in coherent microdiffraction has also been demonstrated (Zhu and Cowley 1983). Other examples of this general approach include coherent microdiffraction studies of diamond platelets (Cowley and Osman 1984), of GP (Guinier–Preston) zones in Al–Cu alloys (Zhu and Cowley 1985), of metal–ceramic interfaces (Lodge and Cowley 1984), and of a CdS quantum dot nanocrystal (Huang et al. 2009). The method has also been applied to the study of twinning and defects in the small metal particles used for catalysis (Roy et al. 1981).

372 Associated techniques

The remarkable ability of carbon nanotubes to trap species of interest can be used to produce electron diffraction patterns from single molecules. In this way, coherent nanodiffraction patterns of C60 molecules have been recorded inside a single-walled carbon nanotube (Ran et al. 2012).

In all this work, the ability to record the weak microdiffraction fine structure efficiently using an axial parallel detection system and image intensifier has proved crucial. Microdiffraction area-detector systems are discussed in Cowley and Spence (1979), Cowley and Ou (1989), Spence and Zuo (1992), and Hembree et al. (2003).

The use of computed Fourier transforms of small regions of HREM images as a microdiffraction technique is discussed in Tomita et al. (1985). This method is subject to all the limitations outlined in Section 3.5 with regard to non-linearities.

13.4. Cathodoluminescence in STEM

The analysis of the electronic and atomic structure of individual defects in crystals requires the development of a technique which is: (1) sensitive to the very low concentrations of impurities which may be electrically important (usually present in concentrations too low to be detected by EDS, EELS, or HREM); (2) capable of sufficient spatial resolution to isolate individual defects; and (3) able to provide sufficiently high spectral energy resolution to study the electronic states of interest. At present the CL technique in STEM provides the most favourable combination of these properties. (Single-atom EELS is a special case—you have to know where the atom is. If we calculate the atomic concentration of foreign atoms along an atomic column within which there is an atom detected by EELS, that concentration would be very high, not low.) In the CL method, the optical emission excited by the electron beam in passing through a thin sample is collected by a small mirror and passed to a conventional optical spectrometer for analysis, as shown in Fig. 13.12. By forming a small electron probe and plotting the CL intensity within a small spectral range as a function of the position of the electron probe, a scanning monochromatic CL image may also be obtained, giving a spatial map of the impurity or electronic state of interest.

The usefulness of the STEM–CL technique for the study of semiconductor defects was first demonstrated by the pioneering work of Petroff and others on GaAs (Petroff et al. 1980): in this work they obtained a correlation between the electrical activity and type of dislocation. The use of a STEM rather than a SEM for CL improves spatial resolution.

The spatial resolution possible in STEM–CL depends on the specimen thickness for thin samples. In general it might be expected to be given by

$$d_\mathrm{r} = \left(d_\mathrm{p}^2 + d_\mathrm{g}^2 + d_\mathrm{D}^2\right)^{1/2} \tag{13.11}$$

where d_p is the electron probe diameter, d_g is the electron–hole pair generation volume, and d_D the carrier diffusion length (Pennycook 1981). For very thin transmission samples, d_g (the 'beam spreading' discussed in connection with EDS) and d_p are small compared with d_D, which is dominated by surface recombination. As a rough approximation one then obtains (M. Brown, personal communication, 1984; see also Shockley (1953))

$$d_\mathrm{r} = t/\sqrt{6} \tag{13.12}$$

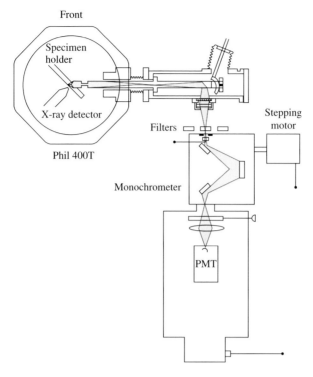

Figure 13.12 Diagram of the experimental arrangement used to observe cathodoluminescence from samples in a TEM instrument at low temperature. The optical path from the ellipsoidal mirror is shown stippled. (From Yamamoto et al. (1984).)

for a thin specimen of thickness t. For such very thin specimens, the emission intensity is very small, requiring efficient light-collection optics. The spatial resolution of a monochromatic scanning CL image is not influenced by the optical resolution limit of the light-collection mirror or lens; however, this does affect the amount of stray light which is collected, and hence the signal-to-noise ratio in the image. From such a small volume (it may be as small as a few thousand cubic angstroms) a spectral resolution of perhaps 10 Å may be obtained from very low concentrations of impurities, and this spectral information may be collected together with the corresponding HREM image. In most applications, however, spectra are recorded from a region of about 1 µm in diameter in combination with diffraction-contrast images, and this ability to correlate CL spectra with high-resolution STEM or TEM images is the most important advantage of the STEM–CL technique over conventional SEM–CL work, and over the PL method. Much lower temperatures, however, can be obtained in PL, with a corresponding gain in emission intensity.

In addition to the semiconductor work referred to above, there have been several studies of MgO and diamond by this method. Using a tapered silver tube as a light-collection element fitted to the Vacuum Generators HB5 STEM, Pennycook et al. (1980) obtained CL images of individual dislocations in type IIb diamond in correlation with their transmitted electron images. In this material, almost all the optical luminescence arises from dislocations, but not all dislocations are luminescent. Both 60° and screw dislocations were found

374 *Associated techniques*

to be luminescent. By subtracting a spectrum recorded from the defect of interest from one recorded nearby, the effects of stray luminescence, filament light, and impurity luminescence may be eliminated. In development of this work, Yamamoto *et al.* (1984) have obtained weak-beam images and spectra from individual dislocations of known type in diamond. The apparatus (shown in Fig. 13.12) has been fitted to a Philips EM 400 TEM–STEM instrument. This apparatus allows spectra to be collected over a range of temperatures down to about 25 K. Results from this work are shown in Fig. 13.13, where the TEM image of a single line defect and its associated optical emission spectrum are shown together with the

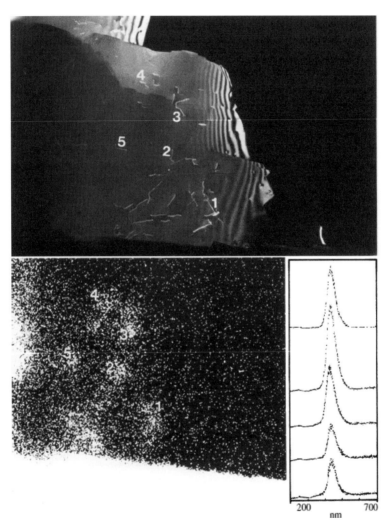

Figure 13.13 Scanning monochromatic CL images (lower panel) and corresponding dark field (220) images (upper panel) of dislocations in type IIb diamond recorded at 89 K with the apparatus of Fig. 13.12. Inset are the optical emission spectra taken from each of the individual dislocations shown. The dislocation types are (1) screw, (2) edge, (3) screw, (4) edge, and (5) screw. (From Yamamoto *et al.* (1984).)

scanning monochromatic CL image used to identify the defect. The spectrum was recorded at 89 K. In most materials, CL is found to increase in intensity at low temperatures, and the resulting fine structure observed may be more readily correlated with theoretical models. In this work, in addition, the polarization of the CL was measured from individual dislocations of known type and degree of dissociation. The dislocation emission (which occurs at a wavelength of 435 nm) is confirmed to be polarized along the direction of the dislocation line (rather than along its Burgers vector), and to form a broad band whose width is 0.412 eV. No clear correlation between CL activity and dislocation type or degree of dissociation was found, although a one-dimensional 'donor–acceptor' model for dislocation core CL is suggested as consistent with all the experimental findings. A high kink density was found to reduce the CL intensity.

This technique shows considerable promise in many areas of microscopy including mineralogy, catalysis, quantum dots, and semiconductor studies. For example, it is likely that measurements of CL polarization could be related to the crystallographic site of impurities, while the extension of CL detectors into the infrared region has opened up the possibility of work on silicon and other small-bandgap semiconductors. As an example, Fig. 13.14 shows the infrared CL emission spectrum obtained at 25 K from a group of straight dislocations in silicon, using an apparatus similar to that shown in Fig. 13.12. An intrinsic germanium detector was used, cooled to 77 K. The interpretation of these CL lines, and their relationship to those observed from similar samples by PL, is discussed in the literature (Graham

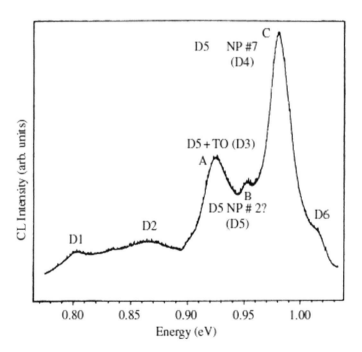

Figure 13.14 Infrared CL spectrum recorded at 23 K from a group of straight dislocations in silicon using a similar apparatus to that shown in Fig. 13.12, but with an intrinsic germanium detector. The transverse optical (TO) and no-phonon (NP) lines are indicated. (From Graham et al. (1986).)

13.5. Environmental HREM, imaging surfaces, holography of fields, and magnetic imaging with twisty beams

The vision of an atomic-resolution 'laboratory in a microscope' has steadily been approached in recent decades, and represents perhaps the most exciting frontier of HREM. By constructing a controlled-atmosphere environmental cell within either the objective lens pole-piece or the sample holder, for example, it is possible to induce oxidation and other reactions at elevated pressures and temperatures for study by HREM. Beyond that, the instrument becomes a complete nanolaboratory for synthesis and characterization, providing atomic-scale imaging and spectroscopy, if EELS and EDX facilities are also fitted. (The discovery of the nanotube by HREM imaging, for example, highlights the importance of direct observation at the atomic scale of the products of novel synthesis.) The literature now abounds with studies of oxidation and reduction, of nitridation, intercalation reactions, polymerization and hydroxylation reactions, and is well reviewed in the text by Gai (1997). Differentially pumped apertures may be used in the pole-piece, or thin carbon films across a cell in the holder, to accommodate the pressure difference. Recent cell designs are described in Lee *et al.* (1991), Boyes and Gai (1997), and Sharma (2001), while the use of a simple graphene envelope is described by Yuk *et al.* (2012). In this remarkable work, a liquid-filled graphene envelope is found to prevent evaporative cooling of the liquid, which does not freeze in the vacuum as a result. The very small sealed volume, however, prevents the supply of liquids under controlled conditions, and reactions with the trapped liquid will run quickly to exhaustion of the species in the cell.

For conventional cells, the larger pole-piece gap available in high-voltage machines is advantageous, but most recent work has used medium-voltage machines to minimize radiation damage. Video recordings of these dynamic processes may be obtained. Near-atomic resolution images (e.g. 0.2 nm) are obtainable in such a cell at pressures as high as 5 Torr (Hansen *et al.* 2001) (1 Torr = 133.3 Pa, 760 Torr is atmospheric pressure). The method has proven invaluable for the study of atomic mechanisms in catalytic reactions (Boyes and Gai 1997) and, for example, the direct observation by bright-field aberration-corrected TEM of the reconstruction of the surface facets on gold nanocrystals at atomic resolution, when they are exposed to adsorbed CO_2 molecules (Yoshida *et al.* 2012). Figure 13.15 shows the *in situ* partial oxidation of a thin crystal of niobium tungsten oxide, in which the oxidation is seen to occur as a result of the formation of domains of different sizes, whose crystal structures may be identified from the images (Sayagues and Hutchison 1999). Using the white-line ratios in EELS spectra, it has been possible to monitor the local oxidation state of cerium in CeO_2 during reduction (Sharma and Crozier 1999). Similar cells have been used to directly observe the process of chemical vapour deposition (Drucker *et al.* 1995). A series of remarkable papers has revealed the mechanism by which silicon nanowires can be grown by the vapour–liquid–solid process, utilizing a Au–Si eutectic cap, and the whole field, including this area, has been exhaustively reviewed by Ross (2006). This ability to directly observe controlled synthesis at the atomic level, combined with nanoscale analytic capabilities, is one of the most exciting areas of future development for HREM.

Figure 13.15 Solid state chemical reactions in action. HREM image along [001] of the *in situ* oxidation product of $Nb_4W_{13}O_{47}$, oxidized in an environmental cell under 20 mbar of oxygen. Domains of different structure appear as a result of the oxidation. Microdomains of WO_3, tetragonal tungsten bronze, and disordered bronze structures can be identified. The instrument used was a modified JEOL 4000EX. (See Krumeich *et al.* (1999).)

The study of phase transitions by *in situ* HREM has deepened our understanding of these processes at the atomic level. Near-atomic resolution has been obtained at temperatures well above 600 °C (Sinclair and Konno 1995), allowing the atomic mechanisms involved in the motion of interfaces to be deduced. Comparative studies have established that surface diffusion does not dominate these *in situ* thin film observations. For the highest temperatures, a heating holder consisting a single current-carrying wire running across the grid has been used, with nanoparticles attached to it. In this way it has been possible to obtain lattice images of silicon at 1400 °C (the melting point of silicon is 1414 °C) (Mori *et al.* 1994). Remarkably, it is also possible to image at high resolution through liquids. A simple, easily manufactured liquid cell has been described (Williamson *et al.* 2003), in which the electrochemical deposition of Cu from a copper sulphate solution is studied by TEM. Dynamic observations resolve the events occurring during the progressive nucleation of clusters, allowing the growth of individual copper clusters to be followed. The liquid cell geometry lends itself to the study of a wide variety of wet processes, including anodization, corrosion, growth of crystals from solution, liquid crystal switching, and alignment, as well

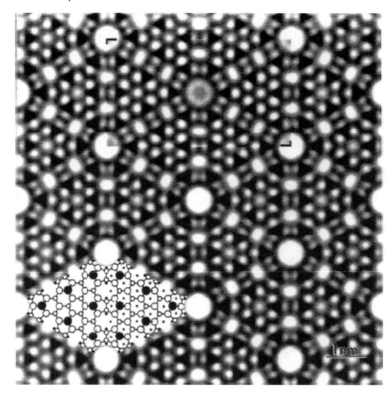

Figure 13.16 Direct imaging of surface atoms. Transmission image of a thin (111) silicon crystal obtained in an ultrahigh-vacuum HREM instrument, showing the 7 × 7 surface reconstruction, with atomic structure inset. Unlike STM images, this HREM image shows all the atoms in the top three layers in projection (Bengu *et al.* 1996).

as biological objects and processes. By extension of this work, the use of lithographically fabricated cells for *in situ* TEM will be an exciting growth area for HREM in the future.

The solution of the atomic structure of the silicon (111) 7 × 7 surface by ultrahigh-vacuum (UHV) electron diffraction using a modified TEM (Takayanagi 1984) solved a long-standing problem in the field of surface science just prior to the appearance of the scanning tunnelling microscope (STM). This has provided a great incentive for further development. Figure 13.16 shows that images may now be formed from these transmission diffraction patterns, projected through a thin crystal with reconstructed surfaces. Unlike the STM images, which image the adatoms only, the UHV HREM imaging shows all the atoms in the top three layers in projection, including the dimers (Bengu *et al.* 1996). (The image has been translationally and rotationally averaged to improve contrast.) For a review of UHV TEM see Gibson and Lanzerotti (1989), Yagi (1993), and Kienzle and Marks (2012).

In addition to the atomic-resolution electron holography described in Section 7.3, which followed Gabor's original aim of using electron holography to improve resolution, electron holography has been found invaluable for the mapping of magnetic and electric fields

(Matteucci et al. 1991). Dramatic images of superconducting vortices and of magnetic domains in nanoparticles may be found in Tonomura (1993), while the electric fields associated with ferroelectric materials (Spence et al. 1993) and field-emission tips (Cumings et al. 2002) have also been studied in this way. Mankos et al. (1994) describe holographic imaging of the multilayers used to produce the giant magnetoresistance effect. Using this technique it is possible to map out magnetic fields with nanometre spatial resolution, and a sensitivity of 1500 Bohr magnetons (about 500 cobalt atoms), which is more sensitive than a superconducting quantum interference device (SQUID) magnetometer. At low temperatures, where flux in superconductors is quantized, experiments may be arranged so that each holographic fringe corresponds to a single quantum of flux (2×10^{-15} Wb). The field is reviewed in Voelkl (1998). Particularly exciting is the three-dimensional reconstruction of electron fields within and around a biased semiconductor p–n junction by off-axis holography (Twitchett-Harrison et al 2007). This field of holographic tomography for the mapping of magnetic and electric fields offers great potential for the future.

The *in situ* manipulation of nanostructures during HREM observation presents great challenges. Straining stages have been used for many years for dynamic observation of

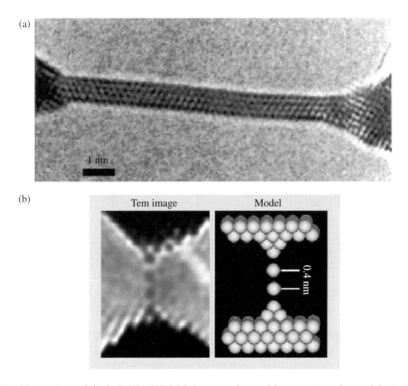

Figure 13.17 Nanowires. (a) A UHV HREM image of a gold nanowire, formed between a gold surface and a gold STM tip, suspended in a vacuum and oriented normal to the TEM beam. The wire is four atoms thick. (b) The one atom thick wire which results from stretching the wire in (a). The quantization of the wire's electrical conductance has been correlated with its structure (Kondo and Takayangi 1997).

strains during deformation using lower-resolution diffraction contrast. HREM imaging of the approach of an operating STM tip to a surface was achieved in 1992 (Lo and Spence 1993), leading to the development of improved UHV HREM instruments with this capability. Figure 13.17 shows an image of a gold nanowire consisting of four strands of atoms (Kondo and Takayangi 1997). With further stretching using a piezo motion, the wire is reduced to a single line of gold atoms, with a variety of novel intermediate structures. In a development of this work, the quantized resistance of these wires can be correlated with their atomic structure, as seen in these images. The conductance changes by $2e^2/h$ with the loss of each atomic layer.

The formation of vortex laser beams which carry orbital angular momentum $m\hbar$ per photon (with m the topological charge) has a long history (Beth 1936), and such beams have been used as an optical wrench to rotate a small particle in a liquid, amongst other applications. Similar states for electron beams have also now been created for use in STEM (Uchida and Tonomura 2010, Verbeek et al. 2010). Since the resulting dipole moment couples to the spin system of a sample, this may offer a way to map out magnetic fields in thin samples with high spatial resolution. These 'twisty beams' are formed using a diffraction grating in the illumination system to create the vortex state. The grating may contain a terminating fringe or form a spiral. Alternatively, the magnetic effects may be observed in the energy-loss spectrum by the method of electron energy-loss magnetic circular dichroism (EMCD), which preserves the high spatial resolution of the STEM (Schattschneider, 2012). By using a broader more collimated beam, it would seem possible to combine this method with the EELS ALCHEMI technique described at the end of Section 13.1.

The limits of spatial resolution in SEM have improved steadily over the years, especially with the development of the field-emission source for UHV SEM imaging (see Drucker et al. (1993) for analysis of contrast mechanisms). Using thin samples, it is now possible to demonstrate atomic resolution using the secondary electron signal, and to compare this directly with the transmitted HAADF-STEM image (Inada et al. 2011). This provides a valuable comparison of surface and bulk structure.

References

Ankudinov, A., Ravel, B., Rehr, J., and Conradson, S. (1998). Real-space multiple scattering calculation of XANES. *Phys. Rev.* **B58**, 7565.

Arslan, I. and Browning, N. D. (2002). Intrinsic electronic structure of threading dislocations in GaN. *Phys. Rev.* **B65**. 073510.

Batson, P. E. (1993). Simultaneous STEM imaging and ELS with atomic column sensitivity. *Nature* **366**, 727.

Bengu, E., Plass, R., Marks, L., Ichihashi, T., Ajayan, P., and Iijima, S. (1996). Imaging the dimers in Si(111)-(7 × 7). *Phys. Rev. Lett.* **77**, 4226.

Beth, R. A. (1936). Mechanical detection and measurement of the angular momentum of light. *Phys. Rev.* **50**, 115.

Bethe, H. (1930). Zur Theorie des Durchgangs schneller Korpuskularstrahlen durch Materie. *Ann. Phys. (Leipzig)* **5**, 325.

Blackman, M. (1939). On the intensities of diffraction rings. *Proc. R. Soc. Lond., Ser. A***173**, 68.

Blake, R. G., Jostsons, A., Kelly, P. M., and Napier, J. G. (1978). The determination of extinction distances by STEM. *Phil. Mag.* **A37**, 1.

Botton, G., Lazar, S., and Dwyer, C. (2010). Elemental mapping at the atomic scale using low accelerating voltages. *Ultramicroscopy* **110**, 926.

Bourdillon, A. J. (1984). Localisation and X-ray emission. *Phil. Mag.* **A50**, 839.

Bourdillon, A. J., Self, P. G., and Stubbs, W. M. (1981). Crystallographic orientation effects in energy dispersive X-ray analysis. *Phil. Mag.* **44**, 1335.

Bourret, A. and Colliex, C. (1982). Combined HREM and STEM microanalysis on decorated dislocation cores. *Ultramicroscopy* **9**, 183.

Boyes, E. and Gai, P. (1997). Environmental high resolution electron microscopy and applications to chemical science. *Ultramicroscopy* **67**, 219.

Brydson, R. (2001). *Electron energy loss spectroscopy*. Springer, New York.

Carpenter, R. and Spence, J. C. H. (1982). Three dimensional strain information from CBED patterns. *Acta Crystallogr.* **A38**, 55.

Cherns, D., Howie, A., and Jacobs, M. H. (1973). Characteristic X-ray production in thin crystals. *Z. Naturforsch.* **28a**, 565.

Christensen, K. K. and Eades, J. A. (1986). On 'parallel' illumination conditions for ALCHEMI. *Proc. 44th Ann. EMSA Meeting* (ed. G. Bailey), p. 622. San Francisco Press, San Francisco.

Ciston, J., Deng, B., Marks, L. D., Own, C. S., and Sinkler, W. (2008). A quantitative analysis of cone-angle dependence in precession electron diffraction. *Ultramicroscopy* **108**, 514.

Colliex, C. (2011). From electron energy-loss spectroscopy to multi-dimensional and multi-signal electron microscopy. *J. Electron Microsc.* **60** (Suppl. 1), S161.

Cosgriff, E. C., Oxley, M., Allen, L., and Pennycook, S. (2005). The spatial resolution of imaging using core-loss spectroscopy in the scanning transmission electron microscope. *Ultramicroscopy* **102**, 317.

Cowley, J. M. (1967). *Crystal structure determination by electron diffraction*. Progress in Materials Science, Vol. 13, No. 6, pp. 269–321. Pergamon Press, Oxford.

Cowley, J. M. (1981). Coherent interference effects in STEM and CBED. *Ultramicroscopy* **7**, 19.

Cowley, J. M. (1992). *Techniques of transmission electron diffraction*. Oxford University Press, New York.

Cowley, J. M. and Osman, A. (1984). Nanodiffraction from platelets in diamond. *Ultramicroscopy* **15**, 311.

Cowley, J. M. and Ou, H. J. (1989). Observation of microdiffraction patterns with STEM. *J. Electron Microsc. Tech.* **11**, 143.

Cowley, J. M. and Spence, J. C. H. (1979). Innovative imaging and microdiffraction in STEM. *Ultramicroscopy* **3**, 433.

Cowley, J. M. and Spence, J. C. H. (1981). Convergent beam electron microdiffraction from small crystals. *Ultramicroscopy* **6**, 359.

Craven, A. J. and Colliex, C. (1977). The effect of energy-loss on phase contrast. *Inst. Phys. Conf. Ser. 36*, p. 271. Institute of Physics, Bristol.

Craven, A. J., Gibson, J. M., Howie, A., and Spalding, D. R. (1978). Study of single-electron excitations by electron-microscopy. *Phil. Mag.* **A38**, 519.

Cumings, J., Zettl, A., McCartney, M. R., and Spence, J. C. H. (2002). Electron holography of field-emitting carbon nanotubes. *Phys. Rev. Lett.* **88**, 056804.

DeGraef, M. (2003). *Introduction to conventional transmission electron microscopy*. Cambridge University Press, New York.

Disko, M., Abn, C., and Fulz, B. (eds) (1992). *Transmission ELS in materials science*. Minerals, Metals and Materials Society (TMS), Warrendale, PA.

Dorset, D. L. (1995). *Structural electron crystallography*. Plenum, New York.

Drucker, J., Scheinfein, M., Liu, J., and Weiss, J. K. (1993). Electron coincidence spectroscopy studies of secondary and Auger electron generation mechanisms. *J. Appl. Phys.* **74**, 7329.

Drucker, J., Sharma, R., Kouvetakis, J., and Weiss, K. (1995). *In situ*, real time observation of Al CVD on SiO_2. *J. Appl. Phys.* **77**, 2846.

Duscher, G., Browning, N., and Pennycook, S. J. (1998). Atomic column resolved electron energy loss spectroscopy. *Phys. Status Solidi (a)* **166**, 327.

Eades, J. A. (1984). Zone-axis diffraction patterns by the Tanaka method. *J. Electron Microsc. Tech.* **1**, 279.

Eades, J. A. (1988). Glide planes and screw axes in CBED: the standard procedure. *Microbeam analysis*, ed. D. E. Newbury, p. 75. San Francisco Press, San Francisco.

Eades, J. A., Shannon, M. D., and Buxton, B. F. (1983). In *Scanning electron microscopy 1983*, Vol. III, ed. O. Johari, p. 1051. A. M. F. O'Hare, Chicago.

Egerton, R. F. (2009). Electron energy loss spectroscopy in the TEM. *Rep. Prog. Phys.* **72**, 1.

Egerton, R. F. (2011). *Electron energy-loss spectroscopy in the electron microscope*. Third edition. Springer, New York.

Eggeman, A. S., White, T. A., and Midgley, P. A. (2010). Is precession diffraction kinematical? *Ultramicroscopy* **110**, 777.

Etheridge, J., Lazar, S., Dwyer, C., and Botton, G. A. (2011). Imaging high energy electrons propagating in a crystal. *Phys. Rev. Lett.* **106**, 160802.

Fathers, D. and Rez, P. (1979). A transport equation theory of electron backscattering. *Scanning electron microscopy 1979*, Vol. II, ed. O. Johari, p. 55. A. M. F. O'Hare, Chicago.

Gai, P. (ed.) (1997). *In-situ microscopy in materials research*. Boston, Kluwer Academic.

Garratt-Reed, A. J. and Bell, D. C. (2003). *Energy dispersive X-ray analysis in the electron microscope*. BIOS Scientific, London.

Gibson, J. M. and Lanzerotti, M. (1989). Silicon oxidation by *in-situ* TEM. *Ultramicroscopy* **31**, 29.

Gjønnes, J., Hansen, V., Berg, B. S., Runde, P., Cheng, Y. F., Gjønnes, K., Dorset, D. L., and Gilmore, C. J. (1998). Structure model for the phase Al_mFe derived from three-dimensional electron diffraction intensity data collected by a procession technique. Comparison with convergent-beam diffraction. *Acta Crystallogr.* **A54**, 306.

Goodman, P. (1975). A practical method of 3-dimensional space-group analysis using convergent-beam electron-diffraction. *Acta Crystallogr.* **A31**, 804.

Goodman, P. and Warble, C. (1987). The top–bottom N-beam phase contrast effect from finite crystals. Use of the effect in HREM studies of surface. *Phil. Mag.* **B56**, 15.

Graham, R. J., Spence, J. C. H., and Alexander, H. (1986). Infrared cathodoluminescence studies from dislocations in silicon in TEM, a Fourier transform spectrometer for CL in TEM and ELS/CL coincidence measurements of lifetimes in semiconductors. *MRS Proc.* **82**, 235 (doi: 10.1557/PROC-82-235).

Hansen, T., Wagner, J., Hansen, P., Dahl, S., Topsoe, H., and Jacobsen, J. (2001). Atomic-resolution *in situ* transmission electron microscopy of a promoter of a heterogeneous catalyst. *Science* **294**, 1508.

Hembree, G., Weierstall, U., and Spence, J. (2003). A quantitative nanodiffraction recording system for STEM. *Microsc. Microanal.* **9**, 468.

Howie, A. (1963). Inelastic scattering of electrons by crystals: (I) The theory of small-angle inelastic scattering. *Proc. R. Soc. Lond.* **271A**, 268.

Howie, A. (1979). Image contrast and localised signal selection techniques. *J. Microsc.* **117**, 11.

Huang, W. J., Zuo, J. M., Jiang, B., Kwon, K. W., and Shim, M. (2009). Sub-angstrom-resolution diffractive imaging of single nanocrystals. *Nature Phys.* **5**, 129.

Humphreys, C. J. and Spence, J. C. H. (1979). 'Wavons'. *Proc. EMSA 1979*, ed. G. Bailey, p. 554. Claitor's, Baton Rogue, LA.

Inada, H., Su, D., Egerton, R., Konno, M., Wu, L., Ciston, J., Wall, J., and Zhu, Y. (2011). Atomic imaging using secondary electrons in an SEM. *Ultramicroscopy* **111**, 865.

Jiang, N., Qiu, J., Spence, J. C. H., and Humphreys, C. J. (2008). Observation of long-range compositional fluctuations in glasses: implications for atomic and electronic structure. *Micron* **39**, 698.

Johnson, D. W. and Spence, J. C. H. (1974). Determination of the single-scattering probability distribution from plural-scattering data. *J. Phys. D: Appl. Phys.* **7**, 771.
Jones, I. P. (2002) Determining the locations of chemical species in ordered compounds: ALCHEMI. *Adv. Imaging Electron Phys.* **125**, 63.
Joy, D. (1995) *Monte Carlo modelling for electron microscopy and microanalysis*. Oxford University Press, New York.
Joy, D. C., Romig, A. D., and Goldstein, J. I. (eds) (1986). *Principles of analytical electron microscopy*. Plenum Press, New York.
Kainuma, Y. (1955). The theory of Kikuchi patterns. *Acta. Crystallogr.* **8**, 247.
Kienzle, D. M. and Marks, L. D. (2012). Surface transmission electron diffraction for $SrTiO_3$ surfaces. *Cryst. Eng. Commun.* **14**, 7833.
Kim, K.-H. and Zuo, J.-M. (2013). Symmetry quantification and mapping using convergent-beam electron diffraction. *Ultramicroscopy* **124**, 71.
Kohl, H. and Rose, H. (1985). Theory of image formation by inelastically scattered electrons in the electron microscope. *Adv. Imaging Electron Phys.* **65**, 173.
Kondo, Y. and Takayangi, K. (1997). Atom bridges to nanowires. *Phys. Rev. Lett.* **79**, 3455.
Krivanek, O. L., Ursin, J. P., Bacon, N. J., Corbin, G. J., Delby, N., Hrncirik, P., Murfitt, M. F., Own, C. S., and Szilagy, Z. S. (2009). High energy-resolution monochromator for aberration-corrected STEM. *Phil Trans. R. Soc. A* **367**, 3683.
Krumeich, F., Hutchison, J. L., and Sayagues, M. J. (1999). *In-situ* oxidation products of $Nb_4W_{13}O_{47}$: a high-resolution transmission microscopy study. *Z. Anorg. Allgem. Chem.* **625**, 755.
Lee, T., Dewald, D., Eades, J., Robertson, I., and Birnbaum, H. (1991). An environmental cell TEM. *Rev. Sci. Instrum.* **62**, 1438.
Lo, W. and Spence, J. C. H. (1993). Investigation of STM image artifacts by *in-situ* reflection electron microscopy. *Ultramicroscopy* **48**, 433.
Lodge, E. A. and Cowley, J. M. (1984). The surface diffusion of silver on MgO. *Ultramicroscopy* **13**, 215.
Lovejoy, Q., Ramasse, M., Falke, M., Kaeppel, A., Terborg, R., Zan, R., Dellby, N., and Krivanek, O. L. (2012). Single atom identification by energy dispersive x-ray spectroscopy. *Appl. Phys. Lett.* **100**, 154101.
Lynch, D. F., Moodie, A. F., and O'Keefe, M. (1975). The use of the charge-density approximation in the interpretation of lattice images. *Acta Crystallogr.* **A31**, 300.
McKeown, J. and Spence, J. C. H. (2009). The kinematic convergent beam method for solving nanocrystal structures. *J. Appl. Phys.* **106**, 074309.
Mankos, M., Yang, Z. J., Scheinfein, M. R., and Cowley, J. M. (1994). Absolute magnetometry of thin cobalt films and Co/Cu multilayer structures at nanometer spatial resolution. *IEEE Trans. Magn.* **30**, 4497.
Matteucci, G., Missiroli, G., Nichelatti, E., Migliori, A., Vanzi, M., and Pozzi, G. (1991). Electron holography of long range electric and magnetic fields. *J. Appl. Phys.* **69**, 1835.
Mori, S., Yasuda, H., and Kamino, T. (1994). Si lattice image at 1400 °C. *Phil. Mag. Lett.* **69**, 279.
Morniroli, J.-P. (2002). *Large-angle convergent-beam electron diffraction*. Society of French Microscopists, Paris.
Mory, J. Colliex, C., and Cowley, J. M. (1987). Optimum defocus for STEM. *Ultramicroscopy* **21**, 171.
Mugnaioli, E., Gorelik, T., Stewart, A., and Kolb, U. (2012). *Ab initio* structure solution of nanocrystalline minerals and synthetic materials by automated electron tomography. *Minerals as advanced materials II*, p. 41. Springer, Berlin.

Muller, D. A., Sorsch, T., Moccio, S., Baumann, F. H., Evans-Lufferodt, K., and Timp, G. (1999). The electronic structure at the atomic scale of ultrathin gate oxides. *Nature* **399**, 758.

Muller, D. A., Kourkoutis, L. F., Mufitt, M., Song, J. H., Hwang, H. Y., Silcox, J., Dellby, N., and Krivanek, O. L. (2008). Atomic-scale chemical imaging of composition and bonding by aberration-corrected microscopy. *Science* **319**, 1073.

Nufer, S., Marinopoulos, A. G., Gemming, T., Elsässer, C., Kurtz, W., Köstlmeier, S., and Rühle, M. (2001). Quantitative atomic-scale analysis of interface structures: transmission electron microscopy and local density functional theory. *Phys. Rev. Lett.* **86**, 5066.

O'Keeffe, M. and Spence, J. C. H. (1994). On the average Coulomb potential and constraints on the electron density in crystals. *Acta Crystallogr.* **A50**, 33.

Own, C. S. and Marks, L. D. (2005). Electron precession: a guide for implementation. *Rev Sci. Instrum.* **76**, 033703.

Own, C. S., Sinkler, W., and Marks, L. D. (2007). Prospects for aberration corrected electron precession. *Ultramicroscopy* **107**, 534.

Petroff, P., Logan, R. A., and Savage, A. (1980). Nonradiative recombination at dislocation in III-V. *Phys. Rev. Lett.* **44**, 287.

Pennycook, S. J. (1981). Investigation of the electronic effects of dislocations by STEM. *Ultramicroscopy* **7**, 99.

Pennycook, S. J. (1987). Impurity lattice and sublattice location by electron channeling. *Scanning electron microscopy 1987*, ed. O. Johari, p. 217. A. M. F. O'Hare, Chicago.

Pennycook, S. J., Brown, L. M., and Craven, A. J. (1980). Observation of cathodoluminescence at single dislocation by STEM. *Phil. Mag.* **A41**, 589.

Pinsker, Z. G. (1949). *Electron diffraction* (translated by J. A. Spink and E. Feigl). Butterworths, London.

Radtke, G. and Botton, G. (2011). Energy loss near edge structures. *Scanning transmission electron microscopy: imaging and analysis*, ed. S. Pennycook and P. Nellist, p. 207. Springer, New York.

Ramesse, Q., Seabourne, C., Kepaptsoglou, D., Zan, R., Bangert, U., and Scott, A. (2013). Probing the bonding and electronic structure of single dopants in graphene with EELS. Nanoletters. in press.

Ran, K., Zuo, J.-M., Chen, Q., and Shi, Z. (2012). Electrons for single-molecule diffraction and imaging. *Ultramicroscopy* **119**, 72.

Reese, G. M., Spence, J. C. H., and Yamamoto, N. (1984). Coherent bremsstrahlung from kilovolt electrons in zone axis orientations. *Phil. Mag.* **A49**, 697.

Reimer, L. (ed.) (1995). *Energy-filtering TEM*, Springer Optical Sciences, Vol. 71. Springer, New York.

Rez, P., Alvarez, J., and Pickard, C. (1999). Calculation of near edge structure. *Ultramicroscopy* **78**, 175.

Ross, F. (2006). In situ transmission electron microscopy. *The science of microscopy*, ed. P. Hawkes and J. C. H. Spence, Ch. 6. Springer, New York.

Rossouw, C. J. and Maslen, V. W. (1987). Localisation and ALCHEMI for zone axis orientations. *Ultramicroscopy* **21**, 277.

Rossouw, C. J. and Miller, P. R. (1999). Analysis of incoherent channelling patterns formed by X-ray emission from host lattice species and interstitial Cr in mullite. *J. Electron Microsc.* **48**, 849.

Rossouw, C. J., Forwood, C. T., Gibson, M. A., and Miller, M. (1996). Statistical ALCHEMI. *Phil. Mag.* **A74**, 57.

Roy, R. A., Messier, R., and Cowley, J. M. (1981). Fine structure of gold particles in thin films prepared by metal–insulator co-sputtering. *Thin Solid Films* **79**, 207.

Saldin, D. and Rez, P. (1987) Theory of the excitation of atomic inner-shells in crystals by fast electrons. *Phil. Mag.* **B55**, 481.

Sayagues, M. and Hutchison, J. (1999). A new niobium tungsten oxide as a result of an *in-situ* reaction in a gas reaction cell TEM. *J. Solid State Chem.* **143**, 33.

Schattschneider, P. (2012). *Linear and circular dichroism in the electron microscope*. Pan Stanford, Singapore.

Scheu, C., Stein, W., and Rühle, M. (2000). ELNES of a Cu/Al_2O_3 interface. *Phys. Status Solidi (b)* **222**, 199.

Self, P. G. and Buseck, P. R. (1983). Low-energy limit to channelling effects in the inelastic scattering of fast electrons. *Phil. Mag.* **A48**, L21.

Shah, A. B., Ramesse, Q. M., Wen, J. G., Bhattacharya, A., and Zuo, J. M. (2011). Practical spatial resolution of electron energy loss spectroscopy in aberration-corrected scanning transmission electron microscopy. *Micron* **42**, 539.

Shah, A. B., Nelson-Cheeseman, B. B., Subramanian, G., Bhattacharya, A., and Spence, J. C. (2012). Structurally induced magnetization in an $La_{2/3}Sr_{4/3},MnO_4$ superlattice. *Phys Status Solidi (a)* **209**, 1322.

Sharma, R. (2001). Design and application of environmental cell TEM for *in-situ* observations of gas–solid reactions. *Microsc. Microanal.* **7**, 495.

Sharma, R. and Crozier, P. (1999). In situ electron microscopy of CeO_2 reduction. *Electron Microsc. Anal.* **161**, 569.

Shockley, W. (1953). *Electrons and holes in semiconductors*. Van Nostrand, New York.

Sinclair, R. and Konno, T. (1995). In-situ HREM: application to metal-mediated crystallization. *Ultramicroscopy* **56**, 225.

Sinkler, W., Own, C. S., Ciston, J., and Marks, L. D. (2007). Statistical treatment of precession electron diffraction data with principal components analysis. *Microsc. Microanal.* **13** (Suppl. 2), 954.

Spence, J. C. H. (1978). Approximations for the calculation of CBED patterns. *Acta Crystallogr.* **34**, 112.

Spence, J. C. H. (1985). The structural sensitivity of ELNES. *Ultramicroscopy* **18**, 165.

Spence, J. C. H. (1987). *High resolution electron microscopy and associated techniques*, ed. P. Buseck, J. M. Cowley, and L. Eyring. Oxford University Press, Oxford.

Spence, J. C. H. (2006). Absorption spectroscopy with sub-angstrom beams: ELS in STEM. *Rep. Prog. Phys.* **69**, 725.

Spence, J. C. H. and Carpenter, R. (1986). Electron microdiffraction. *Elements of analytical electron microscopy*, ed. D. Joy, A. Romig, J. Hren, and H. Goldstein, Ch. 3. Plenum Press, New York.

Spence, J. C. H. and Lynch, J. (1982). STEM microanalysis by transmission electron energy loss spectroscopy in crystals. *Ultramicroscopy* **9**, 267.

Spence, J. C. H. and Reese, G. (1986). Pendellösung radiation and coherent bremsstrahlung. *Acta Crystallogr.* **A42**, 577.

Spence, J. C. H. and Taftø, J. (1983). ALCHEMI: a new technique for locating atoms in small crystals. *J. Microsc.* **130**, 147.

Spence, J. C. H. and Zuo, J. M. (1992). *Electron microdiffraction*. Plenum, New York.

Spence, J. C. H., Krivanek, O. L., Taftø, J., and Disko, M. (1981). The crystallographic information in electron energy loss spectra. *Inst. Phys. Conf. Ser. 61*, p. 253. Institute of Physics, Bristol.

Spence, J. C. H., Kuwabera, M., and Kim, Y. (1988). *Ultramicroscopy* **26**, 103.

Spence, J., Cowley, J. M., and Zuo, J. M. (1993). Electron holography of ferroelectric domain walls. *Appl. Phys. Lett.* **62**, 2446.

Steeds, J. and Vincent, R. (1983). Use of high-symmetry zone axes in electron diffraction in determining crystal point and space groups. *J. Appl. Crystallogr.* **16**, 317.

Stout, G. H. and Jensen, L. H. (1968). *X-ray structure determination*. Macmillan, London.

Suenaga, K., Okazaka, T., Okunishi, E., and Matsumua, S. (2012). Detection of photons emitted from single erbium atoms in energy dispersive x-ray spectroscopy. *Nature Photonics* **6**, 545.

Taftø, J. (1984). Absorption edge fine structure study with subunit cell spatial resolution using electron channelling. *Nucl. Instrum. Methods* **B2**, 733.

Taftø, J. and Lehmfuhl, G. (1982). Direction dependence in electron energy loss spectroscopy from single crystals. *Ultramicroscopy* **2**, 287.

Taftø, J. and Liliental, S. (1982). Studies of the cation distribution in $ZnCr_xFe_{2-x}O_4$ spinels. *J. Appl. Crystallogr.* **15**, 260.

Taftø, J. and Spence, J. C. H. (1982). Crystal site location of iron and trace elements in an Mg–Fe-olivine using a new crystallographic technique. *Science* **218**, 49.

Taftø, J. and Zhu, J. (1982). Electron energy-loss near edge structure (ELNES), a potential technique in the studies of local atomic arrangements. *Ultramicroscopy* **9**, 349.

Taftø, J., Krivanek, O. L., Spence, J. C. H., and Honig, J. M. (1982). Is your spinel normal or inverse? *Proc. Tenth Int. Congr. Electron Microsc. (Hamburg)*, Vol. 1, p. 615.

Takayanagi, K. (1984). Si 7 × 7. *J. Microsc.* **136**, 278.

Tomita, M., Hashimoto, H., Ikuta, T., Endoh, H., and Yokota, T. (1985). Improvement and application of the Fourier-transformed pattern from a small area of high resolution electron microscope images. *Ultramicroscopy* **16**, 9.

Tonomura, A. (1993). *Electron holography*. Springer-Verlag, New York.

Twitchett-Harrison, A. C., Yates, T. J. V., Newcomb, S. B., Dunin-Borkowski, R. E., and Midgely, P. A. (2007). High-resolution three-dimensional mapping of semiconductor dopant potentials. *Nano Lett.* **7**, 2020.

Uchida, M. and Tonomura, A. (2010). Generation of electron beams carrying orbital angular momentum. *Nature* **464** 737.

Vainshtein, B. (1964). *Structure analysis by electron diffraction*. Pergamon, London.

Varella, M., Gazquez, J., Pennycook, T., Magen, C., Oxley, M. P., and Pennycook, S. J. (2011). Applications of aberration-corrected scanning transmission electron microscopy and electron energy loss spectroscopy to complex oxide materials. *Scanning transmission electron microscopy: imaging and analysis*, ed. S. Pennycook and P. Nellist, p. 429. Springer, New York.

Vechio, K. S. and Williams, D. B. (1986). Experimental conditions affecting coherent bremsstrahlung in X-ray microanalysis. *J. Microsc.* **147**, 15.

Verbeeck, J., Tian, H., and Schattschneider, P. (2010). Production and application of electron vortex beams. *Nature* **467**, 301.

Vincent, R. and Midgley, P. (1994). Double conical beam-rocking system for measurement of integrated electron diffraction intensities. *Ultramicroscopy* **53**, 271.

Voelkl, E. (ed.) (1998). *Introduction to electron holography*. Plenum, New York.

Wang, F., Egerton, R., and Malac, M. (2009). Fourier-ratio deconvolution techniques for electron energy-loss spectroscopy (EELS). *Ultramicroscopy* **109**, 1245.

Williams, D. B. and Carter, C. B. (1996). *Transmission electron microscopy: a textbook for materials science*. Plenum, New York.

Williams, D. B. and Joy, D. C. (1984). *Analytical electron microscopy 1984*. San Francisco Press, San Francisco.

Williamson, M., Tromp, R., Hull, R., Vereecken, P., and Ross, F. (2003). Dynamic microscopy of nanoscale cluster growth at the solid–liquid interface. *Nature Mater.* **2**, 532.

Witte, C., Findlay, S., Oxley, M. P., Rehr, J. J., and Allen, L. J. (2009). Theory of dynamical scattering in near-edge electron energy loss spectroscopy. *Phys. Rev.* **B80**, 184108.

Wu, J. and Spence, J. C. H. (2002). Kinematic and dynamical CBED for solving the structure of thin organic films at low temperature. *Acta Crystallogr.* **A58**, 580.

Wu, J. and Veblen, D. R. (2010). Characterization of Al–Si ordering in an alkali feldspar using ALCHEMI. *Am. Mineral.* **95**, 41.

Yagi, K. (1993). *In situ* observations of surface processes by UHV HREM. *Dynamic processes on solid surfaces*, ed. K. Tamaru. Plenum, New York.

Yamamoto, N., Spence, J. C. H., and Fathy, D. (1984). Cathodoluminescence and polarization studies from individual dislocations in diamond. *Phil. Mag.* **B49**, 609.

Yoshida, H., Kuwauchi, Y., Jinschek, J., Sun, K., Tanaka, S., Kohyama, M., Shimada, S., Haruta, M., and Takeda, S. (2012). Visualizing gas molecules interacting with supported nanoparticulate catalysts at reaction conditions. *Science* **335**, 317.

Yuk, J. M., Park, J., Ercius, P., Kim, K., Hellebusch, D. J., Crommie, M. F., Lee, J. Y., Zettle, A., and Alivisatos, A. P. (2012). High-resolution EM of colloidal nanocrystal growth using graphene liquid cells. *Science* **336**, 61.

Zagonel, L., Mazzucco, S., Tenc, M., March, K., Bernard, R., Laslier, B., Jacopin, G., Tchernycheva, M., Rigutti, L., Julien, F., Songmuang, R., and Kociak, M. (2011). Nanometer scale spectral imaging of quantum emitters in nanowires and its correlation to their atomically resolved structure. *Nano Lett.* **11**, 568.

Zhang, P., Kisielowski, C., Istratov, C., He, H., Nelson, C., Mardinly, J., Weber, E., and Spence, J. C. H. (2006). Direct strain measurement in a 65 nm node locally strained silicon device by energy-filtered convergent beam diffraction. *Appl. Phys. Lett.* **89**, 161907.

Zhu, J. and Cowley, J. M. (1982). Microdiffraction from antiphase domain boundaries in Cu_3Au. *Acta Crystallogr.* **A38**, 718.

Zhu, J. and Cowley, J. M. (1983). Microdiffraction from stacking faults and twin boundaries in f.c.c. crystals. *J. Appl. Crystallogr.* **16**, 171.

Zhu, J. and Cowley, J. M. (1985). Study of the early stage precipitation in Al–4% Cu by microdiffraction and STEM. *Ultramicroscopy* **18**, 419.

Zou, X., Hovmöller, S., and Oleynikov, P. (2011). *Electron crystallography – electron microscopy and electron diffraction*, IUCr Texts on Crystallography 16. Oxford University Press. New York.

Zuo, J. M. (1993). A new method of Bravais lattice determination. *Ultramicroscopy* **52**, 459.

Zuo, J. M., Spence, J. C. H., and Hoier, R. (1989). Accurate structure-factor phase determination by electron diffraction in noncentrosymmetric crystals. *Phys. Rev. Lett.* **62**, 547.

Zuo, J. M., Kim, M., O'Keeffe, M., and Spence, J. C. H. (1999). Observation of d holes and Cu–Cu bonding in cuprite. *Nature* **401**, 49.

Zvyagin, B. B. (1967). *Electron diffraction analysis of clay minerals*. Plenum, New York.

Appendix 1

The following FORTRAN program calculates values of the defocus f, spherical aberration constant C_s, and astigmatism constants from optical diffractogram ring radii S_n. It is based on the method of Krivanek (1976) and was written by M. A. O'Keefe. As written, *it is restricted to the case of large under-focus settings*, but is easily modified for over-focus (Δf positive) values, as described below.

The method depends on the fact that minima will occur in the diffractogram intensity (given by eqn (3.42)) if

$$\chi(u) = 2\pi[\Delta f \lambda u^2/2 + C_s \lambda^3 u^4/4] = N\pi \tag{A1.1}$$

Here N is any integer (including zero). On dividing by u^2, we find that a plot of N/u^2 against u^2 forms a straight line with slope $(C_s \lambda^3/2)$ and intercept $(\Delta f \lambda)$. The program uses linear regression to find the slope and intercept (and hence C_s and Δf) from the values of N and S_n supplied. The ring radii S_n must be specified in Å^{-1}, and are related to the measured ring radii by eqn (10.5). Input data include a title and the microscope accelerating voltage in kilovolts. Then follows one line of input data for each maximum or minimum, containing the number of the ring, a zero (for a minimum) or a one (for a maximum), and the radius of this maximum or minimum (in Å^{-1}). The method of numbering the rings is indicated by the numbers across the page in Fig. A1.1. These numbers correspond to $-N$ in eqn (A1.1). (The program actually works with a new $n = 2N$ for convenience, evaluated after line 10, so that the slope and intercept are twice the values given above.)

The correct numbering of the rings requires some care. All the worked examples in Krivanek (1976) follow the system indicated in Fig. A1.1, which is correct only for large under-focus (Δf negative). Figure A1.2 shows $\chi(u)/\pi$ plotted for three defocus values with $C_s = 1.8$ mm at 100 kV. The values of N from eqn (A1.1) are indicated on the curves. It is seen that there are essentially three cases: (1) positive focus, for which $N = 1, 2, 3, 4, \ldots$; (2) near Scherzer focus, for which the first value of N must be 0 ($N = 0, 1, 2, 3, \ldots$); and (3) larger negative focus, in which case (for the example shown) $N = -1, -2, -3, -3, -2, -1, 0, 1, 2, 3, 4, \ldots$. The turning point occurs at the stationary phase condition (see eqn (5.66))

$$U_0 = \sqrt{\frac{\Delta f}{C_s \lambda^2}} \tag{A1.2}$$

Here the declining values of N start to increase, and may or may not repeat at the minimum value (e.g. -3 in the last paragraph). The stationary phase condition may *sometimes* be identified on a diffractogram by a broad intense ring; however, if this condition coincides with $\chi(u) = -m\pi$, a broad absence of intensity will result. In practice, as shown

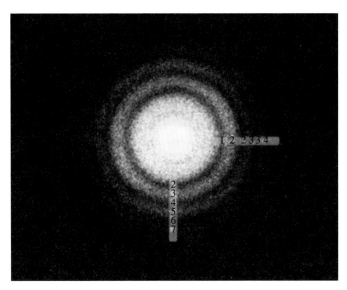

Figure A1.1 Optical diffraction pattern of an electron image of a thin carbon foil suitable for use with the FORTRAN program given. The pattern shows negligible astigmatism and drift (see Section 10.6). Numbers running across the pattern are the numbers NRING needed as input to the program. The maximum of the first (inner) ring has not been used since its position is difficult to determine. The numbers running down the pattern are the quantities $n = 2\text{NRING} - \text{MAXMIN}$ evaluated after line 10 of the program. The intensity in this pattern is given by eqn (3.41), together with the damping effects of partial coherence and chromatic aberration discussed in Chapter 4, Instrumental aberration constants are obtained by the program from patterns such as these.

in Fig. A3.1, the turning point may be driven beyond the resolution of the microscope by choosing from eqn (A1.2)

$$|\Delta f| \geq C_s \lambda^2 u_0^2 \tag{A1.3}$$

where u_0 is the highest spatial frequency in the diffractogram. Then the values of N will increase monotonically by unit increments with negative sign. These are the conditions for which the FORTRAN program is written. An additional test of internal consistency also exists, since the program returns values of the standard deviation for slope and intercept. Because of the method of analysis, these values should *not* be taken as the errors in C_s and Δf. They may, however, be used to confirm that a good straight-line fit has been obtained, and that the values of N have therefore been correctly chosen.

The best way to use the program for the calibration of a new microscope is to record several through-focus series at focus settings satisfying eqn (A1.3) (using the manufacturer's data for C_s and U_0), from which the smallest focal step increment (and C_s) can be deduced. The Scherzer focus required for structure imaging can then be obtained by counting 'clicks' from the minimum contrast position, given by eqn (6.26) (see also Section 12.5).

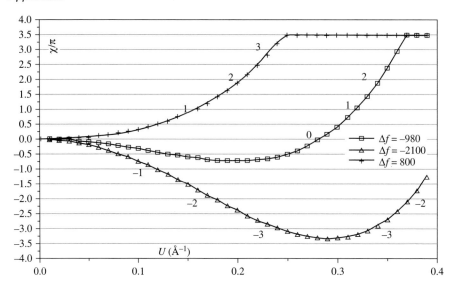

Figure A1.2 The function $\chi(u)$ plotted in units of π for three focus settings $\Delta f = -980$ (Scherzer focus), -2100, and $+800$ Å as indicated. Values of N in eqn (A1.1) are indicated. Here $C_s = 1.8$ mm and the accelerating voltage is 100 kV.

Reference

Krivanek, O. L. (1976). Method for determining coefficient of spherical aberration from a single electron micrograph. *Optik* **45**, 97.

A FORTRAN program for finding defocus and spherical aberration constants from measured optical diffractograms ring radii

```
C     PROGRAM CSEPS
C     TO DETERMINE CS(MM) AND DEFOCUS (ANGSTROMS) GIVEN RING DIAMETERS
C     FROM DIFFRACTOGRAMS OF MICROGRAPHS OF THIN AMORPHOUS MATERIALS
      DIMENSION NAME(13),X(20),Y(20)
C     READ IN VOLTAGE IN KV
C     READ (5,100) NAME,VOLTK
C     CALCULATE WAVELENGTH
      WAVL = 0.387818/SQRT(VOLTK * (1.0 + 0.978459E – 3 * VOLTK))
      WAVCUB = (WAVL * * 3) * 1.0E7
   70 N = 0
   20 CONTINUE
C     READ IN NUMBER OF RING (1 FOR FIRST, 2 FOR SECOND ....),
C     WHETHER RING IS MAXIMUM (MAXIMIN = 1) OR MINIMUM (MAXIMIN = 0)
C     RADIUS OF RING IN RECIPROCAL ANGSTROMS
      READ(5,110)NRING,MAXMIN,URAD
      IF(NRING,NE.0) GO TO 10
      IF(N.EQ.0) GO TO 80
      GO TO 30
   10 CONTINUE
      IF(N.EQ.0) WRITE(6,200) NAME, VOLTK,WAVL
      N = N + 1
      USQ = URAD * URAD
      X(N) = USQ
      Y(N) = –FLOAT(2 * NRING-MAXMIN)/USQ
C     CHECK WHETHER MAXIMUM OR MINIMUM
      IF(MAXMIN.EQ.0) GO TO 25
      IF(MAXMIN.EQ.1) GO TO 35
      WRITE(6,260)
      GO TO 80
   25 WRITE(6,220) NRING, URAD,X(N), Y(N)
      GO TO 20
   35 WRITE(6,210) NRING,URAD,X(N), Y(N)
      GO TO 20
   30 CONTINUE
C     IF N.GT.1 THEN FIT CURVE
      IF(N.FT.1) GO TO 40
      WRITE(6,230)
      GO TO 70
   40 CONTINUE
      SUMX = 0.0
      SUMY = 0.0
      SUMX2 = 0.0
      SUMXY = 0.0
      SUM2 = 0.0
      FN = FLOAT(N)
C     SUM OVER N POINTS
      DO 50 1 = 1,N
      XI = X(I)
      YI = Y(I)
      SUMX = SUMX + XI
```

```
      SUMY = SUMY + YI
      SUMX2 = SUMX2 + XI * XI
      SUMXY = SUMXY + XI * YI
      SUMY2 = SUMY2 + YI * YI
C  CALCULATE CURVE PARAMETERS
   50 CONTINUE
      DELTA = FN * SUMX2 - SUMX * SUMX
      A = (SUMX2 * SUMY - SUMX * SUMXY)/DELTA
      B = (SUMXY * FN - SUMX * SUMY)/DELTA
      EPS = A/(2.0 * WAVL)
      CS = B/WAVCUB
      C = FN - 2.
C  IF NUMBER OF POINTS IS LESS THAN 3 THEN SKIP STD. DEV. CALCULATION
      IF(C.LT.1.E - 3) GO TO 60
      VNCE = (SUMY2 + A*A*FN + B * B * SUMX2 - 2.0 * (A * SUMY + B * SUMXY
     - A * B * SUMX))/C
      SIGA = SQRT(VNCE * SUMX2/DELTA)
      SIGB = SQRT(VNCE * FN/DELTA)
      R = (FN * SUMXY * SUMX * SUMY)/(SQRT(DELTA * (FN * SUMY2 - SUMY * SUMY)))
      SIGEPS = SIGA/(2.0 * WAVL)
      SIGCS = SIGB/WAVCUB
      WRITE(6,240) A,SIGA,B,SIGB,R,EPS,SIGEPS,CS,SIGCS
      GO TO 70
   60 CONTINUE
      WRITE(6,250)A,B,EPS,CS
      GO TO 70
   80 CONTINUE
      CALL EXIT
  100 FORMAT(13AG/F10.5)
  110 FORMAT(2I2,F12.6)
  200 FORMAT(1H1,13AG,/,15HOVOLTAGE (KV) = ,F8.2,10X,16HWAVELENGTH (A) = ,
     $ F9.6,//)
  210 FORMAT(1H0,10X,17HRADIUS OF MAXIMUM,12,2H = ,F7.4,17H (RECIP.ANGSTROM),
     $ 20X,3HX = ,F7.5,5X,3HY = ,F8.2)
  220 FORMAT(1HO,10X,17HRADIUS OF MINIMUM,12,2H = ,F7.4,17H (RECIP.ANGSTROM),
     $ 20X,3HX = ,F7.5,5X,3HY = ,F8.2)
  230 FORMAT(1HO,10X,41HONLY ONE POINT IN DATA SET–DATA IGNORED,/)
  240 FORMAT(1HO,10X,11HINTERCEPT = ,F8.2,2,H(,F5.2,11X,7HSLOPE =
     $ ,F8.2,2H (,F6.2,1H),11X,12HCORR.COEFF. = ,F6.4,///,1HO,50X,20HDEFOCUS
     $ (STD.DEV.) = ,F9.2,2H (,F6.2,6H) ANG.,5X,15HCS (STD,DEV.) = ,F6.3,2H
     $ (,F4.3,5H) MM.,//)
  250 FORMAT(/,1H0,10X,
     $ 11HINTERCEPT = ,F8.2,5X,7HSLOPE = ,F8.2,39X,9HDEFOCUS =
     $ ,F9.2,5H ANG.,5X,4HCS = ,F6.3,4H MM.,//)
  260 FORMAT(1HO,20X,***RING NOT LABELED AS MAXIMUM OR MINIMUM***)
```

Sample data and output from program (100 kV)

TEST OF PROGRAM — O.L.'S DATA
VOLTAGE (KV) = 100.00 WAVELENGTH (A) = 0.037013
  RADIUS OF MAXIMUM 1 = 0.0770 (RECIP.ANGSTROM)   X = 0.00593   Y = −168.66
  RADIUS OF MINIMUM 1 = 0.1120 (RECIP.ANGSTROM)   X = 0.01254   Y = −159.44
  RADIUS OF MAXIMUM 2 = 0.1400 (RECIP.ANGSTROM)   X = 0.01960   Y = −153.06
  RADIUS OF MINIMUM 2 = 0.1660 (RECIP.ANGSTROM)   X = 0.02756   Y = −145.16
  RADIUS OF MAXIMUM 3 = 0.1890 (RECIP.ANGSTROM)   X = 0.03572   Y = −139.97
  RADIUS OF MINIMUM 3 = 0.2150 (RECIP.ANGSTROM)   X = 0.04622   Y = −129.80
  RADIUS OF MAXIMUM 4 = 0.2420 (RECIP.ANGSTROM)   X = 0.05856   Y = −119.53
  RADIUS OF MINIMUM 4 = 0.2890 (RECIP.ANGSTROM)   X = 0.08352   Y = −95.78
  INTERCEPT = −171.83 (0.85) SLOPE = 908.90 (19.57)   CORR.COEFF. = 0.9986
DEFOCUS (STD.DEV.) = −2321.26 (11.50) ANG. CS (STD.DEV.) = 1.792 (0.039)MM

Appendix 2
Use of an absorption function to represent the objective aperture effect

The justification for the use of an absorption function $\phi_i(x, y) = \mu(x, y)t/2\sigma$ in Section 6.1 to account for the exclusion of scattering outside the objective aperture from the image can be seen as follows. The true image amplitude in the 'flat Ewald sphere' approximation (see Section 3.4) is

$$\psi_i(x, y) = [\exp(-i\sigma\phi_R(x, y))] * I(x, y)$$

where the $*$ denotes convolution and $I(x, y)$ is the microscope impulse response (equal to the transform of $A(u, v)$ in Section 3.2). Here $\phi_R(x, y)$ is the unsmoothed specimen electrostatic potential. The use of an absorption function predicts an image amplitude given by eqn (6.2) as

$$\psi'_i(x, y) = \exp(-i\sigma\phi_R(x, y) * I(x, y)) \exp(-\sigma\phi_i(x, y))$$

A comparison of the expansion to second order of these two expressions allows the absorption potential to be defined as

$$\sigma\phi_i(x, y) = \left(\frac{\sigma^2}{2}\right)\phi_R^2(x, y) * I(x, y) - \left(\frac{\sigma^2}{2}\right)[\phi_R(x, y) * I(x, y)]^2$$

which gives the correct limiting behaviour for small and large apertures. I am grateful to Professor J. M. Cowley for a discussion on this point. It is interesting to note that the use of an absorption function common in biological microscopy for the imaging of thick specimens at low resolution is based on a form of 'column approximation'. The image intensity deficit in the neighbourhood of a thin column of specimen is obtained by calculating the scattering from an equivalent specimen whose structure (mass thickness) is everywhere the same as that of the true specimen within this column.

Appendix 3
Resolution-limiting factors and their wavelength dependences

As mentioned in Section 4.2, it is necessary to distinguish two resolution limits in high-resolution electron microscopy. The first, the 'information-resolution limit', gives the ultimate band limit of the instrument and indicates the highest resolution detail which could in principle be extracted by image-processing techniques, or by comparisons with computed images. The second is the familiar 'point-resolution' limit at which images can be simply interpreted. The theory on which these two concepts are based is first reviewed, before discussing their differing wavelength dependences.

For specimen thicknesses at which the central beam is much stronger than any other diffracted beam, the combined effects of electronic instabilities and the use of an extended, incoherent effective disc source result in the imposition of a virtual aperture or 'damping envelope' on the objective lens transfer function of the form (O'Keefe and Anstis, unpublished results; see also references for Section 4.2)

$$A(\mathbf{K}) = \exp\left\{-\pi^2 \lambda^2 \Delta^2 \mathbf{K}^4 / 2\right\} \frac{2\mathrm{J}_1 \left|2\pi\theta_c \left[\Delta f \mathbf{K} + \lambda(\lambda C_s - i\pi\Delta^2)\mathbf{K}^3\right]\right|}{\left|2\pi\theta_c \left[\Delta f \mathbf{K} + \lambda(\lambda C_s - i\pi\Delta^2)\mathbf{K}^3\right]\right|} \quad (A3.1)$$

where the first two terms on the right-hand side of eqn (4.8a) have been omitted. A similar expression is given in Section 4.2 for a Gaussian source. Here λ is the electron wavelength, C_s is the spherical aberration coefficient, Δf is the defocus increment, $\mathbf{K}$ is the scattering vector ($|\mathbf{K}| = \sin\theta/\lambda = (u^2 + v^2)^{1/2}$ where θ is the scattering angle), and θ_c is the illumination semi-angle. The quantity Δ is defined by

$$\Delta = C_c Q = C_c \left[\frac{\sigma^2(V_0)}{V_0^2} + \frac{4\sigma^2(I_0)}{I_0^2} + \frac{\sigma^2(E_0)}{E_0^2}\right]^{1/2} \quad (A3.2)$$

where $\sigma^2(V_0)$ and $\sigma^2(I_0)$ are the variances in the statistically independent fluctuations of accelerating voltage V_0 and objective lens current I_0, respectively. Thus the rms value of the high-voltage fluctuation is equal to the standard deviation $\sigma(V_0) = [\sigma^2(V_0)]^{1/2}$. C_c is the lens chromatic aberration constant and the full width at half maximum height of the energy distribution of electrons leaving the filament is

$$\Delta E = 2.345\sigma(E_0) = 2.345[\sigma^2(E_0)]^{1/2}$$

For high-resolution imaging in which an objective aperture is not used, eqn (A3.1) sets the limit to the highest resolution detail which can be extracted from an electron image. This detail, however, may not be simply related to the specimen structure, since it is obtained using an oscillating transfer function. An 'information-resolution limit' can thus be defined as the **K** value(s) $U_0(i)$ for which $A(\mathbf{K}) = 2J_1(2.6)/2.6 = \exp(-1)$. The solutions to the resulting cubic equation are, for $\Delta = 0$,

$$U_0(1) = S_1 + S_2$$
$$U_0(2) = -U_0(1)/2 + i\sqrt{3}(S_1 - S_2)/2 \qquad (A3.3)$$
$$U_0(3) = -U_0(1)/2 - i\sqrt{3}(S_1 - S_2)/2$$

where

$$S_{1,2} = \left[\left\{b/\theta_c \pm \left(\frac{a/\Delta f^3}{\lambda^2 C_s} + \frac{b}{\theta_c}\right)^{1/2}\right\} \Big/ (\lambda^2 C_s)\right]^{1/3}$$

with

$$a = 1/27 \quad \text{and} \quad b = (26/4\pi)^2$$

The outer cut-off $U_0(1)$ is preceded by one or two crossings of the line $f(\mathbf{K}) = \exp(-1)$ at $U_0(2)$ and $U_0(3)$ if $\text{Im}(U_0(2)) = 0$ as indicated in Fig. A3.3(b). Physically, the effect of partial coherence is seen to attenuate contrast transfer in regions where the slope of $\chi(\mathbf{K})$ (see eqn (3.24)) is large and to preserve good contrast in regions where this slope is small, as shown in Fig. A3.1. The function $\chi(\mathbf{K})$ has a minimum within the contrast transfer intervals which occur between U_1 and U_2 where

$$U_{1,2} = \left[\left(\frac{8n+3}{2}\right)^{1/2} \pm 1\right]^{1/2} C_s^{-1/4} \lambda^{-3/4} = k_n^{1/2} C_s^{-1/4} \lambda^{-3/4} \qquad (A3.4)$$

for a focus setting

$$\Delta f_n = \left(\frac{8n+3}{2}\right)^{1/2} C_s^{1/2} \lambda^{1/2} \qquad (A3.5)$$

as shown in Fig. A3.3(a).

For $n = 0$, eqns (A3.4) and (A3.5) give the Scherzer conditions commonly used for the structure imaging of defects, and this defines the instrumental point-resolution (see eqn (6.17)). The cases $n = 0$ and $n = 4$ are shown for a typical modern instrument in Fig. A3.2. The value of n is equal to the number of minima which precede the passband. We note that the width of these 'passbands' $\Delta U(n) = U_1 - U_2$ depends on C_s, λ, and n as discussed in the next paragraphs. Here the slight effect of α and Δ on $\Delta U(n)$ has been neglected. Solutions similar to those given in eqn (A3.3) can also be obtained for $\Delta \neq 0$; however, the resulting expressions are cumbersome and of little practical use since these resolution-limiting effects can readily be treated separately.

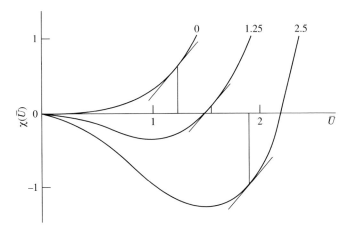

Figure A3.1 The function $\chi(\bar{U})$ drawn for three focus values $\Delta f = d_i C_s^{1/2} \lambda^{1/2}$ where d_i are the numbers indicated on each curve. The ordinate is the generalized spatial frequency U commonly used in the literature, where $|\mathbf{K}| = \theta/\lambda = (C_s \lambda^3)^{-1/4} \bar{U}$. The attenuation of spatial frequencies due to limited spatial coherence is approximately proportional to the gradient of this curve (see eqn (4.8a)). The curves explain how the band limit moves to higher spatial frequencies $\bar{U}$ with increasing defocus. If the attenuation is considered severe for all $\bar{U}$ beyond that for which $\chi(\bar{U})$ has the critical gradient shown by the three tangential line segments indicated, we see that this cut-off gradient occurs at higher spatial frequencies as the defocus is increased (Frank 1976).

For $\theta_c = 0$ and $A(\mathbf{K}) = \exp(-2)$, eqn (A3.1) gives (Fejes 1977)

$$U_0(\Delta) = \left(\frac{2}{\pi \lambda \Delta}\right)^{1/2} \tag{A3.6}$$

as the information-resolution limit due to electronic instabilities alone. We note that it is the *product* $\lambda C_c Q$ which one wishes to minimize for highest 'information-limit' resolution. Assuming that images are recorded at the focus settings given by eqn (A3.5), so that partial coherence effects can be neglected in the neighbourhood of the passband $\Delta U(n)$, eqns (A3.4) and (A3.6) give

$$\Delta = \left(\frac{2}{k_n \pi}\right) C_s^{1/2} \lambda^{1/2} = C_c Q \tag{A3.7}$$

as the electronic stability needed for inclusion of passband $\Delta U(n)$, where $k_n = (8n + 3/2)^{1/2} + 1$. For the commonly used Scherzer condition, this is

$$\Delta = 0.286 C_s^{1/2} \lambda^{1/2} = C_c Q \tag{A3.8}$$

These results indicate that, in the absence of an objective aperture, differing factors will control the resolution of electron images at various accelerating voltages. For example,

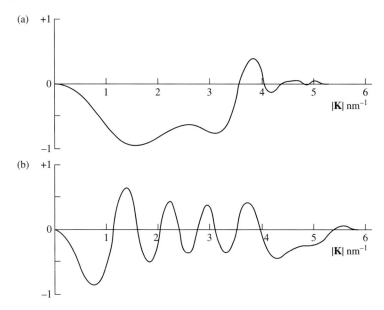

Figure A3.2 Transfer functions for a high-resolution 100 kV instrument with $C_s = 0.7$ mm, beam divergence $\theta_c = 1$ mrad, and $\Delta = 5.4$ nm. The functions shown here and in Fig. 4.3 are actually $A_i(\mathbf{K}) \cos \chi(\mathbf{K}) + A_R(\mathbf{K}) \sin \chi(\mathbf{K})$, where $A_i(\mathbf{K})$ and $A_R(\mathbf{K})$ are the imaginary and real parts of eqn (A3.1), since, for linear imaging, this can be shown to be the coherent transfer function following the method of Section 3.4. In fact the term involving $A_i(\mathbf{K})$ is very small, and the damping envelope can be taken to be the real part of eqn (A3.1) to a good approximation. Curve (a) is drawn for the Scherzer focus $\Delta f = -62.3$ nm while (b) is drawn for $\Delta f = -212.8$ nm ($n = 4$ in eqn (A3.5)). The second curve includes a passband extending beyond 5 nm^{-1}.

at 100 kV with $C_s = 0.7$ mm and $C_c = 1$ mm, eqn (A3.8) gives $Q = 1.5 \times 10^{-5}$ as the stability needed to obtain the Scherzer resolution limit ($n = 0$ in eqns (A3.4) and (A3.5)). This stability is comfortably exceeded by most 100 kV instruments, for which manufacturers typically claim $\sigma(V_0)/V_0 = 2 \times 10^{-6} = \sigma(I_0)/I_0$, and $\sigma(E_0)/E_0 = 5 \times 10^{-6}$ giving $Q = 6 \times 10^{-6}$. Electron energy-loss measurements give 0.7 eV $< \Delta E <$ 2.4 eV as the electron gun-bias setting varies between maximum and minimum for a thermionic filament (see Fig. 9.5). Note that maximum gun bias here corresponds to minimum beam current. Thus, for such an instrument we have from eqn (A3.2)

$$\Delta = 10(4 + 16 + 9)^{1/2} = 5.4 \,\text{nm} \tag{A3.9}$$

for maximum gun bias and

$$\Delta = 10(4 + 16 + 100)^{1/2} = 10.1 \,\text{nm}$$

at minimum gun-bias setting. Since for small illumination semi-angles the effect of partial coherence is negligible in the region of the pass-bands given by eqn (A3.4), the resolution of 100 kV images recorded at the favourable focus settings of eqn (A3.5) is seen to be controlled by the gun-bias setting for small bias settings (high-beam current) and by

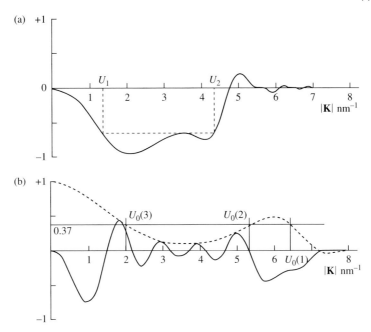

Figure A3.3 Transfer functions for a 300 kV electron microscope. The accelerating voltage chosen (300 kV) is a compromise between the various practical disadvantages of 1 MeV machines and the improvement in point-resolution accompanying a reduction in electron wavelength. Parameters used are: $V_0 = 300$ kV, $C_s = 1.5$ mm, $\Delta = 5.4$ nm, $\theta_c = 1$ mrad, defocus $\Delta f = -66.5$ nm (Scherzer focus) in (a). The electron wavelength is 0.001969 nm. Curve (b) is drawn for similar conditions with $\Delta f = -227.3$ nm ($n = 4$) and includes the damping envelope (shown dotted) given approximately by the real part of eqn (A3.1). This dotted curve clearly illustrates the dip in the damping envelope which occurs for large values of n where the gradient of $\chi(|\mathbf{K}|)$ can be large for $|\mathbf{K}|$ values smaller than the passband. The horizontal line in (b) is drawn for $f(|\mathbf{K}|) = \exp(-1)$ and the points $U_0(1)$, $U_0(2)$, $U_0(3)$, of eqns (A3.3) are indicated. In (a), the dotted lines indicate the method of defining the quantities U_1 and U_2 of eqn (A3.4) which determine the width and position of the passbands.

the objective lens current stability for high bias settings (low-beam current). This broad conclusion is independent of the objective lens aberration coefficients. At maximum gun bias we have, from eqn (A3.6), for such an instrument, $U_0(\Delta) = 0.56$ Å^{-1}, which, by eqn (A3.4) would allow the fifth contrast transfer interval to be included. At minimum gun bias, $U_0(\Delta) = 0.395$ Å^{-1}. Thus, the use of very small illuminating apertures on 100 kV machines would appear to be a less important priority than the minimization of chromatic aberration. For structure images recorded at the Scherzer focus ($n = 0$ in eqn (A3.5)), however, the use of smaller illumination semi-angles will nevertheless introduce higher-resolution detail; however, the use of several images for subsequent analysis recorded at the focus settings of eqn (A3.5) allows the largest illumination aperture to be used (producing the most intense final image) for the smallest resolution penalty. We note also from eqn (A3.5) that for structure imaging under Scherzer conditions ($n = 0$ in eqns (A3.4) and (A3.5)) the rather large illumination semi-angle of $\theta_c = 2.3$ mrad can be used at 100 kV with $C_s = 0.7$ mm if the resolution limit due to partial coherence is to be equal to the Scherzer cut-off ($U_1 = 0.345$ Å^{-1}). Finally, as a matter of practical experience, it is generally found that conscientious monitoring of the vacuum in the gun chamber is needed in order to maintain the manufacturer's quoted high-voltage stability.

At higher voltages the effects of chromatic aberration due to the electron source rapidly become unimportant. The best current values for eqn (A3.2) at an accelerating voltage of 1 MeV appear to be $\sigma(V_0)/V_0 = \sigma(I_0)/I_0 = 5 \times 10^{-6}$ with $\sigma(E_0)/E_0 = 1 \times 10^{-6}$ (S. Horiuchi, personal communication). With $C_c = 4.4$ mm such an instrument has

$$\Delta = 44(25 + 100 + 1)^{1/2} = 49.4 \, \text{nm} \tag{A3.10}$$

giving, from eqn (A3.6), an information-resolution limit of 0.384 Å^{-1}, set chiefly by current instability of the objective lens. This $\lambda\Delta$ value of 4.3 Å^2 is considerably poorer than the value of 2.0 Å^2 for the 100 kV instrument cited in eqn (A3.9) at maximum gun bias. This shows that, if comparable high-voltage stability can be obtained, a field-emission 300 kV instrument with, if possible, improved objective-lens current stability should give the highest information-resolution limit. The chief advantage of high-voltage instruments is thus seen to lie in the extended range of simply interpretable image detail which they can provide.

Practical considerations such as image-viewing convenience, sensitivity of the detector, and viewing phosphor, room height for the microscope, and overall cost suggest that designs for a few hundred kilovolts achieve the best compromise for a machine which is not aberration corrected. For example, a design with $V_0 = 300$ kV, $C_s = 1.5$ mm, and $\Delta = 5.4$ mm gives a Scherzer resolution limit ($n = 0$ in eqn (A3.4)) of 0.21 nm (point-to-point) or an information-resolution limit of $U_0(\Delta) = 0.79$ Å^{-1} (eqn A3.6). The allowed illumination semi-angle consistent with this Scherzer resolution is then 1.9 mrad, producing a conveniently intense final image for observation at high magnification. Transfer functions for this 300 kV design are shown in Fig. A3.3. The design stabilities used are $\sigma(V_0)/V_0 = \sigma(I_0)/I_0 = 2 \times 10^{-6}$, as for 100 kV machines. Unlike 100 kV instruments, however, the quality of the images obtained on such an instrument would be rather insensitive to changes in the electron gun-bias setting, since, at 300 kV, the last figure in the bracket of eqn (A3.9) changes insignificantly between the maximum and minimum gun-bias settings.

Optimum conditions for aberration-corrected instruments are discussed in Section 7.4.

References

Fejes, P. L. (1977). Approximations for the calculation of high-resolution microscope images of thin films. *Acta Crystallogr.* **A33**, 109.

Frank, J. (1976). Determination of source size and energy spread from electron micrographs using the method of Young's fringes. *Optik* **44**, 379.

Appendix 4
What is a structure image?

The term 'structure image' was proposed by J. M. Cowley to describe that restricted set of high-resolution images which may be directly interpreted to some limited resolution in terms of the specimen's projected structure and *which were obtained using instrumental conditions which are independent of the structure*, and so require no a priori knowledge of the crystal structure. This last condition currently excludes all images obtained by 'matching' the microscope transfer function to the crystal reciprocal lattice. At present only the weak-phase object, the projected charge density approximation, and the ADF-STEM incoherent imaging mode provide a prescription for structure-independent instrumental conditions which therefore give structure images.

The development of 'medium-voltage' HREM machines has now made structure imaging routinely possible in many materials. The measured point-resolution of these instruments is about 0.165 nm. Aberration-corrected instruments, with a resolution limit of about 1 Å has finally brought most materials into the domain of atomic-resolution structure imaging.

Appendix 5
The challenge of HREM

This book is intended to teach students and research workers how to record and interpret atomic-resolution transmission electron microscope images. A well-defined aim is the greatest stimulus to progress. The following project therefore provides a thorough test of all the experimental skills taught in this book, together with a severe test of the underlying theory. Any student who completes it may reasonably claim to be an authority on high-resolution electron microscopy!

Magnesium oxide crystals of sub-micrometre dimensions are easily made by burning magnesium ribbon in air (see Section 10.1). A 'holey carbon' grid passed through the smoke will collect particles for examination in the electron microscope. The particles form in perfect cubes, and can be found in the (110) orientation with the electron beam passing along the cube face diagonal. The thickness is therefore known exactly at every point in a lattice image. It is possible to record a structure image of this material on the newest electron microscopes, and to determine the focus setting from diffractogram analysis (see Section 10.6). The image can then be matched as a function of thickness against computed images by the methods described in Sections 5.6 or 5.7. All the thickness-dependent contrast reversals in the lattice fringes due to multiple scattering should be correctly reproduced in the computed images.

Challenging aspects of this project include: (1) the variation of focus along the lower face of the crystal; (2) the possible need for 'absorption' corrections in the calculations for the thick regions (see Goodman and Lehmpfuhl 1967); (3) difficulties in matching the Fresnel fringes along the crystal edge of varying thickness, where a 'profile image' of the MgO surface will be seen; (4) refinement of the atomic scattering factors for ionicity effects; and (5) the need for very accurate alignment of the crystal. To the author's knowledge, a thorough analysis of this problem has yet to be published. Ultimately, the credibility of the dynamical imaging theory given in this book depends on its ability to reproduce these experimental results.

This project represents a summation of practically all the useful knowledge in this book.

Reference

Goodman, P. and Lehmpfuhl, G. (1967). Electron diffraction study of MgO $h00$-systematic interactions. *Acta Crystallogr*. **22**, 14.

Index

A
A.C. field interference, 276
Abbe interpretation of imaging, 53
aberrations, 31, 55
aberration coefficients, 38, 42, 60, 212, 289
aberration correction, 38
absorption, 115
absorption contrast, 154
accelerating voltage, benefits of higher, 165
achromatic circle, 89, 219
ADF-STEM, 243
Alchemi, 348
algorithms, 126
alignment, automated, 213, 339
amplitude contrast, 154
angular magnification, 16
annular aperture, 79, 171
ASIC detector, 283
astigmatism, 36, 38, 58, 295, 297, 309
atom images by HREM, 157
atom location by channelling enhanced microanalysis, 348
atomic columns, bright or dark, 104
atomic position determination, 126, 363
atoms used as lenses, 219
automated focussing, 337
autotuning, 337

B
background, 177
Bayes theorem, 209
beam divergence, 93, 113, 126
bias resistor, 270
biasing, Wehnelt, 270
biprism, 219
Bloch wave methods, 107
Boersch effect, 276
bonding effects in CBED, 363
Born approximation, 61
bound states, 249
brightness, definition and measurement, 268
brightness of sources compared, 267

C
camera length, measurement, 302
carbon films, 340
cardinal planes, 16
cathodoluminescence, 372
cathodoluminescence in STEM, 372
Cathodoluminesence, 372
caustic image, 318
CBED, 219, 363
CCD cameras, 276

channelling radiation, 348
channelling effect on X-ray emission and EELS, 348
channelling in STEM, 249
charge coupled device, 276
chemical mapping at atomic resolution, 357
chromatic aberration, 33, 70, 295
coherence effects in HREM, 70, 81, 93, 299
coherence in STEM, 240
coherence length, 70
coherence width, 76, 299
coherence, complex degree of, 76
coherent bremsstrahlung, 351
coherent detection, 204
column approximation, 117, 124
computer simulation, 122, 126
condenser-objective lenses, 37
condenser lenses, 265
conical illumination, 79
contamination, 321
contrast and atomic number, 171
contrast, dependence on accelerating voltage, 169, 251
contrast transfer intervals, 74, 386
convergent beam diffraction, 363
convex sets, 219, 215
convolution theorem, 47
copper phthalocyanine, 190
correlation functions, 76, 141, 184, 190, 210–1, 246, 299
critical dose, 179
cross-section, elastic, 169
cryo-electron microscopy, 180
cryomicrsocopy, 180
crystal orientation, adjustment, 335
crystallographic point groups, 117

D
damping envelope, 70, 93, 296
dark-field image HREM, 171
dark-field STEM, 243
de Broglie relation, 13
Debye–Waller factor, 116
delocalization, 359
depth of field, 21, 93, 114
detective quantum efficiency (DQE), 177, 276, 280
detectors, 276
detector geometry in STEM, 240
diffraction limit, 50
diffraction, three dimensional, 107
diffractogram tableaux for autotuning, 337
diffractograms, 296, 337

direct methods, 215
dislocation cores, 140
dispersion matrix, surfaces, 107
displacement threshold, 195
DNA, 182, 183, 217
dopant site, determination, 348
dopants imaged, 146, 256
drift, 319
dynamic range of detectors, 276
dynamical electron diffraction, 96, 106, 107
dynamically forbidden reflection, 117

E
EELS, ELNES, 357
effective electron source, 69
eigenvalues, 107
Einstein model for vibration, 243
elastic relaxation, in thin specimens, 104
electron biprism, 219
electron crystallography, 96, 363
electron energy-loss near-edge structure, 357
electron energy loss spectroscopy, 357
electron optics, 13
electron-optical constants, measurement, 289
electron sources, 264
electron source size, 81
electron spin, 107, 376
electron wavelength, 13
electronic instabilities, 33, 70, 315
electrostatic potential, specimen, 13
ELNES, 357
energy filter, imaging, 174
energy-dispersive X-ray spectroscopy, 348
energy-loss analysers, 357
energy-selecting microscopes, 174
energy spread, 33, 70
environmental HREM, 376
environmental requirements for HREM, 315
Ewald sphere, 96
EXAFS, 357
excitation error, 96
EXELFS, 357
extinction distance, 98

F
Fano noise, CCD, 279
Fermat's principle, 31
Feynman path-integral, 15, 122
field-emission sources, 275
filament height, 274
filament life, 264
filament temperature, 274
flux density in lens gap, 24
focus, aberration corrected optimum, 212
focus, dark field, 171
focus increments, measurement, 289
focusing techniques, 337
focus, Scherzer, 74, 160
focus setting, from diffractogram, 296
focus, stationary phase, 74, 138
Fourier coefficients of crystal potential, 96, 107

Fourier images, 93, 95, 114, 136, 221, 334
Fourier transform, 54
Frank method, for resolution measurement, 303
Fraunhofer diffraction, 47
Fresnel diffraction, 47
Fresnel fringe, 300
Fresnel propagator, 48

G
G-value, radiation damage, 192
gaussian focus, 16
genetic code, 183
gold atom, image, 161
goodness-of-fit index, 209
graphene, 146, 160, 194, 340
ground-loops, 315
gun bias, 270

H
hair-pin filaments, 274
hexapole corrector, 38
high-angle dark-field STEM detector, 251
high-voltage instabilities, 318
HIO algorithm, 219
hole count, 348
hole drilling, 321
hollow-cone illumination, 79
holographic Alchemi, 219
holography, in line, 219
holography, off axis electron, 211
holography of fields, 376
Huygens' principle, 50

I
ice-embedding, 180
illumination system, 265
image plate detectors, 281
image reconstruction, 126, 215, 219
image-simulation programs, 126
images of defects, 134
impulse response of an electron microscope, 62
in-situ microscopy, 376
incoherent image, 70, 243
independent Bloch wave approximation, 108
inelastic scattering, 174
information-resolution limit, 73, 204, 214
ionicity, effect on HREM, 145
instabilities, 315
interfaces, studied by HREM, 140
inversion problem, dynamical, 226
irreducible representation, 106

K
kinematic conditions, test for, 363
kinematic theory, 96
kinks, 140
knock-on damage, 195
Kohler illumination, 265, 363

L

lanthanum hexaboride, 274
lattice imaging in HREM, 89
lattice imaging in STEM, 233
Laue circle, 363
Laue geometry, 107
lens current measurement, 289
lens equation, 16
lens field distribution, 29
Lichte focus, 212
linear imaging, 61
logarithmic deconvolution, 357
longitudinal magnification, 21

M

magnetic circular dichroism, 376
magnetic fields, stray, 315
magnification calibration, 293
magnification, angular, 20
magnification, lateral, 20
magnification, longitudinal, 21
many-beam lattice images, 107
mass spectrometer, 321
mass-thickness, 154
matrix optics, 19
mean free path for inelastic scattering, 175
membrane proteins, 188
meridional rays, 23
microanalysis by EELS, 357
microdiffraction, 363
minimum-contrast, 338
minimum-exposure microscopy, 180
Modulation Transfer Function or CCD, 276
Moliere high-energy approximation, 96
multiple elastic scattering, 107
multiple inelastic scattering, 357
multislice method, 106
multiple scattering, dependence on accelerating voltage, 165
multislice theory - polynomial, 106

N

nanoscience, 143
nodal planes, 16
nodal points, 16
noise in an electron image, 177

O

organic crystals, 190
oxygen imaging by HREM in crystals, 141
objective aperture, 20
objective lens, 50
objective-lens focus step, measurement, 289
optical potential, 107
optical theorem, 70
Optical Transfer Function in STEM, 251
optimization, 204
oxidation state, 357

P

parabola method, 207
paraxial ray, 22
paraxial trajectory, 22
partial coherence, 70
partial pressure gauge, 321
passbands, 73, 74
Pendellosung fringes, 99, 333
periodic continuation, 122
Permendur, 29
perturbation theory, 107
phase objects, 2, 61
phase problem, 215, 226
phase-contrast images, 2, 61
phase-grating approximation, 61, 96
phase-plate method, 5
phonon absorption, 115, 251
phonons, frozen, 251
phosphor grain size, 276
Photographic techniques, 282, 343
plasma cleaners, 321
plasmon-scattered electrons, 357
point defects, 146, 256
point group, 363
point spread function, 62
point-resolution, 160
pointed filament, 274
pole-piece, 29
pre-field, 37
principal planes, 16
probe current vs size, 249
probe size, 249–50, 248
probe spreading, 350
projected charge density approximation, 96
projection onto convex sets, 219
projector focal length, 25
projector lenses, 25
ptychography, 219
purple membrane, 188

R

radiation damage, 192, 195
ray diagram, electron microscope, 17
Rayleigh criterion, 219, 303
reciprocity theorem, 233
refractive index, for electrons, 13
relativistic energy, 107
resolution limits, 73, 160, 204, 395–400
resolution of the EDS technique, 348
resolution, dependence on accelerating voltage, 165, 395–400
resolution, measurement, 303
Richardson–Lucy method, 281
robust, statistical, 204
Ronchigrams, 219, 306
Rose equation, 177
Rutherford scattering in STEM, 251

S

scattering matrix, 107
scattering phase, 96

Scherzer focus, 74, 160
Scherzer focus, 73, 74, 160
Scherzer resolution, 160
Schlieren contrast, 2
Schottky emission, 275
Schrodinger equation, 107
seidel aberrations, 38, 55
self-biased guns, 270
sextupole lenses, 38
shadow images, 219
sign conventions, 125
signal-to-noise ratio, 177
simulation of HREM images, 126
single atoms, imaging, 157
single-molecule imaging, 157
single-scattering conditions, 96, 363
Snell's law, 31
space-group, determination, 363
specimen height, 30, 137
specimen movement, 319
spherical aberration, 31, 38, 55
spherical aberration constant, measurement, 291
spin, electron, 107, 376
spread of the electron wavefunction, 122
stability, thermal, 319
standing wave, 348
stationary-phase focus, 74, 138
Stobbs factor, 126
stray fields, 315
structure factor, 96
structure image, 423
structure matrix, 107, 226
substrates, 157
supercells for computation, 122
superesolution, 204
surfaces, imaging, 376
symmetry reduction, 107, 226
symmetry reduction of dispersion matrix, 107
synchrotron brightness, compared to STEM, 264

T
thermal diffuse scattering in STEM, 251
Thickness fringes, 99, 333
thick-phase grating, 96
tilt-series, through, 210
thin-lens formula, 16
three-beam lattice images, 89
three-dimensional imaging, 143, 257
through-focus series, 204

tomography in STEM, 256
tomography in TEM, 143
top-entry stages, 335
transfer function, 61, 70
transfer function for coherent convergent illumination, 113
translational symmetry elements, 363
twin image, 219
two-beam dynamical theory, 96
two-beam lattice fringes, 96

U
Umklapp, 251
unitary transformation, 107

V
vacuum system, 321
van Cittert–Zernike theorem, 76
vibration, 319
video recording, 276
virtual aperture, 50
virtual inelastic scattering, 115
virtual object, 28
virtual source, 265

W
water, cooling, 345
wavelength, electron, 13
wavevector, 107
weak-phase object, 61
Wehnelt, 265
Weiner filter, 204
Wentzel model, 254

X
X-ray absorption fine structure, 357
X-ray absorption spectroscopy, 357
X-ray background, 348

Y
YAG single-crystal screen, 276
Young's slit with electron waves, 78

Z
Z-contrast in STEM, 251
Zernike phase plate, 5
zero-order Laue zone, 363
zone-axis diffraction pattern, 363